Cambridge Studies in Biological and Evolutionary Anthropology 35

Gorilla Biology
A Multidisciplinary Perspective

Gorillas are one of our closest living relatives, the largest of all living primates, and they teeter on the brink of extinction. These fascinating animals are the focus of this in-depth and comprehensive examination of gorilla biology. *Gorilla Biology* combines recent research in morphology, genetics, and behavioral ecology to reveal the complexity and diversity of gorilla populations. The first two sections focus on morphological and molecular variation and underscore the importance of understanding diverse biological patterns at all levels in testing evolutionary and adaptive hypotheses and elucidating subspecies and species diversification. The following section investigates the influence of ecological variables on gorilla social organization, and highlights the surprising behavioral flexibility of this genus. The book ends with discussions of the conservation status of gorillas and the many and increasing threats to their continued survival. Giving insight into the evolutionary biology of these unique primates, this book will be essential reading for primatologists, anthropologists, and evolutionary biologists.

ANDREA B. TAYLOR is Assistant Professor in the Departments of Community and Family Medicine and Biological Anthropology and Anatomy at Duke University Medical Center. For the past 10 years she has studied the comparative anatomy of the African apes, focusing on the ontogeny of the musculoskeletal system to understand better the developmental and evolutionary basis of functional and adaptive differences in morphology.

MICHELE L. GOLDSMITH is Assistant Professor in the Department of Environmental and Population Health at Tufts University School of Veterinary Medicine. As a primatologist, she has spent the past 12 years studying the comparative behavioral ecology of both lowland and mountain gorillas, and most recently is examining the impact of tourism on gorilla behavior.

*Cambridge Studies in Biological and Evolutionary Anthropology*

*Series Editors*

HUMAN ECOLOGY
C. G. Nicholas Mascie-Taylor, University of Cambridge
Michael A. Little, State University of New York, Binghamton
GENETICS
Kenneth M. Weiss, Pennsylvania State University
HUMAN EVOLUTION
Robert A. Foley, University of Cambridge
Nina G. Jablonski, California Academy of Science
PRIMATOLOGY
Karen B. Strier, University of Wisconsin, Madison

*Consulting Editors*
Emeritus Professor Derek F. Roberts

*Selected titles also in the series*
16  *Human Energetics in Biological Anthropology* Stanley J. Ulijaszek
    0 521 43295 2
17  *Health Consequences of 'Modernisation'* Roy J. Shephard & Anders Rode
    0 521 47401 9
18  *The Evolution of Modern Human Diversity* Marta M. Lahr 0 521 47393 4
19  *Variability in Human Fertility* Lyliane Rosetta & C. G. N. Mascie-Taylor (eds.)
    0 521 49569 5
20  *Anthropology of Modern Human Teeth* G. Richard Scott & Christy G. Turner II
    0 521 45508 1
21  *Bioarchaeology* Clark S. Larsen 0 521 49641 (hardback), 0 521 65834 9
    (paperback)
22  *Comparative Primate Socioecology* P. C. Lee (ed.) 0 521 59336 0
23  *Patterns of Human Growth*, second edition Barry Bogin 0 521 56438 7
    (paperback)
24  *Migration and Colonisation in Human Microevolution* Alan Fix 0 521 59206 2
25  *Human Growth in the Past* Robert D. Hoppa & Charles M. FitzGerald (eds.)
    0 521 63153 X
26  *Human Paleobiology* Robert B. Eckhardt 0 521 45160 4
27  *Mountain Gorillas* Martha M. Robbins, Pascale Sicotte & Kelly J. Stewart
    (eds.) 0 521 76004 7
28  *Evolution and Genetics of Latin American Populations* Francisco M. Salzano &
    Maria C. Bortolini 0 521 65275 8

# Gorilla Biology
## A Multidisciplinary Perspective

EDITED BY

## ANDREA B. TAYLOR
*Duke University Medical Center*

and

## MICHELE L. GOLDSMITH
*Tufts University School of Veterinary Medicine*

CAMBRIDGE
UNIVERSITY PRESS

PUBLISHED BY THE PRESS SYNDICATE OF THE UNIVERSITY OF CAMBRIDGE
The Pitt Building, Trumpington Street, Cambridge, United Kingdom

CAMBRIDGE UNIVERSITY PRESS
The Edinburgh Building, Cambridge CB2 2RU, UK
40 West 20th Street, New York, NY 10011-4211, USA
477 Williamstown Road, Port Melbourne, VIC 3207, Australia
Ruiz de Alarcón 13, 28014 Madrid, Spain
Dock House, The Waterfront, Cape Town 8001, South Africa

http://www.cambridge.org

First published 2003

Printed in the United Kingdom at the University Press, Cambridge

*Typeface* Times 11/12.5 pt     *System* LATEX 2ε [TB]

*A catalogue record for this book is available from the British Library*

*Library of Congress Cataloguing in Publication data*

Gorilla biology : a multidisciplinary perspective / edited by Andrea B. Taylor
and Michele L. Goldsmith.
    p.   cm. – (Cambridge studies in biological and evolutionary
anthropology ; 35)
Includes bibliographical references (p.   ).
ISBN 0 521 79281 9
1. Gorilla.  I. Taylor, Andrea B. (Andrea Beth), 1961–  II. Goldsmith,
Michele L. (Michele Lynn), 1963–  III. Series.

QL737. P96 G59   2002
599.884 – dc21   2002023377

ISBN 0 521 79281 9 hardback

## Editors' Dedication

*For Michael Siegel,*
*who made all things possible,*
*and for my family, Serge, Jordan, and Lauren,*
*who make all things worthwhile.*
Andrea B. Taylor

*For my parents, Barbara and George,*
*through fear and worry, you continue to encourage and support.*
*For all of your love and guidance,*
*I thank you.*
Michele L. Goldsmith

## Authors' Dedication

*For the gorillas,*
*past, present, and future*

# Contents

# Contributors

Kate A. Abernethy
Department of Molecular and Biological Sciences
University of Stirling
Stirling FK9 4LA, U.K.

Gene H. Albrecht
Department of Cell and Neurobiology
Keck School of Medicine
University of Southern California
Los Angeles, CA 90089, U.S.A.

Kanyunyi Basabose
Centre de Recherche en Sciences Naturelles
Lwiro, D.S. Bukavu
Democratic Republic of Congo

Richard A. Bergl
PhD Program in Anthropology
CUNY Graduate Center
365 Fifth Avenue
New York, NY 10016, U.S.A.

Mike W. Bruford
Cardiff School of Biosciences
Cardiff University
Cardiff CF1 3TL, U.K.

Stephen L. Clifford
Centre International de Recherches Médicales Franceville (CIRMF)
BP 769 Franceville, Gabon

Amos S. Deinard
School of Veterinary Medicine
University of California–Davis
Davis, CA 95616, U.S.A.

Todd R. Disotell
Department of Anthropology
New York University
25 Waverly Place
New York, NY 10003, U.S.A.

John G. Fleagle
Department of Anatomical Sciences Heath Sciences Center and
Interdepartmental Doctoral Program in Anthropological Sciences
State University of New York
Stony Brook, NY 11794, U.S.A.

Bruce R. Gelvin
Department of Anthropology
California State University
Northridge, CA 91330, U.S.A.

Michele L. Goldsmith
Department of Environmental and Population Health
Tufts University School of Veterinary Medicine
North Grafton, MA 01536, U.S.A.

Colin P. Groves
Department of Archaeology and Anthropology
Australian National University
Canberra, A.C.T. 0200
Australia

Jaqueline L. Groves
Department of Biological Sciences
University of Sussex
Falmer BN1 9RH, U.K.

Jefferson S. Hall
Department of Forestry
Yale University
New Haven, CT 06520, U.S.A.

Alexander H. Harcourt
Department of Anthropology
University of California–Davis
1 Shields Avenue
Davis, CA 95616, U.S.A.

Christopher P. Heesy
Interdepartmental Doctoral Program in Anthropological Sciences
State University of New York
Stony Brook, NY 11794, U.S.A.

Sandra E. Inouye
Department of Anatomy
Chicago College of Osteopathic Medicine
Midwestern University
555 31st Street
Downers Grove, IL 60515, U.S.A.

Michael I. Jensen-Seaman
Human and Molecular Genetics Center
Medical College of Wisconsin
Milwaukee, WI 53226, U.S.A.

William L. Jungers
Department of Anatomical Sciences Heath Sciences Center and
Interdepartmental Doctoral Program in Anthropological Sciences
State University of New York
Stony Brook, NY 11794, U.S.A.

Kiswele Kaleme
Centre de Recherche en Sciences Naturelles
Lwiro, D.S. Bukavu
Democratic Republic of Congo

Kenneth K. Kidd
Department of Genetics
Yale University
New Haven, CT 06520, U.S.A.

Lyle W. Konigsberg
Department of Anthropology
University of Tennessee
Knoxville, TN 37996, U.S.A.

Steven R. Leigh
Department of Anthropology
University of Illinois–Urbana
Urbana, IL 61801, U.S.A.

Joshua M. Linder
PhD Program in Anthropology
CUNY Graduate Center
365 Fifth Avenue
New York, NY 10016, U.S.A.

Kelley L. McFarland
PhD Program in Anthropology
CUNY Graduate Center
365 Fifth Avenue
New York, NY 10016, U.S.A.

Alastair McNeilage
Wildlife Conservation Society
2300 Southern Boulevard
Bronx, NY 10460, U.S.A.

Joseph M.A. Miller
Department of Pathology and Laboratory Medicine
School of Medicine
University of California–Los Angeles
Los Angeles, CA 90095, U.S.A.

John F. Oates
Department of Anthropology
Hunter College
CUNY Graduate Center
365 Fifth Avenue
New York, NY 10016, U.S.A.

Paul B. Park
Department of Anthropology
University of Illinois–Urbana
Urbana, IL 61801, U.S.A.

Andrew J. Plumptre
Wildlife Conservation Society
2300 Southern Boulevard
Bronx, NY 10460, U.S.A.

John D. Polk
Interdepartmental Doctoral Program in Anthropological Sciences
State University of New York
Stony Brook, NY 11794, U.S.A.

John H. Relethford
Department of Anthropology
State University of New York College at Oneonta
Oneonta, NY 13820, U.S.A.

Melissa J. Remis
Department of Sociology/Anthropology
Purdue University
West Lafayette, IN 47907, U.S.A.

Oliver A. Ryder
Center for Reproduction of Endangered Species
Zoological Society of San Diego
San Diego, CA 92112, U.S.A.

Esteban E. Sarmiento
Department of Mammology
American Museum of Natural History
Central Park West and 79th Street
New York, NY 10024, U.S.A.

Rebecca M. Stumpf
Interdepartmental Doctoral Program in Anthropological Sciences
State University of New York
Stony Brook, NY 11794, U.S.A.

Andrea B. Taylor
Departments of Community and Family Medicine
and Biological Anthropology and Anatomy
Duke University Medical Center
Durham, NC 27710, U.S.A.

Caroline E. G. Tutin
Centre International de Recherches Médicales Franceville (CIRMF)
BP 769 Franceville, Gabon

Russell H. Tuttle
Department of Anthropology
University of Chicago
Chicago, IL 60637, U.S.A.

David P. Watts
Department of Anthropology
Yale University
New Haven, CT 06520, U.S.A.

Lee J.T. White
Wildlife Conservation Society
2300 Southern Boulevard
Bronx, NY 10460, U.S.A.

E. Jean Wickings
Centre International de Recherches Médicales Franceville (CIRMF)
BP 769 Franceville, Gabon

Juichi Yamagiwa
Laboratory of Human Evolution Studies
Faculty of Science
Kyoto University
Sakyo, Kyoto 606-8502, Japan

Takakazu Yumoto
Center for Ecological Research
Kyoto University
Otsu, Shiga 520-2113, Japan

# Acknowledgments

The editors gratefully acknowledge the Scientific Program Committee, and especially Mark Teaford, Chair and Program Editor, for their acceptance of the symposium "A revision of the genus *Gorilla*: Seventy years after Coolidge" at the 68th Annual Meeting of the American Association of Physical Anthropologists, Columbus, Ohio, 26 April – 1 May 1999. We thank all who participated in the symposium, and all who contributed a chapter or introductory perspective to this book.

All chapters in this volume were critically reviewed by a minimum of two external reviewers. Although we cannot acknowledge these individuals personally out of a commitment to confidentiality we wish to express our sincere appreciation. Your generous efforts improved the quality of each chapter and, ultimately, the entire book. All authors read and commented on multiple chapters and cross-referenced where appropriate. In this sense, all authors should be viewed as having contributed to the editorial process. We thank Nina Jablonski, Jeffrey Schwartz, and Mark Teaford for advice during the early stages of this project. In addition, a number of individuals read and commented on numerous parts of the book, and otherwise served as personal sounding boards, including Susan Antón, Colin Groves, Matthew Ravosa, Brian Shea, Carel van Schaik, and Richard Wrangham.

We are particularly grateful for the support of Cambridge University Press. A special thanks goes to Dr. Tracey Sanderson, our editor, for her unending support and patience. Thanks also to Anna Hodson, copy-editor, and Carol Miller, production controller, for their excellent work and cheerful dispositions!

Finally, Andrea wishes to thank her husband, Serge, for selflessly taking on the role of both parents, and for allowing me to wax unending about all things related to gorillas, and her children Jordan and Lauren, for relinquishing your mother on many nights and weekends during the two years this book was in preparation. Michele acknowledges her colleagues in the Center for Animals and Public Policy for their support and patience while editing this volume, and thanks her family for their encouragement, even though it meant spending less time together.

# Epigraph

When a young naturalist commences the study of a group of organisms quite unknown to him, he is at first much perplexed to determine what differences to consider as specific, and what as varieties; for he knows nothing of the amount and kind of variation to which the group is subject; and this shows, at least, how very generally there is some variation. But if he confine his attention to one class within one country, he will soon make up his mind how to rank most of the doubtful forms. His general tendency will be to make many species, for he will become impressed, just like the pigeon or poultry-fancier before alluded to, with the amount of difference in the forms which he is continually studying; and he has little general knowledge of analogical variation in other groups and in other countries, by which to correct his first impressions. As he extends the range of his observations, he will meet with more cases of difficulty; for he will encounter a greater number of closely-allied forms. But if his observations be widely extended, he will in the end generally be enabled to make up his own mind which to call varieties and which species; but he will succeed in this at the expense of admitting much variation, – and the truth of this admission will often be disputed by other naturalists.

Charles Darwin
*On the Origin of Species*

# Introduction: Gorilla biology: Multiple perspectives on variation within a genus

It is perhaps widely appreciated in the realm of primatology that until fairly recently, our perspective on gorillas has been an unbalanced one. Our earliest and most comprehensive accounts of gorillas in the wild can be traced almost entirely to studies of the East African gorillas of the Virunga mountains, *Gorilla gorilla beringei* or, as it is now being called by some, *Gorilla beringei beringei*, even though explorers' initial contacts were with the gorillas of West Africa. This is not unlike our early view of the chimpanzee, which for the better half of the twentieth century was based primarily on the chimpanzees of Gombe and the work of Jane Goodall, or our early view of the orangutan, which for years was informed by work conducted at only a few sites along the Bornean coast.

Pioneering work by well known field primatologists, such as George Schaller and Dian Fossey, provided the first systematic accounts of gorillas, cementing in our minds the image of these animals as plant-eating, quadrupedal, terrestrial knuckle-walkers. As a testimonial to the one-sided nature of these early gorilla field studies, the first important attempt to document the behavior of the western lowland gorilla is briefly noted by Schaller in the Preface to the Phoenix Edition of his landmark study *The Mountain Gorilla*, as "an interesting report on the little-known West African gorilla". Ironically, the gorillas of West Africa, that proved to be so difficult to study in the wild, provided the basis for the earliest and most extensive anatomical descriptions, creating a long-standing disconnect between their behavior and morphology that has taken decades to reconcile. It is worth pointing out that the eastern lowland gorillas, subsumed within the subspecies *Gorilla gorilla beringei* by Coolidge in 1929 and classified as such for 40 years, went virtually ignored in the wild until the tail end of the twentieth century.

Research conducted over the past 25 years has begun to redress the imbalance created by unilateral studies of western lowland gorilla morphology on the one hand, and eastern mountain gorilla behavior on the other. Thus, our objectives were twofold. One aim was to fill a notable gap in the primatological literature, resulting from the scattered accumulation of important gorilla research that has been conducted during the past several decades, and to assemble this information in a detailed and integrated framework. Our second objective was to emphasize an interdisciplinary and comparative approach to

1

gorilla biology. This approach provides the critical link between works that have focused exclusively on a single subspecies, and differs from studies of the Great Apes or African apes that have tended to incorporate a single "generic" gorilla subspecies as representative of gorillas. Where at all possible, we introduce multiple perspectives on the same topic, and in some cases, even multiple perspectives on the same data. Comparisons are drawn primarily between two or more groups of gorillas, be they local populations or subspecies, though a number of studies compare similarities and differences between gorillas and their closest living relatives, *Pan*. Placing these studies in a comparative and interdisciplinary framework highlights relatively recent developments that have challenged some previously held and deep-rooted notions of these apes.

We have emphasized areas of primate biology in which some of the greatest strides in our understanding of gorillas have been made; areas that ultimately have the potential to reshape our views of this ape and its relevance to studies of hominoid and hominid evolution, and influence more generally our perspective on interpreting the patterning of variation in nature. The authors provide thoughtful and informed discussions of gorilla phylogeny and taxonomy, the interface between morphology and behavior and its impact on locomotion and mastication, the patterning of morphological, molecular, and ecological variation and its importance for understanding speciation and species diversity, and gorilla demographics, conservation strategies, and the status of gorilla populations in the wild. These areas were chosen because of the natural interface among morphology, ecology, and behavior, and because integration of these types of data with molecular approaches is crucial if we are to advance our understanding of primate phylogeny and evolution, and prolong the existence of natural populations in the wild. Indeed, it is this intrinsic relationship that has provided much of the impetus for the gains we have made in gorilla research during the past decade, a relationship that resulted in our intentional focus on natural gorilla populations. To do justice to experimental research and research conducted in zoological settings, which is of equal importance and interest, simply falls beyond the scope of this volume. Neither do we attempt to tackle the evolutionary history of gorillas as there is scant evidence in the fossil record to address the phylogenetic relationships among fossil and extant gorillas or African apes. Such an exercise must await the discovery of new hominoid fossils.

Most of the chapters included in this book are based on papers originally presented at a symposium in honor of Harold Jefferson Coolidge, which was held at the annual meeting of the American Association of Physical Anthropologists in 1999 in Columbus, Ohio. However, several contributors have been added to round out some of the discussions of various topics. While broader in scope than the symposium, this volume retains as its focus the biology of gorillas. Doubtless, this seemingly narrow focus will engender in the minds of at least some the question of why devote an entire volume to one genus, with only one, or possibly two, species?

Simply stated, primatologists have long been fascinated by gorillas. This fascination began in earnest in the mid nineteenth century, when naturalists and explorers were beginning to discover and document the "elusive giants" of the Virunga mountains. Our interest has not waned over the years but rather, has been renewed, perhaps even intensified. Some of this interest was no doubt sparked by the remarkable finding of Ruvolo and colleagues in the early 1990s that mountain and lowland gorillas exhibit a greater degree of genetic divergence than that observed between chimpanzees and bonobos, the latter long recognized as distinct species of *Pan*. At about the same time that geneticists were beginning to sort out gorilla variation at the molecular level, primatologists were discovering that the well-studied but highly specialized mountain gorilla did not serve well as a "type" specimen for gorilla behavior; though all gorillas are terrestrial, quadrupedal knuckle-walkers and consume herbaceous vegetation, gorillas vary behaviorally and ecologically in ways that relate importantly to differences in social structure. One consequence of this appreciation of gorilla variability is that it has prompted investigators to take a fresh look at gorilla morphology, in light of both the degree of molecular variation, and the comparative data emerging on both their behavior and ecology. Add to these developments the rapidly declining numbers of gorillas in the wild and increased threats to their survival, and there is little mystery behind why our curiosity, along with our sense of urgency, has been heightened towards these apes.

Beyond the study of gorillas for gorillas' sake, the partitioning of variation in nature lies at the heart of evolutionary biology and in our view gorillas are emblematic of the larger challenges all biologists face as we attempt to better understand variation, at different levels and in different systems, and what drives it. Variation was clearly central to Darwin's theory of natural selection, and his observations of the degrees and kinds of variation that characterize everything from domestic pigeons to turnips to his famous Galapagos finches profoundly influenced his ideas on how species originate. Across the animal (not to mention plant) spectrum, long-standing attempts to understand and interpret biological variation continue to characterize a diversity of organisms. *Drosophila*, the well-known fruit fly, neotropical butterflies of the genus *Heliconius* (of which about 25% of species have been found to hybridize), and numerous genera of neotropical bats, are just a handful of examples. Closer to home, baboons (*Papio*) of the *hamadryas–anubis* hybrid zone provide a well-known and persistent example in primates of the difficulty of characterizing variation in biologically meaningful ways. Decades of debate have failed to yield a consensus as to whether *Papio* baboons form a single, highly variable species comprised of multiple subspecies, or whether they represent multiple distinct species. Perhaps because of their close proximity evolutionarily to humans, it has been especially difficult to evaluate and interpret variation in gorillas with the same degree of ojectivity that we might afford other, more distantly related taxa. Placing gorillas within the broader context of comparative evolutionary biology, rather than the more restricted realm of primatology (Great Apes or African apes even more so), may provide a welcome and, in our view, much needed perspective as we continue to deliberate gorilla diversity.

This theme, the partitioning of variation and its meaning in biology, provides the common link between chapters in this volume. We organized this book into four topical sections. The first section, "Gorilla taxonomy and comparative morphology", deals with morphological variation. Craniometrics has historically been at the center of gorilla systematics and has been evaluated repeatedly. Certainly taxonomic classification is important as a means of providing a common language which facilitates comparisons among groups of animals, comparisons that are necessary for addressing fundamental and, arguably, more interesting problems in evolutionary biology. But as Russel Tuttle points out in his introductory perspective, taxonomy is useful when it provides "meaningful" communication and reference. In other words, while the end results of taxonomic classifications may hold less intrinsic interest than the animals that comprise them, the means to the taxonomy are critically important. In this regard, one reason for the absence of a taxonomic consensus among authors, in the context of Tuttle's call that we adopt a uniform subgeneric scheme (based largely on Groves's 2001 book *Primate Taxonomy*; see Groves, this volume), may be that the evidence in support of a taxonomic revision is more clear to some than it is to others.

The first in this series of chapters is by Colin Groves, who, based largely on morphological variation, advises that *G. g. gorilla* and *G. g. beringei* are sufficiently distinct to warrant classification as separate species. This represents a shift in thinking from his one species, three subspecies classification of 1967, and may, in fact, reflect the accumulation of morphological and molecular evidence combined. At the same time, he provides a valuable historical summary of gorilla taxonomy that reminds us all of the importance of adequate samples and an appreciation for variation at all levels. This is the common theme in the chapters that follow by Stumpf *et al.*, Albrecht *et al.*, and Leigh *et al.*, as they consider issues of size adjustment, evaluation of metric vs. nonmetric characters, and assessment of variation at and below the level of the species, all of which lead to diverse interpretations of gorilla phylogeny and taxonomy.

Postcranial morphology has been emphasized in the literature as well in efforts to link anatomical variation to locomotor and positional differences between subspecies of gorillas and more generally among the African apes. The relationship between morphology and diet has been less well investigated, but studies of ontogenetic variation have been relatively neglected. Taylor and Inouye emphasize the importance of ontogenetic allometry in evaluating craniomandibular and postcranial variation, respectively, and show how concordance and discordance in patterns of ontogenetic allometry can influence adaptive arguments as well as clarify issues of taxonomy. Taylor provides a detailed evaluation of the masticatory apparatus in gorillas. She demonstrates for the first time that *G. g. beringei* has the relatively largest face as compared to lowland gorillas. Taylor suggests that increase in facial size, along with other structural modifications of the cranium and mandible, may be linked to their tougher diet. In her comprehensive ontogenetic study of the forelimb, Inouye finds that gorillas show remarkable ontogenetic concordance for most features, but the greater degree of humeral torsion observed in *G. g. beringei* may make sense of recent observations of a greater degree of terrestriality in the mountain as compared to the western lowland gorilla.

The second section, "Molecular genetics", addresses molecular variation between gorilla subspecies and within gorilla populations. There can be little doubt that the impetus to re-evaluate the evolutionary history and taxonomy of gorillas stems directly from a preliminary piece of molecular evidence pointing to a high degree of mitochondrial (mtDNA) variation between eastern and western gorilla subspecies. The application of molecular genetics to evolutionary questions is not new. However, molecular and computer-based technological advancements and, in particular, the complete sequencing of mitochondrial genomes cross-species, have facilitated their use in an unprecedented fashion. Molecular genetics has yielded important insights into the historical origins of diverse groups of animals, gene structure and function, paternity and reproductive success, population demography, and evolutionary history and taxonomy.

Molecular approaches to reconstructing taxonomy, however, have advanced ahead of well-defined and consensus-based criteria for interpreting genetic distances, and gorillas are symbolic of the ubiquitous shifts that seem to be taking place in primate systematics, largely on the basis of mtDNA. Examples include the newly "discovered" subspecies of *Pan, P. troglodytes vellerosus*, and the suggestion that *P. t. verus* may be genetically distinct from other chimpanzee subspecies. The Bornean and Sumatran subspecies of orangutans, it has recently been argued, would likewise more accurately be recognized as separate species on the basis of mtDNA. The recognition of three species of mouse lemur (*Microcebus*) "new to science", as well as the resurrection of two additional species, provides another similar and important example of how mtDNA has come to be viewed by some as a gauge of species diversity. As noted by Ryder in his introductory perspective, the application of molecular genetics to furthering our understanding of the evolutionary history of primates is still young, as is our appreciation for the range and meaning of variation of different parts of the genome, within and between taxa. In many ways, molecular genetics in anthropology is experiencing a stage in its history not unlike that outlined by Groves for morphology in the early twentieth century, when a naïve appreciation for variation led researchers to embrace all variation as equally biologically meaningful.

In the first chapter, Jensen-Seaman *et al.* present a much-needed re-evaluation of genetic variability in western and eastern gorillas using both mtDNA and nuclear DNA markers. Their results reveal discordance between these two types of DNA, but they also provide additional support for previous findings of a deep molecular genetic divergence between eastern and western gorillas. They also caution against a general inclination to use genetic markers for taxonomic purposes, particularly given variation in the evolution of different parts of the genome, and at different loci, both within and between taxa, and without pre-established criteria. This caution is carried over in their judicious approach to the use of *Pan* as a standard reference for characterizing genetic distance in gorillas.

In the second chapter, Clifford *et al.* present the first-ever assessment of genetic variability among local populations of western lowland gorillas, also using both mtDNA and nuclear DNA. Unfortunately, they were unable to obtain reliable results from their analysis of nuclear DNA (for reasons outlined by Ryder and detailed in their chapter), but their mtDNA results show distinct groupings among the western lowland gorillas, adding support for the distinctiveness of the Cross River gorillas. Their finding that the Dzanga group is genetically isolated from other western lowland populations has potential implications for gorilla phylogeography, and warrants further micro-level genetic studies of these gorillas (and possibly micro-level morphological and behavioral studies as well). Nevertheless, their conclusion that molecular differences among western lowland gorilla populations are sufficient to warrant taxonomic restructuring, based as it is exclusively on mtDNA, might be viewed by some as premature, and reflects a less conservative approach to the use of genetic data than that advocated by both Ryder and Jensen-Seaman *et al.* Both studies demonstrate that molecular techniques alone provide no greater resolution to the debate over gorilla phylogeny and taxonomy than morphology. They also underscore the importance of a phylogenetic species concept (*sensu* Cracraft) in systematics, particularly when attempting to assess the relatedness of allopatric taxa.

As we seek to preserve the richness of primate diversity, for both scientific as well as humanistic reasons, the question that is brought to the fore is how molecular data ought to be incorporated in defining and recognizing species, and whether (and how) such data should be used to establish conservation priorities. This question remains, as yet, unanswered, but we should remember that anatomical (and indeed behavioral) characters are also genetically based, and resist the facile temptation to equate "genetic" with "molecular" that seems to be occuring with increasing regularity.

In the section on "Behavioral ecology", years of focused research on the mountain gorilla in its natural habitat is juxtaposed against relatively recent work on natural populations of both western and eastern lowland gorillas. As highlighted by Tutin in her introductory perspective, the relatively recent accumulation of data on geographically diverse gorilla populations has allowed for more refined comparisons, resulting in some notable breakthroughs, particularly in the areas of population structure, sympatry, feeding ecology, and diet. Watts begins this section with a superb, comprehensive overview of gorilla socioecology and shows how comparative data on gorillas, primates and other mammals provide support for the ecological model of primate social relationships advanced by van Schaik. Though the importance of fruit has been highlighted by Tutin (and others), Yamagiwa *et al.* suggest that differences among populations of eastern lowland gorillas may be related to social constraints, such as infanticide and predation. Goldsmith follows these contributions with one of the few studies to date that examines and compares the behavioral ecology of both a highland and lowland gorilla population, thereby limiting some of the confounding variables that may affect inferences drawn from comparisons that have often been conducted on a *post-hoc* basis. In addition, Goldsmith provides interesting findings on perhaps one of the most poorly studied populations, the Bwindi "mountain" gorilla. Remis rounds out this section by providing comparative data on diet and nutrition in the only chapter that includes work on captive zoo gorillas. By combining experimental methods to evaluate taste preferences with dietary data from the field, Remis demonstrates that gorillas are quite selective nutritionally, not only when it comes to foraging on fruits but also on most types of vegetation.

In the final section, "Gorilla conservation", disturbing details are brought to light by these authors as they detail the vulnerability of gorilla populations to habitat destruction, poaching, disease, and warfare. All contributors to this section agree that gorilla populations are threatened. However, they differ, sometimes considerably, in their views of the causal bases for gorilla vulnerability, what can and should be done to minimize it, and what information is most important in efforts to curtail the threat of extinction. Disagreements among these authors are both highlighted in, as well as stimulated by, Sandy Harcourt's introductory perspective. The role of disease as a threat to gorilla populations is one example where authors disagree, with Plumptre *et al.* and Sarmiento emphasizing its importance, and Harcourt calling for a more tempered view. Curiously, various contributors to this volume hint at the importance of molecular genetics for conservation purposes, but Sarmiento is one of the few authors who clearly addresses the ways in which these data have important application, and Sarmiento and Harcourt disagree fairly pointedly as to what sorts of molecular data are most useful. In the final chapter, Oates *et al.* offer an unprecedented, multidisciplinary look at the behavioral ecology, morphology and genetics of the small and poorly known population of gorillas inhabiting the Cross River region on the Nigerian-Cameroon border. If there is a consensus among these authors, beyond their uniform support for gorilla conservation, it comes in the form of reciprocal support for the distinctiveness – in all aspects of their biology – of the Cross River gorillas, and the critical need to actively manage their status in the wild. Of all the local populations of western and eastern gorillas evaluated, there seems to be the most evidence across disciplines to support the Cross River gorillas as a distinct subspecies, but even this evidence must be viewed as preliminary.

Gorillas occupy a unique position in our evolutionary history and their importance in helping us to understand what it means to be uniquely human has not been altered by the recent discovery that chimpanzees and humans are genetically closer to each other than either is to gorillas. Chimpanzees may be our African ape "sisters", but gorillas are our close cousins by a mere 2 million years more. Yet while chimpanzees have historically served as the preferential model for the last common ancestor shared with our earliest hominid ancestors, gorillas have been routinely incorporated as a kind of "calibration tool" in analyses attempting to characterize the degree of morphological variation in fossil hominids.

Compared to other primate genera, such as *Macaca* or *Papio* or even *Pan*, gorillas as a group are fairly specialized, whether recognized as one species or two. It is precisely because they are specialized that contrasts between lowland and mountain gorillas are so important; they enable us to encapsulate the full range of biological differences among extant hominoids – social, ecological, molecular, morphological, cognitive, life history – and a fuller appreciation of the evolutionary bases of our own history depends on this broader hominoid context. Hylobatids, orangutans, chimpanzees, and gorillas are all that remain of a once highly diverse and successful Miocene ape radiation. Hominoids are few in number but differ profoundly in many aspects of their biology, which makes it all the more important for us to understand each to the greatest extent possible.

How much more we will be able to learn about these apes remains to be seen, given that all are listed by the IUCN as endangered, some critically. Gorilla survival depends on the preservation, perhaps even expansion, of gorilla habitats and the maintenance of their genetic diversity. If we wish to extend our study of these apes beyond the twenty-first century, it is incumbent upon all of us to participate in this endeavor. Though the path to success remains uncertain, one thing is clear, and it is a point on which, perhaps belatedly, researchers across disciplines seem to be in agreement: The responsibility for preserving these nonhuman apes is a shared and unequivocally human one.

Andrea B. Taylor and Michele L. Goldsmith

# Part 1
## *Gorilla taxonomy and comparative morphology*

# 1 An introductory perspective: Gorillas – How important, how many, how long?

RUSSELL H. TUTTLE

Gorillas rank highly among elephants, pandas, whales, polar bears, lions, orangutans, and other large mammals as awe-inspiring representatives of *natura naturans* (nature as creative) and *natura naturata* (nature as created). Like unique, imaginative, stimulating literature (Booth, 1988), one cannot encounter them without being changed in ways that are not easily explained. Indeed, attempts to do so can dilute the wonderful effect of having been in their presence. Earth will be a poorer planet if we lose its remaining, already impoverished, continental megafaunae and multifarious smaller beings and their natural habitats. The urgent pedagogical task of ecologists, educators, conservationists, policy-makers, and politicos is to generate appreciation for gorillas and a sense of local and national pride in having gorillas among indigenous peoples upon whom the stewardship of natural diversity is ultimately dependant (Tuttle, 1998).

Gorilla taxonomy is important for meaningful communication and reference, but questions over how many species or subspecies *Gorilla* comprises should be secondary to full descriptions of the morphological, genetic, social, demographic, and ecological diversity of gorilla populations and sample specimens in museums and private collections. The best chance for their survival lies in preserving genetic variety, behavioral plasticity, and a broad geographic distribution in sustainable habitats, some of which should probably be allowed to expand in Africa. This volume is but a small step in the right direction – documenting the diversity of gorillas – but unfortunately this may be the only way that we can progress given the politico-economic status of most countries that are still blessed with gorilla populations.

There is no consensus regarding the number of species and subspecies of *Gorilla* among the authors who analyzed the large data set of cranial features collected by Groves (1967, 1970, 1986; Groves and Humphrey, 1973; Groves and Stott, 1979) and subsets and augmented subsets of it. Inouye provides the only analysis of postcranial features, viz., in the forelimb skeleton, to complement the analyses of cranial skeletal traits. Consequently, the morphological

11

Table 1.1. *Summary of hypotheses on the species and subspecies of* Gorilla *in Chapters 1–6*

| Author | Hypothesis | Method |
|---|---|---|
| Groves | **Two species** (*Gorilla beringei* and *G. gorilla*) and probably **four subspecies** (*Gorilla beringei beringei*, *G. b. graueri*, *G. gorilla gorilla*, *G. g. diehli*) | Canonical discriminant function analyses of male skulls |
| Stumpf *et al.* | **Two species** (*Gorilla beringei* and *G. gorilla*) and noncommittal re subspecies, but do not counter Groves' system | Canonical variates analyses, Mahalanobis $D^2$; and canonical variates analyses on size-adjusted measurements of male and female skulls |
| Albrecht *et al.* | **One species** (*Gorilla gorilla*) and **four subspecies** (*G. gorilla gorilla*, *G. g. diehli*, *G. g. beringei*, *G. g. graueri*) | Principal components analysis of male and female skulls |
| Leigh *et al.* | **One species** (*Gorilla gorilla*) and **three subspecies** (*G. gorilla gorilla*, *G. g. beringei*, *G. g. graueri*), with high diversity among *G. gorilla gorilla* | Wright's $F_{ST}$ and discrete trait analyses on male and female crania |
| Taylor | **One species** (*Gorilla gorilla*) and **three subspecies** (*G. gorilla gorilla*, *G. g. beringei*, *G. g. graueri*) | Principal components and ordinary least-squares regression analyses on log-transformed data from ontogenetic series of male and female skulls |
| Inouye | **One species** (*Gorilla gorilla*) and noncommittal re subspecies | Ontogenetic, allometric analyses of female and male forelimbs |

profiles of subgeneric taxa of *Gorilla* are incomplete, lacking epidermal features and variations in muscles, ligaments, viscera, bodily proportions and much of the postcranial skeleton. In addition to fleshing out this database, it would be useful to have genetic profiles of reliably provenanced museum specimens, based on hair and collagen samples, to compare with samples from living populations.

Groves and Stumpf *et al.* accept two species of *Gorilla*, while Albrecht *et al.*, Leigh *et al.*, Taylor and Inouye subscribe to a monospecific scheme (Table 1.1). Like Butynski (2001), Groves and Stumpf *et al.* acknowledge molecular genetic evidence that support specific status for *Gorilla gorilla* and *Gorilla beringei* (Ruvolo *et al.*, 1994; Garner and Ryder, 1996; Uchida, 1996; Ryder *et al.*, 1999).

Among the authors who comment on subspecific taxa of *Gorilla* (Table 1.1), all accept at least three subspecies, and Groves and Albrecht *et al.* endorse a fourth subspecies (*Gorilla gorilla diehli*).

None of the authors presented a scheme of common names for subgeneric taxa of *Gorilla*. I propose that we adopt the following scheme based on Groves (2001:300–303), and urge editors and authors to employ it henceforth:

| | |
|---|---|
| western gorilla | *Gorilla gorilla gorilla* |
| Cross River gorilla | *Gorilla gorilla diehli* |
| mountain gorilla | *Gorilla gorilla beringei* and *Gorilla beringei beringei* |
| grauer gorilla | *Gorilla gorilla graueri* and *Gorilla beringei graueri* |

The decimation of gorillas due to the bushmeat trade, trophy and subsistence hunting, and deforestation is reducing genetic variation in the genus, and there is little evidence of naturalistic genetic recombination among gorillas in widely dispersed habitats. The less they are variable, the more they might be vulnerable to pandemic diseases, particularly from human repositories as contacts with humans, including indigenous peoples, researchers, conservators and tourists, increase. Sadly, one must wonder how much longer they will survive as natural beings in the twenty-first century.

## References

Booth, W.C. (1988). *The Company We Keep: An Ethics of Fiction*. Chicago, IL: University of Chicago Press.

Butynski, N.T. (2001). Africa's great apes. In *Great Apes and Humans: The Ethics of Coexistence*, eds. B.B. Beck, T.S. Stoinski, M. Huchins, T.L. Maple, B. Norton, A. Rowan, E.F. Stevens, and A. Arluke, pp. 3–56. Washington, D.C.: Smithsonian Institution Press.

Garner, K.J. and Ryder, O.A. (1996). Mitochondrial DNA diversity in gorillas. *Molecular Phylogenetics and Evolution*, **6**, 39–48.

Groves, C.P. (1967). Ecology and taxonomy of the gorilla. *Nature*, **213**, 890–893.

Groves, C.P. (1970). Population systematics of the gorilla. *Journal of Zoology, London*, **161**, 287–300.

Groves, C.P. (1986). Systematics of the great apes. In *Comparative Primate Biology*, vol. 1, *Systematics, Evolution and Anatomy*, eds. D.R. Swindler and J. Erwin, pp. 187–217. New York: A.R. Liss.

Groves, C.P. (2001). *Primate Taxonomy*. Washington, D.C.: Smithsonian Institution Press.

Groves, C.P. and Humphrey, N.K. (1973). Asymmetry in gorilla skulls: Evidence of lateralised brain function? *Nature*, **244**, 53–54.

Groves, C.P. and Stott, K.W., Jr. (1979). Systematic relationships of gorillas from Kahuzi, Tshiaberimu and Kayonza. *Folia Primatologica*, **32**, 161–179.

Ruvolo, M., Pan, D., Zehr, S., Goldberg, T, Disotell, T.R., and von Dornum, M. (1994). Gene trees and hominid phylogeny. *Proceedings of the National Academy of Sciences U.S.A.*, **91**, 8900–8904.

Ryder, O.A., Garner, K.J., and Burrows, W. (1999). Non-invasive molecular genetic studies of gorillas: Evolutionary and systematic implications. *American Journal of Physical Anthropology, Supplement* **28**, 238.

Tuttle, R.H. (1998). Global primatology in a new millennium. *International Journal of Primatology*, **19**, 1–12.

Uchida, A. (1996). What we don't know about great ape variation. *Trends in Ecology and Evolution*, **11**, 163–168.

# 2  *A history of gorilla taxonomy*

COLIN P. GROVES

## Prologue

In the fifth century BC the Carthaginian admiral Hanno was commissioned to sail down the west coast of Africa and found Carthaginian colonies. After dropping off colonists at intervals, he sailed on, and eventually came to a fiery mountain, near which was a bay in which was an island. In the island was a lake, and within this another island, full of savage hairy people whom the interpreters called "gorillas". The Carthaginians tried to catch them; the men escaped and threw stones from the cliffs; but they caught three women and, finding them untameable, killed them, skinned them and took their skins back to Carthage.

Hanno's voyage has been endlessly discussed: did he get to Cameroon? – to Sierra Leone? – just to southern Morocco? Were they really gorillas? – or chimpanzees? – or baboons? – or even Neandertalers? What language was this word "gorillas", and who were these interpreters?

Heuvelmans (1981) was the first to point out what should have been obvious: the account of the voyage which has come down to us is in Greek, not Punic (the language of Carthage), and it was written some centuries after the voyage was said to have taken place. In the interval, who knows how much it has been embellished, abbreviated, and perhaps modified to accord with various accounts of Greek, Phoenician and Egyptian ocean voyages? There seems little hope of ever establishing what those so mercilessly slaughtered "gorillas" actually were, or where they lived. Rightly or wrongly, they live on – in their name.

Real gorillas entered European literature for the first time in an account by Andrew Battell, an English sailor held prisoner by the Portuguese in Angola (almost certainly Cabinda, the enclave north of the Congo River mouth), in the sixteenth century. In his account of the region, he wrote of two human-like "monsters", Pongo and Engeco. Though mixed up with all kinds of misinformation, the description of Pongo is certainly that of a gorilla. Alas, by the late eighteenth century most Great Apes had become inextricably confused under the term "orangutan", with the result that *Pongo* became the generic name for the great red ape of Borneo and Sumatra!

15

## The gorilla becomes known to science

The first scientific description of the gorilla, with its scientific name *Troglodytes gorilla*, was published without fanfare, without fuss of any kind, by Dr. Jeffries Wyman, an anatomist of Boston. Wyman, however, was not the author of the name. The *Proceedings of the Boston Society of Natural History* states that "Dr. J. Wyman read a communication from Dr. Thomas S. Savage, describing the external character and habits of a new species of *Troglodytes* (*T. gorilla*, Savage) recently discovered by Dr. S. in Empongwe, near the river Gaboon, Africa"; this is followed by Dr. Savage's communication, presumably a letter, then a note that Dr. Wyman exhibited some bones, and what seems to be a transcript of Wyman's description of them including the differences from the chimpanzee.

The *International Code of Zoological Nomenclature* (4th edition) states quite clearly, in Art. 50.2:

> **Names in reports of meetings.** If the name of a taxon is made available by publication of the minutes of a meeting, or an account of a meeting, the person responsible for the name, not the Secretary or other reporter of the meeting, is the author of the name.

Therefore the author of the name *T[roglodytes] gorilla* is Savage, not "Savage and Wyman" as commonly cited.

It should be explained that, at that time, the generic name in common use for the chimpanzee was *Troglodytes* (E. Geoffroy St. Hilaire, 1812). It is now realized that this name is preoccupied (i.e., the name had already been proposed for another animal); so the name *Troglodytes* Vieillot, 1806, is the generic name of the wren, and after much struggle we have settled down with *Pan* Oken, 1816, for the chimpanzee.

In December of the same year a fuller description appeared (Savage and Wyman, 1847). Savage gave a full account of its discovery, giving the type locality (the mission house where he had been given the skeletal remains and told of its external appearance and habits) as the Gaboon River, at $0°15'$ N, quoting Hanno and Battell, and explaining that he had adopted Hanno's term as the specific name; followed by an even more detailed account of its osteology by Wyman. Nonetheless the earlier paper, in August of that year, stands as the original description of the name.

Early in the following year Owen (1848*a*) described further specimens, supplied to him by Savage, under the name *Troglodytes savagei*; later (Owen, 1848*b*) he admitted priority to the name given by Savage himself. The specimens brought to the U.S.A. by the Reverend Savage are today in the Museum of Comparative Zoology, Harvard. Those described by Owen are in the Natural History Museum (formerly British Museum (Natural History)), London.

It was not long before the gorilla was awarded a genus of its own. Isidore Geoffroy St. Hilaire (1852), the son of Etienne who had given the chimpanzee its earliest, if already preoccupied, generic name (*Troglodytes*), set it apart in a new genus, *Gorilla*.

### Several species of gorillas? Nineteenth-century thoughts on the matter

No sooner had he accorded the gorilla its own genus, than Isidore Geoffroy received a skull which differed so markedly from those he had previously seen that, though it likewise came from Gabon, he felt sure that it represented a second species: *Gorilla gina* (I. Geoffroy St. Hilaire, 1855). Early in the next decade Slack (1862) described a third species, *Gorilla castaneiceps*, based on the coloured cast of a head in which the crown hair was red, not black. We now know, of course, that the colour of the crown is polymorphic in gorilla populations. The cast studied by Slack is still in the Academy of Natural Sciences, Philadelphia; ironically, when I saw it, it had been repainted – completely black.

More enduring was the name given by Alix and Bouvier (1877) to a supposed dwarf gorilla, *Gorilla mayema*. The small female gorilla on which this name was based differed from *Gorilla gorilla* in osteological details and in its pelage, most notably in having a "collar" of hair round the face and the back covered with long, thick hair. The type locality was given as the village of King Mayêma, on the banks of the Quilo at 4° 35′ S, but was later changed by Famelart (1883), on what evidence we do not know, to Conde, near Landana.

Elliot (1913) could not find Alix and Bouvier's specimen when he searched for it in the Muséum National d'Histoire Naturelle, Paris. I found a female skull in the Laboratoire d'Anatomie Comparée of the Museum, which I judged might have been the type of *G. mayema*; it is the smallest adult gorilla skull I have seen.

In the last years of the century an enormous gorilla, said to be 207 cm long (how measured?!) and with a span of 280 cm, was shot, in the uncompassionate way of those days, by H. Paschen at Yaounde, Cameroon, and figured and named *Gorilla gigas* by the notorious Ernst Haeckel in 1903.

### Early studies on the gorilla's anatomy

The type and follow-up descriptions of the gorilla (Wyman, 1847; Savage and Wyman, 1847) themselves contained detailed descriptions of the gorilla's osteology. Further detailed descriptions were given by Owen (1848*a*,*b*,*c*, 1862*a*,*b*).

By the time Owen was writing his second series of papers, in the 1860s, the Darwinian controversy was in full swing, and Owen was especially anxious to use his studies of the gorilla and other apes to combat "the hypothesis of transmutation". Already (Owen, 1857) he had made it clear that, for him, characters of the brain set humans clearly apart from other mammals, as sole representatives of a new subclass, Archencephala. Only the human brain, he claimed, had a special posterior lobe of the cerebrum, covering the cerebellum; had a posterior horn of the lateral ventricle; and had a hippocampus minor. He repeated these assertions in several other places (see, for example, Owen, 1861), despite their accuracy being increasingly challenged by Huxley. It was the examination of a gorilla brain that led Huxley (1867) to contradict Owen fully and finally, and show that "Man" was far from unique in those features: Owen had been caught letting his prejudices get in the way of his objectivity, and the Darwinian revolution became that much more firmly entrenched. The whole story, of the irrational behavior and downright dishonesty displayed by this great anatomist, is much more complicated (and shaming to Owen) than I have given here; it has been most recently recounted by Wilson (1996).

Subsequent descriptions of the anatomy of the gorilla (Bischoff, 1880; Sonntag, 1924) were much more simply factual. Others (Hartmann, 1886; Keith, 1896; Duckworth, 1915) were more concerned to place the gorilla in evolutionary context, but were equally objective and no one has questioned their accuracy. Small packets of new information were constantly fed into the system as well; thus Neuville (1916) described some skulls of adult males which, unexpectedly, lacked any sagittal crest (and later (Neuville, 1932), in view of the taxonomic prolixity that was to follow, was able to congratulate himself for not having burdened gorilla taxonomy with a new species name for them!).

## Paul Matschie: Let a hundred species bloom

Professor Paul Matschie was an anti-evolutionist, with in addition some highly bizarre ideas about mammalian taxonomy and variability; though quite untrained, he had been appointed curator of mammals in the Berlin Museum. Representatives of each group of animals from each river valley simply had to be different species; it was just a matter of finding the differences. Uncounted scores, perhaps hundreds, of new mammal species were launched into a brief existence under this philosophy; on several occasions a buffalo or antelope would even be described as two separate species (represented by the left and right sides of the animal!) because, being shot on watersheds, they were interpreted as hybrids! Until his eccentricities became so outrageous that they could no longer be ignored, Matschie bestrode mammalian taxonomy for the first quarter of the twentieth century, and his contemporaries could only look on in admiration and try to emulate him.

Two of Matschie's admirers stand out in the present context. The Hon. Walter Rothschild was a millionaire who was passionately interested in natural history, and founded his own private museum for its study, at Tring, north of London. Daniel Giraud Elliot was a general zoologist who worked first in European museums, then in Chicago, and "retired" to an honorary post in the American Museum of Natural History, New York, and in his retirement studied primates, not entirely to the betterment of our understanding. Among them Matschie, Rothschild and Elliot increased the number of putative gorilla species and sub-species nearly fivefold.

As if by accident, the first of the many new gorilla taxa that Matschie described was a good one: the mountain gorilla, *Gorilla beringei* (Matschie, 1903). Captain von Beringe had shot two gorillas at 3000 m on Mt. Sabinio in the Virunga mountains; these differed from *Gorilla gorilla* in their thicker pelage and much stronger beard, and the skull of the male differed from 25 available skulls of *G. gorilla* in several respects:

- narrower, more pointed nasal bones
- palate longer than the distance from its posterior end to the foramen magnum, instead of shorter as in *G. gorilla*
- weaker supraorbital torus, only 8–9 mm thick (cf. more than 11 mm)
- lack of the spur or crest which in West African skulls extends from the lower orbital margin to the infraorbital foramen.

"Ich nenne ihn *Gorilla beringeri*", wrote Matschie (1903:257). As the collector's name was von Beringe, and Matschie himself knew this, this form of the name clearly ranks as an Incorrect Original Spelling under Art. 32.5 of the *Code*, and must be corrected (to *beringei*) while retaining the original author and date (Matschie, 1903). I might add that, of the four skull characters, only the long palate is a good, consistent difference between *G. gorilla* and *G. beringei*.

The type specimen of *G. beringei* could not be discovered when I visited the Zoologisches Museum A. Humboldt in Berlin in the 1960s.

The following year Matschie revised the genus *Gorilla*, recognizing four species: *G. gorilla, G. castaneiceps* (with *G. mayema* a synonym), *G. beringeri* (still spelled incorrectly) and a new species, *G. diehli*, from Dakbe (type locality), Oboni and Basho, on the Cross River in the then-German territory of West Kamerun (Matschie, 1904). The striking difference between *G. diehli* and other gorillas was in the broad, low nuchal surface. Of nine skulls collected by Herr Diehl – most of them from African huts, where they were kept as "fetishes" – eight were of the new species, but one, a female from Basho, differed in no respect from Gabon and southern Cameroon gorillas, and Matschie consequently referred it to *G. gorilla*. All the Cross River skulls are still in the Museum in Berlin.

Rothschild, meanwhile, purchased the Paschen gorilla from the Hamburg Museum for 20 000 marks for his own museum and, apparently unaware that Haeckel had already awarded it a name, described it as *Gorilla gorilla matschiei* (Rothschild, 1905:415). Comparing its skull with four large skulls of the Gabon gorilla, he drew attention to differences in the shape of "the hinder surface of the head", the basioccipital bone, and the condyle and coronoid of the mandible; he did not specify exactly what these differences would be, but from his comparative table the nuchal surface would be longer and wider, and the basioccipital shorter. Rothschild reduced *diehli* to the status of subspecies under *G. gorilla*, and regarded Slack's *castaneiceps* as a mere "casual aberration" of the Gabon gorilla. It must also be stated that Rothschild was actually one of the pioneers of the subspecies concept, and argued very forcibly (1905:439–440) for the trinomial.

The Rothschild Museum, Tring, was incorporated into the then British Museum (Natural History) in 1939, and the type skull of *G. g. matschiei* is now in the Mammal Section of the Natural History Museum; but, the last time I looked, its mounted skin was still on display in Tring.

Matschie (1905) described his third new species, *Gorilla jacobi*, from Lobomouth, on the Dja River, in southern Cameroon. The type skull and the other two skulls are still in Berlin. The distinguishing characters are simple individual variation; but it is worth recording that the type is the largest gorilla skull I have seen.

Matschie was also held responsible for the description of *Gorilla gorilla schwarzi* by Fritze (1912), from "Sogemafarm" (correctly Sogemafam) in southwestern Cameroon, but there is no provision in the *Code* (see especially Art. 50) that anyone but Fritze should be cited as the author of the name. I could not find the type of this supposed subspecies in the Karlsruhe museum. 1913 saw the publication of Elliot's epoch-making, much-maligned *Review of the Primates*. He evidently did not know of *schwarzi*, but he recognized (with admittedly some misgivings) all the taxa that had been described up till then: *gorilla, matschiei, diehli, jacobi, castaneiceps, beringei, mayema* (but added a special touch of his own as far as the last was concerned), and two which he did not name, leaving that task to Matschie. Matschie (1914) took up the challenge, naming the two which Elliot had signalled: *Gorilla hansmeyeri*, from south of the Dume River, on the Assobam road between Mensima and Bimba; and *Gorilla zenkeri*, from Mbiawe on the River Lokundje, six hours upriver from Bipindi (both in southern Cameroon). He also named a second eastern species, *Gorilla graueri*, from 80 km northwest of Boko, on the western shore of Lake Tanganyika, in the present-day Democratic Republic of Congo, formerly Zaïre, and noted its similarity to *G. beringei* (which by this time he was spelling correctly). The types of these three putative species are still in the Berlin museum.

This was Matschie's last paper on gorillas. He had described six supposed new species/subspecies, had been possibly responsible for a seventh, and had had one named after him. His ideas became more and more mystical; eventually he concluded that species' distributions were bounded not by watersheds but by the diagonals of the even degrees of longitude and latitude. He died in 1926, aged only 64.

Only four other species and subspecies of gorillas – if we exclude Frechkop's contribution (below) – were described. Lönnberg (1917), under the impression that mountain gorillas in the Virunga range were confined to individual volcanoes, described *Gorilla beringei mikenensis* from Mt. Mikeno (Matschie's *beringei* was from Mt. Sabinio). Rothschild (1927) described *Gorilla gorilla halli* from Punta Mbouda (correctly Mbonda) in Equatorial Guinea. Finally Schwarz (1927) described another eastern form, *Gorilla gorilla rex-pygmaeorum* from Luofu, west of Lake Edward; whatever one thinks of his taxonomic assessment, he certainly had poetry in his soul (the name means "king of the pygmies"); and in the same paper he mentioned and briefly described three skulls from Djabbir, near Bondo, in the Uelle valley, with a footnote: "Note du Dr. Schouteden: ils s'agit du *G. uellensis* Matschie" (according to the *Code*, Schouteden must be regarded as author of this name). The types of all these taxa are still extant (*mikenensis* in Stockholm, *halli* in London, the other two in the Museum voor Middenafrika, Tervuren, Belgium).

## Pygmy gorillas?

Alix and Bouvier's supposed pygmy gorilla, *Gorilla mayema*, has refused to lie down, almost up to the present day. In his 1905 paper Rothschild indicated that he believed *Gorilla manyema* (*sic!*) to be actually a chimpanzee, but three years later (Rothschild, 1908) he stated that he had received specimens of *Gorilla manyema*, and now saw that it was not a chimpanzee after all but must be the gorilla race of the "South Congo". The skull, he said, is narrow; and a photograph of the whole animal showed, he said, "very sharply defined pale and dark areas".

Now, Manyema (now spelled Maniema) is a district in Democratic Republic of Congo, east of the Lualaba River, where eastern lowland gorillas are found. Did Rothschild misspell *mayema* as *manyema* because that was the source of his specimens, and by "South Congo" did he mean Upper Congo? One of the Rothschild Collection skulls in the Natural History Museum, BM 1939.945, an old male, has "*G. g. manyema*. Upper Congo" written on it, and assorts as an eastern lowland gorilla. In my first publication on gorillas (Groves, 1967), I used the name *manyema* for the eastern lowland gorilla; but Corbet (1967) pointed out that, whatever the circumstances, *manyema* is clearly an inadvertent error for *mayema* and so has no status in nomenclature. And, of course, Rothschild's misreading of the name certainly dissociated it from pygmy gorillas!

Elliot (1913) searched unsuccessfully for Alix and Bouvier's specimen in Paris, and then applied the name, without giving any justification, to a male, female and young (skulls and mounted skins) from "Upper Congo" in the Senckenberg Museum, Frankfurt. Having the impression that these specimens were intermediate between a gorilla and a chimpanzee, he erected a new genus, *Pseudogorilla*. Compared to *Gorilla*, the genus *Pseudogorilla* could be distinguished by its small size, lack of sagittal and nuchal crests, more rounded forehead, and other features.

In 1943, Frechkop gave it as his opinion that the Frankfurt specimens were not, in fact, the same as Alix and Bouvier's, so he renamed them *Pseudogorilla ellioti*.

Actually, two years after Elliot's book was published, Miller (1915:6, n. 1) revealed the true nature of the specimens: "an immature male with all the teeth in place but with the basal suture open and the temporal ridges separate ... and a mature female with the basal suture closed and the temporal ridges joined." I have seen the specimens, and I concur. Maybe Elliot had never seen subadult male skulls, or even those of adult females; or maybe he had seen them but taken little notice of them. Presumably the skins, mounted and on display to this day in the Senckenberg Museum, belong with the skulls; they are, at any rate, of male and female of corresponding ages (plus an infant). Heuvelmans (1981), inclined to accept the real existence of pygmy gorillas, drew attention to the long, sharp canines visible in the open mouth of the mounted male skin; but the canines are made of wood (Dieter Kock, personal communication).

Throughout the century the ghost of the pygmy gorilla has been periodically resurrected. I hope I have finally laid it to rest (Groves, 1985).

### Order out of chaos

Schwarz (1928) was perhaps the first to try to reduce the plethora of described species and subspecies to manageable proportions. He recognized just one species, *Gorilla gorilla*, with the following subspecies:

*G. g. gorilla* (synonyms *castaneiceps, mayema, halli*)
*G. g. matschiei* (synonyms *jacobi, schwarzi, hansmeyeri, zenkeri*)
*G. g. diehli*

*G. g. uellensis*
*G. g. rex-pygmaeorum*
*G. g. graueri*
*G. g. beringei* (synonym *mikenensis*)

He gave no descriptions of these subspecies, and the "revision" is probably more in the nature of a conveninent disposal – for example, all the southern Cameroon taxa are lumped into one subspecies, the Gabon and Equatorial Guinea taxa into another.

The first major formal revision, which became standard for 40 years, was that of Coolidge (1929). This can be criticized, and it has been (Haddow and Ross, 1951), but probably Coolidge did the best that could be done with the restricted amount of material available to him. Coolidge put all gorillas into one species, as had Schwarz, and recognized only two subspecies: *G. g. gorilla* for all western gorillas, and *G. g. beringei* for all eastern ones. The narrow cranium and long palate of the latter were the most striking differences. He noted that the skulls from Bondo (Schouteden's *uellensis*) were indistinguishable from *G. g. gorilla*, and suggested that they had originated further west and had ended up in the Bondo region by trade, but he later (Coolidge, 1936) acknowledged that the evidence was against this, and that gorillas were (or had been) native to the region.

Schultz (1934) argued that Coolidge had underestimated the differences between western and eastern gorillas, and that they were actually distinct species. He listed 20 differences between them. Looking over his list, some of the differences are cogent, some not, but the major criticism is that for the eastern ("mountain") gorillas only Virunga and Kahuzi specimens were available to him; the skeletal differences related mainly to Virunga, the external differences to Kahuzi, none to any of the other eastern populations. Groves and Stott (1979) tried to determine how far Schultz's features could be applied to other eastern populations – with varying results.

A little-known paper by Rzasnicki (1936) used indices taken from human craniometry to sort out gorillas, using Coolidge's data. He regarded six "complexes" of skull measurements as originating in different parts of the distribution and spreading out to different populations so that, just as it was customary until mid-century to regard different human populations as consisting 40% of one race, 25% of another, 20% of a third, and so on, so it was with gorillas. The three centres-of-origin of these "complexes" which he elucidated were made subspecies: these were the Cross River (*G. g. diehli*), the rest of the western area (*G. g. gorilla*), and the eastern area (*G. g. beringei*).

Vogel (1961) studied mandibles. He separated *G. beringei* as a distinct species from *G. gorilla*, and within it recognized two subspecies: *G. g. beringei*

(Virunga) and *G. g. graueri* (rest of the eastern region). He thus deserves credit as the first person to distinguish the eastern lowland gorilla as a category from the true mountain gorilla, although as pointed out by Schaller (1963) some of his localities for the eastern region are incorrectly identified and, in a study based on 38 jaws (divided, of course, into male and female), any misallocation matters.

The definitive revision that finally took over from Coolidge's and became standard for 30 years was that of Groves (1967). The advantages that had accrued in the interval included more material, from a wider range of localities; better understanding of gorilla ecology and distribution; and, in particular, new analytic methods especially multivariate analysis. I measured and took notes on 469 male and 278 female skulls, and allocated the crania to 19 circumscribed geographic regions and the mandibles (of which there were fewer specimens than of crania) to ten such regions, with at least 11 specimens in each. I took up to 45 measurements on each skull. As it is important as a general statistical rule of thumb that the number of specimens per sample should exceed the number of variables, I ran correlation coefficients and made ten intercorrelated groups of cranial variables and six of mandibular variables; I then selected one from each group and ran a series of Mahalanobis generalized distance analyses.

The results enabled me to pool the 19 cranial and 10 mandibular samples into eight major regions ("demes"), as follows: (a) Western: Coast (and including the whole of Gabon), Cameroon Plateau, Sangha River, and Cross River; (b) Eastern: Utu (i.e., lowlands east of the Lualaba), Tshiaberimu, Itombwe ("Mwenga-Fizi" region of Schaller, 1963), and Virunga. Unallocated specimens were from Bondo (*G. g. uellensis* Schouteden, 1927), Lubutu (north of Utu), Kahuzi, and Kayonza (now preferably called Bwindi-Impenetrable) Forest. All western samples, even when grouped into the four regions, were very close – the Cross River sample somewhat more distinctive than the rest – and I considered them to represent a single subspecies, *Gorilla gorilla gorilla* (Savage and Wyman, 1847, *recte* Savage, 1847); and the Bondo specimens fell well within their range of variability. Among the Eastern regions, however, the Virunga sample fell well away from the others, and I restricted *G. g. beringei* to the Virunga population; the others I grouped into a third subspecies, which at first (Groves, 1967) was called *G. g. manyema* (Rothschild, 1908), but after criticism by Corbet (1967) I called it *G. g. graueri* (Matschie, 1914). The few Kahuzi specimens seemed rather closer to *beringei*, the Bwindi ones to *graueri*.

Groves and Stott (1979) returned to the Kahuzi and Bwindi question; a little new information about them had meanwhile become available. The Kahuzi gorilla was now shown quite clearly to be *graueri*, while the Bwindi gorilla was still equivocal, but perhaps closer to *beringei* after all.

### Recent understanding of the anatomy of the gorilla:
### Normal and pathological

The landmark in anatomy, that brought order out of chaos as Coolidge (1929) had done for taxonomy, was the volume edited by Gregory (1950), the Henry Cushier Raven Memorial Volume, *The Anatomy of the Gorilla*. Featuring papers by some of the leading human and primate anatomists of the mid-century era, S.L. Washburn, H. Elftman, W.B. Atkinson, W.L. Straus and A.H. Schultz, the centerpiece was a profusely illustrated *tour de force* on regional anatomy by the dedicatee himself, H.C. Raven. It is perhaps disappointing that the papers are in the main exclusively descriptive, only Straus and Schultz and to a lesser degree Washburn having much in the way of a comparative approach. But the data were now, for the very first time, laid out in an accessible, detailed form, in a single volume.

The era of gross anatomy was coming to an end. Cave (1959) published a study on the nasal fossa of a gorilla fetus, and later (1961) on the frontal sinus; he confirmed that, like *Pan* and *Homo, Gorilla* has a true frontal sinus of ethmoid origin. Cave had specifically examined frontal sinuses from a phylogenetic point of view, and from now on anatomical studies would be problem-oriented in approach: pathology, function and so on.

The first problem studied by Angst (1967, 1970) was sagittal and nuchal cresting. He confirmed the general view that the nuchal crest and the posterior part of the sagittal crest depend predominantly on jaw length and associated m. temporalis size, whereas the anterior part of the sagittal crest is associated with shorter jaws; but the matter is more complicated than this. For example, two adult male gorilla skulls lacking a sagittal crest had a smaller cranial capacity than any other males. Later the same author (Angst, 1976) published what is probably the most complete survey to date on cranial capacity. In 38 adult males measured by him, cranial capacity averaged 565.2 cm$^3$, with a range of 440–672 cm$^3$ (but he noted an unusually large value of 752 cm$^3$ published by Schultz (1962)); in 27 adult females, 482.6 (403–583) cm$^3$. These values would seem to be achieved around the time of eruption of the canines and third molars.

Pathology has been another concern of more recent studies. The skulls of gorillas reared in captivity in the pre-war period showed characteristic differences from those obtained from the wild (Angst and Storch, 1967). The pre-war reared gorillas Bobby (Berlin), Gargantua (from a circus) and Bushman (Chicago) had very dorsoventrally low orbits and prominent supraorbital tori. Frequently the incisors are unusually procumbent. Cranial capacity seems to be small. Bobby had a strikingly vertically oriented planum nuchale. The skull of Abraham, who died in Frankfurt Zoo in 1967, was much more "natural" than this, reflecting favorably on the improvements in captive husbandry that had taken place in the meantime.

Pathology in the wild was studied for mountain gorillas by Lovell (1990). Dental abscesses were frequent, as was antemortem tooth loss, often associated with extensive alveolar bone loss. Lovell discussed this phenomenon and its association with the heavy build-up of calculus; it is not a result of the destruction of interdental alveolar bone by bamboo "packing", as was hypothesized by Colyer (1936). Traumatic injuries were recorded, but the most extensive skeletal pathology was arthritis.

Groves and Humphrey (1973), studying cranial asymmetry, drew attention to the grossly deformed skull of an adult male, found at the foot of Mt. Karissimbi, which was donated by the late Dr. Dian Fossey to this author. The left ascending ramus and mandibular condyle are almost undeveloped, making the whole skull extremely lopsided, perhaps due to destruction in infancy of the left m. temporalis (unsuccessful attempted infanticide?). The only detailed description of extreme skeletal trauma in a western gorilla (from the Sangha River) had been by Maly (1939), who described the results of a severe wound, presumably by a spear, to the pelvis, spinal column and leg bones, and another wound to the skull knocking out the front teeth and damaging part of the upper jaw.

Sarmiento (1994) made a study of the functional anatomy of the hands and feet, documenting in detail for the first time the extent of the gorilla's terrestrial adaptations. Functional anatomy is very much the current concern, and studies like Sarmiento's are clearly the future of the field.

## Epilogue

The genetic revolution has done much to revitalize systematics, and has stimulated a great deal of research on interrelationships within the Hominoidea. Meanwhile conservation concern has renewed attention on two outlying populations – Bwindi and the Cross River.

Garner and Ryder (1996) have analyzed a 250-bp sequence from the "hypervariable" region of the D-loop of mitochondrial DNA. They found a large difference between western and eastern gorillas, and a smaller difference between *graueri* and *beringei*; there was no concordance with geographic origin within either western gorillas or mountain gorillas (the latter included both Virunga and Bwindi), but some of the divisions among western gorillas were very deep (implying a demic population structure: Simon Easteal, personal communication). Ruvolo *et al.* (1994) had previously analysed the COII sequences of several hominoids; though their gorilla representation was smaller (only four western, one *graueri*, one *beringei*) they were at any rate able to make comparisons: thus the *graueri* vs. *beringei* difference was the same as the difference between modern human races, and the western vs. eastern difference nearly twice as great (but less than half that between common and pygmy chimpanzees).

Sarmiento *et al.* (1996) have found that, as far as the admittedly limited evidence goes, the Bwindi gorillas fall in many morphological characters beyond or at the edge of the range of those from Virunga. Eventually, they suggest, the Bwindi gorillas will have to be recognized as a separate subspecies. A suggestion was made (Anon., 1996) that this proposal is at odds with the findings of Garner and Ryder (1996) that Virunga and Bwindi mitochondrial lineages are intermixed; an interpretation to which Butynski *et al.* (1996) took exception, pointing out that Ryder himself had commented that there was no necessary conflict.

I might note here that there is an unfortunate tendency to regard genetic data as "the answer"; if morphologically defined subspecies do not appear to be characterized by unique DNA lineages (especially mtDNA), then they are somehow not different, that the evidence of the eyes and the calipers is illusory. Concerning another case of conservation concern (sable antelopes) I wrote:

> Mitochondrial lineage sorting just has not occurred yet among sable subspecies . . . which is reasonable when you consider that what is characteristic of subspecies is that they do not have fixed allelic differences between them, unlike species . . . Unless we take the phenotypic differences to be merely environmental – due to different conditions of rearing, no more – then the existence of such differences is *ipso facto* evidence that [subspecies] differ genetically. This genetic differentiation does not reside in the control region of mtDNA, that's all: and, as Avise himself, the founder of the phylogeography concept, has shown, lineage sorting is not the same thing as population separation. (Groves, 1998)

Phenotypically, eastern and western gorillas are very different. Their skulls would probably always be distinguishable; certainly externally they are distinct in colour and other characters, such as nose form (Cousins, 1974). It is on this basis that I think that there are, after all, two full species: *Gorilla gorilla* and *Gorilla beringei* (Groves, 2000, 2001). Within the latter, for the moment both recognized subspecies, *G. b. beringei* and *G. b. graueri*, should stand as homogeneous entities, but further enquiry may well establish that the Bwindi population should be separated from *beringei* and one or more of the Utu, Tshiaberimu and Kahuzi populations should be separated from *graueri*.

Among western gorillas, there has been a move to reinvestigate the status of *diehli*, the gorillas of the Cross River (now in Nigeria). Stumpf *et al.* (1998) used size-adjusted cranial data to show that this population differs much more strongly than the raw data of Groves (1967, 1970) had suggested.

I made my craniometric data available to Rebecca Stumpf, who put them on a datafile and kindly sent me a copy. I ran the male skull data on SPSS,

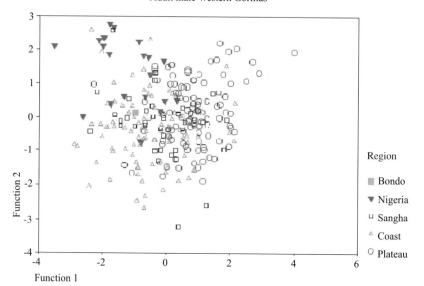

Fig. 2.1. *Gorilla gorilla* male skulls, grouped by major region. The Bondo skull was entered as an unknown. The first discriminant function accounts for 65.0% of the variance, the second for 30.4%. Standardized canonical discriminant function coefficients are as follows:

|  | Function 1 | Function 2 |
|---|---|---|
| Braincase length | −0.521 | 0.841 |
| Palatal breadth at $M^1$ | 0.298 | 0.001 |
| Facial height | −0.346 | 0.346 |
| Interorbital breadth | 0.345 | −0.062 |
| Breadth of nuchal surface | −0.569 | −1.052 |
| Biporionic breadth | 0.708 | 0.628 |
| Basal length | 0.228 | −0.487 |
| Maximum cranial length | 0.328 | −0.408 |
| Palatal length | 0.400 | 0.007 |
| Basion to inion | 0.159 | 0.267 |

Discriminant, just as a preliminary test. In five minutes I recreated the analyses that had taken six months in 1966. As an experiment, I tested the differentiation of the Bwindi population (further data kindly supplied by T. Butynski) and the Cross River sample. The results are shown in Figs. 2.1–2.3.

In Fig. 2.1 we see that the Cross River sample occupies its own cluster, in a way which the other regional groups of *G. g. gorilla* do not – even the single skull from Bondo is buried deep in the general Cameroon/Gabon/Congo scatter. If discriminant analysis of raw data can suggest such a separation, it is evident that size-adjusted data would have much greater potential to reveal the validity of *Gorilla gorilla diehli*.

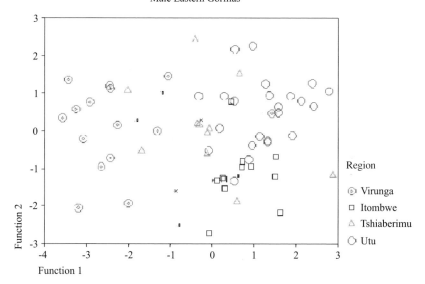

Fig. 2.2. *Gorilla beringei* skulls. Ungrouped cases: B = Bwindi; K = Kahuzi. The first function accounts for 77.0% of the variance, the second for 16.9%. Standardized canonical discriminant function coefficients are as follows:

|  | Function 1 | Function 2 |
|---|---|---|
| Braincase length | −0.150 | −0.278 |
| Palatal breadth at M$^1$ | 0.530 | 0.051 |
| Facial height | 0.624 | 0.304 |
| Interorbital breadth | 0.314 | −0.060 |
| Breadth of nuchal surface | −0.436 | 0.995 |
| Biporionic breadth | 0.270 | −0.651 |
| Basal length | −0.410 | 1.137 |
| Maximum cranial length | −0.439 | −0.366 |
| Palatal length | −0.218 | −1.266 |
| Basion to inion | 0.070 | 0.865 |

In Fig. 2.2, the separation of *G.b. beringei* and *G.b. graueri* is clear, but the four *graueri* samples are somewhat separated. The Tshiaberimu sample – for which the name *rex-pygmaeorum* is available if required – approaches *beringei* and overlaps with it, and the three Kahuzi and four Bwindi specimens (all entered ungrouped) also approach *beringei* somewhat.

In Fig. 2.3, all male skulls are compared, keeping the (now four) presumed subspecies separate. There is a clear separation between *G. gorilla* and *G. beringei*, with just a little overlap; quite unexpectedly, two of the four Bwindi fall within *G. g. gorilla* (this does not mean that they are somehow geographically displaced western gorillas, only that they are different from *beringei* in

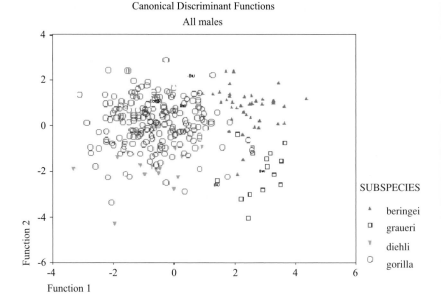

Fig. 2.3. Skulls of the genus *Gorilla*, grouped by subspecies. 19 of the 210 *G. gorilla* skulls (9%, all *G. g. gorilla*) were misclassified as *G. beringei*, but there was no reciprocal misclassification. Within *G. gorilla* only two of the 16 *diehli* skulls (12.5%) are misclassified as *gorilla*; 27 of the 191 gorilla *skulls* (12.9%) are misclassified as *diehli*. Within *G. beringei* only two of the 42 *graueri skulls* (4.8%) are misclassified as *beringei*; one of the 14 *beringei* skulls (7.1%) is misclassified as *graueri*.
Bw = Bwindi; the other ungrouped skull (inside the *gorilla* dispersion) is a skull from Bondo. The first discriminant function accounts for 69.7% of the variance, the second for 22.8%. Standardized canonical discriminant function coefficients are as follows:

|  | Function 1 | Function 2 |
|---|---|---|
| Maximum cranial length | −0.172 | 0.014 |
| Braincase length | −0.031 | −0.054 |
| Palatal breadth at $M^1$ | 0.083 | 0.334 |
| Facial height | −0.176 | 0.231 |
| Palatal length | 1.246 | 0.125 |
| Interorbital breadth | −0.149 | 0.442 |
| Breadth of nuchal surface | −0.254 | −1.116 |
| Basal length | −0.001 | −0.344 |
| Opisthion to inion | 0.027 | −0.425 |
| Biporionic breadth | −0.156 | 0.773 |
| Basion to inion | −0.263 | 0.509 |

some of the same ways in which *G. gorilla* is). Within each of the species, *graueri* is more distinct from *beringei* than *diehli* is from *gorilla*.

As new methods of investigation become available to us, levels of analysis can be conducted: nuances undreamed of by Wyman, Matschie, Rothschild, even Coolidge. Science has advanced, but human behavior has not. People still hunt gorillas for food or trophies, and still cut down their forests; but now those same advances in science also enable forests to be cut down more efficiently, gorillas to be hunted more efficiently, human populations to increase ever faster and press in on the remaining habitat, so that our second-closest relative is threatened with disappearing for ever. More and more, the work of taxonomists and other biologists must be put at the service of conservation.

## References

Alix, E. and Bouvier, A. (1877). Sur un nouvel anthropoïde (*Gorilla mayéma*). *Bulletin de la Société Zoologique de France*, **11**, 2, 488–490.

Anon. (1996). Bwindi's gorillas in the midst of taxonomic dispute. *African Wildlife Update*, Feb. 1996, 7.

Angst, R. (1967). Beitrag zum Formwandel des Craniums der Ponginen. *Zeitschrift für Morphologie und Anthropologie*, **58**, 109–151.

Angst, R. (1970). Über die Schädelkämme der Primaten. *Natur und Museum*, **100**, 293–302.

Angst, R. (1976). Das Endocranialvolumen der Pongiden (Mammalia: Primates). *Beiträge der naturkunliche Forschung in Südwest Deutschland*, **35**, 181–188.

Angst, R. and Storch, G. (1967). Bemerkungen über den Schädel des Gorilla Abraham aus dem Frankfurter zoologischen Garten. *Natur und Museum*, **97**, 417–420.

Bischoff, T.L.W. (1880). Beiträge zur Anatomie des Gorilla. *Abhandlungen des bayerischen Akademie der Wissenschaft*, **13**, 3, 1–48.

Butynski, T., Sarmiento, E.E., and Kalina, J. (1996). More on the Bwindi gorillas. *African Wildlife Update*, Apr. 1996, 6.

Cave, A.J.E. (1959). The nasal fossa of a foetal gorilla. *Proceedings of the Zoological Society of London*, **133**, 73–77.

Cave, A.J.E. 1961. The frontal sinus of the gorilla. *Proceedings of the Zoological Society of London*, **136**, 359–373.

Colyer, F. (1936). *Variations and Diseases of the Teeth of Animals*. London: John Bale & Sons.

Coolidge, H.J. (1929). A revision of the genus *Gorilla*. *Memoirs of the Museum of Comparative Zoology, Harvard*, **50**, 291–381.

Coolidge, H.J. (1936). Zoological results of the George Vanderbilt African Expedition of 1934. IV: Notes on four gorillas from the Sanga River region. *Proceedings of the Academy of Natural Sciences, Philadelphia*, **88**, 479–501.

Corbet, G.B. (1967). Nomenclature of the "Eastern Lowland Gorilla". *Nature*, **215**, 1171–1172.

Cousins, D. (1974). Classification of captive gorillas *Gorilla gorilla*. *International Zoo Yearbook*, **14**, 155–159.

Duckworth, W.L.H. (1915). *Morphology and Anthropology*. Cambridge, U.K.: Cambridge University Press.

Elliot, D.G. (1913). *Review of the Primates*. New York: American Museum of Natural History.

Famelart, L. (1883). Observations sur un jeune gorille. *Bulletin de la Société Zoologique de France*, **8**, 149–152.

Frechkop, S. (1943). *Exploration du Parc National Albert, Mission S. Frechkop (1937–8)*. Brussels: Institut des Parcs Nationaux du Congo Belge.

Fritze, A. (1912). Kleinere Mitteilungen. I: *Gorilla gorilla* Schwarzi Matschie. *Jahrbuch der Provinzialmuseum, Hannover*, **1912**, 113.

Garner, K.J. and Ryder, O.A. (1996). Mitochondrial DNA diversity in gorillas. *Molecular Phylogenetics and Evolution*, **6**, 39–48.

Geoffroy St. Hilaire, I. (1852). Sur les rapports naturels du gorille: Remarques faites à la suite de la lecture de M. Duvernoy. *Comptes rendus de l'Académie des Sciences, Paris*, **36**, 933–936.

Geoffroy St. Hilaire, I. (1855). Description des mammifères nouveaux ou imparfaitement connus de la collection du Muséum d'Histoire Naturelle et remarques sur la classification et les caractères des mammifères. *Archives du Muséum d'Histoire Naturelle, Paris*, **10**, 1–102.

Gregory, W.K. (ed.) (1950). *The Anatomy of the Gorilla*. New York: Columbia University Press.

Groves, C.P. (1967). Ecology and taxonomy of the gorilla. *Nature*, **213**, 890–893.

Groves, C.P. (1970). Population systematics of the gorilla. *Journal of Zoology, London*, **161**, 287–300.

Groves, C.P. (1985). The case of the pygmy gorilla: A cautionary tale for cryptozoology. *Cryptozoology*, **4**, 37–44.

Groves, C.P. (1998). Of course the Giant Sable is a valid subspecies. *Gnusletter*, **17**, 1, 4.

Groves, C.P. (2000). What, if anything, is taxonomy? *Gorilla Journal*, **21**, 12–15.

Groves, C.P. (2001). *Primate Taxonomy*. Washington, D.C.: Smithsonian Institution Press.

Groves, C.P. and Humphrey, N.K. (1973). Asymmetry in gorilla skulls: Evidence of lateralized brain function? *Nature*, **244**, 53–54.

Groves, C.P. and Stott, K.W., Jr. (1979). Systematic relationships of gorillas from Kahuzi, Tshiaberimu and Kayonza. *Folia Primatologica*, **32**, 161–179.

Haddow, A.J. and Ross, R.W. (1951). A critical review of Coolidge's measurements of gorilla skulls. *Proceedings of the Zoological Society of London*, **121**, 43–54.

Hartmann, R. (1886). *Anthropoid Apes*. New York: Appleton & Co.

Haeckel, E. (1903). *Anthropogenie*, 5th edn, vol. 1, *Keimesgeschichte des Menschen*. Leipzig: Engelmann.

Heuvelmans, B. (1981). *Les Bêtes Humaines d'Afrique*. Paris: Pion.

Huxley, T.H. (1867). On the zoological relations of Man with the lower animals. *Natural History Review*, 1867, **1**, 67–84.

*International Code of Zoological Nomenclature, 4th edn*, adopted by the XX General Assembly of the International Union of Biological Sciences. London: International Trust for Zoological Nomenclature, in association with British Museum (Natural History).

Keith, A. (1896). An introduction to the study of the anthropoid apes. I: The gorilla. *Nature and Science*, **9**, 26–37.

Lönnberg, E. (1917). Mammals collected in Central Africa by Captain E. Arrhenius. *Proceeding, Kunglisk Svenska Vetensk-Akademisk Handlingar*, **58**, 2, 1–110.

Lovell, N.C. (1990). Skeletal dental pathology of free-ranging mountain gorillas. *American Journal of Physical Anthropology*, **81**, 399–412.

Maly, J. (1939). Kostra ranèné gorily. *Anthropologie*, **17**, 21–36.

Matschie, P. (1903). Über einen Gorilla aus Deutsch-Ostafrika. *Sitzungsberichte des Gesellschaft naturforschender Freunde, Berlin*, **1903**, 253–259.

Matschie, P. (1904). Bemerkungen über die Gattung *Gorilla*. *Sitzungsberichte des Gesellschaft naturforschender Freunde, Berlin*, **1904**, 45–53.

Matschie, P. (1905). Merkwürdige Gorilla-Schädel aus Kamerun. *Sitzungsberichte des Gesellschaft naturforschender Freunde, Berlin*, **1905**, 277–283.

Matschie, P. (1914). Neue Affen aus Mittelafrika. *Sitzungsberichte des Gesellschaft naturforschender Freunde, Berlin*, **1914**, 323–342.

Miller, G.S. (1915). The jaw of the Piltdown Man. *Smithsonian Miscellaneous Collections*, **65**, 12, 7–31.

Neuville, H. (1916). Remarques sur la variabilité de la crête sagittale de crane des gorilles. *Bulletin du Muséum d'Histoire Naturelle, Paris*, **22**, 2–7.

Neuville, H. (1932). La classification des gorilles. *L'Anthropologie*, **42**, 330–337.

Owen, R. (1848*a*). On a new species of chimpanzee (*T. savagei*). *Proceedings of the Zoological Society of London*, **1848**, 27–35.

Owen, R. (1848*b*). Supplementary note on the great chimpanzee (*Troglodytes gorilla, Trogl. savagei* Owen). *Proceedings of the Zoological Society of London*, **1848**, 53–56.

Owen, R. (1848*c*). Osteological contributions to the natural history of the chimpanzees (*Troglodytes*, Geoffroy) including the description of the skull of a large species (*Troglodytes gorilla*, Savage) discovered by Thomas Savage, M.D., in the Gabon country, West Africa. *Transactions of the Zoological Society of London*, **3**, 381–422.

Owen, R. (1857). On the characters, principles of division and primary groups of the class Mammalia. *Proceedings of the Linnean Society*, **1857**, 2, 1–37.

Owen, R. (1861). On the cerebral characters of Man and the ape. *Annals and Magazine of Natural History*, **7**, 3, 456–458.

Owen, R. (1862*a*). Osteological contributions to the natural history of the chimpanzees (*Troglodytes*) and orangs (*Pithecus*). IV: Description of the cranium of an adult male gorilla from the River Danger, West coast of Africa, indicative of a variety of the Great Chimpanzee (*Troglodytes gorilla*), with remarks on the capacity of the cranium and other characters shown by sections of the skull, in the orangs (*Pithecus*), chimpanzees (*Troglodytes*), and in different varieties of the human race. *Transactions of the Zoological Society of London*, **4**, 75–88.

Owen, R. (1862*b*). Osteological contributions to the natural history of the chimpanzees (*Troglodytes*) and orangs (*Pithecus*). V: Comparison of the lower jaw and vertebral column of the *Troglodytes gorilla, Troglodytes niger, Pithecus satyrus,* and in different varieties of the human race. *Transactions of the Zoological Society of London*, **4**, 89–116.

Rothschild, W. (1905). Notes on anthropoid apes. *Proceedings of the Zoological Society of London*, **1905**, 2, 413–440.

Rothschild, W. (1908). Note on *Gorilla gorilla diehli* (Matschie). *Novitates Zoologicae*, **15**, 391–392.

Rothschild, W. (1927). On a new race of Bongo and of Gorilla. *Annals and Magazine of Natural History*, **19**, 9, 271.

Ruvolo, M., Pan, D., Zehr, S., Goldberg, T., Disotell, T.R., and von Dornum, M. (1994). Gene trees and hominoid phylogeny. *Proceedings of the National Academy of Sciences U.S.A.*, **91**, 8900–8904.

Rzasnicki, A. (1936). Versuch einer Rassenunteilung der Gorillas. *Annales Musei Zoologici Polonici*, **15**, 11, 1–32.

Sarmiento, E.E. (1994). Terrestrial traits in the hands and feet of gorillas. *American Museum Novitates*, **3091**, 1–56.

Sarmiento, E.E., Butynski, T.M., and Kalina, J. (1996). Gorillas of Bwindi-Impenetrable Forest and the Virunga Volcanoes: Taxonomic implications of morphological and ecological differences. *American Journal of Primatology*, **40**, 1–21.

Savage, T.S. and Wyman, J. (1847). Notice of the external characters and habits of *Troglodytes gorilla*, a new species of orang from the Gaboon River; Osteology of the same. *Boston Journal of Natural History*, **5**, 417–442.

Schaller, G.B. (1963). *The Mountain Gorilla: Ecology and Behavior.* Chicago, IL: University of Chicago Press.

Schultz, A.H. (1934). Some distinguishing characters of the mountain gorilla. *Journal of Mammalogy*, **15**, 51–61.

Schultz, A.H. (1962). Die Schädelkapazität männlicher Gorillas und ihr Höchstwert. *Anthropologischer Anzeiger*, **25**, 197–203.

Schwarz, E. (1927). Un gorille nouveau de la forêt de l'Ituri, *Gorilla gorilla rex-pygmaeorum* subsp. n. *Revue de Zoologie et Botanique Africaine*, **14**, 333–336.

Schwarz, E. (1928). Die Sammlung afrikanischer Affen in Congo-Museum. *Revue de Zoologie et Botanique Africaine*, **16**, 2, 1–48.

Slack, J.H. (1862). Note on *Gorilla castaneiceps. Proceedings of the Academy of Natural Sciences, Philadelphia*, **1862**, 159–160.

Sonntag, C.F. (1924). *The Morphology and Evolution of the Apes and Man.* London: John Bale, Sons & Danielson.

Stumpf, R.M., Fleagle, J.G., Jungers, W.L., Oates, J.F., and Groves, C.P. (1998). Morphological distinctiveness of Nigerian gorilla crania. *American Journal of Physical Anthropology Supplement*, **1998**, 213.

Vogel, C. (1961). Zur systematischen Untergliederung der Gattung *Gorilla* anhand von Untersuchungen der Mandibel. *Zeitschrift für Säugetierkunde*, **26**, 2, 1–12.

Wilson, L.G. (1996). The gorilla and the question of human origins: The brain controversy. *Journal of the History of Medicine and Allied Sciences*, **51**, 184–207.

Wyman, J. (1847). A communication from Dr. Thomas S. Savage, describing the external character and habits of a new species of *Troglodytes* (*T. gorilla*, Savage,) recently discovered by Dr. S. in Empongwe, near the river Gaboon, Africa. *Proceedings of the Boston Society of Natural History*, **2**, 245–7.

# 3 Patterns of diversity in gorilla cranial morphology

REBECCA M. STUMPF, JOHN D. POLK, JOHN F. OATES,
WILLIAM L. JUNGERS, CHRISTOPHER P. HEESY,
COLIN P. GROVES, AND JOHN G. FLEAGLE

## Introduction

Gorillas, perhaps because of their size, always seem to be the subject of spectacularly divergent interpretations. Views of their behavior have ranged from the rapacious, vicious giant ape of the nineteenth century, through Robert Ardrey's (1961) view of them as lethargic, depressed, evolutionary dead ends, to the current view of them as gentle giants, albeit with infanticidal tendencies. Views of gorilla systematics have been no less diverse over the past century and a half (see review by Groves, this volume). Beginning with the initial description of *Gorilla gorilla* by Savage and Wyman in 1847 through the 1920s, ten separate species of *Gorilla* were described by systematists from all over the world, often from a single skull. The modern systematics of *Gorilla* stems from the work of Harold Jefferson Coolidge in 1929. Coolidge reviewed all of the previously described species and provided measurements and graphs of 213 specimens from seven major geographic regions. He placed all gorillas in a single species, *Gorilla gorilla*, in accordance with others such as Rothschild (1906), Elliot (1913), and Schwarz (1928), but went even further in identifying only two subspecies – *Gorilla gorilla gorilla* for the gorillas of western and central Africa, and *Gorilla gorilla beringei* for gorillas from the Virunga mountains and adjacent regions.

More recently, *Gorilla* systematics has derived almost exclusively from Groves's (1970, 1986; also Groves and Stott, 1979) study of 747 skulls and over 100 skeletons. Groves largely followed Coolidge (1929) in recognizing a single species, but recognized three subspecies: *Gorilla gorilla gorilla*, the western lowland gorilla; *G. g. beringei*, the mountain gorilla from the Virunga volcanos; and *G. g. graueri*, the eastern lowland gorilla from "the Eastern Congo lowlands and the mountains to the west of Lakes Tanganyika and Edward" (Groves, 1970:298). Subsequently, there have been various questions concerning the relationships among eastern gorillas. Are the eastern lowland gorillas (*G. g. graueri*) more closely related to western lowland gorillas (*G. g. gorilla*) or to the mountain gorillas (*G. g. beringei*)? For several local populations of eastern gorillas, should they be placed in the subspecies *graueri* or *beringei* (e.g., Groves and Stott, 1979) or identified as a distinct species (Sarmiento *et al.*, 1996)? Finally, in the last few years there has been increasing focus on the endangered sample of gorillas adjacent to the Cross River in Nigeria and northern Cameroons, which is geographically and morphologically distinct from other western gorillas and may deserve separate subspecies status (Stumpf *et al.*, 1997, 1998; Sarmiento and Oates, 2000; Oates *et al.*, this volume).

We examine the morphological diversity of gorillas using multivariate analysis of cranial morphology. There is considerable variation in body size among gorillas, both within samples as a result of sexual dimorphism, and among samples perhaps related to climatic or other environmental factors. Therefore, a major goal of our study is to examine the extent to which the morphological differences among gorilla crania in different samples are mainly differences in size or whether they also reflect differences in shape. Thus, using analyses of raw cranial measurements as well as analyses of "size-adjusted" data (Jungers *et al.*, 1995), we examine patterns of morphological diversity among gorillas. In particular we address three specific questions:

1. Is the major morphological separation among gorillas between western (*G. g. gorilla*) and eastern samples (*G. g. graueri* and *G. g. beringei*), or between "lowland gorillas" (*G. g. gorilla* and *G. g. graueri*) and mountain gorillas (*G. g. beringei*)?'
2. Is the Cross River sample of gorillas (Nigerian gorillas) distinct from other western gorillas?
3. What is the influence of body size (using cranial size as a surrogate) on patterns of gorilla cranial morphometrics? Does analysis of size-adjusted (shape) cranial data yield different patterns of morphological diversity from analysis of cranial data without any size-adjustment?

## Materials and methods

The data used in this study are a subset of Groves's (1970) original data set (see also chapters by Groves, Leigh *et al.*, and Albrecht *et al.*, this volume). The complete data consisted of 30 linear cranial measurements and 10 linear mandibular measurements from 747 gorilla crania. Because we planned to use a size-adjustment for part of our analysis, we needed a data set of specimens with complete measurements. As many mandibles were missing, we used only crania for this analysis. In addition, there were many measurements that were taken on only part of the sample. In order to maximize our sample size while concurrently ensuring a robust set of measurements for each specimen, this study is based on 19 cranial measurements (Table 3.1) that were available for the majority of

Table 3.1. *Cranial measurements used in this study*

| |
|---|
| 1. Cranial length (prosthion to opisthocranion) |
| 2. Neurocranial length (glabella to opisthocranion) |
| 3. Biorbital breadth (external) |
| 4. Bimolar breadth ($M^1 - M^1$ breadth) |
| 5. Breadth of both nasal bones |
| 6. Premaxillary length (pyriform aperture to prosthion) |
| 7. Bicanine breadth (alveolar) |
| 8. Facial height (supraorbital torus to prosthion) |
| 9. Height of one orbit |
| 10. Width of one orbit |
| 11. Postorbital (constriction) breadth |
| 12. Pyriform aperture breadth |
| 13. Interorbital breadth |
| 14. Supraorbital torus height |
| 15. Breadth of nuchal surface |
| 16. Neurocranial width (biporionic breadth) |
| 17. Maxillary tooth row ($P^3$ to $M^3$ inclusive) |
| 18. Palatal length (prosthion to staphylion) |
| 19. Lateral facial height (length of a perpendicular from infraorbital foramen to maxillary alveoli) |

*Source:* Groves (1970).

specimens. Specimens with an incomplete set of these 19 measurements were also excluded from the study, leaving a total of 504 gorilla crania (307 males and 197 females).

All specimens in this sample are of known geographic origin (Fig. 3.1). Following Groves (1970), individual specimens were assigned to restricted geographic locations reflecting their place of origin, usually to within 100 square miles. Exceptions are three samples from single localities (Ouesso, Ebolowa, and Abong Mbang) and specimens from a 10 000 square mile area of the eastern Congo lowland forests referred to as Utu (Groves, 1970). Of the 504 specimens, there are male crania from 19 geographic locations and female crania from 14 (Table 3.2). In lieu of any information about gene flow, we treated these geographic samples equally and did not combine them into populations or any other larger groupings. Because gorillas show extreme sexual size dimorphism, with almost no overlap in size between males and females (Albrecht *et al.*, this volume), male and female crania were analyzed separately.

Canonical variates analyses, using the mainframe version of the statistical program SAS (SAS Institute, 1985), were used to summarize morphological differences among gorilla samples from different localities, to depict graphically the clustering of samples, and to determine the influence of different variables on the morphological differences observed. Generalized or Mahalanobis distances ($D^2$) were calculated between centroids of samples from individual localities

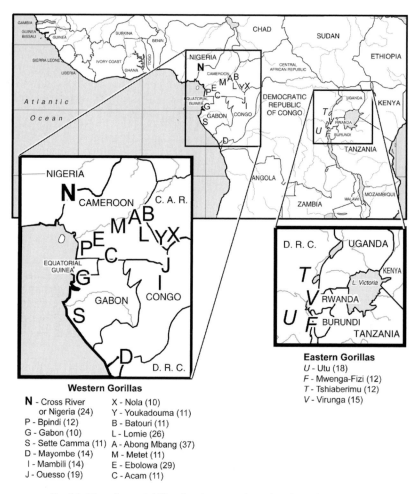

**Western Gorillas**

**N** - Cross River
      or Nigeria (24)
P - Bpindi (12)
G - Gabon (10)
S - Sette Camma (11)
D - Mayombe (14)
I - Mambili (14)
J - Ouesso (19)

X - Nola (10)
Y - Youkadouma (11)
B - Batouri (11)
L - Lomie (26)
A - Abong Mbang (37)
M - Metet (11)
E - Ebolowa (29)
C - Acam (11)

**Eastern Gorillas**
*U* - Utu (18)
*F* - Mwenga-Fizi (12)
*T* - Tshiaberimu (12)
*V* - Virunga (15)

Fig. 3.1. Map of central Africa showing approximate locations of the samples of gorilla crania analyzed in this study (after Groves, 1970).

in order to quantify the morphological differences among gorilla samples and to evaluate their significance.

In order to evaluate the influence of size on cranial differences, we also performed the same analyses using size-adjusted measurements (Jungers, *et al.*, 1995). Following Mosimann (1970), Falsetti *et al.* (1993) and Jungers *et al.* (1995), indices of morphological shape were created by dividing all individual measurements by a generalized size factor – an estimated volume of each cranium, calculated as the cube root of the product of cranial length (prosthion to opisthocranion), cranial width (biporionic breath) and facial height (supraorbital torus to prosthion).

Table 3.2. *Cranial samples used in this study by location, sex and subspecies*

| | Males | | | Females | | |
|---|---|---|---|---|---|---|
| | Subspecies | Abbreviation | Sample (n) | Subspecies | Abbreviation | Sample (n) |
| | **Western lowland gorilla** | N | Cross River or Nigeria (24) | **Western lowland gorilla** | N | Cross River or Nigeria (24) |
| | | P | Bpindi (12) | | P | Bpindi (4) |
| | | G | Gabon (10) | | S | Sette Camma (11) |
| | | S | Sette Camma (11) | | I | Mambili (9) |
| | | D | Mayombe (14) | | Y | Youkadouma (10) |
| | | I | Mambili (14) | | B | Batouri (22) |
| | | J | Ouesso (19) | | L | Lomie (19) |
| | | X | Nola (10) | | A | Abong Mbang (28) |
| | | Y | Youkadouma (11) | | M | Metet (9) |
| | | B | Batouri (11) | | E | Ebolowa (18) |
| | | L | Lomie (26) | | | |
| | | A | Abong Mbang (37) | | | |
| | | M | Metet (11) | | | |
| | | E | Ebolowa (29) | | | |
| | | C | Acam (11) | | | |
| | **Eastern lowland gorilla** | U | Utu (18) | **Eastern lowland gorilla** | U | Utu (10) |
| | | F | Mwenga-Fizi (12) | | F | Mwenga-Fizi (9) |
| | | T | Tshiaberimu (12) | | T | Tshiaberimu (14) |
| | **Mountain gorilla** | V | Virunga (15) | **Mountain gorilla** | V | Virunga (10) |

## Results

### *Analysis of raw measurements*

The results of the canonical variates analysis of male crania from all 19 geographic locations using the raw data are illustrated in Fig. 3.2. The first two
canonical axes account for 63% of the information included in the data.
Although this plot clearly documents overlap among the gorilla samples in
the first two axes (see Albrecht *et al.*, this volume), there are nevertheless some
major groupings and separations. The major separation on axis 1 is between
western samples to the left and eastern samples (Utu (U), Mwenga-Fizi (F),
Tshiaberimu (T) and Virunga (V)) on the right. The second axis separates the
Cross River (Nigerian) sample (N) and the Virunga sample (V) from the other
western and eastern samples, respectively. The correlations between individual

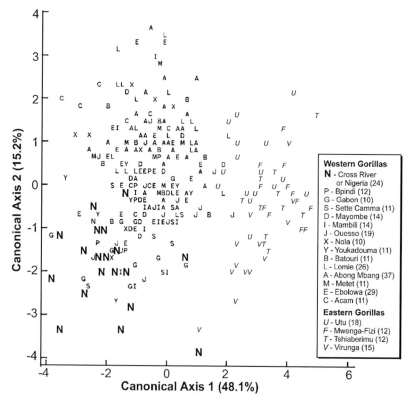

Fig. 3.2. First two canonical axes in the analysis of 19 samples of male gorilla crania
from western and eastern Africa. Some individuals may not be visible due to overlap.

Table 3.3. *Correlations (structure coefficients) between individual cranial measurements and canonical axes 1 and 2 in the analysis of male crania from western and eastern gorillas*

| Cranial measurement | Canonical axis 1 | Canonical axis 2 |
|---|---|---|
| 1. Cranial length | 0.431912 | 0.611032 |
| 2. Neurocranial length | −0.154511 | 0.573552 |
| 3. Biorbital breadth | −0.316672 | 0.702041 |
| 4. Bimolar breadth | 0.591907 | 0.765188 |
| 5. Breadth of both nasal bones | 0.852296 | −0.078789 |
| 6. Premaxillary length | 0.790360 | 0.400218 |
| 7. Bicanine breadth | 0.599808 | 0.509714 |
| 8. Facial height | 0.601290 | 0.519322 |
| 9. Height of one orbit | 0.184274 | −0.269272 |
| 10. Width of one orbit | −0.542470 | −0.240535 |
| 11. Postorbital breadth | −0.592348 | 0.433336 |
| 12. Pyriform aperture breadth | 0.860615 | −0.194665 |
| 13. Interorbital breadth | −0.007250 | 0.848769 |
| 14. Supraorbital torus height | −0.798573 | −0.099672 |
| 15. Breadth of nuchal surface | −0.302103 | 0.059558 |
| 16. Neurocranial width | 0.055677 | 0.713552 |
| 17. Maxillary tooth row | 0.936502 | −0.002560 |
| 18. Palatal length | 0.907835 | 0.107938 |
| 19. Lateral facial height | −0.319016 | 0.785945 |

morphological features and the canonical axes ("structure coefficients") are presented in Table 3.3. The first axis separates western samples from the eastern samples on the basis of their smaller values for palatal length, maxillary tooth row length, nasal aperture breadth, nasal breadth, lateral facial height, and their greater supraorbital torus height. Axis 2 separates the Virunga mountain gorillas and the Nigerian gorillas from the rest of the sample on the basis of their relatively smaller interorbital breadth, smaller lateral facial height, and narrower palate.

Table 3.4 is a matrix of Mahalanobis $D^2$ distances between the multivariate centroids of individual gorilla samples and the significance levels for those distances based on an approximate $F$ test. In general, the eastern samples are all significantly different from the western samples. Among the western samples, the Cross River (Nigerian) sample stands out as significantly different from all other samples whereas the remaining western samples are rarely significantly different from one another.

Canonical variates analysis of the female crania from 14 geographic localities yields a roughly similar separation of eastern and western samples on axis 1 (Fig. 3.3) as found in the analysis of male crania. Similarly the second axis separates the Cross River (Nigerian) samples from other western gorillas, in

Table 3.4. *Mahalanobis D² distances between sample centroids and probabilities for the analysis of male crania from western and eastern gorillas*

Mahalanobis $D^2$ distances between sample centroids

| Location | Abong Mbang A | Batouri B | Acam C | Mayombe D | Ebolowa E | Mwenga-Fizi F | Gabon G | Mam-bili I | Ouesso J | Lomie L | Metet M | Nigeria N | Bpindi P | Sette Camma S | Tshiab-erimu T | Utu U | Virunga V | Nola X | Youka-douma Y |
|---|---|---|---|---|---|---|---|---|---|---|---|---|---|---|---|---|---|---|---|
| Abong Mbang A | — | 3.53 | 3.13 | 5.05 | 2.40 | 22.03 | 12.23 | 4.93 | 3.66 | 1.71 | 2.52 | 12.07 | 6.33 | 8.72 | 18.65 | 14.03 | 18.77 | 4.03 | 3.89 |
| Batouri B | 0.0937 | — | 5.89 | 8.89 | 6.02 | 26.16 | 13.71 | 7.94 | 8.49 | 3.71 | 5.81 | 13.63 | 7.97 | 12.35 | 20.85 | 20.10 | 17.62 | 5.34 | 7.50 |
| Acam C | 0.1764 | 0.0562 | — | 4.70 | 3.68 | 31.36 | 8.91 | 5.62 | 5.76 | 3.84 | 4.69 | 11.20 | 4.37 | 9.94 | 29.84 | 21.25 | 29.46 | 3.88 | 5.23 |
| Mayombe D | 0.0006 | 0.0002 | 0.1131 | — | 3.85 | 22.69 | 7.18 | 1.69 | 2.48 | 5.99 | 7.46 | 12.60 | 3.06 | 3.49 | 22.40 | 14.64 | 26.98 | 7.75 | 6.32 |
| Ebolowa E | 0.0128 | 0.0013 | 0.1041 | 0.0236 | — | 24.09 | 9.18 | 4.27 | 2.56 | 3.94 | 2.64 | 8.66 | 5.37 | 9.01 | 20.50 | 14.62 | 21.09 | 4.33 | 5.03 |
| Mwenga-Fizi F | 0.0001 | 0.0001 | 0.0001 | 0.0001 | 0.0001 | — | 31.70 | 23.71 | 20.93 | 22.25 | 25.38 | 40.71 | 26.11 | 24.29 | 2.87 | 5.82 | 14.19 | 30.83 | 28.68 |
| Gabon G | 0.0001 | 0.0001 | 0.0019 | 0.0001 | 0.0001 | 0.0001 | — | 5.06 | 6.67 | 13.46 | 11.00 | 9.62 | 5.19 | 6.84 | 31.41 | 26.05 | 27.57 | 9.44 | 6.40 |
| Mambili I | 0.0008 | 0.0010 | 0.0345 | 0.0063 | 0.0092 | 0.0001 | 0.1015 | — | 2.48 | 7.41 | 6.99 | 9.72 | 2.80 | 3.03 | 21.99 | 17.58 | 21.80 | 7.30 | 5.02 |
| Ouesso J | 0.0022 | 0.0001 | 0.0096 | 0.4793 | 0.1033 | 0.0001 | 0.0040 | 0.4755 | — | 4.64 | 4.62 | 7.29 | 5.12 | 2.86 | 19.13 | 13.00 | 20.07 | 4.58 | 3.62 |
| Lomie L | 0.1894 | 0.1178 | 0.0984 | 0.0002 | 0.0003 | 0.0001 | 0.0001 | 0.0001 | 0.0006 | — | 2.82 | 12.79 | 8.15 | 9.46 | 18.96 | 14.08 | 21.50 | 3.19 | 5.57 |
| Metet M | 0.3974 | 0.0616 | 0.2000 | 0.0023 | 0.4176 | 0.0001 | 0.0001 | 0.0047 | 0.0584 | 0.3766 | — | 10.45 | 8.75 | 11.09 | 20.81 | 17.97 | 19.32 | 4.16 | 4.49 |
| Nigeria N | 0.0001 | 0.0001 | 0.0001 | 0.0001 | 0.0001 | 0.0001 | 0.0001 | 0.0001 | 0.0001 | 0.0001 | 0.0001 | — | 9.08 | 11.20 | 33.91 | 30.98 | 27.81 | 8.64 | 7.58 |
| Bpindi P | 0.0001 | 0.0024 | 0.0002 | 0.0001 | 0.0025 | 0.0001 | 0.0001 | 0.5933 | 0.0175 | 0.0001 | 0.0007 | 0.0001 | — | 5.93 | 24.48 | 18.12 | 23.69 | 9.23 | 5.51 |
| Sette Camma S | 0.0001 | 0.2259 | 0.0002 | 0.3933 | 0.4884 | 0.0001 | 0.0001 | 0.5583 | 0.4830 | 0.0001 | 0.0001 | 0.0001 | 0.0394 | — | 24.18 | 17.53 | 24.82 | 9.85 | 6.40 |
| Tshiaberimu T | 0.0001 | 0.0001 | 0.0001 | 0.0001 | 0.0001 | 0.6452 | 0.1277 | 0.0001 | 0.0001 | 0.0001 | 0.0001 | 0.0001 | 0.0001 | 0.0001 | — | 6.88 | 9.46 | 27.86 | 24.50 |
| Utu U | 0.0001 | 0.0001 | 0.0001 | 0.0001 | 0.0001 | 0.0063 | 0.0001 | 0.0001 | 0.0001 | 0.0001 | 0.0001 | 0.0001 | 0.0001 | 0.0001 | 0.0009 | — | 15.85 | 22.86 | 20.91 |
| Virunga V | 0.0001 | 0.0001 | 0.0001 | 0.0001 | 0.0001 | 0.0001 | 0.0001 | 0.0001 | 0.0001 | 0.0001 | 0.0001 | 0.0001 | 0.0001 | 0.0001 | 0.0001 | 0.0001 | — | 25.80 | 22.20 |
| Nola X | 0.0644 | 0.1359 | 0.4581 | 0.0001 | 0.0588 | 0.0001 | 0.0017 | 0.0053 | 0.0920 | 0.3144 | 0.3761 | 0.0001 | 0.0007 | 0.0005 | 0.0001 | 0.0001 | 0.0001 | — | 4.89 |
| Youkadouma Y | 0.0493 | 0.0074 | 0.1170 | 0.0130 | 0.0096 | 0.0001 | 0.0439 | 0.0762 | 0.2216 | 0.0047 | 0.2411 | 0.0001 | 0.0658 | 0.0306 | 0.0001 | 0.0001 | 0.0001 | 0.2063 | — |

Probabilities that centroids are the same

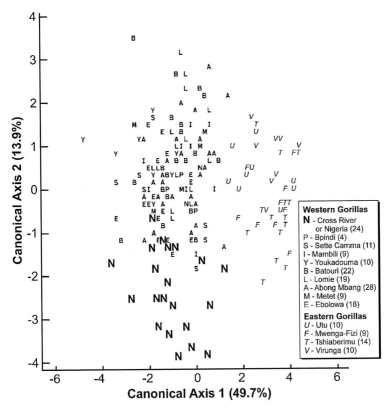

Fig. 3.3. First two canonical axes in the analysis of 14 samples of female gorilla crania from eastern and western Africa. Some individuals may not be visible due to overlap.

this case more completely than among the male crania. However, in contrast with the results from analysis of male crania, the female gorillas from the Virungas overlap extensively with other eastern samples on these first two axes. The morphological features most highly correlated with axis 1, separating eastern and western females, are similar to those separating eastern and western males (Table 3.5). They include palatal length, maxillary tooth row length, breadth of pyriform aperture and nasal bones, lateral facial height, and overall cranial length, the last of which was not important in separating eastern and western males. The second axis, which mainly separates the Cross River (Nigerian) females from all others, is not highly correlated with any individual measurements; the best correlations are with the length and breadth of the braincase, breadth across the molars and canines, and lateral facial height.

The $D^2$ values (Table 3.6) for the female crania show a result similar to that for males, with highest values between eastern and western samples. Among the

Table 3.5. *Correlations (structure coefficients) between individual cranial measurements and canonical axes 1 and 2 in the analysis of female crania from western and eastern gorillas*

| Cranial measurement | Canonical axis 1 | Canonical axis 2 |
|---|---|---|
| 1. Cranial length | 0.852419 | 0.379512 |
| 2. Neurocranial length | 0.598582 | 0.651150 |
| 3. Biorbital breadth | −0.306742 | 0.298530 |
| 4. Bimolar breadth | 0.330996 | 0.621060 |
| 5. Breadth of both nasal bones | 0.608063 | −0.503107 |
| 6. Premaxillary length | 0.679326 | 0.558693 |
| 7. Bicanine breadth | 0.374467 | 0.542454 |
| 8. Facial height | 0.350784 | 0.276219 |
| 9. Height of one orbit | 0.201393 | −0.181297 |
| 10. Width of one orbit | −0.394496 | 0.199206 |
| 11. Postorbital breadth | −0.542635 | 0.278372 |
| 12. Pyriform aperture breadth | 0.773892 | −0.297121 |
| 13. Interorbital breadth | 0.178643 | 0.241217 |
| 14. Supraorbital torus height | −0.416172 | 0.052826 |
| 15. Breadth of nuchal surface | 0.134847 | −0.048931 |
| 16. Neurocranial width | 0.202660 | 0.640816 |
| 17. Maxillary tooth row | 0.902312 | 0.369374 |
| 18. Palatal length | 0.959714 | 0.056103 |
| 19. Lateral facial height | −0.604550 | 0.433328 |

eastern samples, the Virunga gorillas are most distinct, though this is not evident in the plot of the first two canonical axes. Higher axes contribute substantially to the separation of Virunga populations. Among the western gorillas, the Cross River sample shows the highest distances from other western samples with the Gabon samples also exhibiting relatively high values. These are also the most geographically separated samples among the western gorillas.

When the analysis is limited to only the western gorilla samples, the main separation in the canonical variates analyses of both male and female crania is between the Cross River (Nigerian) sample and all others (Figs. 3.4 and 3.5). In the analysis of male crania, the Cross River (Nigerian) sample is separated from the other samples on both axes 1 and 2. The morphological features most highly correlated with axis 1 are measurements of cranial length, neurocranial length, bimolar breadth, bicanine breadth, interorbital breadth, neurocranial width, palatal length as well as lateral facial height and facial height (Table 3.7). The highest correlations on the second axis are with orbit height and breadth of the nuchal surface. In the analysis of female crania axis 1 separates the Cross River sample from the others. The morphological features most highly correlated with this axis include some of the same measurements that separated the Cross River (Nigerian) males, such as lateral facial height, but other factors

Table 3.6. *Mahalanobis* $D^2$ *distances between sample centroids and probabilities for the analysis of female crania from western and eastern gorillas*

| | | | | | | | Mahalanobis $D^2$ distances between sample centroids | | | | | | | |
|---|---|---|---|---|---|---|---|---|---|---|---|---|---|---|
| Location | Abong Mbang A | Batouri B | Ebolowa E | Mwenga-Fizi F | Mambili I | Lomie L | Metet M | Nigeria N | Bpindi P | Sette Camma S | Tshiab-erimu T | Utu U | Virunga V | Youka-douma Y |
| Abong Mbang | — | 3.63 | 2.03 | 18.76 | 5.29 | 1.78 | 5.07 | 8.33 | 9.45 | 7.28 | 18.98 | 18.19 | 22.40 | 6.97 |
| Batouri | 0.0062 | — | 4.92 | 22.77 | 4.46 | 2.72 | 4.22 | 9.98 | 7.33 | 8.30 | 21.39 | 22.67 | 24.36 | 5.90 |
| Ebolowa | 0.4013 | 0.0025 | — | 25.63 | 5.07 | 3.07 | 4.64 | 6.16 | 7.81 | 5.53 | 27.63 | 24.39 | 30.58 | 6.35 |
| Mwenga-Fizi | 0.0001 | 0.0001 | 0.0001 | — | 20.82 | 19.94 | 24.34 | 25.40 | 28.58 | 31.33 | 6.71 | 10.53 | 16.23 | 35.63 |
| Mambili | 0.0391 | 0.1574 | 0.1136 | 0.0001 | — | 3.09 | 4.09 | 9.52 | 7.58 | 7.18 | 21.90 | 17.38 | 21.62 | 9.12 |
| Lomie | 0.514 | 0.1780 | 0.1623 | 0.0001 | 0.5878 | — | 3.13 | 10.36 | 8.94 | 7.49 | 20.41 | 18.99 | 22.31 | 6.99 |
| Metet | 0.0523 | 0.2027 | 0.1756 | 0.0001 | 0.6167 | 0.5736 | — | 9.91 | 5.97 | 8.38 | 25.04 | 21.22 | 22.90 | 8.78 |
| Nigeria | 0.0001 | 0.0001 | 0.0001 | 0.0001 | 0.0001 | 0.0001 | 0.0001 | — | 9.07 | 9.78 | 24.09 | 26.50 | 30.12 | 11.37 |
| Bpindi | 0.0693 | 0.2824 | 0.2521 | 0.0001 | 0.4681 | 0.1322 | 0.7234 | 0.1003 | — | 7.94 | 26.35 | 26.82 | 26.98 | 10.60 |
| Sette Camma | 0.0003 | 0.0001 | 0.0273 | 0.0001 | 0.0428 | 0.0011 | 0.0125 | 0.0001 | 0.3493 | — | 27.95 | 23.60 | 29.69 | 8.37 |
| Tshiaberimu | 0.0001 | 0.0001 | 0.0001 | 0.0337 | 0.0001 | 0.0001 | 0.0001 | 0.0001 | 0.0001 | 0.0001 | — | 12.96 | 10.84 | 31.96 |
| Utu | 0.0001 | 0.0001 | 0.0001 | 0.0019 | 0.0001 | 0.0001 | 0.0001 | 0.0001 | 0.0001 | 0.0001 | 0.0001 | — | 18.19 | 32.03 |
| Virunga | 0.0001 | 0.0001 | 0.0001 | 0.0001 | 0.0001 | 0.0001 | 0.0001 | 0.0001 | 0.0001 | 0.0001 | 0.0001 | 0.0001 | — | 32.30 |
| Youkadouma | 0.0013 | 0.0152 | 0.0143 | 0.0001 | 0.0086 | 0.0048 | 0.0122 | 0.0001 | 0.1155 | 0.0074 | 0.0001 | 0.0001 | 0.0001 | — |

Probabilities that centroids are the same

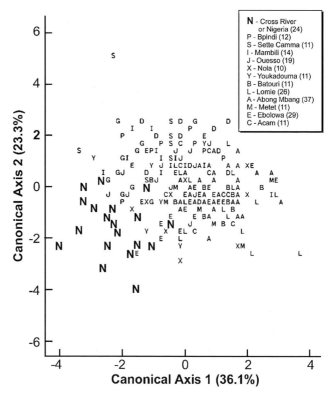

Fig. 3.4. First two canonical axes in the analysis of 13 samples of male gorilla crania from western central Africa. Some individuals may not be visible due to overlap.

are also important, including breadth of nasal bones, orbital width, pyriform aperture breadth, and breadth of the nuchal surface (Table 3.8).

The $D^2$ values for analyses of both male and female crania (Tables 3.9 and 3.10) show the Cross River (Nigerian) samples to be significantly different from almost all other western samples. The only other sample with relatively large $D^2$ values separating it from other groups is the Gabon males. There were too few females from the Gabon sample for a reliable analysis.

### *Analysis of size-adjusted measurements (shape)*

Canonical variates analysis of male crania from both the eastern and western samples using size-adjusted measurements shows essentially the same patterns that were found in the analyses of raw, unadjusted measurements (Fig. 3.6; Tables 3.11 and 3.12). The eastern samples form a distinct cluster separated on

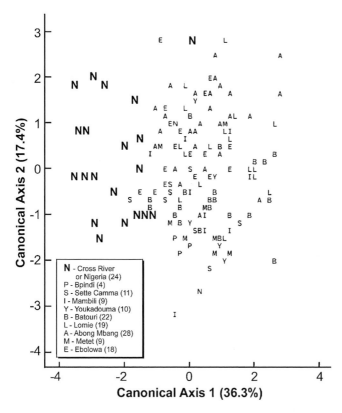

Fig. 3.5. First two canonical axes in the analysis of 10 samples of female gorilla crania from western central Africa. Some individuals may not be visible due to overlap.

the first canonical axis from the western samples, with many individuals from the Virunga sample separating from other eastern gorillas on axis 2. Even after size adjustment, many of the variables most highly correlated with axis 1 that separate western and eastern gorillas are the same as those in analysis of the raw data. These variables include relative maxillary tooth row length, palatal length, and premaxillary length. Axis 1 shows high negative correlations with several breadth factors, including relative neurocranial width, relative postorbital breadth, relative biorbital breadth, and relative orbital width, as well as height measurements, including relative lateral facial height and relative supraorbital torus height. Axis 2, which separates the Virunga sample from other eastern gorillas, is most highly correlated with relative breadth of the nuchal surface.

The western samples overlap extensively in the analysis of size-adjusted data, but as in the analysis of unadjusted data, the Nigerian (Cross River) sample is the most distinctive, lying at the edge of the cloud of western individuals.

Table 3.7. *Correlations (structure coefficients) between individual cranial measurements and canonical axes 1 and 2 in the analysis of male crania from western gorillas*

| Cranial measurement | Canonical axis 1 | Canonical axis 2 |
|---|---|---|
| 1. Cranial length | 0.921067 | −0.022201 |
| 2. Neurocranial length | 0.808694 | 0.152341 |
| 3. Biorbital breadth | 0.767841 | −0.449342 |
| 4. Bimolar breadth | 0.958140 | −0.054163 |
| 5. Breadth of both nasal bones | −0.046741 | 0.442226 |
| 6. Premaxillary length | 0.914576 | −0.057820 |
| 7. Bicanine breadth | 0.895322 | −0.233886 |
| 8. Facial height | 0.690119 | 0.343934 |
| 9. Height of one orbit | −0.391376 | 0.777705 |
| 10. Width of one orbit | −0.310598 | −0.635577 |
| 11. Postorbital breadth | 0.426792 | −0.497801 |
| 12. Pyriform aperture breadth | 0.219958 | −0.138009 |
| 13. Interorbital breadth | 0.790808 | −0.390690 |
| 14. Supraorbital torus height | −0.351077 | −0.299453 |
| 15. Breadth of nuchal surface | 0.341987 | −0.760705 |
| 16. Neurocranial width | 0.896278 | −0.146261 |
| 17. Maxillary tooth row | 0.607187 | 0.642821 |
| 18. Palatal length | 0.827953 | −0.367222 |
| 19. Lateral facial height | 0.824748 | −0.025498 |

Table 3.8. *Correlations (structure coefficients) between individual cranial measurements and canonical axes 1 and 2 in the analysis of female crania from western gorillas*

| | Canonical axis 1 | Canonical axis 2 |
|---|---|---|
| 1. Cranial length | 0.883787 | −0.274590 |
| 2. Neurocranial length | 0.913579 | −0.012386 |
| 3. Biorbital breadth | 0.569018 | 0.552517 |
| 4. Bimolar breadth | 0.821780 | 0.674694 |
| 5. Breadth of both nasal bones | −0.522437 | 0.564234 |
| 6. Premaxillary length | 0.866034 | 0.476479 |
| 7. Bicanine breadth | 0.725407 | 0.680475 |
| 8. Facial height | 0.477039 | −0.301040 |
| 9. Height of one orbit | −0.225736 | −0.622306 |
| 10. Width of one orbit | 0.318273 | −0.017645 |
| 11. Postorbital breadth | 0.464291 | 0.424911 |
| 12. Pyriform aperture breadth | −0.406287 | 0.443783 |
| 13. Interorbital breadth | 0.547715 | 0.662632 |
| 14. Supraorbital torus height | −0.147134 | −0.169809 |
| 15. Breadth of nuchal surface | 0.124284 | 0.097259 |
| 16. Neurocranial width | 0.849307 | 0.462317 |
| 17. Maxillary tooth row | 0.835402 | 0.112237 |
| 18. Palatal length | 0.371329 | 0.543554 |
| 19. Lateral facial height | 0.489531 | 0.279290 |

Table 3.9. *Mahalanobis* $D^2$ *distances between sample centroids and probabilities for the analysis of male crania from western gorillas*

Mahalanobis $D^2$ distances between sample centroids

| Location | Abong Mbang A | Batouri B | Acam C | Mayombe D | Ebolowa E | Gabon G | Mambili I | Ouesso J | Lomie L | Metet M | Nigeria N | Bpindi P | Sette Camma S | Nola X | Youka-douma Y |
|---|---|---|---|---|---|---|---|---|---|---|---|---|---|---|---|
| Abong Mbang A | — | 3.39 | 2.89 | 4.65 | 2.55 | 11.54 | 4.52 | 3.58 | 1.71 | 2.55 | 12.48 | 5.98 | 8.26 | 3.91 | 3.88 |
| Batouri B | 0.1310 | — | 5.41 | 8.53 | 5.87 | 12.79 | 7.44 | 8.27 | 3.82 | 5.62 | 13.34 | 7.39 | 12.01 | 5.03 | 7.01 |
| Acam C | 0.2667 | 0.1083 | — | 4.49 | 3.73 | 8.82 | 5.41 | 5.68 | 3.51 | 4.52 | 11.67 | 4.11 | 9.72 | 3.80 | 5.19 |
| Mayombe D | 0.0022 | 0.0006 | 0.1580 | — | 3.86 | 7.17 | 1.72 | 2.39 | 5.54 | 6.99 | 13.05 | 3.18 | 3.42 | 7.45 | 6.24 |
| Ebolowa E | 0.0090 | 0.0024 | 0.1076 | 0.0277 | — | 8.85 | 4.33 | 2.58 | 3.90 | 2.46 | 8.60 | 5.53 | 9.08 | 4.04 | 5.12 |
| Gabon G | 0.0001 | 0.0001 | 0.0029 | 0.0082 | 0.0001 | — | 4.92 | 6.65 | 13.17 | 10.45 | 9.92 | 5.09 | 7.21 | 9.06 | 5.87 |
| Mambili I | 0.0031 | 0.0031 | 0.0530 | 0.9156 | 0.0101 | 0.1312 | — | 2.54 | 7.12 | 6.71 | 10.33 | 2.82 | 3.13 | 7.05 | 4.81 |
| Ouesso J | 0.0039 | 0.0002 | 0.0136 | 0.5400 | 0.1113 | 0.0054 | 0.4693 | — | 4.52 | 4.39 | 7.59 | 5.37 | 2.82 | 4.37 | 3.71 |
| Lomie L | 0.2086 | 0.1121 | 0.1741 | 0.0010 | 0.0005 | 0.0001 | 0.0001 | 0.0012 | — | 2.79 | 13.11 | 7.83 | 9.17 | 3.19 | 5.79 |
| Metet M | 0.4029 | 0.0865 | 0.2520 | 0.0061 | 0.5189 | 0.0003 | 0.0091 | 0.0917 | 0.4074 | — | 10.34 | 8.45 | 10.77 | 3.90 | 4.59 |
| Nigeria N | 0.0001 | 0.0001 | 0.0001 | 0.0001 | 0.0001 | 0.0001 | 0.0001 | 0.0001 | 0.0001 | 0.0001 | — | 10.00 | 11.91 | 8.34 | 7.85 |
| Bpindi P | 0.0004 | 0.0071 | 0.3064 | 0.4643 | 0.0024 | 0.1549 | 0.6020 | 0.0139 | 0.0001 | 0.0016 | 0.0001 | — | 6.21 | 8.90 | 5.35 |
| Sette Camma S | 0.0001 | 0.0001 | 0.0005 | 0.4359 | 0.0001 | 0.0207 | 0.5367 | 0.5148 | 0.0001 | 0.0001 | 0.0001 | 0.0328 | — | 9.69 | 6.41 |
| Nola X | 0.0884 | 0.1983 | 0.4994 | 0.0055 | 0.1027 | 0.0035 | 0.0096 | 0.1321 | 0.3331 | 0.4699 | 0.0001 | 0.0016 | 0.0009 | — | 4.71 |
| Youkadouma Y | 0.0578 | 0.0172 | 0.1346 | 0.0176 | 0.0102 | 0.0890 | 0.1103 | 0.2142 | 0.0041 | 0.2378 | 0.0001 | 0.0898 | 0.0356 | 0.2609 | — |

Probabilities that centroids are the same

Table 3.10. *Mahalanobis* D² *distances between sample centroids and probabilities for the analysis of female crania from western gorillas*

Mahalanobis $D^2$ Distances between Sample Centroids

| Location | Abong Mbang A | Batouri B | Ebolowa E | Mambili I | Lomie L | Metet M | Nigeria N | Bpindi P | Sette camma S | Youkadouma Y |
|---|---|---|---|---|---|---|---|---|---|---|
| Abong Mbang | — | 3.72 | 1.99 | 5.44 | 1.76 | 5.46 | 8.13 | 9.88 | 6.84 | 6.89 |
| Batouri | 0.0077 | — | 4.59 | 4.28 | 2.76 | 4.23 | 9.10 | 7.02 | 7.72 | 5.29 |
| Ebolowa | 0.4624 | 0.0083 | — | 4.95 | 2.91 | 4.66 | 5.84 | 8.00 | 5.15 | 6.15 |
| Mambili | 0.0429 | 0.2236 | 0.1542 | — | 3.35 | 3.97 | 9.31 | 7.19 | 6.38 | 8.53 |
| Lomie | 0.5625 | 0.1972 | 0.2381 | 0.5332 | — | 3.61 | 9.72 | 8.96 | 6.72 | 6.56 |
| Metet | 0.0421 | 0.2344 | 0.2038 | 0.6763 | 0.4486 | — | 9.47 | 5.35 | 7.64 | 8.51 |
| Nigeria | 0.0001 | 0.0001 | 0.0004 | 0.0003 | 0.0001 | 0.0003 | — | 8.98 | 9.28 | 10.75 |
| Bpindi | 0.0678 | 0.3645 | 0.2629 | 0.563 | 0.1571 | 0.8306 | 0.1295 | — | 7.61 | 10.26 |
| Sette Camma | 0.0014 | 0.0008 | 0.0608 | 0.114 | 0.0063 | 0.0375 | 0.0001 | 0.4344 | — | 8.16 |
| Youkadouma | 0.0028 | 0.0492 | 0.0271 | 0.0228 | 0.0138 | 0.0232 | 0.0001 | 0.1643 | 0.0144 | — |

Probabilities that centroids are the same

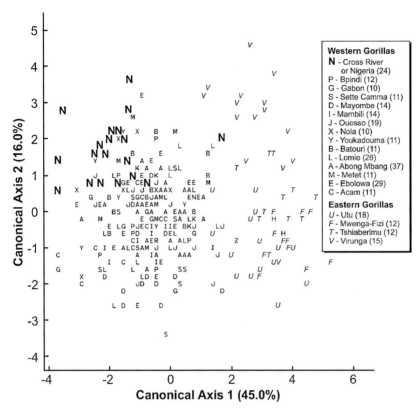

Fig. 3.6. First two canonical axes in the analysis of 19 samples of male gorilla crania from western and eastern Africa using size-adjusted measurements. Some individuals may not be visible due to overlap.

Comparison of $D^2$ values between sample centroids (Table 3.12) shows that the Nigerian (Cross River) sample is the most distinct from other western groups and is significantly different from all other samples. The separation of the Cross River (Nigerian) sample is primarily on axis 2, most highly correlated positively with relative nuchal breadth.

Analysis of the female crania using size-adjusted (shape) measurements yields a similar pattern yet again (Fig. 3.7; Tables 3.13 and 3.14). Western and eastern samples separate along the first canonical variates axis. Although there is considerable overlap among the eastern samples on canonical axes 1 and 2, the centroids of the samples from the Virungas, Tshiaberimu, Utu and Mwenga-Fizi are all significantly distinct from one another except Tshiaberimu and Mwenga-Fizi. Among the western samples, the Cross River (Nigerian) sample is the most distinct, being separated primarily along axis 2. The centroid of the Cross River

Table 3.11. *Correlations (structure coefficients) between individual cranial measurements and canonical axes 1 and 2 in the analysis of male crania from western and eastern gorillas using size-adjusted data*

| Cranial measurement | Canonical axis 1 | Canonical axis 2 |
|---|---|---|
| 1. Cranial length | 0.166706 | 0.197455 |
| 2. Neurocranial length | −0.605550 | 0.036313 |
| 3. Biorbital breadth | −0.692038 | 0.093285 |
| 4. Bimolar breadth | 0.437969 | −0.578982 |
| 5. Breadth of both nasal bones | 0.661064 | 0.005953 |
| 6. Premaxillary length | 0.851997 | 0.076594 |
| 7. Bicanine breadth | 0.541162 | 0.122401 |
| 8. Facial height | 0.393780 | −0.295493 |
| 9. Height of one orbit | −0.013345 | 0.004830 |
| 10. Width of one orbit | −0.582079 | 0.613549 |
| 11. Postorbital breadth | −0.716069 | 0.231228 |
| 12. Pyriform aperture breadth | 0.698867 | 0.484720 |
| 13. Interorbital breadth | −0.182665 | −0.523533 |
| 14. Supraorbital torus height | −0.759760 | 0.376164 |
| 15. Breadth of nuchal surface | −0.564772 | 0.750737 |
| 16. Neurocranial width | −0.609820 | 0.207706 |
| 17. Maxillary tooth row | 0.863213 | 0.211044 |
| 18. Palatal length | 0.905281 | 0.273313 |
| 19. Lateral facial height | −0.674147 | −0.408344 |

(Nigerian) sample is significantly different from all other samples except the tiny, widely scattered sample from Bpindi ($n = 4$). However, it is notable that in the analysis of male crania using size-adjusted measurements, the Virunga sample overlaps with the Cross River (Nigerian) sample on axis 2, but in the analysis of size-adjusted female crania these two samples are separated by this second axis.

The variables most highly correlated with the canonical variates axes in the size-adjusted analysis of female crania are similar to those in the analyses of raw data. Axis 1, which separates eastern and western gorillas, is most highly and positively correlated with relative palatal length, maxillary tooth row length and cranial length, and negatively correlated with breadth measurements, including relative neurocranial width, postorbital breadth, and biorbital breadth, as well as relative lateral facial height. As in the shape analysis of males, axis 2, which separates the Cross River (Nigerian) sample from other samples, is most highly correlated with relative breadth of the nuchal area. Although not shown here, analyses of male and of female crania from the western samples alone using size-adjusted data exhibit a similar pattern of overlap among the western samples with the Nigerian sample being the most distinct graphically and statistically.

Table 3.12. *Mahalanobis $D^2$ distances between sample centroids and probabilities for the analysis of male crania from western and eastern gorillas using size-adjusted data*

Mahalanobis $D^2$ distances between sample centroids

| Location | Abong Mbang A | Batouri B | Acam C | Mayombe D | Ebolowa E | Mwenga-Fizi F | Gabon G | Mambili I | Ouesso J | Lomie L | Metet M | Nigeria N | Bpindi P | Sette Camma S | Tshiab-erimu T | Utu U | Virunga V | Nola X | Youka-douma Y |
|---|---|---|---|---|---|---|---|---|---|---|---|---|---|---|---|---|---|---|---|
| Abong Mbang A | — | 3.19 | 3.21 | 4.99 | 2.43 | 21.35 | 11.62 | 4.86 | 3.40 | 1.49 | 2.46 | 11.01 | 6.57 | 8.36 | 17.73 | 13.02 | 19.19 | 3.96 | 3.67 |
| Batouri B | 0.1600 | — | 5.51 | 8.38 | 5.72 | 25.17 | 12.02 | 7.19 | 7.56 | 3.51 | 5.50 | 11.39 | 7.93 | 11.24 | 20.03 | 19.03 | 18.30 | 4.52 | 6.60 |
| Acam C | 0.1550 | 0.0836 | — | 5.00 | 4.31 | 30.71 | 8.78 | 5.72 | 6.02 | 3.66 | 5.00 | 10.70 | 5.57 | 10.08 | 28.91 | 20.18 | 30.60 | 4.09 | 5.50 |
| Mayombe D | 0.0007 | 0.0005 | 0.0779 | — | 3.97 | 22.18 | 6.60 | 1.68 | 2.23 | 5.73 | 7.30 | 11.67 | 3.30 | 3.01 | 21.49 | 13.81 | 27.17 | 7.81 | 6.14 |
| Ebolowa E | 0.0113 | 0.0025 | 0.0364 | 0.0181 | — | 23.55 | 8.91 | 4.54 | 2.40 | 3.69 | 2.50 | 7.85 | 5.44 | 8.81 | 19.46 | 13.63 | 20.84 | 4.38 | 4.90 |
| Mwenga-Fizi F | 0.0001 | 0.0001 | 0.0001 | 0.0001 | 0.0001 | — | 30.27 | 22.88 | 20.09 | 21.58 | 24.47 | 38.80 | 25.74 | 23.34 | 2.71 | 5.34 | 14.78 | 30.22 | 27.59 |
| Gabon G | 0.0001 | 0.0001 | 0.0001 | 0.0001 | 0.0001 | 0.0001 | — | 4.97 | 7.00 | 11.93 | 10.33 | 10.36 | 5.59 | 7.24 | 28.90 | 23.06 | 26.70 | 9.91 | 6.58 |
| Mambili I | 0.0009 | 0.0035 | 0.0304 | 0.0142 | 0.0050 | 0.0001 | 0.1113 | — | 2.66 | 6.89 | 6.98 | 9.41 | 3.63 | 3.10 | 20.69 | 15.99 | 22.28 | 7.78 | 5.29 |
| Ouesso J | 0.0052 | 0.0004 | 0.0064 | 0.5979 | 0.1469 | 0.0001 | 0.0023 | 0.3931 | — | 3.75 | 4.23 | 7.41 | 5.45 | 2.90 | 17.56 | 11.16 | 19.80 | 4.89 | 3.68 |
| Lomie L | 0.3263 | 0.1591 | 0.1276 | 0.0004 | 0.0007 | 0.0001 | 0.0023 | 0.0001 | 0.0075 | — | 2.58 | 10.55 | 7.95 | 8.38 | 18.32 | 13.47 | 22.02 | 2.52 | 4.74 |
| Metet M | 0.4276 | 0.0874 | 0.1483 | 0.0029 | 0.4808 | 0.0001 | 0.0003 | 0.0001 | 0.1021 | 0.4804 | — | 9.35 | 8.63 | 10.64 | 19.66 | 16.59 | 19.23 | 4.11 | 4.26 |
| Nigeria N | 0.0001 | 0.0001 | 0.0001 | 0.0001 | 0.0001 | 0.0001 | 0.0003 | 0.0001 | 0.0001 | 0.0001 | 0.0001 | — | 9.28 | 11.59 | 30.58 | 27.22 | 26.20 | 8.71 | 7.57 |
| Bpindi P | 0.0001 | 0.0001 | 0.0001 | 0.0001 | 0.0021 | 0.0001 | 0.0829 | 0.2956 | 0.0098 | 0.0001 | 0.0009 | 0.0001 | — | 6.19 | 23.18 | 16.83 | 22.87 | 9.84 | 5.80 |
| Sette Camma S | 0.0001 | 0.0001 | 0.0002 | 0.5641 | 0.0001 | 0.0001 | 0.0163 | 0.5305 | 0.4632 | 0.0001 | 0.0001 | 0.0001 | 0.0285 | — | 22.35 | 15.23 | 24.39 | 10.41 | 6.54 |
| Tshiaberimu T | 0.0001 | 0.0001 | 0.0001 | 0.0001 | 0.0001 | 0.7045 | 0.0001 | 0.0001 | 0.0001 | 0.0001 | 0.0001 | 0.0001 | 0.0001 | 0.0001 | — | 6.55 | 9.72 | 20.69 | 22.61 |
| Utu U | 0.0001 | 0.0001 | 0.0001 | 0.0001 | 0.0001 | 0.0001 | 0.0001 | 0.0001 | 0.0001 | 0.0001 | 0.0001 | 0.0001 | 0.0001 | 0.0001 | 0.0016 | — | 16.11 | 21.27 | 18.92 |
| Virunga V | 0.0001 | 0.0001 | 0.0001 | 0.0001 | 0.0001 | 0.0001 | 0.0001 | 0.0001 | 0.0001 | 0.0001 | 0.0001 | 0.0001 | 0.0001 | 0.0001 | 0.0001 | 0.0001 | — | 25.84 | 21.65 |
| Nola X | 0.0720 | 0.2855 | 0.3956 | 0.0025 | 0.0537 | 0.0001 | 0.0009 | 0.0026 | 0.0607 | 0.5849 | 0.3895 | 0.0001 | 0.0003 | 0.0002 | 0.0001 | 0.0001 | 0.0001 | — | 5.15 |
| Youkadouma Y | 0.0736 | 0.0237 | 0.0874 | 0.0168 | 0.0125 | 0.0001 | 0.0361 | 0.0542 | 0.2067 | 0.0222 | 0.2959 | 0.0001 | 0.0463 | 0.0258 | 0.0001 | 0.0001 | 0.0001 | 0.1628 | — |

Probabilities that centroids are the same

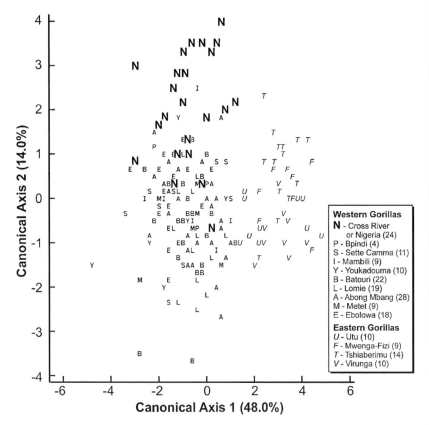

Fig. 3.7. First two canonical axes in the analysis of 14 samples of female gorilla crania from western and eastern Africa using size-adjusted measurements. Some individuals may not be visible due to overlap.

## Discussion

Results of this study largely accord with those reported by Groves (1970) and with other studies of gorilla cranial morphology in this volume using different parts of this same data set (e.g., Albrecht *et al.*, Groves, Leigh *et al.*, this volume). However, our analyses also address several issues that have received less attention in other studies.

In all analyses of samples from throughout Africa, the major distinction observed in this study is between the western samples normally recognized as the western lowland gorilla, *Gorilla gorilla gorilla*, and the eastern samples, including those normally recognized as eastern lowland gorillas (*G. g. graueri*: samples from Utu, Tshiaberimu, and Mwenga-Fizi), mountain gorillas (*G. g. beringei*: samples from the Virunga mountains). Nevertheless, based on

Table 3.13. *Correlations (structure coefficients) between individual cranial measurements and canonical axes 1 and 2 in the analysis of female crania from western and eastern gorillas using size-adjusted data*

| Cranial measurement | Canonical axis 1 | Canonical axis 2 |
|---|---|---|
| 1. Cranial length | 0.892396 | −0.111676 |
| 2. Neurocranial length | 0.078751 | −0.310233 |
| 3. Biorbital breadth | −0.761118 | 0.263099 |
| 4. Bimolar breadth | −0.133933 | −0.311833 |
| 5. Breadth of both nasal bones | 0.461737 | 0.623455 |
| 6. Premaxillary length | 0.586859 | −0.496782 |
| 7. Bicanine breadth | −0.072956 | −0.231686 |
| 8. Facial height | −0.517297 | 0.526449 |
| 9. Height of one orbit | −0.180480 | 0.394058 |
| 10. Width of one orbit | −0.744521 | 0.260695 |
| 11. Postorbital breadth | −0.710345 | 0.106819 |
| 12. Pyriform aperture breadth | 0.558421 | 0.467371 |
| 13. Interorbital breadth | −0.056920 | −0.051437 |
| 14. Supraorbital torus height | −0.476530 | 0.059219 |
| 15. Breadth of nuchal surface | −0.515216 | 0.659983 |
| 16. Neurocranial width | −0.583514 | −0.353017 |
| 17. Maxillary tooth row | 0.873473 | −0.256163 |
| 18. Palatal length | 0.967378 | 0.099077 |
| 19. Lateral facial height | −0.793602 | −0.131066 |

$D^2$ distances, the Virunga gorillas are the most distinctive among the eastern samples. The Virunga gorillas also differ from other eastern gorillas in limb proportions (Groves and Stott, 1979), and in some craniomandibular (Taylor, 2002, this volume) and dental proportions (Uchida, 1996, 1998). In analyses of raw data, there was often some overlap among the eastern samples, but group distinctions were greater in the analyses of size-adjusted data.

Based on the cranial data, these results suggest that the major dichotomy among gorilla samples is the east–west separation, and that *G. g. graueri* is more closely related to *G. g. beringei* than to *G. g. gorilla* (the western lowland subspecies), an alignment that has gained increasing support since Groves's original (1970) study (see, for example, Groves and Stott, 1979; Ruvolo *et al.*, 1994; Garner and Ryder, 1996; Groves, this volume). Likewise, the distinctiveness of the mountain gorilla sample seems well supported. The evolutionary relationships among the other eastern samples are unresolved by our data. There are several possible reasons why our data do not provide a clear analysis of the relationships among eastern gorillas. These relationships are more complex than previously appreciated (see also Groves and Stott, 1979; Leigh *et al.*, this volume). Resolution of these complex relationships clearly requires input from

Table 3.14. *Mahalanobis* D² *distances between sample centroids and probabilities for the analysis of female crania from western and eastern gorillas using size-adjusted data*

Mahalanobis $D^2$ distances between sample centroids

| Location | Abong Mbang A | Batouri B | Ebolowa E | Mwenga-Fizi F | Mambili I | Lomie L | Metet M | Nigeria N | Bpindi P | Sette Camma S | Tshiab-erimu T | Utu U | Virunga V | Youka-douma Y |
|---|---|---|---|---|---|---|---|---|---|---|---|---|---|---|
| Abong Mbang | — | 3.67 | 1.93 | 18.10 | 5.01 | 1.70 | 5.16 | 7.73 | 8.64 | 7.36 | 16.61 | 17.94 | 21.68 | 7.41 |
| Batouri | 0.0055 | — | 4.77 | 22.30 | 4.06 | 2.80 | 4.16 | 9.51 | 6.74 | 8.82 | 19.00 | 22.46 | 23.88 | 6.16 |
| Ebolowa | 0.4582 | 0.0035 | — | 24.94 | 4.10 | 2.79 | 4.48 | 6.03 | 7.47 | 6.18 | 24.10 | 23.69 | 29.41 | 6.73 |
| Mwenga-Fizi | 0.0001 | 0.0001 | 0.0001 | — | 20.15 | 18.90 | 24.40 | 24.57 | 27.52 | 30.11 | 5.91 | 10.49 | 15.48 | 35.77 |
| Mambili | 0.0567 | 0.2395 | 0.2912 | 0.0001 | — | 3.05 | 3.57 | 7.55 | 5.59 | 6.45 | 20.18 | 17.32 | 21.41 | 8.62 |
| Lomie | 0.5679 | 0.1577 | 0.2423 | 0.0001 | 0.6031 | — | 3.23 | 9.33 | 7.85 | 7.27 | 18.12 | 18.68 | 21.39 | 7.36 |
| Metet | 0.0462 | 0.2170 | 0.2053 | 0.0001 | 0.7481 | 0.5388 | — | 9.55 | 5.39 | 8.91 | 22.75 | 21.22 | 22.84 | 8.94 |
| Nigeria | 0.0001 | 0.0001 | 0.0001 | 0.0001 | 0.0021 | 0.0001 | 0.0001 | — | 9.30 | 10.60 | 19.57 | 25.36 | 28.36 | 11.55 |
| Bpindi | 0.1166 | 0.3730 | 0.2970 | 0.0001 | 0.7786 | 0.2383 | 0.8074 | 0.0864 | — | 7.83 | 22.32 | 25.17 | 25.09 | 10.59 |
| Sette Camma | 0.0003 | 0.0001 | 0.0108 | 0.0001 | 0.0862 | 0.0016 | 0.0071 | 0.0001 | 0.3648 | — | 24.35 | 22.39 | 27.53 | 9.86 |
| Tshiaberimu | 0.0001 | 0.0001 | 0.0001 | 0.0795 | 0.0001 | 0.0001 | 0.0001 | 0.0001 | 0.0001 | 0.0001 | — | 12.62 | 10.67 | 29.13 |
| Utu | 0.0001 | 0.0001 | 0.0001 | 0.0020 | 0.0001 | 0.0001 | 0.0001 | 0.0001 | 0.0001 | 0.0001 | 0.0001 | — | 18.38 | 32.29 |
| Virunga | 0.0001 | 0.0001 | 0.0001 | 0.0001 | 0.0001 | 0.0001 | 0.0001 | 0.0001 | 0.0001 | 0.0001 | 0.0001 | 0.0001 | — | 32.36 |
| Youkadouma | 0.0006 | 0.0103 | 0.0084 | 0.0001 | 0.0143 | 0.0028 | 0.0104 | 0.0001 | 0.1162 | 0.0013 | 0.0001 | 0.0001 | 0.0001 | — |

Probabilities that sample centroids are the same

other sources of data including skeletal and dental anatomy, soft structures and genetic analyses. Moreover, the eastern samples used in this study were in some cases drawn from very large geographic areas.

Compared with the eastern samples, samples of western lowland gorillas analyzed in this study showed both a larger overall variation and greater overlap among samples. Thus on the first two canonical variates axes, individuals from the western samples occupy a greater multivariate space than the eastern samples (see also Albrecht *et al.*, this volume). At the same time, there is so much overlap among the different samples that individuals from a single sample may be dispersed over essentially the same area covered by all of the western samples (see also Albrecht *et al.*, this volume). Among the western samples, gorillas from the Nigerian sample are invariably the most distinctive (Stumpf *et al.*, 1998). In all analyses of samples from both western and eastern Africa, or western Africa only, whether based on raw or size-adjusted data, the Cross River (Nigerian) sample is graphically separated from other western samples; the generalized distances between the centroid of the Cross River sample and centroids of other samples are the largest, and the centroid of the Cross River sample is significantly different from almost all other centroids. By contrast, in all analyses, the western samples excluding Nigerian gorillas tend to overlap extensively; they have lower generalized distances between centroids and their centroids are often not significantly different from one another.

The distinctiveness of the Cross River gorillas was first noted by Matschie (1904), who diagnosed it as a new species, *Gorilla dielhi*, on the basis of its short cranial length, short molar tooth row length, palate morphology, and nuchal morphology. In his major revision of gorillas, Coolidge (1929) subsumed this taxon within *Gorilla gorilla gorilla*, but noted that it was characterized by a peculiar form of the posterior part of the cranium, and that compared to other western gorillas "*G. dielhi*" also had a relatively large intercoronoid width and interorbital width. Coolidge (1929:348) also noted that of all the western gorillas, those from Western Cameroon (assigned to *G. dielhi*) were the most distinctive (Fig. 3.8), but felt they did not justify separate taxonomic status. It is worth noting that Coolidge's caution in this instance was well justified, as univariate comparisons generally fail to consistently distinguish the Cross River from other western gorillas. However, multivariate approaches are more successful in emphasizing the distinctive size and shape differences characteristic of this sample (see also Sarmiento and Oates, 2000). In Groves's 1970 study of gorilla crania, the $D^2$ values between the Cross River and other western samples are generally higher compared with distances among other samples, but Groves did not emphasize this distinction. Current attention to this sample has largely come from increased realization of their endangered status due to their dwindling numbers and geographic isolation (e.g., Gonder *et al.*, 1997; Oates, 1998; Stumpf *et al.*, 1998; Oates *et al.*, this volume), as well

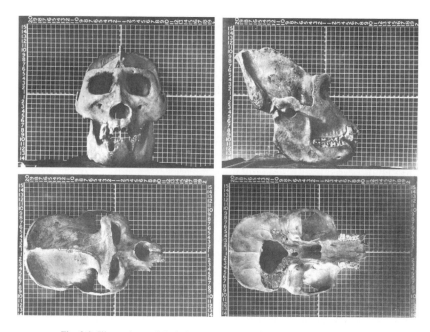

Fig. 3.8. Illustrations of the holotype specimen of *Gorilla diehli* (Matschie, 1904) from Coolidge (1929). Clockwise from upper left: anterior, lateral, inferior, and superior views. (Illustrations courtesy of Coolidge Archives, Stony Brook, NY. Reprinted with permission from Harvard Museum of Comparative Zoology.)

as the recent finding that chimpanzees from the same region are genetically distinct from other species of *Pan* (e.g., Gonder *et al.*, 1997). Results of this study accord well with other recent studies in supporting the distinctiveness of the Nigerian gorillas from other western gorillas.

The initial goal of this study was to determine the extent to which results of previous analyses of gorilla cranial morphology may have reflected size differences. Size has played a large part in taxonomic descriptions of gorilla systematics (see Groves, this volume) and, indeed, it has been argued that many morphological differences among African apes reflect uniform, and perhaps developmentally simple, allometric changes associated with differences in body size (e.g., Shea, 1985). Previous studies have not attempted to correct for the influence of overall size in evaluating cranial differences among gorilla populations.

In our study, results of analyses of size-adjusted data were very similar to those obtained using unadjusted data. The same taxa were distinctive in both sets of analyses and the generalized distances among taxa showed the same magnitude and distribution. These similarities suggest that many of the proportion differences among taxa are size-related or "allometric" in a general sense.

However, in some cases, size-adjustment permitted a clearer separation of size and shape differences. For example, in the analyses of both male and female crania using raw measurements, the second canonical axis separating the Cross River gorillas showed a broad pattern of correlation with several cranial variables. By contrast, size-adjustment of the variables using an estimate of cranial size identifies one variable in particular, breadth of the nuchal surface, as most highly correlated with this axis. Not surprisingly, it is the nuchal region that has been repeatedly identified in descriptive studies as being the most distinctive aspect of the cranium of the Cross River gorillas (e.g., Matschie, 1904; Coolidge, 1929; cf. Groves, 1970).

In summary, this study has identified several patterns of morphological distinctiveness among the gorillas of western and eastern central Africa based on multivariate analysis of geographic samples of cranial measurements. The greatest distinction remains between western and eastern gorillas. Among the eastern gorillas, the mountain gorillas of the Virunga region are the most distinct. Among western gorillas, the sample from the Cross River (Nigeria) is the most distinctive. However, appropriate taxonomic designations for the different populations of gorillas in eastern and western central Africa should be based on a broader analysis of all available data. Moreover, a more phylogenetic approach, rather than the purely phenetic studies conducted thus far, would probably be of considerable value in clarifying the evolutionary significance of the various similarities and differences among all of the gorilla populations.

### Acknowledgments

This paper is based on results initially presented as a poster at the annual meetings of the American Association of Physical Anthropologists in 1997. We thank Gene Albrecht and Bruce Gelvin for thoughtful advice and encouragement during discussion of this work at that time. We thank Luci Betti for producing and correcting seemingly endless versions of the figures and Stephen Nash, Curator of the Coolidge Archives, for permission to reproduce Fig. 3.8. Finally we thank Michele Goldsmith and Andrea Taylor for inviting us to contribute to this volume and for a remarkable display of patience and good humor in putting up with our tardiness.

### References

Ardrey, R. (1961). *African Genesis*. New York: Dell Publishing.
Coolidge, H.J., Jr. (1929). A revision of the genus *Gorilla*. *Memoirs of the Museum of Comparative Zoology, Harvard*, **50**, 291–381.

Elliot, D.G. (1913). *A Review of the Primates*. New York: American Museum of Natural History.

Falsetti, A.B., Jungers, W.L., and Cole, T.M. III (1993). Morphometrics of the callitrichid forelimb: A case study in size and shape. *International Journal of Primatology*, **14**, 551–571.

Garner, K.J. and Ryder, O.A. (1996). Mitochondrial DNA diversity in gorillas. *Molecular Phylogeny and Evolution*, **6**, 39–48.

Gonder, M.K., Oates, J.F., Disotell, T.R., Forstner, M.R.J., Morales, J.C., and Melnick, D.J. (1997). A new West African chimpanzee subspecies? *Nature* **388**, 337.

Groves, C.P. (1970). Sample systematics of the gorilla. *Journal of Zoology, London*, **161**, 287–300.

Groves, C.P. (1986). Systematics of the great apes. In *Comparative Primate Biology*, vol. 1, *Systematics, Evolution and Anatomy*, eds. D.R. Swindler and J. Erwin, pp. 187–217. New York: Alan R. Liss.

Groves, C.P. and Stott, K.W., Jr. (1979). Systematic relationships of gorillas from Kahuzi, Tshiaberimu and Kayonza. *Folia Primatologica*, **32**, 161–179.

Jungers, W.L., Falsetti, A.B., and Wall, C.E. (1995). Shape, relative size, and size-adjustments in morphometrics. *Yearbook of Physical Anthropology*, **38**, 137–161.

Matschie, P. (1904). Bemerkungen über die Gattung *Gorilla*. *Sitzungsberichte des Gesellschaft naturforschender Freunde, Berlin*, **1904**, 45–53.

Mosimann, J.E. (1970). Size allometry: Size and shape variables with characterizations of the log-normal and gamma distributions. *Journal of the Americal Statistical Association*, **56**, 930–945.

Oates, J. (1998). The gorilla population in the Nigeria–Cameroon border region. *Gorilla Conservation News*, **12**, 3–6.

Rothschild, W. (1906). Further notes on anthropoid apes. *Proceedings of the Zoological Society of London*, **2**, 465–468.

Ruvulo, M., Pan, D., Zehr, S., Goldberg, T., Disotell, T., and von Dornum, M. (1994). Gene trees and hominid phylogeny. *Proceedings of the National Academy of Sciences U.S.A.*, **91**, 8900–8904.

Sarmiento, E.E. and Oates, J.F. (2000). The Cross River gorillas: A distinct subspecies, *Gorilla gorilla diehli* Matschie 1904. *American Museum Novitates*, **3304**, 1–55.

Sarmiento, E.E., Butynski, T.M., and Kalina, J. (1996). Gorillas of the Bwindi-Impenetrable Forest and Virunga Volcanoes: Taxonomic implications of morphological and ecological differences. *American Journal of Primatology*, **40**, 1–20.

SAS Institute Inc. (1985). *SAS User's Guide: Statistics*, version 5 edition. Cary, NC: SAS Institute Inc.

Savage, T.S. and Wyman, J. (1847). Notice of the external characters and habits of *Troglodytes gorilla*, a new species of orang from the Gaboon River; Osteology of the same. *Boston Journal of Natural History*, **5**, 417–441.

Schwarz, E. (1928). Die Sammlung afrikanischer Affen in Congo-Museum. *Revue de Zoologie et Botanique Africaine*, **16**, 2, 1–48.

Shea, B.T. (1985). Ontogenetic allometry and scaling: A discussion based on the growth and form of the skull in African apes. In *Size and Scaling in Primate Biology*, ed. W. L. Jungers, pp. 175–205. New York: Plenum Press.

Stumpf, R.M., Fleagle, J.G., and Groves, C.P. (1997). Sexual dimorphism and geographic variation among subspecies of *Gorilla*. *American Journal of Physical Anthropology, Supplement* **24**, 223–224.

Stumpf, R.M., Fleagle, J.G., Jungers, W.L. Oates, J.F., and Groves, C.P. (1998). Morphological distinctiveness of Nigerian gorilla crania. *American Journal of Physical Anthropology, Supplement* **26**, 213.

Taylor, A.B. (2002). Masticatory form and function in the African apes. *American Journal of Physical Anthropology*, **117**, 133–156.

Uchida, A. (1996). *Craniodental Variation among the Great Apes*, Peabody Museum Bulletin no. 4. Cambridge, MA: Harvard University Press.

Uchida, A. (1998). Variation in tooth morphology of *Gorilla gorilla*. *Journal of Humam Evolution*, **34**, 55–70.

# 4 The hierarchy of intraspecific craniometric variation in gorillas: A population-thinking approach with implications for fossil species recognition studies

GENE H. ALBRECHT, BRUCE R. GELVIN, AND
JOSEPH M. A. MILLER

## Introduction

Understanding the nature of intraspecific craniometric variation in *Gorilla gorilla* is of major importance in paleoanthropology since gorillas are often used to model what variation might have looked like in a fossil hominid species. The use of modern species analogs, like gorillas, is a critical factor in fossil species recognition studies that seek to determine whether a sample of fossils represents one or more species. In such studies, a decision must be made as to whether the variation in the fossil sample is representative of intraspecific or interspecific variation. Consequently, accurately characterizing intraspecific variation in modern taxa is the foundation upon which such decisions rest.

The concept of "population thinking" should have a central role in fossil species recognition studies. Mayr (1963:5–6) states: "the replacement of typological thinking by population thinking is perhaps the greatest conceptual revolution that has taken place in biology" (see also, among many others: Mayr, 1942, 1969, 1976, 1999; Mayr *et al.*, 1953; Simpson, 1953, 1961; Mayr and Ashlock, 1991). By population thinking, we mean the theory and practice of population systematics in which biological species are thought of as aggregates of interbreeding natural populations comprising individuals that vary genetically and phenetically. Questions have been raised about the extent to which biological anthropologists have embraced population thinking (e.g., Fuller & Caspari, 2001). While paleoanthropologists may be familiar with the theory, fossil species recognition studies have not fully incorporated population thinking into their practical design or interpretation of results. Any study dependent on making comparisons of intraspecific variation between modern species and fossil samples must take into account the nature of biological species as defined by population taxonomy and population structure.

We use a population-thinking approach to demonstrate the kinds, degrees, and patterns of intraspecific variation in the craniofacial morphology of gorillas (see Gelvin *et al.*, 1997, for our preliminary results). We will illustrate an approach to characterizing intraspecific variation that reflects the population taxonomy and structure of a species. The goal is to obtain an accurate, well-controlled, comprehensive picture of intraspecific variation that reflects infraspecific species structure in which individuals are aggregated into populations, which in turn are aggregated into the species. The result is a "yardstick" of intraspecific variation in a modern analog species that is both well calibrated, in terms of measuring fossil differences against overall species-level variation, and flexible, in terms of comparing fossil differences to the various infraspecific levels of species structure. Our ultimate objective is to use the yardstick of intraspecific variation in gorillas to interpret the meaning of variation among early *Homo* crania (see Miller *et al.*, 1997).

### *Gorillas in paleoanthropological fossil species recognition studies*

Our interest in gorillas derives from their common use as a modern species analog for interpretation of morphological variation in fossil hominoid samples. For example, Oxnard *et al.* (1985) compared patterns of sexual dimorphism in dental variability in *Ramapithecus* and *Sivapithecus* to a sample of 84 *Gorilla gorilla*. Stringer (1986) used 29 *G. gorilla* to estimate ranges of intraspecific variability to test the credibility of *Homo habilis* as a single species. Lieberman *et al.* (1988) used 40 *G. gorilla* to determine if sexual dimorphism can explain craniometric variation between specimens of *H. habilis*. Wood (1991:230–254, 1993) measured 64 *G. g. gorilla* as a comparative sample in univariate and multivariate craniometric studies of species number in *H. habilis*. Kelley (1993) compared odontometric variation in *Lufengpithecus* to samples of as many as 44 *G. g. gorilla*. Martin and Andrews (1993) investigated species recognition in Middle Miocene hominoids based on tooth metrics of as many as 40 *G. gorilla*. Kramer *et al.* (1995) included 61 *G. gorilla* as a comparative analog to construct tests of the single species hypothesis for *H. habilis*. Grine *et al.* (1996) used 46 *G. g. gorilla* and 4 *G. g. beringei* to assess phenetic affinities among early *Homo* crania from East and South Africa. Silverman *et al.* (2001) measured 22 *G. g. gorilla* and 12 *G. g. beringei* mandibles to test if morphological variability in *Paranthropus boisei* exceeds what might be expected for a single species.

Teaford *et al.* (1993) used 40 *G. gorilla* skeletons to examine postcranial metric variation with respect to species discrimination in *Proconsul* from Kenya. These papers, and many others not cited here, illustrate the importance of understanding intraspecific variation in gorillas as a prerequisite for interpreting variation in fossil hominoid samples.

Paleoanthropological fossil species recognition studies, such as those cited above, typify the traditional approach to characterizing intraspecific variation in a modern analog species. The traditional approach does not fully characterize intraspecific variation in gorillas and other primates that are used as the comparative analogs for interpreting morphological variation among fossils. One problem is that relatively small sample sizes (e.g., 29–84 gorillas) are inadequate for investigating questions about the degree and pattern of variation within a species. Large numbers of specimens are needed to fully characterize the potential morphological diversity of gorillas or other sexually dimorphic, polytypic, geographically wide-ranging species. Consider, for example, that relying on a sample of 80 specimens means that each sex of each of the four gorilla subspecies (including *G. g. diehli*: Sarmiento and Oates, 2000) would be characterized by an average of only 10 individuals representing many natural populations spread across hundreds to thousands of square kilometers.

The second deficiency of traditional fossil species recognition studies is not fully characterizing the population taxonomy and population structure of gorillas. Typically, fossil species recognition studies are based on gorilla specimens that are available at a few museums. There is no systematic sampling of the species except for differentiating females and males because gorilla sexual dimorphism is known to be substantial. None of the cited fossil species recognition studies include specimens of all three historically recognized subspecies. Often specimens are not identified by subspecies rendering it impossible to know what the study's sample actually represents *vis-à-vis* the infraspecific taxonomic diversity of gorillas. If all specimens come from a single locality, this would provide little information about the nature of variation in a polytypic species.

Whatever the samples of traditional fossil species recognition studies represent, they do not characterize the full scope of intraspecific variation in gorillas. Thus, when examining the results of fossil species recognition studies, it is difficult to know for certain whether differences among fossils correspond to intrasexual, intersexual, geographic, subspecific, or interspecific variation. Questions involving the distinctions between intraspecific and interspecific variation can only be answered using an intraspecific yardstick constructed from large, well-controlled, representative samples that fully characterize the population taxonomy and population structure of gorillas.

*Population thinking, intraspecific variation, and the hierarchy of morphological variation*

Population thinking takes into account the population taxonomy and the population structure of a species (Mayr, 1942, 1963, 1969; Mayr *et al.*, 1953; Mayr and Ashlock, 1991). Population taxonomy is the study of the biological organization of species that recognizes species as geographically variable aggregates of populations. Polytypic species are naturally divided into subspecies, which are aggregates of phenotypically similar local populations that inhabit part of the species' geographic range and that differ taxonomically from other such groups (e.g., Mayr, 1969:41). However, "infraspecific categories, such as the subspecies, are by no means the final 'atoms' of the species. They are still composite units consisting of small local populations, each of which differs noticeably or statistically from other local populations" (Mayr, 1999:98). The local population, though not a formally recognized taxonomic category, is the basic taxonomic unit of the modern systematist (e.g., Mayr, 1999:7). A local population, or deme, is generally defined as the group of potentially interbreeding females and males at a locality (e.g., Mayr, 1963:136). Thus, the taxonomic organization of a polytypic biological species can be understood as a natural hierarchy of individuals belonging to local populations that are grouped into taxonomically recognized subspecies.

Population structure refers to the geographic arrangement of local populations across the species' range. Population structure can be described in terms of three phenomena: the population continuum, geographic isolates, and zones of secondary intergradation (hybrid zones) (e.g., Mayr and Ashlock, 1991). The population continuum is that part of the species' range where there is continuity of contact among local populations, some of which may be recognized as subspecies if sufficiently differentiated. A geographically isolated population is an incipient species that may differentiate genetically and phenetically because it is no longer in contact with the species continuum. A zone of secondary intergradation forms where a former geographic isolate, which attained some level of differentiation from the species continuum, re-establishes contact with the population continuum thus forming a hybrid zone. The population structure of the species is an aspect of the biological organization of a species that overlaps and complements its population taxonomy.

The population taxonomy and population structure of a polytypic species provide the conceptual framework for understanding morphological variation within a species (i.e., intraspecific variation). There is a natural hierarchical arrangement in which a polytypic species comprises subspecies that are comprised of local populations that are comprised of male and female individuals at one or more localities. Albrecht and Miller (1993) introduced the concept

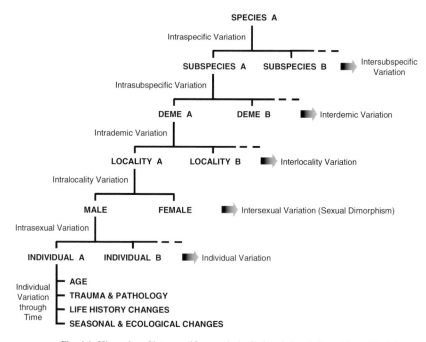

Fig. 4.1. Hierarchy of intraspecific morphological variation (adapted from Fig. 1 in Albrecht and Miller, 1993).

of the hierarchy of morphological variation (Fig. 4.1), which corresponds to the population taxonomy and population structure of a species. In Fig. 4.1, each infraspecific level of species organization, shown as a nested hierarchy on the left, corresponds to a different "kind" of morphological variation listed on the right. At the highest level, intraspecific variation is comprised of intersubspecific differences among subspecies. Intrasubspecific variation is comprised of interdemic differences among local populations. Intrademic variation is comprised of interlocality differences among groups of individuals at different collecting localities. These three levels together make up geographic variation, which is defined as the "occurrence of differences among spatially segregated populations of a species" (Mayr, 1963: 297). Intralocality variation is comprised of intersexual differences between females and males (sexual dimorphism). Intrasexual variation is comprised of interindividual differences among members of the same sex (individual variation). Finally, individuals may exhibit variation across time related to age, trauma or pathology, seasonal changes, or other life-history events (e.g., pregnancy or lactation).

There are practical considerations in implementing population thinking as reflected in the hierarchy of morphological variation. One consideration is that collections of primate specimens housed in museum collections rarely correspond to the demes of population biology. In a practical sense, with respect to population taxonomy, museum specimens can be organized according to: (1) collecting localities, as noted in museum records, where specimens are sampled from an undefined but limited geographic area; (2) groups of localities that can be organized on the basis of geographic proximity into "demes" that approximate local breeding populations; and (3) subspecies "named only if they differ 'taxonomically,' that is, by sufficient diagnostic morphological characters" (Mayr, 1969:42).

With respect to population structure, it is necessary to sample across the entire species range including representative samples of the population continuum, geographic isolates, and any hybrid zones. This sampling requires large numbers of specimens from many localities across the species range. Additionally, depending on the purpose of the study, there should be appropriate controls on sex, age, trauma/pathology, and other life-history parameters. Thus, the broad-scale sampling needed to adequately characterize intraspecific variation at all its hierarchical infraspecific levels requires paleoanthropologists to adopt the attitude of population systematists:

> They appreciate the fact that all organisms occur in nature as members of populations and that specimens cannot be understood and properly classified unless they are treated as samples of natural populations. As a consequence, they attempt to collect statistically adequate samples, which in the case of variable species often amount to hundreds or thousands of specimens, in order to be able to undertake a study of individual and geographic variation with the help of the best biometric and statistical tools.
>
> (Mayr and Ashlock, 1991:52–53)

## Materials and methods

Groves's (1967, 1970) extensive craniometric survey of gorillas from known collecting localities across the species' range in central Africa exemplifies the kind of data required to characterize the intraspecific variation in a modern species analog. Including some additional specimens measured by us, gorillas are represented by 657 adults from 196 localities grouped into 12 demes of four subspecies. We combine population thinking with a graphical quantitative approach to illustrate how morphometric data is structured infraspecifically according to the hierarchy of morphological variation in a manner that reflects population taxonomy and population structure. A graphical approach provides a strong visual impression of what intraspecific variation actually looks like in terms of its infraspecific organization.

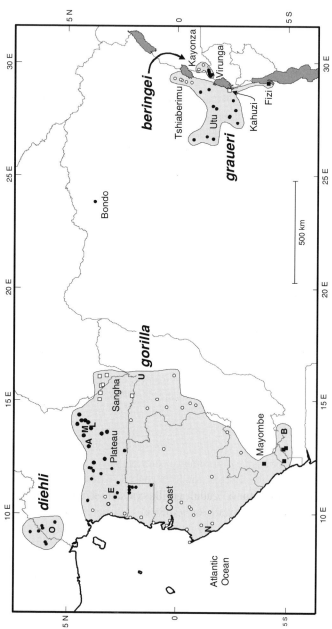

Fig. 4.2. Map of central Africa showing gorilla localities, demes, and subspecies. Subspecies ranges are shaded, demes are labeled, and localities are shown as either symbols or single letters. *Gorilla gorilla gorilla* demes: Bondo (separated from main subspecies range), Coast (open circles; N = Lake Nkomi), Mayombe (solid squares; B = Bamba), Plateau (solid circles; A = Abong Mbang, E = Ebolowa, L = Batouri-Lomie, M = Mayos), and Sangha (open squares; U = Ouesso); *G. g. diehli* (solid circles; O = Ossing); *G. g. graueri* demes: Mwenga-Fizi (solid square), Kahuzi (open square), Tshiaberimu (open circles), and Utu (solid circles); and *G. g. beringei* demes: Kayonza (open circles) and Virunga (solid circles). Boundaries are country borders.

## Gorilla systematics

The history of gorilla taxonomy is reviewed by Groves (this volume). We treat gorillas as a single species divided into four subspecies (Fig. 4.2): *Gorilla gorilla gorilla* (the western lowland gorilla), *G. g. diehli* (the Cross River gorilla), *G. g. graueri* (the eastern lowland gorilla), and *G. g. beringei* (the mountain gorilla). We follow Groves (1967, 1970, 1986; see Jenkins (1990) for a concise summary of taxonomy and synonymy), who separated western gorillas from eastern gorillas and, among the eastern forms, differentiated lowland from mountain gorillas. We include the Cross River gorilla as a separate subspecies in recognition of accumulating evidence that this geographic isolate may be phenotypically and genetically differentiated from the main population continuum of western gorillas (see Stumpf *et al.*, 1998, this volume; Sarmiento and Oates, 2000; Suter and Oates, 2000; Oates *et al.*, this volume; Groves, this volume). Clearly, questions remain about the subspecific affinities of some local populations of eastern gorillas (e.g., Casimir, 1975; Groves and Stott, 1979; Sarmiento *et al.*, 1996; Clifford *et al.*, this volume).

We retain all gorillas in a single species because the morphological and genetic evidence seems equivocal for recognizing western and eastern subspecies as separate species as suggested by some workers (e.g., see Groves, this volume, and Jensen-Seaman *et al.*, this volume, for reviews of this issue). It may be that the western and eastern gorillas are semispecies defined as "borderline cases between species and subspecies . . . that have acquired some but not all of the attributes of species rank" (Mayr and Ashlock, 1991:428). We follow Mayr and Ashlock (1991:105) in preferring "to treat allopatric populations of doubtful rank as subspecies, because the use of trinomial nomenclature conveys two important pieces of information: (1) closest relationship and (2) allopatry."

## Sampling strategy

Our total sample is comprised of 657 adult gorilla crania (Table 4.1). Most of our data (605 specimens) are derived from Groves's data set. Groves worked at eight American and 25 European institutions in the 1960s to assemble a comprehensive survey of available museum specimens for his original systematic studies of gorillas (Groves, 1967, 1970). We obtained Groves's data some years ago from copies of his original data sheets that he deposited at the Natural History Museum in London, England. Some of Groves's specimens are omitted because of missing measurements and/or unacceptable uncertainty about where they were collected in Africa.

Table 4.1. *Numbers of localities and specimens by sex, deme, and subspecies of* Gorilla gorilla

| Subspecies | Deme[a] | Localities | Females | Males | Totals |
|---|---|---|---|---|---|
| *Gorilla gorilla gorilla* | Bondo | 1 | 0 | 2 | 2 |
| | Coast | 53 | 36 | 84 | 120 |
| | Mayombe | 5 | 4 | 14 | 18 |
| | Plateau | 52 | 112 | 135 | 247 |
| | Sangha | 22 | 15 | 46 | 61 |
| | unassigned[b] | 4 | 3 | 20 | 23 |
| | Subtotal | 137 | 170 | 301 | 471 |
| *Gorilla gorilla diehli* | Cross River | 10 | 25 | 26 | 51 |
| *Gorilla gorilla graueri* | Mwenga-Fizi | 4 | 10 | 12 | 22 |
| | Kahuzi | 3 | 1 | 2 | 3 |
| | Tshiaberimu | 6 | 15 | 12 | 27 |
| | Utu | 14 | 10 | 21 | 31 |
| | unassigned[c] | 0 | 1 | 1 | 2 |
| | Subtotal | 27 | 37 | 48 | 85 |
| *Gorilla gorilla beringei* | Kayonza | 2 | 1 | 2 | 3 |
| | Virunga | 15 | 16 | 24 | 40 |
| | unassigned[d] | 0 | 1 | 0 | 1 |
| | Subtotal | 17 | 18 | 26 | 44 |
| *Gorilla gorilla graueri / beringei* | unassigned | 5 | 1 | 5 | 6 |
| | Total | 196 | 251 | 406 | 657 |

[a]See Fig. 4.2 for mapping of demes.
[b]Includes two females and 17 males that do not have precise locality information but can be reasonably assigned to *G. g. gorilla*.
[c]Includes one female and one male that do not have precise locality information but can be reasonably assigned to *G. g. graueri*.
[d]Includes one female that does not have precise locality information but can be reasonably assigned to *G. g. beringei*.

We supplemented Groves's data with 52 additional adult specimens measured by one of us (J.M.A.M.) at the U.S. National Museum in 1995. These more recently measured specimens include a sample of 18 mountain gorilla skulls (*G. g. beringei*) donated by Dian Fossey after Groves visited Washington, DC in the 1960s.

All specimens used in our analyses are adults based on full dental eruption (third molars in occlusion) and closed sphenooccipital synchondrosis (C. Groves, personal communication). Presumably, limiting specimens to adults controls for the confounding effects of age at the lowest level of the hierarchy of variation shown in Fig. 4.1. There may be age-related changes in gorillas, especially males, whose skulls as young adults with unworn dentitions seem relatively smooth and gracile compared to the often rugose, robust skulls of old adults with very worn dentitions (e.g., see photographs in Coolidge, 1929). However, age-related changes in adult gorilla skulls are only a minor component of morphometric variation compared to the broad range of individual differences. In any case, age-related changes throughout adulthood are part of the individual variation that should be characterized in a study of intraspecific variation.

The total sample includes 251 females and 406 males. There are 471 specimens of *G. g. gorilla*, 51 *G. g. diehli*, 85 *G. g. graueri*, and 44 *G. g. beringei*. Six specimens, from localities that might be attributed to either *G. g. graueri* or *G. g. beringei*, are included only when analyzing overall species variation and excluded at lower hierarchical levels (specimens are from five localities: Babault Mission, Lake Kivu; Kivu Forest; southwest of Kivu; mountains north of Lake Kivu; and Masisi).

The 657 specimens that comprise our sample are wild-caught individuals from 196 different collecting localities (Table 4.1). We carefully reviewed and revised Groves's locality information using Groves's notes, museum records, a variety of maps, the country gazetteers published by the U.S. Board of Geographic Names, and the Geographic Names Database on the GEOnet Names Server maintained by the National Imagery and Mapping Agency. This time-consuming process was necessary to assure geographically well-controlled samples consistent with a population-thinking approach for analyzing morphometric variation. We could reliably determine latitude and longitude for 133 localities (Fig. 4.2). Precise coordinates could not be determined for the other 63 localities but there was sufficient information to assign each of them to one of the demes shown in Fig. 4.2.

Most localities are known from just a single individual or a few animals. However, there are eight localities with samples of at least 10 females and/or 10 males (Fig. 4.2 and Table 4.1). Of particular interest are the two localities with the largest samples representing *G. g. gorilla* from Cameroon – Abong Mbang with 28 females and 37 males and Ebolowa with 15 females and 24 males. Field-collected primates attributed to a single locality like Abong Mbang or Ebolowa were actually collected, in most cases, not from that precise location but from a broader geographic area in the neighborhood of the named place. In other cases, such as *G. g. beringei* in the Virunga mountains, localities may be rather precisely defined and close enough together to approximate the geographic collecting area of a single locality elsewhere, such as either Abong Mbang or Ebolowa.

The 196 collecting localities are divided into the same geographic subgroups indicated in Groves's archived data based on geographic and ecological criteria thought to be relevant for gorilla systematics (Groves, 1967, 1970). We call these subgroups "demes" to indicate their intermediate position between local groups (localities) and subspecies as a first approximation to the deme of population biology. The demes are shown in Fig. 4.2 and their sample sizes are given in Table 4.1. Two males from Bondo on the Uele River in central Zaïre were treated as a deme of *G. g. gorilla* (Fig. 4.2). These specimens, which are usually attributed to *G. g. gorilla*, are enigmatic and problematic because Bondo is about halfway between the ranges of western and eastern lowland gorillas, hundreds of kilometers from the closest known distribution of either subspecies (Coolidge, 1929:358–359). A number of other specimens, listed in Table 4.1 as "unassigned," can be ascribed to a subspecies but are from localities that cannot be reliably assigned to any one deme.

### Measurements

Our study is based on the 16 craniofacial dimensions listed in Table 4.2. These are a subset of the 30 cranial measurements taken by Groves. Groves (1967, 1970) actually used only 10 cranial dimensions for his multivariate analyses (the seven indicated in Table 4.2 plus tooth row length, palatal length, and nuchal surface length). We selected the 16 measurements because they could be reliably taken on the gorilla skulls we measured at the U.S. National Museum as well as on the most complete crania of fossil hominids used in our fossil

Table 4.2. *Sixteen craniofacial measurements used in this study*

| |
| --- |
| Greatest length of skull (prosthion to opisthocranion) |
| Cranial length (glabella to opisthocranion)[a] |
| Skull breadth (biporionic breadth)[a] |
| Nuchal breadth (breadth of nuchal surface in mastoid region) |
| Biorbital breadth[a] |
| Palatal breadth (outside first molars)[a] |
| Upper nasal breadth (greatest breadth of nasal bones) |
| Nasal height (top of piriform aperture to prosthion) |
| Bicanine breadth (across canine alveoli)[a] |
| Facial height (top of supraorbital torus to prosthion)[a] |
| Orbital height (inside) |
| Orbital breadth (inside one orbit) |
| Postorbital breadth (at postorbital constriction)[a] |
| Lower nasal breadth (greatest breadth of piriform aperture) |
| Interorbital breadth |
| Supraorbital thickness (normal thickness of supraorbital torus) |

[a]Seven of 10 measurements used by Groves (1967, 1970).

species recognition study of *Homo habilis* (Miller *et al.*, 1997). We did not use Groves's mandibular data because there are no mandibles associated with the fossil crania in which we are interested. The 16 craniofacial dimensions generally characterize the overall size and shape of gorilla skulls.

### Principal components analysis

Principal components analysis (PCA) is used as a simple, straightforward ordination technique to display graphically the craniometric relationships among the 657 gorilla specimens. All of our PCA plots are based on a single, covariance-based PCA of all 657 specimens. The different plots shown in Figs. 4.3–4.6 are based on the same PCA but simply highlight different subsets of gorillas. Means and standard deviations for localities, demes, subspecies, and sexes of gorillas are listed in Table 4.3.

PCA is a well-known, robust, multivariate technique ideally suited to our purpose of efficiently describing multivariate variation in graphical form (see Albrecht, 1977, 1978, 1980, 1992, for graphical approaches to understanding multivariate techniques and the procedural relationships among them). Covariance-based PCA is an ideal technique for investigating within-group structure without transformations, assumptions, or constraints that affect the character of the raw data. Imagine that the 657 gorilla specimens are plotted in a multivariate data space defined by 16 orthogonally constructed, intersecting axes which correspond to the original 16 craniofacial dimensions. When applied to a covariance matrix, PCA rotates these orthogonally drawn axes rigidly around the origin to a new position that displays as much information (variability) as possible along a single primary axis, the first principal component (PC1). The subsequent 15 PC axes are constructed to be mutually orthogonal, they are statistically independent (uncorrelated), and they are successively decremental in the amount of variation displayed. Typically, the first two PCs account for a large portion of the overall variation previously dispersed among the original dimensions.

As an ordination technique, an important property of covariance-based PCA is that it preserves Euclidean distances among specimens in the multivariate data space. Since PCA effects a rigid rotation of orthogonal axes without rescaling or otherwise transforming the original axes, there is no distortion of the metric relationships in shifting from the original variables to the PC axes.

Using PCA as a descriptive technique for ordering, summarizing, and displaying morphometric relationships among gorillas can be thought of as "the quantification of the classical methodology of comparative anatomy" (Albrecht, 1980:681). In this view, the measurement values for an individual gorilla can

Table 4.3. *Summary statistics by sex, locality, and by demic, subspecies, and species levels of intraspecific variation for* Gorilla gorilla

| Sex, subspecies, deme, locality[a] | n | Mean PC1 | Mean PC2 | SD PC1 | SD PC2 | PC$_{1-2}$ area[b] | Total variance[c] |
|---|---|---|---|---|---|---|---|
| Female *Gorilla gorilla* | 251 | −57.4 | −0.3 | 14.6 | 7.3 | 106.1 | 441.7 |
| *Gorilla gorilla gorilla* | 170 | −58.9 | 0.2 | 13.6 | 7.1 | 97.0 | 404.3 |
| Coast deme | 36 | −66.2 | −1.6 | 14.5 | 5.8 | 83.8 | 412.2 |
| Plateau deme | 112 | −56.1 | 0.4 | 12.8 | 7.0 | 89.3 | 373.5 |
| Abong Mbang | 28 | −56.0 | 2.5 | 12.7 | 7.8 | 99.2 | 366.7 |
| Batouri-Lomie | 15 | −53.9 | −2.4 | 11.7 | 6.2 | 72.8 | 337.0 |
| Ebolowa | 16 | −58.8 | 0.6 | 10.4 | 6.4 | 66.2 | 281.1 |
| Mayos | 10 | −57.6 | −2.3 | 15.0 | 6.2 | 93.1 | 442.0 |
| Sangha deme | 15 | −62.7 | 1.0 | 9.4 | 9.3 | 87.4 | 322.0 |
| *Gorilla gorilla diehli* | 25 | −68.8 | 1.0 | 12.2 | 6.0 | 73.7 | 339.3 |
| Ossing[d] | 16 | −72.1 | 0.4 | 12.3 | 6.1 | 75.4 | 337.8 |
| *Gorilla gorilla graueri* | 37 | −48.4 | −4.3 | 14.2 | 6.7 | 95.2 | 377.9 |
| Mwenga-Fizi deme | 10 | −51.7 | −4.2 | 7.7 | 5.9 | 45.8 | 209.9 |
| Tshiaberimu (Lubero) deme[e] | 15 | −41.1 | −3.7 | 13.7 | 7.7 | 106.0 | 346.2 |
| Utu deme | 10 | −57.3 | −4.6 | 15.8 | 6.7 | 105.5 | 405.0 |
| *Gorilla gorilla beringei* | 18 | −48.1 | 2.2 | 11.2 | 8.5 | 95.2 | 326.0 |
| Virunga deme | 16 | −48.6 | 1.6 | 11.6 | 8.8 | 102.0 | 335.1 |
| Male *Gorilla gorilla* | 406 | 35.5 | 0.2 | 22.7 | 12.0 | 271.9 | 986.8 |
| *Gorilla gorilla gorilla* | 301 | 36.5 | 1.7 | 23.5 | 11.3 | 266.1 | 1003.8 |
| Coast deme | 84 | 26.0 | 0.0 | 22.1 | 10.0 | 220.9 | 926.5 |
| Lake Nkomi | 12 | 16.8 | −0.7 | 25.3 | 9.0 | 227.4 | 921.0 |
| Mayombe deme | 14 | 34.3 | −3.3 | 17.1 | 9.1 | 155.4 | 615.5 |
| Bamba | 10 | 35.8 | −1.1 | 17.0 | 9.9 | 168.3 | 591.0 |
| Plateau deme | 135 | 45.4 | 1.7 | 21.2 | 11.4 | 241.4 | 886.6 |
| Abong Mbang | 37 | 45.8 | 0.6 | 19.6 | 11.8 | 231.3 | 796.0 |
| Ebolowa | 25 | 36.8 | 1.3 | 13.8 | 11.2 | 155.2 | 598.1 |
| Sangha deme | 46 | 33.6 | 5.2 | 23.8 | 13.0 | 309.3 | 1043.1 |
| Ouesso | 19 | 25.8 | 1.6 | 23.3 | 12.1 | 282.7 | 949.3 |
| *Gorilla gorilla diehli* | 26 | 13.3 | 6.0 | 20.6 | 10.4 | 212.9 | 775.2 |
| Ossing[d] | 11 | 7.9 | 4.2 | 17.1 | 6.4 | 108.4 | 636.2 |
| *Gorilla gorilla graueri* | 48 | 35.8 | −11.1 | 13.4 | 9.5 | 127.0 | 541.8 |
| Mwenga-Fizi deme | 12 | 34.4 | −14.0 | 16.0 | 6.2 | 100.0 | 463.6 |
| Tshiaberimu deme | 12 | 40.7 | −8.0 | 12.6 | 12.8 | 161.1 | 761.7 |
| Utu deme | 21 | 34.0 | −11.6 | 12.7 | 8.6 | 108.6 | 455.3 |
| *Gorilla gorilla beringei* | 26 | 43.6 | −0.6 | 17.6 | 12.0 | 210.7 | 723.9 |
| Virunga deme | 24 | 42.7 | −0.5 | 17.4 | 12.4 | 215.6 | 706.3 |
| *Gorilla gorilla* (male and female) | 657 | 0.0 | 0.0 | 49.4 | 10.4 | 515.5 | 2820.4 |

[a] Table includes only localities and demes whose sample sizes are greater than or equal to 10 specimens for males and/or females.
[b] PC$_{1-2}$ area = SD$_{PC1}$ × SD$_{PC2}$ (a proportionate measure of a group's area of dispersion in the bivariate plot of the first two PC variates).
[c] Total variance = sum of the variances for PC1–16 (a measure of a group's total dispersion in the 16 dimensional multivariate data space).
[d] Ossing is given as Oddinge on Groves's data sheets and as Ossidinge by Sarmiento and Oates (2000).
[e] All female specimens of the Tshiaberimu deme are from a single locality (Lubero).

be used to plot a single point in a 16-dimensional data space that summarizes the overall morphology of that animal's skull. The position of any individual in the multidimensional space is an objective, empirical representation of the craniofacial morphology of that one gorilla. Nearness to others in the multi-dimensional space means similarity in morphology while distance means dissimilarity as described by the 16 raw measurements. A gorilla may be unique in its skull morphology and occupy a distant part of the data space, or groups of gorillas may look alike and be found close together.

Our choice of PCA can be further understood by explaining why we do not use other well-known multivariate techniques. Group-finding techniques (i.e., cluster or mixture analysis) are appropriate for determining group structure when the classification of the specimens is unknown. However, for gorillas, the identity of each specimen is known as to age, sex, locality, deme, and subspecies. Among-group techniques like canonical variates analysis (CVA; or discriminant functions or generalized distances) are appropriate for investigating relationships among groups whose membership is known and whose within-group variability is homogeneous. Canonical variates can be thought of graphically and computationally as the sequential application of covariance-based PCA, first to the dispersion within groups and then to the dispersion among groups (Albrecht, 1980, 1992). While CVA is a natural extension of PCA ordination, it portrays among-group relationships in a multivariate data space that has been distorted by standardizing the pooled within-group variation. Our purpose is not to discover previously unrecognized groups nor to look only at among-group relationships, but to use the known taxonomy to characterize the hierarchical infraspecific structure of variation within and among gorillas.

The bivariate plots of PC1–2 provide a visual impression of the dispersion within and among the sexes, subspecies, demes, and localities of gorillas. We confirm the visual impressions of variation with a simple quantitative measure of bivariate dispersion ($PC_{1-2}$ area; see Table 4.3). $PC_{1-2}$ area is calculated as the product of the standard deviations of the first two PCs ($PC_{1-2}$ area $= SD_{PC1} \times SD_{PC2}$). It is not a measure of the actual dispersion size (area) of a group's dispersion, but is proportionate to a group's variability shown in the PCA plots.

A second measure of variation estimates the overall variation of localities, demes, subspecies, or sexes (total variance; see Table 4.3 and Fig. 4.7). Total variance is used as a measure of the amount (volume) of the 16–dimensional multivariate data space occupied by a group of individuals (Van Valen, 1974; Albrecht, 1978). Total variance for a group of individuals is easily calculated as the sum of the univariate variances of the 16 craniofacial dimensions or, equivalently, as the sum of variances (eigenvalues) of the 16 PC variates. Like univariate variances, total variance may vary widely for small samples but become increasingly stable as sample sizes increase.

### Graphics versus statistics

Some readers may question why we choose a graphical approach, rather than inferential statistical hypothesis testing, to assess differences and similarities among individuals, sexes, localities, demes, and subspecies of gorillas. Our primary purpose is to illustrate the application of population thinking to the analysis of intraspecific variation, not to test degrees of differences and similarities. Familiarity with the data gained through graphical exploration is an essential, often overlooked prerequisite to insightful statistical analysis. Chambers *et al.* (1983:1) begin their book on graphical techniques by observing: "our eye–brain system is the most sophisticated information processor ever developed, and through graphical displays we can put this system to good use to obtain deep insight into the structure of data." Wilson (1984:261) expresses a similar sentiment about the utility of graphical approaches:

> In many situations involving the analysis of physical anthropological data, the techniques which are needed are informal exploratory techniques rather than formal confirmatory techniques, such as tests of specific hypotheses. Graphical techniques are well-suited for this purpose since they tend to be less formal and confining, and enable the investigator to gain 'insight' into the structure of his data, without imposing restrictive constraints.

In no way should our use of graphic ordinations of multivariate data be construed as a recommendation to eschew statistical analysis. Rather, we choose here to emphasize the descriptive use of multivariate methods to ordinate and summarize intraspecific variation in gorillas. In applying a graphical approach, care must be taken to emphasize clear differences, obvious similarities, and broad patterns consistent with the limitations of such an approach. The uncovering of apparent differences graphically should serve as working hypotheses to be tested statistically in future studies.

## Results

### Overall species level variation

The PCA of all 657 gorillas is shown in Fig. 4.3A. The scatter of plotted points represents the overall variation in *G. gorilla*, which incorporates all levels of the hierarchy of intraspecific variation. The first two PCs account for 90.4% of the variation originally distributed among the 16 craniofacial variables. The first PC orders gorillas by increasing overall skull size from the smallest skulls on the left to the largest skulls on the right. There appear to be two distinct clusters of specimens, with the right-hand cluster being more dispersed than the one on the left.

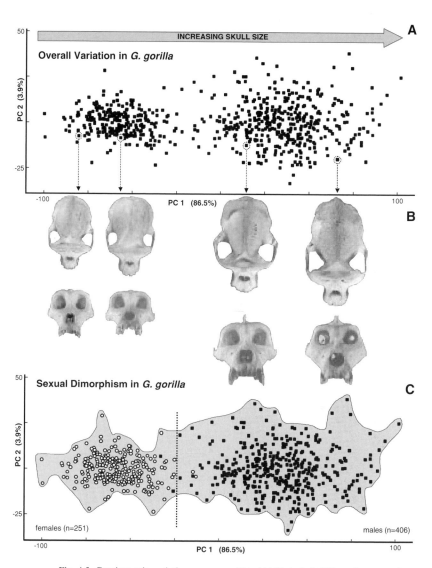

Fig. 4.3. Craniometric variation among gorillas. (A) Plot of all 657 specimens on the first two PC axes which account for 90.4% of the total variation in the original 16 craniofacial variables. (B) Dorsal and facial views of gorilla skulls corresponding to the four individual gorilla specimens identified in (A). (C) Sexual dimorphism in gorilla craniometrics. The plot is identical to that shown in (A) except the 251 females and the 406 males are identified by sex (females = circles; males = squares). The dotted vertical line is positioned to maximally separate (discriminate) females from males. The shaded envelope is representative of the overall distribution of all 657 gorilla specimens.

The four skulls shown in Fig. 4.3B are mapped onto the distribution of specimens in Fig. 4.3A to "put a face" on the craniometric variation of gorillas. This juxtaposition of individual skulls and their plotted points illustrates how metric variability on PC1–2 represents actual morphological similarities and differences among the skulls themselves, keeping in mind that 9.6% of overall skull variation is represented in higher PC variates. The metric distances between points reflect morphological affinities – specimens close together look alike while specimens far apart look very different in terms of skull size and shape. As is obvious by the photographs, the variation across the spectrum of individuals represents substantive morphological differences in their skulls.

Figure 4.3A shows most (90%) of the total overall variation that exists among all individuals of a large, reasonably comprehensive sample of a polytypic primate species. Figure 4.3A also shows overall morphological variation without superimposing any biological information about the individuals depicted. Two clusters of specimens are apparent with an intermediate region of lesser density. There is no apparent structure within either of the clusters. It is only when further biological information is superimposed on the plotted points that the significance of the metric relationships among specimens becomes apparent. Thus, it is necessary to identify individuals by sex, locality, deme, and subspecies to see the infraspecific structure within the overall species level variation portrayed in Fig. 4.3A.

Figures 4.3C–4.6 show a series of plots successively representing variation by sex, subspecies, deme, and locality. The points of each graph are plotted on a shaded background envelope representing the overall species level variation of Fig. 4.3A. Means and standard deviations for PC1–2, PC$_{1-2}$ areas, and total variances are listed in Table 4.3. Total variances are plotted in Fig. 4.7.

### Sexual dimorphism

This section considers the amount and pattern of sexual dimorphism at successive infraspecific levels from the species level down through the locality level. Subsequently, we control for sexual dimorphism by treating females and males separately when further examining variation at the subspecific, demic, and locality levels.

Sexual dimorphism at the overall species level is shown in Fig. 4.3C, which is identical to Fig. 4.3A except specimens are identified by sex. The two clusters of specimens seen in Fig. 4.3A now become evident as females and males. There is extensive sexual dimorphism between large-skulled males on the right and small-skulled females on the left. The sexes are so dimorphic in skull size that there is only minimal overlap between females and males. A simple "by eye" discriminant function, constructed as a straight vertical line, separates the sexes such that only two of 406 males and two of 251 females are misclassified as the wrong sex based on PC1 scores. Only the very largest females might be mistaken for the very smallest males, and vice versa, using the 16 craniometric dimensions employed here. However, it is important to remember that species-level differentiation of the sexes may be confounded by geographic variation comprising intersubspecific, interdemic, and interlocality differences.

Sexual dimorphism at the subspecies level is shown in the four panels of Fig. 4.4. In *G. g. gorilla* (Fig. 4.4A), the highly dimorphic females and males minimally overlap in their distribution. The other three subspecies (*G. g. diehli*, *G. g. graueri*, and *G. g. beringei* shown in Fig. 4.4B–D) exhibit mean sex differences that are broadly comparable to *G. g. gorilla*. However, the sexes are well separated from one another in their distributions. Thus, these three subspecies appear to be more dramatically dimorphic than *G. g. gorilla* because there are no specimens with intermediate morphologies. Apparently, the minimal overlap between the sexes seen at the overall species level (Fig. 4.3C) is attributable to only one gorilla subspecies (*G. g. gorilla* in Fig. 4.4A). It is important to remember that sexual dimorphism at the subspecies level may be confounded by interdemic and interlocality variation.

Sexual dimorphism at the demic level is shown in the four panels of Fig. 4.5. The three demes of *G. g. gorilla* (Fig. 4.5A–C) exhibit mean differences between the sexes that are about the same as seen for the subspecies as a whole (Fig. 4.4A). However, at the demic level, the sexes are more distinct in their distributions, especially the Sangha deme, which displays a considerable gap between females and males. For the Mwenga-Fizi deme of *G. g. graueri* shown in Fig. 4.5D, the mean difference between the sexes approximates the mean sexual difference at the subspecies level (Fig. 4.4C), but there is a greater morphological gap at the demic level than the subspecies level. Here, again, differences between the sexes observed at the demic level may be confounded by interlocality variation.

Sexual dimorphism at the locality level is shown in the four panels of Fig. 4.6, which include two localities for the Plateau deme of *G. g. gorilla* (Ebolowa in Fig. 4.6A and Abong Mbang in Fig. 4.6B) and one locality each for *G. g. diehli* (Ossing in Fig. 4.6C) and *G. g. graueri* (Baraka of the Mwenga-Fizi deme in Fig. 4.6D). For each locality, the mean difference between females and males is similar to mean sex differences of their respective demes and subspecies (Figs. 4.4 and 4.5). However, the sexes at the locality level exhibit a distinct morphological gap that separates the largest female from the smallest male. In general, the sexes appear progressively more distinct as geographic variation is more tightly controlled (e.g., compare Figs. 4.3B, 4.4A, 4.5A, and 4.6A for the Ebolowa locality of the Plateau deme of *G. g. gorilla*).

In the preceding discussion, sexual dimorphism is judged by: (1) how far apart the sexes are in terms of distance (i.e., mean difference), and (2) how

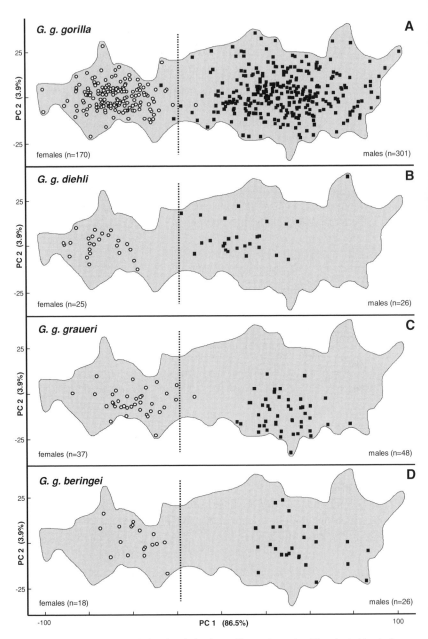

Fig. 4.4. Subspecific level variation in gorilla craniometrics. The plot is identical to Fig. 4.3A except the specimens of each subspecies are separated into similarly scaled panels for (A) *Gorilla gorilla gorilla*, (B) *G. g. diehli*, (C) *G. g. graueri*, and (D) *G. g. beringei* (females = open circles; males = closed squares). Other conventions are the same as Fig. 4.3B.

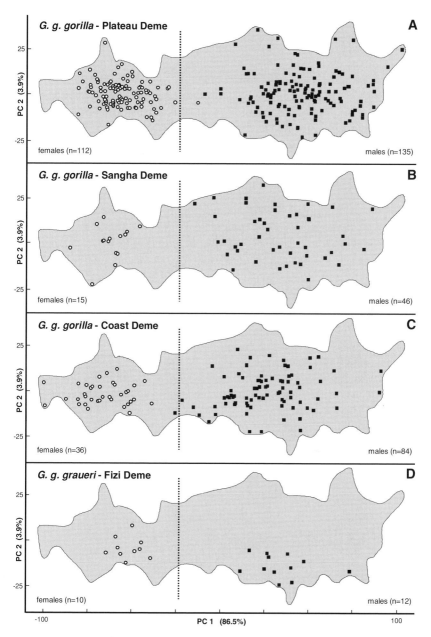

Fig. 4.5. Demic level variation in gorilla craniometrics. The plot is identical to Fig. 4.3A except the specimens of representative demes are separated into similarly scaled panels for (A) the Plateau deme of *Gorilla gorilla gorilla*, (B) the Sangha deme of *G. g. gorilla*, (C) the Coast deme of *G. g. gorilla* and (D) the Mwenga-Fizi deme of *G. g. graueri* (females = open circles; males = closed squares). Other conventions are the same as Fig. 4.3B.

81

distinct the sexes are in terms of the amount of overlap or separation from one another. An additional consideration is variance dimorphism, which refers to how variable the sexes are in comparison to each other. At the species level (Fig. 4.3C), the "area" (or volume in a multidimensional sense) of the cloud of male points is easily double that of the female cloud, which indicates much greater morphological variation among males than among females. The greater variability in males applies equally to size differences along PC1 and non-size-related shape differences along PC2. This disparity in within-sex dispersions applies variably at other hierarchical levels (see Table 4.3 for PC1–2 standard deviations, PC1–2 area, and total variance). At the subspecies level, the amount of variance dimorphism ranges from a large difference for *G. g. gorilla* (Fig. 4.4A) comparable to that seen at the species level (Fig. 4.3C), to intermediate differences for *G. g. diehli* and *G. g. beringei* (Figs. 4.4B and 4.4D, respectively), to a small difference between the size of the male and female clouds for *G. g. graueri* (Fig. 4.4C). The variance dimorphism in the three demes of *G. g. gorilla* (Coast, Sangha, and Plateau demes in Figs. 4.5A–C) approximates the same 2:1 disparity seen at the subspecies level (Fig. 4.4A). In contrast, the Mwenga-Fizi deme of *G. g. graueri* (Fig. 4.5D), like the subspecies (Fig. 4.4C), exhibits only a small degree of variance dimorphism. At the Ebolowa and Abong Mbang localities of the Plateau deme of *G. g. gorilla* (Figs. 4.6A–B), the variance dimorphism is evident but less dramatic than at the subspecies and demic levels (Figs. 4.4A and 4.5A, respectively). There is little if any variance dimorphism for the Ossing locality of *G. g. diehli* (Fig. 4.6C) and the Baraka locality of *G. g. graueri* (Fig. 4.6D). However, the small sample sizes make it difficult to draw any definitive conclusion about variance dimorphism at these two localities. Figure 4.7 demonstrates that the variance dimorphisms seen in the bivariate plots of Figs. 4.3C–4.6 hold equally for total variance in the full PCA data space.

### Subspecific level variation

Craniometric variation within and among the four gorilla subspecies is shown in the four panels of Fig. 4.4. In this and the next two sections, females and males are considered separately to control for the substantial sexual dimorphism discussed above.

For each sex in Fig. 4.4, the four subspecies exhibit considerable overlap in their distributions that together form the species envelope shown by the shaded region. For the large sample of *G. g. gorilla* (Fig. 4.4A), the subspecies clouds for females and males both substantially correspond in distribution to the overall species distribution (Fig. 4.3C). The smaller number of specimens for *G. g. diehli* (Fig. 4.4B), *G. g. graueri* (Fig. 4.4C), and *G. g. beringei*

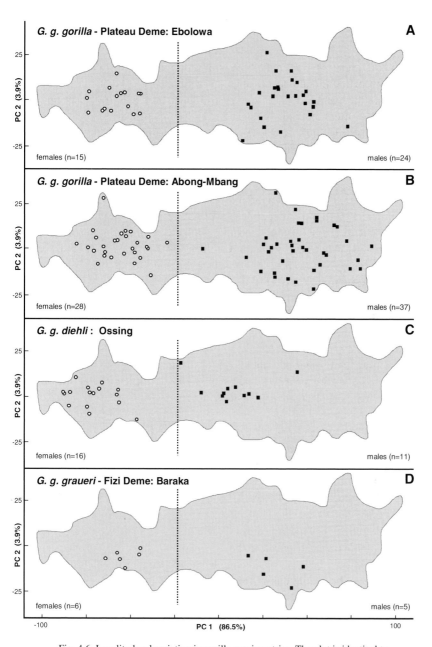

Fig. 4.6. Locality level variation in gorilla craniometrics. The plot is identical to
Fig. 4.3A except the specimens of representative localities are separated into similarly
scaled panels for (A) Ebolowa in the Plateau deme of *Gorilla gorilla gorilla*,
(B) Abong Mbang in the Plateau deme of *G. g. gorilla*, (C) Ossing in *G. g. diehli*, and
(D) Baraka in the Mwenga-Fizi deme of *G. g. graueri* (females = open circles;
males = closed squares). Other conventions are the same as Fig. 4.3B.

83

(Fig. 4.4D) broadly overlap the *G. g. gorilla* sample but occupy different regions of the overall species envelope. Thus, since the subspecies differ in their relative positions within the overall species envelope, they must differ in the mean morphology of their skulls (Table 4.3).

The subspecies differ in the amount of variation expressed on PC1–2 (Fig. 4.4 and Table 4.3). As inferred from the size (area) of the PC1–2 subspecies clouds, males of *G. g. gorilla* are more variable than males of the other three subspecies. Males of *G. g. diehli* and *G. g. beringei* are intermediate in their variability, and *G. g. graueri* has low male variability. The total variance statistics for males confirm this pattern of disparities in male subspecific variability (Table 4.3 and Fig. 4.7). Thus, just as variance dimorphism is a component of sexual dimorphism, geographic variation includes variance "polymorphism" among the male subspecies. These differences in relative variabilities of males are evident along both PC1 (size) and PC2 (shape) axes.

Females display a different pattern of subspecific variance polymorphism. The bivariate plots of Fig. 4.4 show some differences in the size of the female subspecies clouds. However, when overall PCA variation is considered, as measured by total variance (Table 4.3 and Fig. 4.7), female subspecies samples are much more homogeneous in their intrasubspecific dispersions than males. Note that females of *G. g. graueri* do not show the same restricted level of variability as their male counterparts (Fig. 4.4C).

### Demic level variation

Examples of demic variation within and among gorilla subspecies are shown in the four panels of Fig. 4.5. For *G. g. gorilla*, the Plateau (Fig. 4.5A), Sangha (Fig. 4.5B), and Coast (Fig. 4.5C) demes are broadly overlapping for each sex, but differ slightly in mean position (Table 4.3). The male samples of the three *G. g. gorilla* demes appear to be less variable than the subspecies cloud (Fig. 4.4A) only because of the smaller demic sample sizes. Total variances indicate that the overall dispersions of the male demic samples of *G. g. gorilla* are similar to one another and minimally, if at all, less diverse than the subspecies as a whole (Table 4.3 and Fig. 4.7). The Mayombe deme of *G. g. gorilla* is considerably less variable than the Sangha, Coast, and Plateau demes of this subspecies (Table 4.3 and Fig. 4.7), but the Mayombe sample consists of only 14 specimens, 10 of which are from a single locality. For male *G. g. graueri* (Fig. 4.5D), the Mwenga-Fizi deme occupies only part of the subspecies cloud (Fig. 4.4C). Figure 4.7 indicates that the Utu and Mwenga-Fizi demes of *G. g. graueri* are both restricted in their male variability. Most notable, however, is the limited male intrasexual variability in the Mwenga-Fizi deme compared to the Sangha, Coast, and Plateau demes of *G. g. gorilla* (Figs. 4.5 and 4.7).

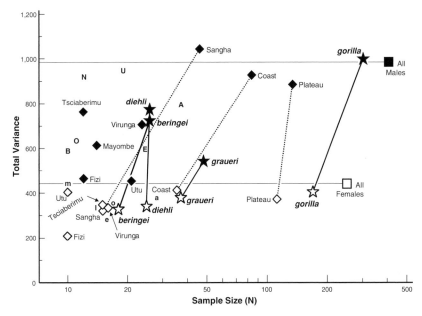

Fig. 4.7. Total variances of localities, demes, and subspecies of each sex of *Gorilla gorilla*. The graph is based on the same data and same groups listed in Table 4.3 (sample size of 10 or more specimens per group). The horizontal axis representing sample size is logarithmically scaled only to reduce the length of the axis. Male samples are plotted as solid symbols and females as open symbols. The overall variation for each sex is plotted as squares, subspecies as stars, and demes as diamonds. Localities are marked by uppercase (males) or lowercase (females) letters: A,a = Abong Mbang; B = Bamba; E,e = Ebolowa; l = Batouri–Lomie; m = Mayos; N = Lake Nkomi; O,o = Ossing; U = Ouesso (see Table 4.3 for subspecific and demic associations of the these localities). A solid line joins the sexes of each subspecies and a dashed line joins the sexes of each deme.

For females, there is little variance polymorphism among the demes of the species once sample sizes are taken into account (Table 4.3; Figs. 4.5 and 4.7). The one exception is the Mwenga-Fizi deme of *G. g. graueri* which exhibits substantially less intrademic variability than female samples of other demes (see Fig. 4.5D). However, in contrast to males, the females of the Utu deme of *G. g. graueri* do not have the same low total variance as in the Mwenga-Fizi deme.

### Locality level variation

Examples of locality variation within and among the demes and subspecies of gorillas are illustrated in the four panels of Fig. 4.6. The Ebolowa (Fig. 4.6A) and Abong Mbang (Fig. 4.6B) localities of the Plateau deme of *G. g. gorilla* have the largest locality sample sizes in our data set. Males from these two localities broadly overlap in their distributions, but there are slight differences in their mean morphology (Table 4.3). The two localities differ in the relative size of their male dispersions with Ebolowa overlapped almost entirely by the more broadly dispersed Abong Mbang cloud of points. The total variances for these localities indicate that males from Abong Mbang are about one-third more variable than males from Ebolowa (Table 4.3 and Fig. 4.7). The variation at each of these localities is much less than that of the Plateau deme (Fig. 4.5A), the subspecies (Fig. 4.4A), and the species (Fig. 4.3C) to which they belong (Table 4.3 and Fig. 4.7). For the Ossing locality of *G. g. diehli* (Fig. 4.6C), male intrasexual variability is less than that seen in the subspecies (Fig. 4.4B), whereas female variability is more comparable. For the Baraka locality of *G. g. graueri* (Fig. 4.6D), the small samples for females and males appear to be slightly less variable than intrasexual variability at the demic level (Fig. 4.5D) but substantially less variable than at the subspecies level (Fig. 4.4C).

## Discussion

### *Craniometric variation in gorillas*

Our study differs in its design and intent from the work of previous investigators who have examined craniometric characters of gorillas (e.g., Keith, 1927; Schultz, 1927, 1934; Coolidge, 1929; Randall, 1943, 1944; Haddow and Ross, 1951; Ashton, 1957; Groves, 1967, 1970; Casimir, 1975; Hursh, 1975, 1976; Wood, 1975, 1976; Groves and Stott, 1979; Shea, 1983, 1986; Schmid and Stratil, 1986; Preuschoft, 1989; O'Higgins and Dryden, 1993; Sarmiento *et al.*, 1996; Stumpf *et al.*, 1997, 1998, this volume; Leigh *et al.*, 1998, this volume; Humphrey *et al.*, 1999; Oates *et al.*, 1999, this volume; Park *et al.*, 1999; Sarmiento and Oates, 2000; Taylor, 2002, this volume). Accordingly, our results are not directly comparable to previous studies that, for the most part, addressed taxonomic, systematic, ontogenetic, or functional questions using metric features of the skulls of one or more kinds of gorillas. Our discussion focuses on the significance of major patterns of morphological variability that have been revealed by our hierarchical analysis of intraspecific variation.

### *Sexual dimorphism*

Gorillas are well known to display a high degree of sexual dimorphism in their craniofacial morphology (e.g., see Wood, 1975, 1976, for early references on gorilla sexual dimorphism; see the fossil species recognition studies cited in our "Introduction"; Schmid and Stratil, 1986; Wood *et al.*, 1991; O'Higgins and Dryden, 1993; Stumpf *et al.*, 1997). We have systematically examined the degrees and patterns of sexual dimorphism at each infraspecific level of the hierarchy of morphological variation. The degree of dimorphism is approximately the same at every infraspecific level in terms of mean differences between the sexes. However, the gap between the sexes becomes progressively greater as one moves from the species level to the locality level (Figs. 4.3C–4.6). At the species level (Fig. 4.3C), there is slight morphological overlap with the largest females similar to the smallest males, while at the locality level (Fig. 4.6) the sexes are clearly distinct from one another with a substantial morphological gap between females and males. Thus, the proximity of the male and female clouds of points at the species level (Fig. 4.3C) is an artifact resulting from the confounding effects of geographic variation. For example, the small amount of overlap of the sexes for *G. g. gorilla* (Fig. 4.4A) is due to comparing females of the Plateau deme (Fig. 4.5A) with males of the Coast deme (Fig. 4.5C).

Comparisons of females and males from the same locality provide the best true estimate of sexual dimorphism in a species (Wood, 1975; Jenkins and Albrecht, 1990). Such an approach is simply good comparative anatomy by trying to control for geographic variation among localities, demes, and subspecies in order to eliminate extraneous sources of variation that confound estimates of sexual dimorphism. Ebolowa and Abong Mbang (Figs. 4.6A–B), as the two best-sampled gorilla localities, provide the most realistic, biologically relevant estimates of how dramatically different female and male gorillas actually are in their craniofacial morphology.

In general, when characterizing sexual dimorphism, it is inadequate to consider only mean differences while ignoring the relative distributions of females versus males. For example, the index of dimorphism (male mean/female mean) provides an estimate of the degree of difference between the sexes but does not provide important information about relative dispersions and degree of overlap that may exist. Looking at mean sexual difference without also considering degree of dispersion is like trying to test for mean differences (e.g., a *t*-test) between two samples without also estimating variances. Consider the theoretical case of taxa A and B that are identical in the mean differences between the sexes (Fig. 4.8). These two taxa would have the same indices of dimorphism. In taxon A, the distributions of females and males extensively overlap such that some small males are similar to an average female and some large females are similar to an average male. In taxon B, the distributions of the sexes are nonoverlapping with a morphological gap between them. Despite similar indices of dimorphism, taxon B is actually more dimorphic than taxon A when the degree of overlap (or separation) between the sexes is taken into account. We contend that sexual dimorphism in taxon A versus taxon B represents situations of different biological significance.

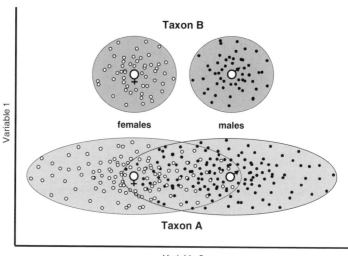

Fig. 4.8. Simple model illustrating the importance of considering distributional parameters when studying sexual dimorphism. The plot shows the distributions of females (open circles) and males (closed circles) of hypothetical taxa A and B, whose sex means (centroids) are shown as female and male symbols. The degree of sexual dimorphism is identical based on mean values for each sex but taxon A exhibits wide overlap in the ranges of the sexes while taxon B exhibits a discontinuity between the female and male distributions.

Another aspect of sexual dimorphism that must be considered is variance dimorphism (Wood, 1976; Leutenegger and Cheverud, 1982, 1985; Wood *et al.*, 1991; Plavcan, 2000; Gordon, 2001). That males are more variable than females is evident from the bivariate PCA plots (Figs. 4.3C–4.6), which show substantially larger dispersions for males as compared to females. The greater variability among males applies equally to size differences along PC1 and non-size-related shape differences along PC2. Furthermore, on subsequent variates (PC3–16), males are more variable than females for the overwhelming number of intersexual comparisons at the locality, demic, and subspecific levels. Total variance (Fig. 4.7) demonstrates that males are more variable than females at every infraspecific level of the hierarchy of morphological variation. At any infraspecific level, there is a range of differences in the relative amount of variation among females compared to males. For subspecies, the length of the lines connecting females and males in Fig. 4.7 indicates that the disparity in total variance is largest for *G. g. gorilla*, intermediate for *G. g. beringei* and *G. g. diehli*, and smallest for *G. g graueri*. The well-sampled demes of *G. g. gorilla* (Sangha, Coast, and Plateau) seem to reflect the variance dimorphism of the subspecies as a whole. Unfortunately, it is difficult to see any obvious, consistent pattern of variance dimorphism among the localities and demes of the other three subspecies because sample sizes are smaller and the number of replicate samples of localities and demes within any one subspecies are few.

Variance dimorphism has been proposed as a key factor in the evolution of primate sexual dimorphism. The theoretical model of Leutenegger and Cheverud (1982, 1985) holds that the evolution of sexual dimorphism involves differential selective responses for body size increase when male characters are more variable than female characters. There has been controversy regarding this model based on studies of the teeth, crania, and postcrania of primates (e.g., Pickford and Chiarelli, 1986; Plavcan and Kay, 1988; Plavcan, 2000; Gordon, 2001). It is true that gorillas are characterized by large body size, a high degree of sexual dimorphism, and, as shown here, males that are more variable than females. However, variance dimorphism differs substantially among the subspecies of gorillas such that male *G. g. graueri* are about 50% more variable than females while male *G. g. gorilla* are about 250% more variable (Table 4.3 and Fig. 4.7). Evolutionary models of sexual dimorphism need to incorporate the complex pattern of intraspecific sex differences that may exist in primates.

Sexual dimorphism is thought to have many correlates to various aspects of primate biology including intra- and intersexual selection, niche expansion, energy strategies, predator defense, and other behavioral and ecological life-history parameters (e.g., Leutenegger and Kelly, 1977; Leutenegger and Cheverud, 1982, 1985; Clutton-Brock, 1985; Pickford, 1986; Oxnard, 1987). In general, hypotheses about the biological significance of sexual dimorphism relate to mean sexual size differences. However, as we have shown, sexual dimorphism should also be characterized by the amount of distributional overlap or separation of the sexes (sexual distinctiveness) and the relative within-sex variabilities (variance dimorphism). If the gorilla sexes exhibit differences in their biology related to their extreme sexual dimorphism in size, then we would also expect to see biological correlates of the similarly substantial sexual distinctiveness and variance dimorphism of gorillas. The greater variation in the craniofacial morphology of male gorillas compared to females (or vice versa) must have biological explanations related to the breadth of dietary adaptations, strength of sexual selection, social organization, use of habitat, or other aspects of gorilla biology.

*Intraspecific variation within sexes*
Geographic variation in craniometric morphology among the localities, demes, and subspecies of gorillas is well known (Coolidge, 1929; Groves, 1967, 1970, this volume; Groves and Stott, 1979; Albrecht and Miller, 1993; Sarmiento *et al.*, 1996; Stumpf *et al.*, 1997, 1998, this volume; Leigh *et al.*, 1998, this

volume; Sarmiento and Oates, 2000). Our results are generally consistent with multivariate results of other studies despite differences in methodological approaches. Other multivariate studies rely on methods that maximize mean differences among groups after standardizing for within-group variation (i.e., canonical variates, discriminant functions, and generalized distances), whereas our PCA approach ordinates the data to maximize overall variation without presumptions about group membership. Our PCA shows a hierarchy of mean differences within each sex (Table 4.3): among subspecies of gorillas (Fig. 4.4), among demes of the same subspecies (Fig. 4.5), and among localities of the same deme of the same subspecies (Fig. 4.6). Since PC1 is dominated by sexual dimorphism and the large sample of *G. g. gorilla*, some mean differences involving the other three subspecies are apparent only on the higher principal component variates (PC3–16) not displayed here.

Variance polymorphism is an important component of intraspecific variation in gorillas. Figure 4.7 and Table 4.3 demonstrate variance polymorphism at the different hierarchical levels. Variation in females appears to be constrained to relatively narrow limits compared to males. For females, the four gorilla subspecies differ little in their overall intraspecific variation (total variance ranges from 326 to 404). Female variation within localities and demes is comparable to the dispersions of the four subspecies given the broad range of sample sizes. In contrast, males display a range of within-group variances that is three to four times greater than for females. For males, the most variable subspecies, *G. g. gorilla*, has an intrasubspecific total variance that is twice that of the least variable subspecies, *G. g. graueri*. Intrademic variation in the Sangha, Coast, and Plateau demes of *G. g. gorilla* is comparable to the overall variability of the subspecies as a whole. The greater range of intrademic variation in *G. g. graueri* compared to *G. g. gorilla* may be a sampling effect of the small numbers of specimens for the Tshiaberimu, Mwenga-Fizi, and Utu demes. The total variance values for the Virunga deme and its parent subspecies (*G. g. beringei*) are so similar because they are based on nearly the same specimens. Sample sizes are too small and the number of localities too few to make any meaningful interpretation of variance polymorphism among localities.

One interesting question with respect to within-sex variance polymorphism relates to the association between craniometric and geographic variation. Specimens of *G. g. beringei* are about as variable as specimens of *G. g. gorilla* from single localities (the N, B, A, E, and U localities of Fig. 4.7). This is not a surprising result for *G. g. beringei*, which is geographically restricted to a small mountainous area of East Africa. The mountain gorilla's geographic range is probably comparable in size to the collecting area of a single locality of *G. g. gorilla* (e.g., Ebolowa and Abong Mbang undoubtedly include specimens from rather substantial geographic regions rather than specific places). However, for males of *G. g. graueri*, there is an incongruity between the geographic range of the subspecies and its intraspecific variability. This subspecies inhabits thousands of square kilometers in central Africa but its craniometric variation is less than that seen at a single locality of other subspecies. The fact that females do not display a similarly low variability adds an interesting element of variance dimorphism to the picture of craniometric variation in *G. g. graueri*.

There is increasing evidence of geographic differentiation in the ecology, behavior, social structure, and other adaptations of gorillas (see Dixson, 1981; Watts, 1991, this volume; Sarmiento *et al.*, 1996; Uchida, 1996; Doran and McNeilage, 1998, 1999; Goldsmith *et al.*, 1999; Remis, 1999, this volume; Goldsmith, this volume). Morphological studies are also increasingly associating morphological differentiation among gorillas to various aspects of their natural history (e.g., Groves, 1970; Sarmiento, 1994; Taylor, 1995, 1997, 2002, this volume; Sarmiento *et al.*, 1996; Uchida, 1996; Sarmiento & Oates, 2000; Inouye, this volume). Thus, it seems likely that the restricted craniometric variation in *G. g. graueri* might relate to one or more of the following: ecological homogeneity, biogeographical events, evolutionary history, functional adaptations, genetic drift, selective forces, and/or restricted gene flow. More generally, the diversity of degrees and patterns of intraspecific differences at the various infraspecific levels demonstrated in our PCA results may have adaptive correlates yet to be determined.

### Population thinking and species structure

We have shown how a population-thinking approach, as embodied in the hierarchy of morphological variation, can be used to characterize the kinds, degrees, and patterns of intraspecific craniometric variation in a polytypic species. Our results provide a comprehensive picture of what intraspecific variation actually looks like when infraspecific species structure is investigated using large numbers of specimens. A profile of intraspecific variation can be constructed when specimens are identified by sex, subspecies, deme, and locality. The intraspecific profile will be an amalgam of mean differences (i.e., how much distance separates groups) and dispersional differences (i.e., how much variation is present within the groups) among the various hierarchical levels.

For example, Fig. 4.9 shows a nested hierarchy of envelopes for male gorillas: (1) locality level variation (Baraka) within demic level variation (Mwenga-Fizi deme), (2) demic variation within subspecific level variation (*G. g. graueri*), and (3) subspecific variation within the overall species envelope. The envelope of variation for Baraka is one of four envelopes that could be drawn for the Mwenga-Fizi deme, each of which is for a different locality that

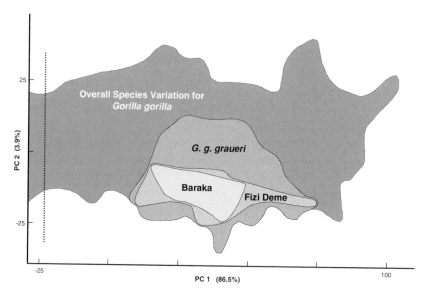

Fig. 4.9. Nested hierarchy of variation in gorillas. The plot shows the male (right) "half" of the PC plots with nested, shaded envelopes representing the Baraka locality (Fig. 4.6D) within the Mwenga-Fizi deme (Fig. 4.5D) of *Gorilla gorilla graueri* (Fig. 4.4C) within the male species cloud (Fig. 4.3C).

may vary in position and size from the others. The composite variation of these four localities forms the envelope of variation for the Mwenga-Fizi deme. Likewise, the Mwenga-Fizi deme is but one of three demes, each of which differs in position and size of their envelopes, whose combined variation forms the *G. g. graueri* subspecies envelope of variation. In turn, the *G. g. graueri* envelope combines with those of the other three subspecies to form the overall species envelope of craniometric variation.

Nested envelopes of variation, like those of Fig. 4.9, exemplify the hierarchical nature of intraspecific morphological variation in a manner that reflects the population taxonomy and population structure of a polytypic species. The hierarchy of morphological variation provides the analytic framework for studying polytypic species. However, each species will have varying numbers of subspecies, demes, and localities that differ in the degrees and patterns of their variation. The composite picture of intraspecific variation will differ from species to species. To reveal that picture for any one species requires samples of specimens large enough to adequately characterize each infraspecific level of variation (i.e., population taxonomy) across the geographic range of the species (i.e., population structure).

The traditional approach of relying on relatively few specimens to characterize intraspecific variation runs contrary to population thinking. Rather, the traditional approach begins with a concern for acquiring a comparative sample large enough to satisfy the requirements of statistical testing (typically 30–80 specimens). However, by using such limited samples, there is an implicit decision to ignore the biological reality that species are complex in their population taxonomy and population structure. As is clear from gorillas, such small numbers of specimens cannot possibly represent the complexities of variation among the sexes, localities, demes, and subspecies that comprise intraspecific variation in a sexually dimorphic, geographically variable, polytypic species.

In contrast, the population-thinking approach begins with an appreciation for the biological organization of species as they exist in nature. For morphometric studies, the population taxonomy and population structure of a species give rise to an analytic framework that we call the hierarchy of morphological variation (Fig. 4.1). This hierarchical framework is used to develop a sampling strategy that will adequately characterize the complexity of the species. The same hierarchical framework also provides a structure for both analyzing and interpreting the data in terms of comparisons within and among individuals, sexes, localities, demes, subspecies, and any other levels that may be relevant to the questions being investigated. The population thinking approach, in contrast to the traditional approach, provides an appropriate model that most closely approximates the biological complexity of polytypic species.

Detailed knowledge of infraspecific species structure is important for comparative studies of gorillas and other primates. Such an emphasis is fitting at a time when studies are increasingly focused on testing fine-grained hypotheses about the comparative anatomy, behavior, genetics, and systematics of primates. Comprehensive knowledge of intraspecific variation allows greater insight into the nature of biological variation through comparisons within and across levels of the infraspecific hierarchy. In making such comparisons, it is important to understand that the significance of variation at any one hierarchical level relies on interpreting the magnitude and pattern of variation at other levels – analogous to a nested analysis of variance in statistics.

### *Implications for interpreting fossil variation*

A population-thinking approach to the study of intraspecific variation has implications for fossil species recognition studies in paleoanthropology. These implications include: (1) a way of thinking about intraspecific variation in fossil species that is consistent with the biology of polytypic species; (2) an appropriate sampling strategy for constructing adequate yardsticks of intraspecific variation in modern analog species; and (3) a strategy of comparison for using well-calibrated yardsticks to interpret fossil sample variation.

*Population thinking and fossil species variation*
Just as the principles of population thinking should be integrated into studies of intraspecific variation of modern species such as gorillas, so too should these principles be applied to evaluating intraspecific variation of fossil species. For example, when considering a fossil hominid species, it is reasonable to assume that it is polytypic given that modern humans, great apes, and most other modern primates are polytypic. Thus, a fossil hominid species is likely to exhibit all the infraspecific levels of variation indicated in the hierarchy of morphological variation (Fig. 4.1). Certainly, every fossil specimen represents an individual at a locality who was a member of a group, which belonged to a deme, which belonged to a subspecies of a polytypic species, which had a certain population structure across the geographic range of the species. Local groups, demes, and subspecies are "ephemeral" and change through evolutionary time, and are unlikely to be preserved in the fossil record as a whole. Nevertheless, at one point in time, each individual has a certain morphology that reflects the variation of the infraspecific groups to which it belongs. Any individual who might be preserved as a fossil must carry the morphological attributes of its hierarchical group membership. Consequently, paleoanthropologists should think about intraspecific variation in a fossil species in the same terms that apply to intraspecific variation of modern polytypic species.

Time is another important factor to consider with respect to intraspecific variation in fossil species. When studying a modern polytypic species, intraspecific variation is viewed from the vantage point of a single "time slice" or "moment" in geological time. Most certainly, intraspecific variation at each infraspecific level can change differentially through time, whether morphology oscillates back and forth ("stasis") or shifts directionally. For a fossil sample that spans a considerable time depth, the result is expanded "envelopes" of variation at the various infraspecific levels and, thus, at the species level. Therefore, intraspecific variation of a fossil species sampled across geologic time will be greater than in a modern polytypic species observed at a single geologic "instant". Just as population thinking provides the biological context for understanding intraspecific variation in modern polytypic species, it also provides the context for understanding the evolution of a species.

*Sampling strategy for constructing yardsticks*
The population-thinking approach leads to a sampling strategy for constructing the yardsticks of morphological variation in modern species that are used to "measure" the significance of variation in fossil samples. This study has shown that intraspecific variation is a composite of variation at different infraspecific levels that reflects the population taxonomy and population structure of polytypic species. The sampling strategy for constructing the yardstick derives from

the hierarchy of morphological variation (Fig. 4.1). The hierarchy provides a biologically appropriate framework for characterizing the various infraspecific components of intraspecific variation.

In practical terms, this means measuring large numbers of wild-caught specimens of known age, sex, locality, and taxonomy. While sex and age are usually readily determined from museum records or the specimens themselves, considerably more effort is required to check locality information and confirm taxonomic identities of each individual specimen. With precise locality information, specimens can be assigned to demes and subspecies based on the current knowledge of the species taxonomy. Sample sizes should be sufficiently large so that each infraspecific hierarchical level is well characterized from the locality level to the species level. For a polytypic species, the analog sample must be represented by all known subspecies. Additionally, it is necessary to sample as widely from the species range as possible to account for the various kinds of geographic variation. This sampling strategy will provide the best estimates of the kinds, degrees, and patterns of intraspecific variation for modern comparative species since it is consistent with their known population biology. The result will be a robust, appropriate yardstick of intraspecific variation that can be used effectively to interpret fossil sample variation.

Given the nature of intraspecific variation in polytypic species, it is inappropriate to use comparative samples that do not reflect the complexity of that variation. For example, intraspecific variation in gorillas cannot be accurately characterized by samples of only a single subspecies such as *G. g. beringei* or *G. g. graueri*, or by specimens of only one deme or locality. Such sampling would clearly result in an underestimate of intraspecific variation. Yet, it is not uncommon for fossil species recognition studies to use small, comparative analog samples that only comprise a single subspecies or even a small number of specimens whose subspecific designation is unidentified. Clearly, this "traditional" practice is inconsistent with the principles of modern population biology.

The two yardsticks shown in Fig. 4.10 illustrate the contrast between the population-thinking approach and the traditional approach to characterizing intraspecific variation. For the population-thinking approach, the hierarchical levels of infraspecific variation represented in Fig. 4.1 can be reconfigured as a well-calibrated yardstick, the ends of which represent the limits of intraspecific morphometric variation. One can imagine that this yardstick is double-sided so that the scale for males is represented along one edge while the scale for females is represented on the other edge. This double-edged ruler is necessary because males and females may differ in their variation at each respective hierarchical level. The smallest gradations of measurement are provided by individuals, which are represented by the thinnest, shortest lines. The next higher unit of

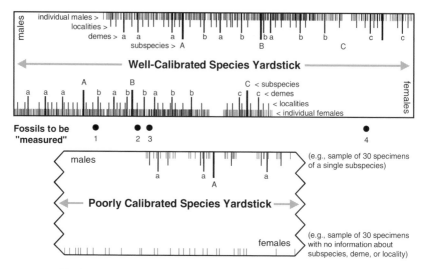

Fig. 4.10. The hierarchical levels of intraspecific variation (Fig. 4.1) portrayed as a species yardstick of morphological variation. The measurement values of individuals are shown as short, thin lines along the male and female edges of the yardstick representing some morphological variable. Mean values for localities, demes (lowercase letters), and subspecies (uppercase letters) are indicated by successively longer and thicker lines. The upper yardstick is well calibrated on the basis of several hundred specimens that represent a reasonable sampling of the localities and demes of the four subspecies. The lower yardstick is poorly calibrated using only 30 specimens of each sex, which is typical of the sample size in most studies of primate morphometrics. In fossil species recognition studies, the yardstick can be used to "measure" the significance of variability among fossils such as the four specimens shown between the yardsticks.

measurement is provided by locality means, which are represented by the next longest tick marks in Fig. 4.10. Demic means are represented by the next longest marks (labeled with lowercase letters) and the grossest of the graduated units are subspecies means (labeled with uppercase letters).

As opposed to a real ruler that is graduated with regular, standardized, iso-metric units of measurement, the species yardstick will vary in terms of the length and scale of each hierarchical infraspecific level, depending on the ana-log species used to construct the yardstick. A well-calibrated yardstick provides the best picture of what intraspecific variation actually looks like in a species when each infraspecific level is characterized by relatively large samples from known localities. For each level of calibration, the yardstick provides infor-mation about mean positions, limits of variation, distributional properties, and patterns of variation.

Such a finely graduated, well-calibrated yardstick contrasts with the in-complete picture of intraspecific variation typical of fossil species recognition studies (see "Introduction"). The traditional approach attempts to portray intraspecific variation of an entire species with a poorly calibrated yardstick constructed with inadequate samples often selected without regard for species structure. Such a poorly calibrated yardstick is aberrant in its length, and the ticks representing the various hierarchical levels of infraspecific variation are mostly or entirely missing (Fig. 4.10). However, even if some infraspecific hierarchical levels are demarcated, sample sizes are typically so small that the measurement scale is unlikely to be well defined. Clearly, decisions about the meaning of fossil sample variation should be based on a well-calibrated yardstick that best represents the actual biological reality of intraspecific variation within the chosen analog species.

*Strategy for using well-calibrated yardsticks*
The population-thinking approach not only allows for proper yardsticks of intraspecific variation to be constructed according to meaningful biological criteria, but it also provides a strategy for using those yardsticks. This strategy is characterized by comparing variation in a fossil sample to each infraspecific hierarchical level of variation of the modern analog species. For example, as is traditionally done, differences among fossils could be compared to overall species variation (e.g., as represented in Fig. 4.3 for gorillas). However, the population-thinking approach allows more refined comparisons of variation to be made at any infraspecific level: the subspecific level (e.g., Fig. 4.4), the demic level (e.g., Fig. 4.5), the locality level (e.g., Fig. 4.6), or within-sex variation at any level (e.g., Figs. 4.4–4.6) (see Miller *et al.*, 1997, 1998). These comparisons increase the interpretative power of fossil species recognition studies in a manner not attainable by the traditional approach. Such a strategy of comparison also provides a meaningful biological perspective and context for understanding fossil sample variation.

**Concluding remarks**

This study uses a population-thinking approach to demonstrate the kinds, degrees, and patterns of intraspecific variation in gorillas. This approach characterizes intraspecific variation by using the hierarchy of morphological variation as a conceptual framework that corresponds to the population taxonomy and population structure of a polytypic species. The result of applying this approach to gorillas is a portrait of variation at every infraspecific level from the species level down to the locality level. These portraits demonstrate that overall intraspecific variation is a composite of variation within and among the various infraspecific levels. An essential component of the population-thinking approach is a sampling strategy that adequately characterizes each infraspecific level across the geographic range of the species.

Population thinking has important implications for paleoanthropologists involved in fossil species recognition studies. For the investigator seeking to distinguish intraspecific from interspecific variation in fossil samples, the population-thinking approach provides: (1) a strategy for evaluating intraspecific variation in a fossil species that is consistent with the biology of polytypic species; (2) a sampling strategy for constructing well-calibrated yardsticks of intraspecific variation in modern analog species; and (3) an analytic strategy of more interpretative power that uses well-calibrated yardsticks for assessing the significance of fossil sample variation.

Mayr and Ashlock (1991:55) remind us: "the populations of most species of animals contain several different phena [morphs] as a result of sexual dimorphism, age variation, seasonal variation, polymorphism, and other causes". The critical task for fossil species recognition studies is to interpret correctly the biological meaning of the phena that comprise fossil samples. For modern species, Mayr and Ashlock (1991:57) observe:

> Differences between phena thus may reflect either a species difference or intraspecific variation. A full understanding of intraspecific variation is therefore necessary before we can make the probabilistic statement that phenon B belongs to a species different from phenon A. This is the reason for the immense importance of a thorough understanding of individual and geographic variation.

These comments apply equally to fossil species recognition studies that seek to interpret differences between fossils as reflecting either a species difference or intraspecific variation. A well-calibrated yardstick of intraspecific variation, established by applying the principles of population thinking in the form of the hierarchy of morphological variation, is a prerequisite for understanding variation in fossil samples.

## Acknowledgments

We thank Colin Groves (Australian National Museum) for archiving his gorilla data and for his help with clarifying some of the locality information. We thank Richard Thorington for providing access to gorilla specimens in the U.S. National Museum of Natural History. We thank two anonymous reviewers, who provided useful suggestions to improve our paper. Andrea Taylor and Michele Goldsmith are to be highly commended for their efforts and patience in organizing and editing this volume.

# References

Albrecht, G.H. (1977). Methodological approaches to morphological variation in primate populations: The Celebesian macaques. *Yearbook of Physical Anthropology*, **20**, 290–308.

Albrecht, G.H. (1978). The craniofacial morphology of the Sulawesi macaques: Multivariate approaches to biological problems. *Contributions to Primatology*, **13**, 1–151.

Albrecht, G.H. (1980). Multivariate analysis and the study of form, with special reference to canonical variate analysis. *American Zoologist*, **20**, 679–693.

Albrecht, G.H. (1992). Assessing the affinities of fossils using canonical variates and generalized distances. *Human Evolution*, **7**, 49–69.

Albrecht, G.H. and Miller, J.M.A. (1993). Geographic variation in primates: A review with implications for interpreting fossils. In *Species, Species Concepts, and Primate Evolution*, eds. W.H. Kimbel and L.B. Martin, pp. 123–161. New York: Plenum Press.

Ashton, E.H. (1957). Age changes in dimensional differences between the skulls of male and female apes. *Proceedings of the Zoological Society of London*, **128**, 259–265.

Casimir, M.J. (1975). Some data on the systematic position of the Eastern gorilla population of the Mt. Kahuzi region (Zäire). *Zeitschrift für Morphologie und Anthropologie*, **66**, 188–201.

Chambers, J.M., Cleveland, W.S., Kleiner, B., and Tukey, P.A. (1983). *Graphical Methods for Data Analysis*. Pacific Grove, CA: Wadsworth & Brooks/Cole.

Clutton-Brock, T.H. (1985). Size, sexual dimorphism, and polygyny in primates. In *Size and Scaling in Primate Biology*, ed. W.L. Jungers, pp. 51–60. New York: Plenum Press.

Coolidge, H.J. (1929). Revision of the genus *Gorilla*. *Memoirs of the Museum of Comparative Zoology, Harvard*, **50**, 291–381.

Dixson, A.F. (1981). *The Natural History of the Gorilla*. New York: Columbia University Press.

Doran, D.M. and McNeilage, A. (1998). Gorilla ecology and behavior. *Evolutionary Anthropology*, **6**, 120–131.

Doran, D.M. and McNeilage, A. (1999). Diet of western lowland gorillas in south-west Central African Republic: Implications for subspecific variation in gorilla grouping and ranging patterns. *American Journal of Physical Anthropology, Supplement* **28**, 121.

Fuller, K. and Caspari, R. (2001). Population thinking: Dead or alive. *American Journal of Physical Anthropology, Supplement* **24**, 25.

Gelvin, B.R., Albrecht, G.H., and Miller, J.M.A. (1997). The hierarchy of craniometric variation among gorillas. *American Journal of Physical Anthropology, Supplement* **24**, 117.

Goldsmith, M.L., Nkurunungi, J.B., and Stanford, C.B. (1999). Gorilla behavioral ecology: Effects of altitudinal changes on highland/lowland populations. *American Journal of Physical Anthropology, Supplement* **28**, 137.

Gordon, A.D. (2001). Sexual size dimorphism in primates: Considerations of relative variation between sexes. *American Journal of Physical Anthropology, Supplement* **32**, 72.

Grine, F.E., Jungers, W.L., and Schultz, J. (1996). Phenetic affinities among early *Homo* crania from East and South Africa. *Journal of Human Evolution*, **30**, 189–225.

Groves, C.P. (1967). Ecology and taxonomy of the gorilla. *Nature*, **213**, 890–893.

Groves, C.P. (1970). Population systematics of the gorilla. *Journal of Zoology, London*, **161**, 287–300.

Groves, C.P. (1986). Systematics of the great apes. In *Comparative Primate Biology*, vol. 1, *Systematics, Evolution, and Anatomy*, eds. D.R. Swindler and J. Erwin, pp. 187–217. New York: A.R. Liss.

Groves, C.P. and Stott, K.W., Jr. (1979). Systematic relationships of gorilla from Kahuzi, Tshiaberimu, and Kayonza. *Folia Primatologica*, **32**, 161–179.

Haddow, A.J. and Ross, R.W. (1951). A critical review of Coolidge's measurements of gorilla skulls. *Proceedings of the Zoological Society of London*, **121**, 43–54.

Humphrey, L.T., Dean, M.C., and Stringer, C.B. (1999). Morphological variation in great ape and modern human mandibles. *Journal of Anatomy*, **195**, 491–513.

Hursh, T.M. (1975). A multivariate study of chimpanzee and gorilla crania. PhD thesis, Harvard University, Cambridge, MA.

Hursh, T. (1976). Multivariate analysis of allometry in crania. *Yearbook of Physical Anthropology*, **18**, 111–120.

Jenkins, P.D. (1990). *Catalogue of Primates in the British Museum (Natural History) and Elsewhere in the British Isles*, Part V, *The Apes, Superfamily Hominoidea*. London: Natural History Museum Publications.

Jenkins, P.D. and Albrecht, G.H. (1990). Sexual dimorphism and sex ratios in Madagascan primates. *American Journal of Primatology*, **24**, 1–14.

Keith, A. (1927). Cranial characteristics of gorillas and chimpanzees. *Nature*, **120**, 914–915.

Kelley, J. (1993). Taxonomic implications of sexual dimorphism in *Lufengpithecus*. In *Species, Species Concepts, and Primate Evolution*, eds. W.H. Kimbel and L.B. Martin, pp. 429–458. New York: Plenum Press.

Kramer, A., Donnelly, S.M., Kidder, J.H., Ousley, S.D., and Olah, S.M. (1995). Craniometric variation in large-bodied hominoids: Testing the single-species hypothesis for *Homo habilis*. *Journal of Human Evolution*, **29**, 443–462.

Leigh, S.R., Relethford, J.H., and Groves, C. (1998). Morphological differentiation of gorilla subspecies. *American Journal of Physical Anthropology, Supplement* **26**, 149.

Leutenegger, W. and Kelly, J. (1977). Relationships of sexual dimorphism in canine size and body size to social, behavioral and ecological correlates in anthropoid primates. *Primates*, **18**, 117–136.

Leutenegger, W. and Cheverud, J. (1982). Correlates of sexual dimorphism in primates: Ecological and size variables. *International Journal of Primatology*, **3**, 387–402.

Leutenegger, W. and Cheverud, J. (1985). Sexual dimorphism in primates: The effects of size. In *Size and Scaling in Primate Biology*, ed. W.L. Jungers, pp. 32–50. New York: Plenum Press.

Lieberman, D.E., Pilbeam, D.R., and Wood, B.A. (1988). A probabilistic approach to the problem of sexual dimorphism in *Homo habilis*: A comparison of KNM-ER 1470 and KNM-ER 1813. *Journal of Human Evolution*, **17**, 503–511.

Martin, L.B. and Andrews, P. (1993). Species recognition in Middle Miocene hominoids. In *Species, Species Concepts, and Primate Evolution*, eds. W.H. Kimbel and L.B. Martin, pp. 393–428. New York: Plenum Press.

Mayr, E. (1942). *Systematics and the Origin of Species from the Viewpoint of a Zoologist*. New York: Columbia University Press.

Mayr, E. (1963). *Animal Species and Evolution*. Cambridge, MA: Belknap Press of Harvard University Press.

Mayr, E. (1969). *Principles of Systematic Zoology*. New York: McGraw-Hill.

Mayr, E. (1976). Typological versus population thinking. In *Evolution and the Diversity of Life: Selected Essays*, pp. 26–29, Cambridge, MA: Belknap Press of Harvard University Press.

Mayr, E. (1999). *Systematics and the Origin of Species from the Viewpoint of a Zoologist*. (Reprint of Mayr (1942) with a new introduction by the author.) Cambridge, MA: Harvard University Press.

Mayr, E. and Ashlock, P.D. (1991). *Principles of Systematic Zoology*, 2nd ed. New York: McGraw-Hill.

Mayr, E., Linsley, E.G., and Usinger, R. (1953). *Methods and Principles of Systematic Zoology*. New York: McGraw-Hill.

Miller, J.M.A., Albrecht, G.H., and Gelvin, B.R. (1997). An hierarchical analysis of craniofacial variation in *Homo habilis* using a *Gorilla* analog. *American Journal of Physical Anthropology, Supplement* **24**, 170.

Miller, J.M.A., Albrecht, G.H., and Gelvin, B.R. (1998). A hierarchical analysis of craniofacial variation in *Homo habilis* compared to a modern human analog. *American Journal of Physical Anthropology, Supplement* **26**, 163.

Oates, J.F., McFarland, K.L., Stumpf, R.M., Fleagle, J.G., and Disotell, T.R. (1999). New findings on the distinctive gorillas of the Nigeria–Cameroon border region. *American Journal of Physical Anthropology, Supplement* **28**, 213–214.

O'Higgins, P.O. and Dryden, I.L. (1993). Sexual dimorphism in hominoids: Further studies of craniofacial shape differences in *Pan, Gorilla*, and *Pongo*. *Journal of Human Evolution*, **24**, 183–205.

Oxnard, C.E. (1987). *Fossils, Teeth, and Sex: New Perspectives on Human Evolution*. Seattle, WA: Seattle University Press.

Oxnard, C.E., Lieberman, S.S., and Gelvin, B.R. (1985). Sexual dimorphisms in dental dimensions of higher primates. *American Journal of Physical Anthropology*, **8**, 127–152.

Park, P.B., Leigh, S.R., and Konigsberg, L.W. (1999). Craniometric discrete trait variation in *Gorilla gorilla*. *American Journal of Physical Anthropology, Supplement* **28**, 218.

Pickford, M. (1986). On the origins of body size dimorphism in primates. In *Sexual Dimorphism in Living and Fossil Primates*, eds. M. Pickford and B. Chiarelli, pp. 77–91. Florence, Italy: Il Sedicesimo.

Pickford, M. and Chiarelli, A.B. (1986). Sexual dimorphism in primates: Where are we and where do we go from here. In *Sexual Dimorphism in Living and Fossil Primates*, eds. M. Pickford and B. Chiarelli, pp. 1–5. Florence, Italy: Il Sedicesimo.

Plavcan, J.M. (2000). Variance dimorphism in primates. *American Journal of Physical Anthropology, Supplement* **30**, 251.

Plavcan, J.M. and Kay, R.F. (1988). Sexual dimorphism and dental variability in platyrrhine primates. *International Journal of Primatology*, **9**, 169–178.

Preuschoft, H. (1989). Quantitative approaches to primate morphology. *Folia Primatologica*, **53**, 82–100.

Randall, F.E. (1943). The skeletal and dental development and variability of the gorilla. *Human Biology*, **15**, 236–337.

Randall, F.E. (1944). The skeletal and dental development and variability of the gorilla (concluded). *Human Biology*, **16**, 23–76.

Remis, M.J. 1999. Nutritional aspects of diet of gorillas at Bai Hokou, Central African Republic with interpopulation and interspecific comparisons. *American Journal of Physical Anthropology, Supplement* **28**, 231.

Sarmiento, E.E. (1994). Terrestrial traits in the hands and feet of gorillas. *American Museum Novitates*, **3091**, 1–56.

Sarmiento, E.E. and Oates, J.F. (2000). The Cross River gorillas: A distinct subspecies, *Gorilla gorilla diehli* Matschie 1904. *American Museum Novitates*, **3304**, 1–55.

Sarmiento, E.E., Butynski, T.M., and Kalina, J. (1996). Gorillas of the Bwindi-Impenetrable Forest and the Virunga Volcanoes: Taxonomic implications of morphological and ecological differences. *American Journal of Primatology*, **40**, 1–21.

Schmid, P. and Stratil, Z. (1986). Growth changes, variations and sexual dimorphism of the gorilla skull. In *Selected Proceedings of the 10th Congress of the International Primatological Society*, vol. 1, *Primate Evolution*, eds. J. G. Else and P. C. Lee, pp. 239–247. Cambridge, U.K.: Cambridge University Press.

Schultz, A.H. (1927). Studies on the growth of *Gorilla* and of other higher primates, with special reference to a fetus of *Gorilla*, preserved in the Carnegie Museum. *Memoirs of the Carnegie Museum, Pittsburgh*, **11**, 1–86.

Schultz, A.H. (1934). Some distinguishing characters of the mountain gorilla. *Journal of Mammalogy*, **15**, 51–61.

Shea, B.T. (1983). Allometry and heterochrony in the African apes. *American Journal of Physical Anthropology*, **62**, 275–289.

Shea, B.T. (1986). Ontogenetic approaches to sexual dimorphism in the anthropoids. *Human Evolution*, **1**, 97–110.

Silverman, N., Richmond, B., and Wood, B. (2001). Testing the taxonomic integrity of *Paranthropus boisei sensu stricto*. *American Journal of Physical Anthropology*, **115**, 167–178.

Simpson, G.G. (1953). *The Major Features of Evolution*. New York: Columbia University Press.

Simpson, G.G. (1961). *Principles of Animal Taxonomy*. New York: Columbia University Press.

Stringer, C.B. (1986). The credibility of *Homo habilis*. In *Major Topics in Primate and Human Evolution*, eds. B. Wood, L. Martin, and P. Andrews, pp. 266–294, Cambridge, U.K.: Cambridge University Press.

Stumpf, R.M., Fleagle, J.G., and Groves, C.P. (1997). Sexual dimorphism and geographic variation among subspecies of *Gorilla*. *American Journal of Physical Anthropology, Supplement* **24**, 223–224.

Stumpf, R.M., Fleagle, J.G., Jungers, W.L., Oates, J.F., and Groves, C.P. (1998). Morphological distinctiveness of Nigerian gorilla crania. *American Journal of Physical Anthropology, Supplement* **26**, 213.

Suter, J. and Oates, J.F. (2000). Sanctuary in Nigeria for possible fourth subspecies of gorilla. *Oryx*, **34**, 71.

Taylor, A.B. (1995). Effects of ontogeny and sexual dimorphism on scapula morphology in the mountain gorilla (*Gorilla gorilla beringei*). *American Journal of Physical Anthropology*, **98**, 431–445.

Taylor, A.B. (1997). Relative growth, ontogeny, and sexual dimorphism in *Gorilla* (*Gorilla gorilla gorilla* and *G. g. beringei*): Evolutionary and ecological considerations. *American Journal of Primatology*, **43**, 1–33.

Taylor, A.B. (2002). Masticatory form and function in the African apes. *American Journal of Physical Anthropology*, **117**, 133–156.

Teaford, M.F., Walker, A., and Mugaisi, G.S. (1993). Species discrimination in *Proconsul* from Rusinga and Mfangano Islands, Kenya. In *Species, Species Concepts, and Primate Evolution*, eds. W. H. Kimbel and L. B. Martin, pp. 373–392. New York: Plenum Press.

Uchida, A. (1996). What we don't know about great ape variation. *Trends in Ecology and Evolution*, **11**, 163–168.

Van Valen, L. (1974). Multivariate structural statistics in natural history. *Journal of Theoretical Biology*, **45**, 235–247.

Watts, D.S. (1991). Comparative socioecology of gorillas. In: *Great Ape Societies*, eds. W.C. McGrew, L.F. Marchant, and T. Nishida, pp. 16–28. Cambridge, U.K.: Cambridge University Press.

Wilson, S.R. (1984). Towards an understanding of data in physical anthropology. In *Multivariate Statistical Methods in Physical Anthropology*, eds. G.N. van Vark and W.W. Howells, pp. 261–282. Boston, MA: D. Reidel.

Wood, B.A. (1975). An analysis of sexual dimorphism in primates. PhD thesis, University of London.

Wood, B.A. (1976). The nature and basis of sexual dimorphism in primates. *Journal of Zoology, London*, **180**, 15–34.

Wood, B.A. (1991). *Koobi Fora Research Project*, vol. 4. Oxford, U.K.: Clarendon Press.

Wood, B.A. (1993). Early *Homo*: How many species? In *Species, Species Concepts, and Primate Evolution*, eds. W.H. Kimbell and L.B. Martin, pp. 485–522. New York: Plenum Press.

Wood, B.A., Li, Y., and Willoughby, C. (1991). Intraspecific variation and sexual dimorphism in cranial and dental variables among higher primates and their bearing on the hominid fossil record. *Journal of Anatomy*, **174**, 185–205.

# 5 Morphological differentiation of Gorilla subspecies

STEVEN R. LEIGH, JOHN H. RELETHFORD, PAUL B. PARK,
AND LYLE W. KONIGSBERG

## Introduction

Classic analyses of gorilla systematics by Coolidge (1929) and Groves (1970) form a vital foundation for virtually all contemporary studies of gorillas (see also important morphological studies by Albrecht *et al.*, this volume; Groves, this volume; Haddow and Ross, 1951; Vogel, 1961; Sarmiento *et al.*, 1996; Stumpf *et al.*, 1998, this volume). Investigations of *Gorilla* subspecies diversity also inform many other fields such as paleoanthropology (cf. Albrecht *et al.*, this volume; Lieberman *et al.*, 1988). Despite a good understanding of variation within this species, theoretical advances and new analytical techniques point to a need to reassess our current understanding of gorilla variation. Measures of subspecific variation have important implications for studies of microevolution, ecomorphology, population dynamics, and evolutionary history. Concomitantly, increasing pressures on wild populations underscore the need for a clearer understanding of the structure of subspecific variability in a conservation context (Suter and Oates, 2000).

Therefore, the goal of this analysis is to investigate morphological diversity in gorillas through new measures of variation below the species level. More specifically, this analysis evaluates overall levels of variation within the *Gorilla* species through application of Wright's $F_{ST}$ to craniometric data (Wright, 1951; 1969; Relethford, 1994). This approach, typically used to measure genetic microdifferentiation, calibrates the degree to which subdivision within populations or species departs from a quantitative expectation of no substructure. In contrast, contemporary statistical approaches, like those normally applied to analyses of gorilla craniometric variation, simply describe whether or not distinctions among groups or operational taxonomic units (OTUs) are statistically perceptible. In essence, $F_{ST}$ employs an explicit biological model to assess the degree of subdivision in a group, population, or species.

Analyses based on Wright's $F_{ST}$ enable several extensions of our current understanding of *Gorilla* subspecies variation. Most broadly, $F_{ST}$ can help answer two important questions about the general nature of variation in *Gorilla*. First,

104

despite a number of excellent studies documenting distinctions among *Gorilla* subspecies (see below) we lack a quantitative understanding of the degree to which *Gorilla* subspecies differ. Simply put, it is difficult to interpret *how* different *Gorilla* subspecies are from one another based on traditional statistical techniques. Second, $F_{ST}$ allows interspecific comparisons of subspecies diversity because it is a standardized measure of variation (Templeton, 1999). Templeton shows that $F_{ST}$-based measures of subspecies variation can be compared among species, providing a valuable foundation for investigating the evolutionary history of *Gorilla* subspecies.

Investigations of these general problems provide a basis for addressing a number of more specific questions regarding subspecific variation in *Gorilla*. First, we use $F_{ST}$ to investigate patterns of concordance among measures of subspecific taxonomic divergence. Concordance refers to the degree to which measures of subspecific diversity (e.g., molecular, morphological, ecological, and behavioral) agree in their estimates of divergence (Diamond, 1994). Discrepancies among estimates of divergence within hominoid species are emerging as our understanding of these taxa increases. Second, we use $F_{ST}$ estimates to generate hypotheses about population histories and dynamics (Relethford and Harpending, 1994). We base hypotheses about changes in historical population sizes on comparisons between observed and theoretically expected patterns of variation. Third, we suggest that $F_{ST}$ estimates have roles to play in conservation efforts. Measures of relative craniometric variation can help define patterns of significant biological diversity.

We address these problems through two different kinds of craniometric analyses. First, Colin Groves's craniometric data set (Groves, 1966, 1970) is utilized to measure $F_{ST}$. This data set forms the basis of contemporary classifications, and includes the vast majority of crania available for study. Second, we analyze discrete trait variation among *Gorilla* subspecies using a relatively new analytical protocol. Although the data for these particular analyses are extremely limited, discrete traits have significant potential for improving our understanding of subspecific variation (Braga, 1995a,b,c). The analytical techniques that we apply to discrete traits have been used in studies of human population variation (Konigsberg, 1990a,b) and have considerable potential for studies of nonhuman primate variation.

### *Partitioning of* Gorilla *subspecific variation*

#### *Morphology*

The formal study of morphological variation within *Gorilla* has an interesting and dynamic history. Groves (1966, this volume) details this history, beginning with Savage and Wyman's description of the species in 1847. Groves notes that taxonomic names proliferated through the early 1900s, mainly because of a pervasive typological perspective (cf. Matschie, 1904, 1905; Rothschild, 1905, 1906). Realizing the problems of a typological perspective, Harold Coolidge, in his "Revision of the genus *Gorilla*", established the foundation for contemporary taxonomies (1929). Groves's subsequent revision thus proposes a single gorilla species divided into three subspecies (Groves, 1966, 1967, 1970), including the western lowland gorilla (*G. g. gorilla*), the eastern lowland gorilla (*G. g. graueri*), and the mountain gorilla (*G. g. beringei*). Groves's interpretations are based on detailed multivariate analysis of hundreds of gorilla crania, evaluation of cranial nonmetric trait variation, study of postcranial skeletal material, and various soft-tissue features. Reanalyses of Groves's database support his hypothesis, confirming division of the species into three subspecies (Albrecht and Miller, 1993; Gelvin *et al.*, 1997; Stumpf *et al.*, 1998; Albrecht *et al.*, this volume).

Analyses of discrete trait variation document further differences among gorilla subspecies. Specifically, Braga's application of divergence measures and parsimony analysis to frequency data for 25 discrete cranial traits illustrates distinctions among subspecies (1995*a,c*). Divergence measures suggest fairly consistent distinctions among subspecies, but nominal distances vary slightly. For example, an analysis based on bilateral traits shows proximity between eastern lowland gorillas and mountain gorillas, while analyses of unilateral or midline traits indicates slightly greater similarity between western lowland and mountain gorillas. Principal components analyses follow the pattern seen for midline traits, but parsimony analyses do not separate eastern lowland gorillas and mountain gorillas. Interestingly, Braga's analyses of chimpanzees clearly differentiate *Pan troglodytes verus* from other subspecies (1995*a,b*; see also Morin *et al.*, 1994; Leigh and Konigsberg, 1996).

Distinctions recognized by craniometric analyses are paralleled by studies that investigate the dentition and postcrania. For example, Uchida notes that *G. g. gorilla* is highly variable in terms of dental morphology. Furthermore, *G. g. graueri* exhibits distinctive dental features, and cannot be considered as "intermediate" between the other subspecies (Uchida, 1996, 1998). Contrasts in scapular morphology have long been known in gorillas. Specifically, the vertebral border of the mountain gorilla scapula is more sinuous than that of western gorillas (Schultz, 1934). Overall scapular shape differences between these subspecies are also present (Taylor, 1997*b*). Based on a small sample, Sarmiento and colleagues recognize a variety of anatomical distinctions between Bwindi-Impenetrable Forest gorillas and mountain gorillas (*G. g. beringei*). The differences observed in the limited sample of Bwindi gorillas include smaller size of the Bwindi gorillas and differences in postcranial proportions (Sarmiento *et al.*, 1996). In all, dental and postcranial data are compatible with the hypothesis that variation in *Gorilla* can be partitioned into three subspecies. Craniometrically and odontometrically, differences among subspecies are often statistically significant (Groves, 1970; Uchida, 1998; Taylor, 2002, this volume). The statistical significance of other distinctions is difficult to calibrate, but these differences appear consistently and are often interpretable with respect to functional models (e.g., Sarmiento *et al.*, 1996; Taylor, 1997a,b, 2002).

We can note that subspecific morphological variability in gorillas may be more thoroughly documented than morphological variability in other large-bodied hominoid species. Studies of other hominoids have tended to use smaller samples than have analyses of gorillas (e.g., Shea and Coolidge (1988) for *Pan*, $n = 360$; and Groves *et al.* (1992) for *Pongo*, $n =$ approximately 200). This may account for what we perceive as a broad appreciation of morphological variation in gorillas.

*Molecular variation*

Molecular data with which to evaluate gorilla subspecies variability are still quite limited, at least when compared to the amount of data available for common chimpanzees (*Pan troglodytes*: cf. Morin *et al.*, 1994). Ruvolo *et al.*'s (1994) analysis of mitochondrial DNA (mtDNA) variation illustrates comparatively deep taxonomic divisions within *Gorilla*. Although their study is based on a very small sample size, Ruvolo *et al.* detect distinctions between western and eastern gorillas that are comparable to the level of mtDNA variation separating the two generally recognized chimpanzee species, *P. troglodytes* and *P. paniscus*. Garner and Ryder (1996) confirm this finding by suggesting that the most divergent mtDNA control region sequences in gorillas resemble distances that distinguish between chimpanzee species. Garner and Ryder also find that *G. g. graueri* and *G. g. beringei* sequences are quite similar, but show some distinctions. These two subspecies may present low levels of mitochondrial variability (Garner and Ryder, 1996). More recently, Seaman *et al.* (1998) find that *G. g. graueri* and *G. g. beringei* have distinct mtDNA D-loop haplotypes. However, now-allopatric populations of *G. g. graueri* share certain haplotypes, implying fragmentation of a formerly sympatric *G. g. graueri* population (Seaman *et al.*, 1998). Analyses that include *G. g. gorilla* indicate a substantial difference between western gorillas and a combined sample of *G. g. graueri* and *G. g. beringei* (Saltonstall *et al.*, 1998; Seaman *et al.*, 2000), along with important variation within *G. g. gorilla* (Clifford *et al.*, this volume). Seaman *et al.*, (2000) demonstrate low levels of mtDNA diversity in *G. g. graueri*. They also record virtually no mtDNA differences between Bwindi-Impenetrable Forest and Virunga samples, contrasting with Sarmiento *et al.*'s (1996) preliminary taxonomic inferences. Nuclear haplotypes are shared frequently among subspecies, and suggest less divergence than mitochondrial regions (Jensen-Seaman *et al.*, this volume).

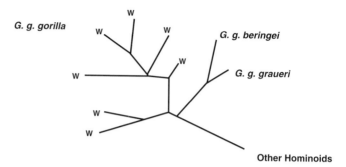

Fig. 5.1. Unrooted phylogram (after topiary pruning) for *Gorilla* based on mitochondrial DNA (adapted from Gagneaux *et al.*, 1999). All branches at the left of the figure identified with W belong to the *G. g. gorilla* subspecies.

Gagneaux *et al.* (1999) illustrate very high levels of mtDNA diversity in *Gorilla* (Fig. 5.1). Their phylograms reveal deep divergences among OTUs, with remarkable diversity in the western lowland subspecies. Branch lengths for the entire gorilla species are comparable to the amount of diversity in *Pan*. Estimates of pairwise sequence divergences suggest that mtDNA diversity in *Gorilla* exceeds diversity in all other hominid species, with maximum divergences in the range 20–25%. Furthermore, their analysis links *G. g. graueri* and *G. g. beringei* on close branches. Gagneaux *et al.* (1999) relate the high degree of molecular variation in *Gorilla* to restricted gene flow among populations and subspecies. They suggest that isolation is largely a product of ecological specialization (relative to chimpanzees) and habitat fragmentation.

*Ecology and social behavior*

Recent field studies reveal a variety of ecological and social distinctions among *Gorilla* subspecies. Several excellent contributions document geographic and taxonomic differences in behavior and ecology (see Doran and McNeilage, 1998). In general, western lowland gorilla diets include more fruits and are more diversified than mountain gorilla diets (Tutin and Fernandez, 1985). The latter subspecies selects foods that are widely distributed and high in protein (Doran and McNeilage, 1998). The diet of eastern lowland gorillas seems to covary closely with altitude (Watts, 1996). Dietary differences between *G. g. beringei* and other subspecies seem to exceed differences between *G. g. graueri* and *G. g. gorilla*. Some of these differences correspond to masticatory (Taylor, 2002, this volume) and dental (Uchida, 1998) contrasts that ultimately stem from altitude-driven habitat differences between mountain gorillas and other subspecies (Groves, 1971).

Social organization seems to be similar among subspecies, with most groups including one or two adult males along with several females and their offspring

(Doran and McNeilage, 1998). Group sizes vary among subspecies, as do home range sizes, and these differences are difficult to explain (Doran and McNeilage, 1998). It can also be expected that variation in sexual selection contributes to differences among populations and subspecies (Fisher, 1930; Lande, 1981). Although differences in sexual selection among gorilla subspecies have not been discussed, sexual selection via female mate choice can accelerate the evolution of differences among OTUs (Lande, 1981). Sexual selection in gorillas is obviously quite intense (Plavcan and van Schaik, 1992; Leigh and Shea, 1995; Watts, 1996), so differences in the degree and type of sexual selection could contribute to distinctions among taxa. Differences between eastern gorillas (mountain and eastern lowlands) and western lowland gorillas in the border between silverback pelage and surrounding hair (Groves, 1966) may reflect the evolution of differences due to mate choice. Certainly, sexual selection may be a factor in the differentiation of gorilla subspecies in comparison to chimpanzees, in which levels of sexual selection are much lower.

### Concordance

Variable levels of concordance among characters can complicate subspecific taxonomies (Diamond, 1994). In other words, taxonomic schemes may vary depending upon which characters are analyzed. Theoretically, analyses between species should show greater concordance than analyses within species (although homoplasy among species may reduce concordance (Tattersall, 1993)). At the subspecific level, concordance is important because it provides insight into the dynamics of divergence. If all characters or character sets (e.g., morphological, molecular, and behavioral) are concordant, then subspecies divergence can be unambiguously recognized. However, low concordance may reflect adaptations as a result of differential rates of evolution among characters.

Morphological, mitochondrial, social, and ecological measures of subspecific variation provide concordant measures of variation in *Gorilla*. In other words, these lines of evidence, particularly morphological and mitochondrial data, denote clear distinctions among subspecies. It should be noted that although measures of craniometric and mitochondrial distinctions among *Gorilla* subspecies are generally concordant, analyses of mandibular dimensions suggest some overlap between eastern and western lowland gorillas (Groves, 1970: Fig. 2). Moreover, nuclear diversity may be patterned differently than mitochondrial variation (Jensen-Seaman *et al.*, this volume).

### Gorillas and subspecies

Gorillas provide excellent opportunities to address questions regarding the general significance of subspecies variation. At the most basic level, variation in the degree of morphological differentiation among gorilla subspecies can help us understand evolutionary processes that generate and maintain clusters of variation within species. These clusters of variation are theoretically important because they provide direct insight into the general phenomenon of evolutionary divergence (Futuyma, 1987, 1992; Gould and Eldredge, 1993; Jolly, 1993; Kimbel and Martin, 1993). More specifically, the evolution of clusters of variation within a species serves as the basis for speciation when reproductive isolation sequesters reservoirs of existing variation, triggering speciation while retaining "characters that evolved in a localized context" (Futuyma, 1987:467). Gould and Eldredge stress that this subspecific process can account for associations between morphological change and speciation (1993:226). When such an association occurs, it may be influenced by the structure of variation within an ancestral species.

Although subspecific variation potentially plays a role in structuring diversity during speciation events, differences among subspecies may also erode through recombination. Thus, several authors have characterized subspecific variation as "potentially ephemeral" variation "that could ... comfortably be ignored" (Tattersall, 1993:174). Similarly, Kimbel has viewed the long-term significance of subspecific variation as "largely ephemeral" (1991:362). These contributions have been critiqued previously (Shea *et al.*, 1993), and it is worth reiterating that they offer very little empirical evidence to support their cases. We emphasize that studying intraspecific variation in gorillas can help us understand the processes that produce evolutionary divergence, and thus should not be dismissed *a priori* (see also Jolly, 1993; Shea *et al.*, 1993; Smith *et al.*, 1997).

## Materials and methods

### Data

#### Craniometrics

The present analysis includes 571 individuals measured by Colin Groves (1966, 1967, 1970) (Table 5.1). A total of 18 measurements are analyzed (Table 5.2). These measures mainly reflect facial measures, and are derived from adult skulls

Table 5.1. *Sample sizes for craniometric analyses*

|  | *Gorilla gorilla beringei* | *Gorilla gorilla gorilla* | *Gorilla gorilla graueri* | Total |
|---|---|---|---|---|
| **Female** | 9 | 180 | 34 | 223 |
| **Male** | 17 | 286 | 45 | 348 |
| **Total** | 26 | 466 | 79 | 571 |

Table 5.2. *Craniometric measurements analyzed*

| 1. | GRLEN | Greatest skull length, prosthion to opisthocranion |
|----|---------|-----------------------------------------------------------------|
| 2. | ZYGBR | Bizygomatic breadth |
| 3. | CRLEN | Cranial length, glabella to opisthocranion |
| 4. | BIORB | Biorbital breadth at frontozygomatic suture |
| 5. | UM1UM1 | Outside palatal breadth, across first molars |
| 6. | NASBR | Combined nasal breadth |
| 7. | PYRPRO | Lower facial height, from top of pyriform aperture to prosthion |
| 8. | BICAN | Bicanine breadth across alveoli |
| 9. | SOTPR | Facial height, from top of supraorbital torus to prosthion |
| 10. | ORBHT | Inside orbital height |
| 11. | ORBBR | Inside orbital breadth |
| 12. | POBR | Breadth of postorbital constriction |
| 13. | PYAPBR | Breadth of pyriform aperture |
| 14. | IOB | Interorbital breadth |
| 15. | SOTHT | Height of supraorbital torus at thinnest part |
| 16. | BIPORL | Biporionic breadth |
| 17. | PALLEN | Length of palate, from prosthion to hind border (estimated if incomplete) |
| 18. | BRNPTR | Breadth across medial pterygoids |

(based on presence of wear on third molars and closure of the basilar suture (Groves, 1970)). It should be noted that gorilla males have very long growth periods (Leigh and Shea, 1995), raising the possibility that some variation among individuals in this sample is a consequence of differences in age.

*Discrete traits*

Skeletal material for discrete trait analyses consists of 142 crania of mixed age and sex. Operational taxonomic units (OTUs) were composed of individuals from each of the three currently recognized subspecies (Table 5.3). Each cranium was scored for the presence or absence of eight nonmetric traits by a single observer (SRL) (Table 5.4). These traits were initially selected based on analyses of comparable traits in humans (Konigsberg *et al.*, 1993). This list of traits was modified based on comparative study of nonhuman cranial material. For the present study, only left-side traits are analyzed. In order to maximize sample sizes, we pooled sexes and age groups, then tested for sex, age, and

Table 5.3. *Sample sizes for discrete trait analyses*

|  | *Gorilla gorilla beringei* | *Gorilla gorilla gorilla* | *Gorilla gorilla graueri* | Total |
|-----------|------|------|------|-------|
| **Female** | 5 | 29 | 36 | 70 |
| **Male** | 2 | 21 | 35 | 58 |
| **Unknown** | 3 | 0 | 11 | 14 |
| **Total** | 10 | 50 | 82 | 142 |

Table 5.4. *Discrete traits analyzed*

| |
|---|
| 1. Infraorbital suture evident at zygomaxillary articulation |
| 2. Frontomaxillary contact in medial orbital wall |
| 3. Divided hypoglossal canal |
| 4. Foramen spinosum and ovale not separated |
| 5. Postglenoid tubercle extends beyond inferior rim of auditory canal |
| 6. Clival foramen at vomerosphenoid contact |
| 7. Obelionic foramen present |
| 8. Sagittal ossicle present |

sex-by-age interactions using an analysis of covariance design. No significant interactions for these traits were observed, suggesting that these variables are not significant covariates.

### $F_{ST}$ and morphological variation

The value of Wright's $F_{ST}$ lies in its ability to provide a measure of relative variation, where variation *among* subspecies is expressed relative to the *total* variation in a species expected under random breeding. In Wright's notation, $F_{ST}$ refers to the degree by which subdivisions ($S$) of a total population ($T$) are isolated, as measured by $F$ (the "fixation index", a measure of genetic homogeneity) (Wright, 1951, 1969). More specifically, $F_{ST}$ is the proportion of total variation due to variation among groups (in this case, gorilla subspecies). The remainder ($1 - F_{ST}$) is referred to as $P$ or the panmictic index. $F_{ST}$ provides a way to compare variation across taxonomic units, including comparisons among species (cf. Templeton, 1999). Divergence can also be compared among variables (e.g., genetic $F_{ST}$ vs. morphological $F_{ST}$ (see Relethford, 1994)). Neither kind of comparison is possible with traditional $F$-ratio based statistical techniques such as discriminant functions, which produce results that are specific to the number of groups and sample sizes analyzed.

A primary advantage of $F_{ST}$ is that it is a "model-bound" approach, meaning that it measures the partitioning of variation relative to a theoretical expectation of random breeding (Wright, 1951; Relethford, 1994). In contrast, traditional statistical techniques are "model-free." One result of a model-free approach is that it may overemphasize differences among populations, particularly given the specificity of results of traditional statistical analyses. A second advantage is that $F_{ST}$ is sensitive to variation in population sizes. This property helps control for differences in morphological variation that might be expected as a result of subgroup sample size differences. However, incorporation of population estimates also introduces some uncertainties when population sizes are poorly known.

### $F_{ST}$ estimation

$F_{ST}$ is easily defined in terms of allele frequencies. These procedures utilize a $g$-by-$g$ matrix of standardized variances and covariances of populations (an **R** matrix) around average allele frequencies, where $g$ is the number of populations in an analysis. For any given allele, the elements of the **R** matrix provide a measure of the genetic similarity of populations $i$ and $j$:

$$r_{ij} = \frac{(p_i - \bar{p})(p_j - \bar{p})}{\bar{p}(1 - \bar{p})} \tag{5.1}$$

where $p_i$ and $p_j$ are the allele frequencies in populations $i$ and $j$ respectively, and $\bar{p}$ is the mean allele frequency derived from all populations in the analysis. The final **R** matrix is computed by averaging Equation 5.1 over all alleles (Harpending and Jenkins, 1973). The **R** matrix expresses variation among groups (the numerator) relative to the total variation expected under panmixia (the denominator). The mean allele frequencies must be weighted by population size in order to accurately reflect panmixia, which would occur if all individuals belonged to a single random mating population (since larger populations would have a greater genetic impact if all groups were pooled into a panmictic group). Mean allele frequencies are thus computed as:

$$\bar{p} = \sum_{i=1}^{g} w_i p_i \tag{5.2}$$

where $w_i$ is the relative population size of population

$$w_i = N_i \Big/ \sum N_i \tag{5.3}$$

$F_{ST}$ can be estimated as the average diagonal of the **R** matrix, weighted by population size:

$$F_{ST} = \sum_{i=1}^{g} w_i r_{ii} \tag{5.4}$$

Thus, $F_{ST}$ is a measure of the dispersion of populations around the mean allele frequencies, giving a measure of among-group variation relative to a panmictic population.

$F_{ST}$ can also be estimated from morphological data (Williams-Blangero and Blangero, 1989; Relethford and Blangero, 1990; Relethford, 1994). Given $t$ traits, a $g$-by-$g$ codivergence matrix, **C**, is computed as:

$$\mathbf{C} = \Delta \mathbf{G}_w^{-1} \Delta' \tag{5.5}$$

where $\Delta$ is a $g$-by-$t$ matrix containing the deviations of group means from the pooled mean over all groups, $\Delta'$ is its transpose, and $\mathbf{G}_w$ is the $t$-by-$t$ pooled within-group genotypic variance–covariance matrix. Note that both the pooled means over all groups and $\mathbf{G}_w$ are averaged over all groups weighted by relative population size ($w_i$). The $\mathbf{C}$ matrix is related to the $\mathbf{R}$ matrix as

$$\mathbf{R} = \frac{\mathbf{C}(1 - F_{ST})}{2t} \tag{5.6}$$

and $F_{ST}$ is therefore estimated as:

$$F_{ST} = \sum_{i=1}^{g} w_i r_{ii} = \frac{\sum_{i=1}^{g} w_i c_{ii}}{2t + \sum_{i=1}^{g} w_i c_{ii}} \tag{5.7}$$

The pooled within-group genotypic variance–covariance matrix ($\mathbf{G}_w$) can be estimated from the pooled within-group *phenotypic* covariance matrix ($\mathbf{P}_w$) by multiplying the latter by an estimate of average heritability:

$$\mathbf{G}_w = h^2 \mathbf{P}_w \tag{5.8}$$

Two extensions to $F_{ST}$ are used here: (1) correction for sampling bias, and (2) computation of the standard error (Relethford *et al.*, 1997).

### *Population size, morphometric size adjustment, and heritability*

Calculation of $F_{ST}$ involves estimates of population sizes. Obviously, reliable census data for gorillas are difficult to obtain. Despite the role of population size in $F_{ST}$ algorithms, we note that $F_{ST}$ values tend to be fairly robust to difficulties with population size estimates (Relethford and Harpending, 1994; Relethford and Jorde, 1999). The method is especially resilient when there are major size differences among subgroups because it uses proportional population sizes to weight estimates (see Equation 5.3).

*Gorilla* population sizes have probably diminished radically during the last century. Therefore, current population size estimates may not accurately reflect long-term demographic history, and the relative population sizes of the sub-species may have been different in the past. These questions can be addressed by applying Relethford and Blangero's (1990) method of comparing observed and expected variances within each group. The method was originally developed as a means of detecting differences in long-range gene flow passing into the populations under analysis. When applied to groups within an entire species, there is no external gene flow, and the method can be used to test deviations in observed variation due to misspecification of relative census population size

(Relethford and Harpending, 1994). The same method was used here; different relative population sizes were used repeatedly in a "brute-force" method to find the set of relative weights that provides the best fit of observed and expected variances. This method, as outlined in more detail by Relethford and Harpending (1994), provides an estimate of the long-term relative population sizes for groups within a species.

We use two different population size estimates. First, Doran and McNeilage (1998), following Oates (1996), provide estimates of 110 000 for *G. g. gorilla*, 12 000 for *G. g. graueri*, and 600 for *G. g. beringei*. We also calculate "historical" population sizes by dividing the total subspecies range by the average home range size for each subspecies, then multiply this number by the average number of individuals per group. Group and home range size estimates from Doran and McNeilage (1998) imply a historical population size for *G. g. gorilla* at about 435 000, with *G. g. graueri* at 98 000. We assume a historical population size of 2000 for *G. g. beringei*, which reflects a percentage reduction intermediate between those of the western and eastern lowland subspecies.

Several protocols were applied in the calculation of $F_{ST}$ values. First, we calculated $F_{ST}$s based on raw data. Second, we analyzed log-transformed data. Finally, we transformed data to Darroch and Mosimann (1985) shape variables. Multivariate plots of all variable types revealed differences in shape between sexes. Consequently, we standardized by sex. We also found that matrix operations could not be accomplished on the Darroch and Mosimann shape variables without sex adjustment. We report only values for the first set of analyses (sex adjusted raw data). Results were similar for all forms of data treatment.

Finally, Wright's $F_{ST}$ relies on estimates of heritabilities. Therefore, we first assumed complete trait heritability (e.g., all phenotypic variation in a trait is a consequence of genetic variation): assuming that $h^2 = 1$ returns what is referred to as a "minimum" $F_{ST}$ value (Relethford, 1994). Second, we assumed a heritability of 0.50, which is a realistic value for skeletal dimensions (Cheverud, 1981). Given the uncertainties with census size estimates and heritabilities, we expect a range of calculated $F_{ST}$ values, at least until further research can refine parameter estimates.

### Discrete trait analyses

Subspecies rarely present the kind of taxon-specific, invariant, or monomorphic traits that would serve as the basis for cladistic analysis. Instead, many characters used in studies of subspecies are polymorphic, and vary in their frequencies. Measuring these data as percentages severely limits options for statistical analyses because percentage data violate fundamental assumptions of standard statistical techniques. Despite limitations imposed by standard methods, there are several advantages to analyzing discrete traits, particularly when comparing molecular and morphological measures of divergence. Specifically, polymorphic discrete traits may meet a criterion of selective neutrality more adequately than metric traits (Hauser and De Stefano, 1989; Van Valen, 1990). This property means that estimates of divergence based on discrete traits should be relatively uninfluenced by selection pressures that normally affect cranial dimensions (e.g., lever and load arms). If discrete traits are selectively neutral or nearly so, then we should expect differences among subspecies or OTUs to have evolved via drift, with no selection to alter the degree of divergence among taxa. Thus, differences among OTUs would be linked to divergence time, rather than to either differences in selection intensity or in responses to selection. In this case, we anticipate that discrete traits may parallel molecular measures of divergence, most of which are, theoretically, based on selectively neutral mutations (Kimura, 1968). We can also note that discrete traits are observable on fragmentary materials such as fossils. While this is not a concern for the present study, the methodologies that we develop may have application to fossil problems (see Braga, 1995*a,b,c*).

Discrete trait data are analyzed using the Gibbs sampler, a Monte Carlo Markov chain estimator (Manly, 1997). Essentially, this technique approximates a multivariate distribution by sampling univariate distributions for each trait. These new variables can then be analyzed with both standard multivariate statistics and with $F_{ST}$ methods. Moreover, the Gibbs sampler uses Bayesian inference, which enables it to approximate multivariate distributions by continuously updating probability distributions upon adding each new case (Manly, 1997). We assume that discrete traits are polygenic and dichotomized by the presence of a trait threshold. The Gibbs sampler estimates mean thresholds for each trait in each OTU, then constructs a multivariate distribution from these values.

### Results

#### Craniometric analyses

##### $F_{ST}$ estimation
The minimum $F_{ST}$ estimate based on Groves's craniometric data suggests unexpectedly low levels of variation among gorilla subspecies (Table 5.5). The minimum $F_{ST}$ estimate is 0.107, while a more realistic calculation that assumes heritability at 0.50 returns a value of 0.199. This latter estimate indicates that about 20% of total gorilla phenotypic variation occurs among subspecies, while 80% of the total variation can be found within each subspecies.

Table 5.5. $F_{ST}$ *results for gorillas*

| Population sizes | $F_{ST}$ | SE | $h^2$ | Census weights: *Gorilla gorilla beringei*, *G. g. graueri*, *G. g. gorilla* |
|---|---|---|---|---|
| Equal | 0.107 | 0.007 | 1.00 | equal |
| Equal | 0.199 | 0.008 | 0.50 | equal |
| Estimate A[a] | 0.024 | 0.002 | 1.00 | 0.005, 0.098, 0.897 |
| Estimate A | 0.049 | 0.002 | 0.50 | 0.005, 0.098, 0.897 |
| Estimate B[b] | 0.040 | 0.003 | 1.00 | 0.004, 0.183, 0.813 |
| Estimate B | 0.079 | 0.004 | 0.50 | 0.004, 0.183, 0.813 |
| Best fit | 0.190 | 0.010 | 0.50 | 0.380, 0.010, 0.610 |

[a]Estimate A assumes population sizes of 110 000 for *G. g. gorilla*, 12 000 for *G. g. graueri*, and 600 for *G. g. beringei*.
[b]Estimate B assumes population sizes of 435 000, 98 000, and 2000 for *G. g. gorilla*, *G. g. graueri*, and *G. g. beringei*, respectively.

Adjustment for population size figures yields lower values, reflecting the sensitivity of $F_{ST}$ to census estimates. Thus, the western lowland gorilla population size, which is much larger than the other subspecies, tends to "overwhelm" variation in other subspecies. Consequently, $F_{ST}$ values drop when population sizes are included because the population weights for *G. g. graueri* and *G. g. beringei* are low relative to *G. g. gorilla*. As a preliminary way to minimize problems associated with heritabilities and population sizes, we calculated best-fit weights for each subspecies. Best-fit weights suggest that mountain gorilla weights should be far higher than their census estimates might indicate (Table 5.5). In this case, the weight for *G. g. beringei* indicates a population size slightly more than half the *G. g. gorilla* estimate, but nearly four times the *G. g. graueri* approximation. The $F_{ST}$ value for the best-fit analysis is consistent with the assumption of equal population sizes and 0.50 heritability. Both analyses suggest that about 20% of the total variation occurs between subspecies, while each subspecies holds about 80% of total variation.

These relatively low $F_{ST}$ estimates, ranging from 0.024 to 0.199 (depending upon census and heritability estimates), are unexpected given traditional analyses that emphasize statistically significant morphological differences among subspecies (cf. Groves, 1970). Even our highest craniometric $F_{ST}$ value places 80% of the total variation within each subspecies. Descriptive statistics and inspection of plots for each dimension indicate that absolute variability is extremely high in gorillas, particularly in males (Albrecht *et al.*, this volume). It is important to recall in this context that $F_{ST}$ measures relative variation, effectively adjusting this high level of subspecies variation by the total variation in the species.

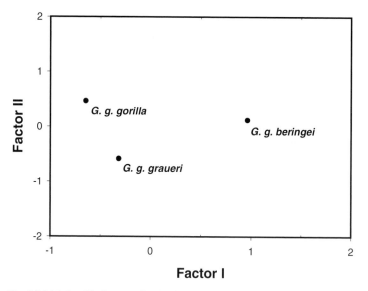

Fig. 5.2. Mahalanobis distances for *Gorilla* discrete traits. These values measure multivariate discrete trait distances among gorilla subspecies.

### *Discrete trait analyses*

Conversion of univariate discrete trait thresholds into interval-level data enables graphic display of multivariate relations among taxa (Fig. 5.2). Specifically, principal coordinates analyses of multivariate thresholds illustrate morphometric distinctions among the three recognized *Gorilla* subspecies. Eastern lowland gorillas are slightly closer to the western lowland gorilla centroid than to the mountain gorilla centroid. Discrete traits suggest an "intermediate" morphology for the eastern lowland subspecies (see also Groves, 1970; Taylor, this volume), and nearly equal divergence separating *G. g. graueri* from other subspecies.

Discrete trait $F_{ST}$ estimates are 0.272 when equal population sizes are assumed. When adjusted for differences in population sizes (Estimate A), the $F_{ST}$ value calculated in discrete trait variation for gorillas is 0.195.

In summary, our results indicate that relative craniometric variation among *Gorilla* subspecies may be somewhat lower than is implied by traditional statistical analyses. Although earlier studies consistently demonstrated statistically significant differences among *Gorilla* subspecies (e.g., Groves, 1970), several of our $F_{ST}$ estimates fall below 0.100, suggesting relatively low levels of divergence among subspecies. Discrete trait analyses, although quite preliminary, reflect moderately high levels of variation in *Gorilla*. Discrete trait estimates are concordant with our upper estimates for craniometric data. On balance, the degree of divergence measured by discrete traits exceeds the level of divergence measured by craniometric $F_{ST}$s.

## Discussion

### *Degree of variation in* Gorilla

Our results illustrate a conflict between traditional craniometrics and craniometric $F_{ST}$s. Traditional craniometrics provide evidence for comparatively deep divergences among *Gorilla* subspecies. However, $F_{ST}$ estimates, which calibrate variation among subspecies relative to total species diversity, suggest reduced variation when compared to traditional craniometrics. The discrepancy between $F_{ST}$ values and traditional craniometric interpretations suggests that these lines of evidence provide different kinds of information. In addition, we find some limitations inherent in the ways that traditional measures account for total variation. Specifically, model-free techniques may be especially problematic in a species with as much total variation as *Gorilla*. In this species, both sexes show high variation for many measurements, and males are especially so for some dimensions (Coolidge, 1929; Groves, 1966).

The exceptional level of total variation in *Gorilla* may reflect age-related changes in the skull, most notably, the development of muscle attachments such as nuchal and sagittal crests. The development of these structures occurs late in ontogeny, especially in males, and possibly beyond the age of third molar eruption. An extended male growth period (Leigh and Shea, 1995) may be partly responsible for the high level of total variation. Thus, the discrepancies between traditional and $F_{ST}$ craniometrics may result from high levels of variation that are associated with sex differences, prolonged growth periods, large size, and geographic variation. The distinctions among subspecies as reflected by traditional discriminant function techniques may be more strongly influenced by this remarkable level of total variation than are $F_{ST}$-based approaches. We suggest that the high level of variation is effectively "controlled" by $F_{ST}$ techniques.

The discrepancies between $F_{ST}$ and standard approaches, particularly those associated with age, should not be dismissed as "noise". Specifically, gorilla subspecies, populations, or demes may have evolved different adult morphologies by following common ontogenetic trajectories to diverse adult endpoints. The high level of craniometric variation may be partly explicable in terms of ontogenetic responses to sexual and natural selection. Thus, discrepancies among methods may imply that ontogenetic processes play important roles in structuring *Gorilla* variation. In effect, ontogenetic allometric processes that drive variation among contemporary African apes (Shea, 1981, 1983) could influence variation below the species level (cf. Shea *et al.*, 1993; Taylor, 1997*a*,*b*, 2002, this volume). It is also important to note that the incongruities that we find provide a cautionary note for fossil studies. Several analyses of fossil variation use gorillas as referential models (Lieberman *et al.*, 1988; Lockwood *et al.*, 1996; Miller, 2000). Our results reinforce the importance of analyzing relative variation in fossil studies, as noted by proponents of coefficient of variation approaches (e.g., Cope, 1993; but see also Polly, 1998).

### Taxonomic implications

The possibility that variation among subspecies of gorillas is lower than might be inferred based on traditional morphometric analyses has several implications. Although these data are not complete enough to re-evaluate the current taxonomy, it is apparent that neither craniometric nor discrete trait analyses provides a basis for advocating the establishment of multiple gorilla species. This contrasts with earlier (and likely preliminary) suggestions of species status for some *Gorilla* subspecies based on mtDNA results (e.g., Ruvolo *et al.*, 1994). Our $F_{ST}$ estimates fluctuate, but even our highest estimates fall well below 30%, leaving more than 70% of total variation within each subspecies. The finding of low to moderate levels of morphometric variation within *Gorilla* also has important implications for debates regarding the relevance of the subspecies concept (Kimbel, 1991; Jolly, 1993; Shea *et al.*, 1993; Tattersall, 1993; Smith *et al.*, 1997). The modest levels of relative craniometric variation that we observe are accompanied both by geographic distinctions and by ecological and social differences (see above). Given these $F_{ST}$ values, relative craniometric variation might be dismissed as ephemeral or unimportant by some researchers. However, it is clear that this decision would preclude insights into gorilla evolutionary history that can be derived by evaluating patterns of concordance among measures of diversity. Futhermore, our findings reinforce the idea that even subtle degrees of morphological variation can be associated with significant molecular, ecological and behavioral distinctions (see Shea *et al.*, 1993; Tattersall, 1993).

### Comparisons to other species

Published $F_{ST}$ estimates for large-bodied mammals vary widely (Templeton, 1999). Gorillas and humans fall towards the low end of this range, which extends

from approximately 0.100 in humans and African buffalo to about 0.850 in
North American gray wolves. Relethford estimates a minimum human craniometric $F_{ST}$ of 0.065 using three geographically defined samples (sub-Saharan
Africa, Europe, Far East); analyses using six regions provide slightly higher
$F_{ST}$ values (0.085) (Relethford, 1994). Our minimum $F_{ST}$s for gorillas, not
accounting for population sizes, suggest that relative variation among gorilla
OTUs exceeds human relative craniometric variation by roughly a factor of two.
When a realistic heritability of 0.50 and equal population sizes are assumed,
the gorilla $F_{ST}$s climb considerably, suggesting much more variation among
*Gorilla* subspecies than among human groups. However, census-adjusted $F_{ST}$
estimates for gorillas are lower (0.025–0.079). Despite low values of census-
adjusted data, our interpretation is that gorillas display greater craniometric
variation among subspecies than do geographically defined human groups. The
precise degree to which relative variation in gorillas exceeds relative human
variation is difficult to establish without better heritability and population size
estimates for each species.

Human craniometric $F_{ST}$s reflect marginally lower amounts of among-group
variation than recent genetic analyses based on geographically defined human
groups (Barbujani *et al.*, 1997). Specifically, Barbujani *et al.*'s study of human
microsatellite and restriction fragment length polymorphism variation across
major geographic regions returns an $F_{ST}$ of about 0.108. This value exceeds
Lewontin's (1972) pioneering estimate of 0.063 based on genetic markers. A
very preliminary estimate of craniometric $F_{ST}$ in Neandertals and early modern
humans returns values at about 0.130 (Donnelly *et al.*, 1998). Finally, discrete
trait $F_{ST}$s with which to compare gorilla values are rare. A discrete trait minimum $F_{ST}$ of about 0.200 for common chimpanzees has been estimated (Leigh &
Konigsberg, 1996). This value falls within our discrete trait range for gorillas
(0.195–0.272).

The level of *Gorilla* craniometric divergence measured by traditional techniques seems to exceed craniometric differences among chimpanzee (*Pan
troglodytes*) subspecies. For instance, Shea and Coolidge find "a surprisingly
low level of metric separation or distinctiveness among these geographic groups
[chimpanzee subspecies]" (1988:679–680). Shea and Coolidge (1988) further
note that Mahalanobis $D^2$ distances among populations within *Gorilla* subspecies are consistent with distances among chimpanzee subspecies.

Piecing together these lines of evidence, the implication is that African apes,
including humans, show comparatively small levels of relative craniometric
variation. Furthermore, even though subspecies variation in African apes is often
statistically significant (Groves, 1970), it may be low relative to other measures
of subspecific divergence, including molecular (Ruvolo *et al.*, 1994; Garner

and Ryder, 1996; Gagneaux *et al.*, 1999; Jensen-Seaman *et al.*, this volume), odontometric (Uchida, 1998), mandibular (Taylor, 2002, this volume), postcranial (Schultz, 1934; Groves, 1966; Taylor, 1997a,b), and ecological (Doran and McNeilage, 1998) characters. Finally, our assessment that other variables show greater levels of variation among subspecies is qualitative; estimates of $F_{ST}$ for these attributes would enable tests of this hypothesis.

## Concordance and variation

Morphological $F_{ST}$s blur the concordance among traits implied by comparing traditional analyses of *Gorilla* craniometrics to mitochondrial, behavioral, and ecological studies. Mitochondrial distinctions among gorilla populations, demes, and subspecies appear to be quite marked (Gagneaux *et al.*, 1999), but mitochondrial diversity seems to be discordant both with nuclear DNA data (Jensen-Seaman *et al.*, this volume) and with gorilla craniometric $F_{ST}$s. This pattern of concordance can be compared to that of chimpanzees and humans, bringing several features of *Gorilla* evolutionary history to light.

In humans (and gorillas), mitochondrial and morphometric views of variation appear to be concordant. Very low levels of human DNA variation parallel minimal levels of relative craniometric variation (Relethford, 1994). $F_{ST}$ values for both character sets are generally below 0.120 (Relethford, 1994; Barbujani *et al.*, 1997).

In chimpanzees, mitochondrial and morphological measures of variation are discordant. Chimpanzee mtDNA (and possibly nuclear DNA) diversity seems sufficiently high to warrant consideration of species designation for some populations (Morin *et al.*, 1994; Gagneaux *et al.*, 1999; Kaessmann *et al.*, 1999). Chimpanzee mitochondrial studies distinguish the extreme western subspecies, *P. t. verus*, from the other subspecies. However, craniometric differences among subspecies are limited (Shea *et al.*, 1993; Shea and Coolidge, 1988), while analysis of discrete trait variation also finds differences between *P. t. verus* and other subspecies (Braga, 1995b; see also Leigh and Konigsberg, 1996). Key aspects of their biology can explain the discordance illustrated by chimpanzees. Specifically, chimpanzees are eurytopic, highly mobile, and utilize large home ranges. Gene flow in a highly mobile taxon may reduce craniometric variation and their wide habitat range may also be reflected by a rather generalized cranial morphology (Shea and Coolidge, 1988).

Patterns of concordance in gorillas seem to roughly parallel the human case in that several measures of diversity are mutually consistent. However, $F_{ST}$ estimates do not seem to accord with mitochondrial, ecological, and traditional craniometric calibrations of subspecies differences. This discrepancy may be resolvable in light of new evidence showing comparatively low levels of nuclear DNA divergence in *Gorilla* (Jensen-Seaman *et al.*, this volume). Low nuclear DNA diversity and low relative craniometric diversity could suggest that gene flow has played a major role in structuring gorilla variability. Therefore, concordance between nuclear DNA and relative craniometric diversity may suggest that migration (particularly among males) and subsequent gene flow have reduced interpopulation differences for both traits. The role of gene flow may have been underappreciated given the apparently large differences among subspecies that are reflected by mitochondrial and traditional morphometric measures of subspecies diversity. Examining our results in light of nuclear DNA analyses implies a major role for gene flow in reducing relative cranial variation. Since Y-chromosome diversity may parallel that of nuclear DNA (Jensen-Seaman *et al.*, this volume), male migration may have been especially important in reducing relative craniometric variation.

Investigating concordance among traditional craniometric, molecular, and $F_{ST}$ measures of *Gorilla* diversity suggests that each way of accounting for variation incorporates a different kind of information. Ideally, molecular data provide a clear indication of how individuals or OTUs relate to one another. Traditional craniometric methods seem especially useful in detecting evolutionarily significant differences in adaptation among taxonomic units. In addition, these techniques, when coupled with molecular and $F_{ST}$ approaches, may provide some indication of rates of morphological evolution. On the other hand, $F_{ST}$ approaches help balance the results of traditional morphological techniques by controlling the effects of total variation. In addition, differences in $F_{ST}$ values for multiple traits provide indications of rates of evolution, and seem to provide information on population dynamics (see below). These are all valuable and complementary approaches, so we suggest that there is no "right" way to calibrate subspecies diversity. Instead, estimates of diversity and inferences about evolutionary histories must be judged according to multiple lines of evidence. Despite this optimistic perspective, we must note that analyses of gorillas (and possibly chimpanzees) distressingly suggest that morphological differences are not especially good indicators of taxonomic divergence below the species level. This inference parallels findings of discrepancies between morphological and molecular measures of divergence at the interspecific level that reflect either homoplasy (see Lockwood, 1999; Lockwood and Fleagle, 1999) or conserved variation. We caution that better $F_{ST}$ estimates are needed for other character complexes before we can fully assess the degree of concordance among traits. We can also suggest that patterns of discrete trait divergence may match estimates of molecular divergence more closely than traditional craniometrics (see also Braga, 1995*a,c*). However, much further testing with discrete traits within and among species is needed before this idea can be confirmed.

### *Evolutionary history*

Analyses of relative craniometric variation have important implications for understanding the recent evolutionary history of gorillas. Specifically, our best-fit analyses, though tentative, provide evidence for a dynamic and interdependent evolutionary history of eastern forms (*G. g. beringei* and *G. g. graueri*). The best fit between actual and expected patterns of variation requires a very high census weight for *G. g. beringei*, which suggests that relative population sizes in *G. g. graueri* and *G. g. beringei* were historically quite different. Specifically, the population of mountain gorillas may have outnumbered eastern lowland gorillas by a considerable margin at some point in the past. We note that Gagneaux *et al.*'s (1999) molecular results may support this hypothesis. Their phylograms seem to indicate long branch lengths within *G. g. beringei* that exceed branch lengths in eastern lowlands (Gagneaux *et al.*, 1999) (Fig. 5.1). However, more recent molecular analyses by Jensen-Seaman and Kidd (2001) suggest very low levels of mtDNA diversity in both *G. g. graueri* and *G. g. beringei*, with slightly higher levels of diversity in the latter subspecies. Jensen-Seaman and Kidd also provide evidence for a genetic bottleneck in each eastern subspecies, which may have occurred at the time of the last glacial maximum.

If our best-fit analyses accurately reflect historical circumstances, then populations of *G. g. beringei* may have contracted while those of *G. g. graueri* expanded. It is conceivable that these events are related, with *G. g. graueri* populations increasing at the expense of *G. g. beringei* populations. We suspect that montane environments currently occupied by *G. g. beringei* were much more extensive during the last glacial maximum (Groves, 1966), resulting in a wider distribution of "mountain" gorillas. Warming since the last glacial maximum would have dramatically reduced these habitats, restricting *G. g. beringei* to their present locations. We can thus suggest that eastern lowland gorilla populations may have expanded rapidly and recently in response to increasing size of lower-elevation forest habitats. These results could imply that *G. g. graueri* evolved from a *G. g. beringei*-like ancestor, although parsimony considerations based on morphological similarities between eastern and western lowland gorillas would suggest that mountain gorillas are derived. In either case, the discordance between molecular (Jensen-Seaman and Kidd, 2001) and morphological (Sarmiento *et al.*, 1996) attributes of Bwindi-Impenetrable Forest gorillas could suggest fairly high evolvability of gorilla morphological differences. This population seems to be indistinguishable from *G. g. beringei* based on mtDNA, but shows some morphological differences from mountain gorillas (unfortunately, Sarmiento *et al.* (1996) did not study *G. g. graueri*).

### Conservation issues

Our investigations of *Gorilla* highlight the important role that morphology can play in defining conservation priorities. The most important conservation implication of this study is that morphological variation within *Gorilla* may be more subtle than has been previously recognized. While there are statistically significant differences in morphology among subspecies, analyses of $F_{ST}$ suggest that this variation is small relative to the total morphological variation. Therefore, $F_{ST}$ results suggest that we must place a premium on understanding and preserving the fullest possible range of *Gorilla* morphological variation.

Analyses of cranial morphologies reveal three more specific conservation issues facing gorillas. First, the possibility that *G. g. beringei* harbors very high levels of relative craniometric variation reinforces the priority of preserving this subspecies. More optimistically, although the current population size for mountain gorillas is tiny, relatively high levels of diversity in this subspecies could offer positive prospects for conservation. These prospects may be especially optimistic if craniometric diversity reflects nuclear DNA diversity.

Our finding of relatively low levels of craniometric diversity in *G. g. graueri* raises an opposing set of issues. Low diversity might make conservation of *G. g. graueri* more difficult than is apparent from census size estimates, particularly if comparatively large population sizes obscure conservation challenges (cf. Templeton, 1994). However, our morphological results are compatible with the idea that *G. g. graueri* has successfully expanded its population, despite relatively low variability. We can speculate that this expansion could be based on newly evolved adaptations that enhance demographic performance. For example, derived life-history attributes, such as early maturation, reduced interbirth intervals, or reduced mortality, may have enabled the population to increase in size in spite of low morphological and molecular diversity. Future field studies might attempt to understand why this subspecies appears to be comparatively successful despite low relative morphological and molecular variation.

Craniometric $F_{ST}$s, along with molecular analyses (Gagneaux *et al.*, 1999), indicate high levels of diversity within *G. g. gorilla*. Molecular results show deep divergences among demes and populations, suggesting that conservation efforts should concentrate on numerous but relatively restricted geographic areas. $F_{ST}$s bolster this idea, suggesting that preservation of multiple populations is critical to retaining morphological and molecular diversity in this subspecies.

## Conclusions

Analyses of craniometric $F_{ST}$s indicate lower levels of morphological variation within *Gorilla* than is implied by traditional craniometric analyses. Investigations of discrete trait variation may indicate slightly higher levels of variation among subspecies, but these results are preliminary. $F_{ST}$ estimates for gorillas fall toward the low to moderate end of a range for other large-bodied species. The discordance between traditional craniometrics and $F_{ST}$ analyses seems to be related to high levels of total variation within the species. Craniometric $F_{ST}$s may reveal discrepancies between morphological and molecular estimates of divergence within the species. These discrepancies could stem from similar responses among subspecies to selection on cranial morphologies, or could indicate conserved morphologies within subspecies. Conservation of morphologies is a strong possibility for gorillas because of their consistent preference for forest habitats.

The evolutionary history of gorillas appears to be complex. Analyses that compare expected and observed variation by relative population weights suggest a complex interaction between *G. g. graueri* and *G. g. beringei* populations. A reversal in relative population sizes, from large to small population size in *G. g. beringei* and the converse in *G. g. graueri*, seems to best explain the relation between these subspecies. Finally, morphological analyses reveal distinct conservation challenges for each subspecies that involve understanding and preserving differing ranges of relative morphological variation.

## Acknowledgments

We thank Colin Groves for making his database publicly accessible, and for tremendous insight into *Gorilla* biology. In addition, several museums facilitated access to gorilla specimens for discrete trait analyses. Dr. Wim VanNeer (Koninklijk Museum voor Midden-Afrika, Tervuren), Derek Howlett (Powell-Cotton Museum), and Dr. David Pilbeam (Peabody Museum, Harvard University) granted access to materials in their care. Dr. Peter Andrews (Natural History Museum, London) made the museum's copies of Groves's database available to us, and Mr. Tom Donnelly transferred the data into computerized form. We also thank Drs. Michael Jensen-Seaman and K.K. Kidd for sharing a manuscript under review. Dr. Tony Goldberg discussed issues in gorilla biogeography, and raised several questions about ideas in the manuscript. We also thank an anonymous referee and Dr. Brian Shea for thorough and critical evaluations of the manuscript. Finally, we thank Drs. Andrea Taylor and Michele Goldsmith for inviting our contribution and for patience during the entire process.

# References

Albrecht, G.H. and Miller, J.M.A. (1993). Geographic variation in primates: A review with implications for interpreting fossils. In *Species, Species Concepts, and Primate Evolution*, eds. W.H. Kimbel and L.B. Martin, pp.123–162. New York: Plenum Press.

Barbujani, G., Magagni, A., Minch, E., and Cavalli-Sforza, L.L. (1997). An apportionment of human DNA diversity. *Proceedings of the National Academy of Sciences U.S.A.*, **94**, 4516–4519.

Braga, J. (1995*a*). Définition de certains caractères discrets crâniens chez *Pongo, Gorilla*, et *Pan*: Perspectives taxonomiques et phylogénétiques. PhD thesis, University of Bordeaux, France.

Braga, J. (1995*b*). Variation squelettique et mesure de divergence chez les chimpanzés: Contribution des caractères discrets. *Comptes Rendus de l'Académie des Sciences de Paris*, series II, **320**, 1025–1030.

Braga, J. (1995*c*). Study of two osseous discrete traits in the occipitocervical region of lowland gorillas. *Folia Primatologica*, **64**, 37–43.

Cheverud, J.M. (1981). Variation in highly and lowly heritable morphological traits among social groups of rhesus macaques (*Macaca mulatta*) on Cayo Santiago. *Evolution*, **35**, 75–83.

Coolidge, H.J. (1929). A revision of the genus *Gorilla*. *Memoirs of the Museum of Comparative Zoology, Harvard*, **50**, 291–381.

Cope, D.A. (1993). Measures of dental variation as indicators of multiple taxa in samples of sympatric *Cercopithecus* species. In *Species, Species Concepts, and Primate Evolution*, eds. W.H. Kimbel and L.B. Martin, pp. 211–238. New York: Plenum Press.

Darroch, J.N. and Mosimann, J.E. (1985). Canonical and principal components of shape. *Biometrika*, **72**, 241–252.

Diamond, J. (1994). Race without color. *Discovery*, **15**(11), 82–89.

Donnelly, S.M., Konigsberg, L.W., and Stringer, C.B. (1998). Interpretation of population structure when group structure is unknown. *American Journal of Physical Anthropology, Supplement* **26**, 106.

Doran, D.M. and McNeilage, A. (1998). Gorilla ecology and behavior. *Evolutionary Anthropology*, **6**, 120–131.

Fisher, R.A. (1930). *The Genetical Theory of Natural Selection*. Oxford, U.K.: Clarendon Press.

Futuyma, D. (1987). On the role of species in anagenesis. *American Naturalist*, **130**, 465–473.

Futuyma, D. (1992). History and evolutionary process. In *History and Evolution*, eds. M.H. Nitecki and D.V. Nitecki, pp. 103–129. Stony Brook, NY: SUNY Press.

Gagneaux, P., Wills, C., Gerloff, U., Tautz, D., Morin, P.A., Boesch, C., Fruth, B, Hohmann, G., Ryder, O.A., and Woodruff, D.S. (1999). Mitochondrial sequences show diverse evolutionary histories of African hominoids. *Proceedings of the National Academy of Sciences U.S.A.*, **96**, 5077–5082.

Garner, K.J. and Ryder, O.A. (1996). Mitochondrial DNA diversity in gorillas. *Molecular Phylogenetics and Evolution*, **6**, 39–48.

Gelvin, B.R., Albrecht, G.H., and Miller, J.M.A. (1997). The hierarchy of craniometric variation among gorillas. *American Journal of Physical Anthropology, Supplement* **24**, 117.

Gould, S.J. and Eldredge, N. (1993). Punctuated equilibrium comes of age. *Nature*, **366**, 223–227.

Groves, C.P. (1966). Variation in the skulls of gorillas with particular reference to ecology. PhD thesis, University of London.

Groves, C.P. (1967). Ecology and taxonomy of the gorilla. *Nature*, **213**, 890–893.

Groves, C.P. (1970). Population systematics of the gorilla. *Journal of the Zoological Society of London*, **161**, 287–300.

Groves, C.P. (1971). Distribution and place of origin of the gorilla. *Man*, **6**, 44–51.

Groves, C.P., Westwood, C., and Shea, B.T. (1992). Unfinished business: Mahalanobis and a clockwork orang. *Journal of Human Evolution*, **22**, 327–340.

Haddow, A.J. and Ross, R.W. (1951). A critical review of Coolidge's measurements of gorilla skulls. *Proceedings of the Zoological Society of London*, **121**, 43–45.

Harpending, H. and Jenkins, T. (1973). Genetic distance among southern African populations. In *Methods and Theories of Anthropological Genetics*, eds. M.H. Crawford and P.L. Workman, pp. 177–200. Albuquerque, NM: University of New Mexico Press.

Hauser, G. and De Stefano, G.F. (1989). *Epigenetic Variants of the Human Skull*. Stuttgart, Germany: Schweitzerbartsche Verlagsbuchhandlung.

Jensen-Seaman, M.I. and Kidd, K.K. (2001). Mitochondrial DNA variation and biogeography of eastern gorillas. *Molecular Ecology*, **10**, 2241–2247.

Jolly, C.J. (1993). Species, subspecies, and baboon systematics. In *Species, Species Concepts, and Primate Evolution*, eds. W.H. Kimbel and L.B. Martin, pp. 67–108. New York: Plenum Press.

Kaessmann, H., Wiebe, V., and Paabo, S. (1999). Extensive nuclear DNA sequence diversity among chimpanzees. *Science*, **286**, 1159–1162.

Kimbel, W.H. (1991). Species, species concepts, and hominid evolution. *Journal of Human Evolution*, **20**, 355–372.

Kimbel, W.H. and Martin, L.B. (1993). Species and speciation: Conceptual issues and their relevance for primate evolutionary biology. In *Species, Species Concepts, and Primate Evolution*, ed. W.H. Kimbel and L.B. Martin, pp. 539–554. New York: Plenum Press.

Kimura, M (1968). Evolutionary rate at the molecular level. *Nature*, **217**, 624–626.

Konigsberg, L.W. (1990*a*). Analysis of prehistoric biological variation under a model of isolation by geographic and temporal distance. *Human Biology*, **62**, 49–70.

Konigsberg, L.W. (1990*b*). Temporal aspects of biological distance: Serial correlation and trend in a prehistoric skeletal lineage. *American Journal of Physical Anthropology*, **82**, 45–52.

Konigsberg, L.W., Kohn, L.A.P., and Cheverud, J.M. (1993). Cranial deformation and nonmetric trait variation. *American Journal of Physical Anthropology*, **90**, 35–48.

Lande, R. (1981). Models of speciation by sexual selection on polygenic characters. *Proceedings of the National Academy of Sciences U.S.A.*, **78**, 3721–3725.

Leigh, S.R. and Konigsberg, L.W. (1996). Intraspecific discrete trait polymorphism in African apes: Implications for variation in the fossil record. *American Journal of Physical Anthropology, Supplement* **22**, 147.

Leigh, S.R. and Shea, B.T. (1995). Ontogeny and the evolution of adult body size dimorphism in apes. *American Journal of Primatology*, **36**, 37–60.

Lewontin, R.C. (1972). The apportionment of human diversity. *Evolutionary Biology*, **6**, 381–398.

Lieberman, D.E., Pilbeam, D.R., and Wood, B.A. (1988). A probabilistic approach to the problem of sexual dimorphism in *Homo habilis*: A comparison of KNM-ER 1470 and KNM-ER 1813. *Journal of Human Evolution*, **17**, 503–511.

Lockwood, C.A. (1999). Homoplasy and adaptation in the atelid postcranium. *American Journal of Physical Anthropology*, **108**, 459–482.

Lockwood, C.A. and Fleagle, J.G. (1999). The recognition and evaluation of homoplasy in primate and human evolution. *Yearbook of Physical Anthropology*, **42**, 189–232.

Lockwood, C.A., Richmond, B.G., Jungers, W.L., and Kimbel, W.H. (1996). Randomization procedures and sexual dimorphism in *Australopithecus afarensis*. *Journal of Human Evolution*, **31**, 537–548.

Manly, B.F.J. (1997). *Randomization, Bootstrap and Monte Carlo Methods in Biology*, 2nd edn. New York: Chapman & Hall.

Matschie, P. (1904). Bemerkungen über die Gattung *Gorilla*. *Sitzungsberichte des Gesellschaft naturforschender Freunde, Berlin*, **1904**, 45–53.

Matschie, P. (1905). Merkurdige Gorilla-Schadel aus Kamerun. *Sitzungsberichte des Gesellschaft naturforschender Freunde, Berlin*, **1905**, 277–283.

Miller, J.M.A. (2000). Craniofacial variation in *Homo habilis*: An analysis of the evidence for multiple species. *American Journal of Physical Anthropology*, **112**, 103–128.

Morin, P.A., Moore, J.J., Chakraborty, R., Lin, L, Goodall, J., and Woodruff, D.S. (1994). Kin selection, social structure, gene flow, and evolution of chimpanzees. *Science*, **265**, 1193–1201.

Oates, J. (1996). *African Primates*. Gland, Switzerland: International Union for Conservation of Nature (IUCN).

Plavcan, J.M. and van Schaik, C.P. (1992). Intrasexual competition and canine dimorphism in anthropoid primates. *American Journal of Physical Anthropology*, **87**, 461–478.

Polly, P.D. (1998). Variability in mammalian dentitions: Size-related bias in the coefficient of variation. *Biological Journal of the Linnean Society*, **64**, 83–99.

Relethford, J.H. (1994). Craniometric variation among modern human populations. *American Journal of Physical Anthropology*, **95**, 53–62.

Relethford, J.H. and Blangero, J. (1990). Detection of differential gene flow from patterns of quantitative variation. *Human Biology*, **62**, 5–25.

Relethford, J.H. and Harpending, H.C. (1994). Craniometric variation, genetic theory, and modern human origins. *American Journal of Physical Anthropology*, **95**, 249–270.

Relethford, J.H. and Jorde, L.B. (1999). Genetic evidence for larger African population size during recent human evolution. *American Journal of Physical Anthropology*, **108**, 251–260.

Relethford J.H., Crawford, M.H., and Blangero, J. (1997). Genetic drift and gene flow in post-famine Ireland. *Human Biology*, **69**, 443–65.

Rothschild, W. (1905). Notes on anthropoid apes. *Proceedings of the Zoological Society of London*, **2**, 413–440.

Rothschild, W. (1906). Further notes on anthropoid apes. *Proceedings of the Zoological Society of London*, **2**, 465–468.

Ruvolo, M., Pan, D., Zehr, S., Goldberg, T., Disotell, T.R., and von Dornum, M. (1994). Gene trees and hominoid phylogeny. *Proceedings of the National Academy of Sciences U.S.A.*, **91**, 8900–8904.

Saltonstall, K., Amato, G., and Powell, J. (1998). Mitochondrial DNA variability in Grauer's gorillas of Kahuzi-Biega National Park. *Journal of Heredity*, **89**, 129–135.

Sarmiento, E.E., Butynski, T.M., and Kalina, J. (1996). Gorillas of the Bwindi-Impenetrable Forest and Virunga Volcanoes: Taxonomic implications of morphological and ecological differences. *American Journal of Primatology*, **40**, 1–21.

Schultz, A.H. (1934). Some distinguishing characters of the Mountain Gorilla. *Journal of Mammology*, **15**, 51–61.

Seaman, M.I., Saltonstall, K., and Kidd, K.K. (1998). Mitochondrial DNA diversity and biogeography of Eastern gorillas. *American Journal of Physical Anthropology, Supplement* **26**, 199.

Seaman, M.I., Deinard, A.S., and Kidd, K.K. (2000). African ape nuclear phylogeography. *American Journal of Physical Anthropology, Supplement* **30**, 276.

Shea, B.T. (1981). Relative growth of the limbs and trunk in African apes. *American Journal of Physical Anthropology*, **56**, 179–202.

Shea, B.T. (1983). Allometry and heterochrony in the evolution of African ape craniodental form. *Folia Primatologica*, **40**, 32–68.

Shea, B.T. and Coolidge, H.J. (1988). Craniometric differentiation and systematics in the genus *Pan*. *Journal of Human Evolution*, **17**, 671–686.

Shea, B.T., Leigh, S.R., and Groves, C.P. (1993). Multivariate craniometric variation in chimpanzees: Implications for species identification. In *Species, Species Concepts, and Primate Evolution*, eds. W.H. Kimbel and L.B. Martin, pp. 265–296. New York: Plenum Press.

Smith, H.M., Chiszar, D., and Montanucci, R.R. (1997). Subspecies and classification. *Herpetological Review*, **28**, 13–16.

Stumpf, R.M., Fleagle, J.G., Jungers, W.L., Oates, J.F., and Groves, C.P. (1998). Morphological distinctiveness of Nigerian gorilla crania. *American Journal of Physical Anthropology, Supplement* **26**, 213.

Suter, J. and Oates, J.F. (2000). Sanctuary in Nigeria for possible fourth subspecies of gorilla. *Oryx*, **34**, 71.

Tattersall, I. (1993). Speciation and morphological differentiation in the genus *Lemur*. In *Species, Species Concepts, and Primate Evolution*, eds. W.H. Kimbel and L.B. Martin, pp. 163–176. New York: Plenum Press.

Taylor, A.B. (1997*a*). Relative growth, ontogeny, and sexual dimorphism in *Gorilla* (*Gorilla gorilla gorilla* and *G. g. beringei*): Evolutionary and ecological considerations. *American Journal of Primatology*, **43**, 1–31.

Taylor, A.B. (1997*b*). Scapula form and biomechanics in gorillas. *Journal of Human Evolution*, **33**, 529–553.

Taylor, A.B. (2002). Masticatory form and function in the African apes. *American Journal of Physical Anthropology*, **117**, 133–157.

Templeton, A.R. (1994). Biodiversity at the molecular genetic level: Experiences from disparate macroorganisms. *Proceedings of the Royal Society of London, Series B*, **345**, 59–64.

Templeton, A.R. (1999). Human races: A genetic and evolutionary perspective. *American Anthropologist*, **100**, 632–650.

Tutin, C.E.G. & Fernandez, M. (1985). Foods consumed by sympatric populations of *Gorilla gorilla gorilla* and *Pan troglodytes troglodytes* in Gabon: Some preliminary data. *International Journal of Primatology*, **6**, 27–43.

Uchida, A. (1996). What we don't know about great ape variation. *Trends in Ecology and Evolution*, **11**, 163–168.

Uchida, A. (1998). Variation in tooth morphology of *Gorilla gorilla*. *Journal of Human Evolution*, **34**, 55–70.

Van Valen, L. (1990). Age changes vs. natural selection in human skeletal traits, and statistics for their study. *International Journal of Anthropology*, **5**, 281–282.

Vogel, C. (1961). Zur systematischen Untergliederung der Gattung *Gorilla* anhand von Untersuchungen der Mandibel. *Zeitschrift für Säugetieren*, **26**, 1–12.

Watts, D.S. (1996). Comparative socioecology of gorillas. In *Great Ape Societies*, eds. W.C. McGrew, L.F. Marchant, and T. Nishida, pp. 16–28. Cambridge, U.K.: Cambridge University Press.

Williams-Blangero, S. and Blangero, J. (1989). Anthropometric variation and the genetic structure of the Jirels of Nepal. *Human Biology*, **61**, 1–12.

Wright, S. (1951). The genetical structure of populations. *Annals of Eugenics*, **15**, 323–354.

Wright, S. (1969). *Evolution and the Genetics of Populations*, vol. 2, *The Theory of Gene Frequencies*. Chicago, IL: University of Chicago Press.

# 6 Ontogeny and function of the masticatory complex in Gorilla: Functional, evolutionary, and taxonomic implications

ANDREA B. TAYLOR

## Introduction

The classic studies of Coolidge (1929) and Groves (1967, 1970a,b, 1986; Groves and Stott, 1979), coupled with more recent investigations (Uchida, 1996, 1998; Taylor, 1998a,b, 1999, 2002), have revealed considerable variability in the cranium, mandible, and dentition among subspecies and even geographic populations of gorillas (Albrecht et al., Leigh et al., Stumpf et al., this volume). Historically, investigators have tended to emphasize the taxonomic implications of this variation (Coolidge, 1929; Groves, 1967, 1970a) and with good reason, as a well-founded taxonomy clearly forms the basis for meaningful studies in other important fields, including morphology, behavioral ecology, and genetics. While there have been some attempts to draw functional inferences from differences in craniomandibular and dental morphology among gorilla subspecies (Vogel, 1961; Groves, 1970a,b; Uchida, 1996, 1998; Taylor, 1998b, 1999, 2000), nevertheless the nature and patterning of variation in jaw form among gorilla subspecies remains unresolved.

I examine masticatory form and function in gorillas to assess whether gorilla subspecies differ predictably in morphology as a function of dietary specialization. I focus particularly on the expected pattern of morphological differentiation in *Gorilla* based on the degree to which these taxa differ in folivory versus frugivory. I use an ontogenetic, allometric approach in an explicitly phylogenetic context to evaluate whether differences among subspecies can be principally attributed to extrapolation or truncation of inherited patterns of ontogenetic allometry (Shea, 1995), or to derived dissociations of ancestral allometries. While differences in body size among gorilla subspecies may be relatively minor, body weight estimates for these taxa are derived from very few animals, and regional and local differences in cranial and postcranial size have been demonstrated, leaving this issue open to debate. I also examine variation during ontogeny as

132

a means of evaluating allometric changes in the masticatory complex during growth that may be functionally important. Such analyses have the potential to yield important insights into gorilla functional anatomy as well as inform biomechanical hypotheses regarding primate mandibular form and function. These data may also be applied indirectly to address both issues of ecophenotypic plasticity and microevolutionary change in gorillas and other primates, and also questions of mastication, diet, and evolution in fossil hominoid and hominid taxa (McCrossin and Benefit, 1993; Taylor, 1998*b*; Ward *et al.*, 1999). More broadly, evaluation of the ecological bases of variation below the level of the species provides an important explanatory framework for understanding what drives speciation events, and how species evolve in nature (Morell, 1999).

### Cranial, dental, and mandibular variation

In his seminal monograph, Coolidge (1929) undertook a comprehensive re-evaluation of morphological variation in the skull of *Gorilla*. Coolidge attributed to idiosyncratic variation much of the variability that had been previously invoked to defend the existing multitude of gorilla species and subspecies designations. The result was the conversion of 15 recognized species or geographical races to a simplified taxonomy comprised of only a single species *Gorilla gorilla*, with two subspecies: *Gorilla gorilla gorilla*, the western lowland gorilla, and *G. g. beringei*, the eastern mountain subspecies. The only modification to this classification, still generally accepted today (cf. Sarmiento and Oates, 2000; Groves, 2001, this volume; Oates *et al.*, this volume), was the subsequent recognition of *G. g. graueri*, the eastern lowland gorilla, as a separate subspecies (Vogel, 1961; Groves, 1967).

Notable differences in skull form between *G. g. gorilla* and *G. g. beringei* that justified their status as subspecies included variation in jaw and condylar dimensions as well as some structural asymmetries of the mandible. Coolidge did not link these differences directly to dietary specialization in *G. g. beringei*. However, he did suggest that at least some of the variation in skull form could reflect a plastic response to feeding behavior during growth:

> In looking at our problem of variation in the gorilla skulls it is impossible to determine how far this variation is caused by muscles developed by their form of living. As in the lions, the feeding habits of gorillas undoubtedly influence the shape of the skull. By way of suggestion: the breaking off of some tough bamboo stalk with the teeth might well be done easier by always pulling in a certain direction and after this was repeated a few times it might become a habit.
>
> (Coolidge, 1929:372)

A

B

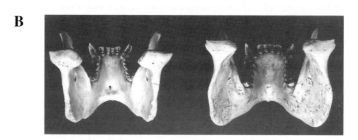

Fig. 6.1. Mandibles of adult *G. g. gorilla* and *G. g. beringei*. (A) Lateral view. Groves noted the absolutely higher mandibular ramus and larger molars of *G. g. beringei* (left) as compared to *G. g. gorilla* (right). (B) Posterior view. Groves also noted the greater bigonial breadth (i.e., "more flared jaw angles") of *G. g. beringei* (right) as compared to *G. g. gorilla* (left). (Photographs courtesy of C.P. Groves).

Groves (1970*b*) later revisited the issue of craniodental variability among gorillas, this time incorporating the newly resurrected eastern lowland subspecies, *G. g. graueri*, in the comparative mix. In his reanalysis of the existing two-subspecies classification, Groves (1970*b*) noted that measurements of the jaw produced a distinct separation among the three taxa. Based largely on length of the tooth row (noted also by Vogel, 1961) and palate, Groves also observed an increase in the size of the jaws associated with the geographic distribution of gorillas from west to east.

Groves (1970*a*) further noted that in comparison to the other gorilla subspecies, *G. g. beringei* exhibited structural features of the jaw complex that appeared to be shared by other taxa such as *Gigantopithecus* and *Theropithecus* (Figs. 6.1A,B). These features included absolutely larger teeth with correspondingly higher crowns, a higher ascending ramus, greater bigonial breadth and tendon-associated bone features that reflect muscle structure and a strongly developed masticatory pattern. Groves, unlike Coolidge, explicitly linked this suite of features to a specialized diet of roots, bark and other fibrous items and suggested the mountain gorilla jaw complex might well serve as a comparative model for evaluating functional interpretations of similar morphologies in fossil taxa. *Gorilla g. graueri*, the eastern lowland subspecies, was observed to be intermediate in jaw form between *G. g. gorilla* and *G. g. beringei*.

Extending the work of Groves, Uchida (1996, 1998) undertook an evaluation of craniodental variability within adult *Gorilla*. Uchida showed that compared to *G. g. gorilla*, the two eastern subspecies exhibit a relatively larger postcanine dentition characterized by molars with higher crowns, sharper cusps and transverse ridges. By contrast, *G. g. gorilla* displays relatively wider upper incisors. To some extent, these results parallel differences in relative size of the anterior versus postcanine dentition noted specifically between *G. g. gorilla* and *Pan troglodytes* (Shea, 1983*a*), observed more generally in broad interspecific primate comparisons (Hylander, 1975), and associated with the differential mechanical demands of fruits versus leaves. Taken together, these data suggest that the African apes are confronted with conflicting mechanical demands depending upon whether they utilize their postcanine dentition to masticate a largely resistant diet of herbaceous vegetation, or consume a more frugivorous diet of ripe fleshy fruits along with other resistant fruit items that may require greater incisal preparation. Indeed, Uchida (1998) speculated that variation in dental morphology among gorilla subspecies may reflect dietary differences.

In addition, Uchida (1996, 1998) also observed that *G. g. beringei* has a uniformly deep and thick mandibular corpus as compared to *G. g. gorilla* and *G. g. graueri*, who together display gradations of a more distally shallow but transversely thicker corpus. Proportion differences in the cranium and dentition were found to be greatest in comparisons between *G. g. gorilla* and *G. g. beringei*, concurring with Groves's cranial results. However, *G. g. graueri* and *G. g. beringei* converged on some features to the exclusion of *G. g. gorilla*, including the presence in the former of larger molars relative to body size, larger distal cusps on $M_2$ and $M_3$, and mesial cusps of equal size in the lower molars (Uchida, 1996). These results suggest that *G. g. graueri* does not consistently present with an "intermediate" morphology.

Taylor (1998*b*, 1999), building on the work of others (Coolidge, 1929; Vogel, 1961; Groves, 1970*a*; Uchida, 1996, 1998), evaluated the morphology of the masticatory complex in *Gorilla*. Taylor demonstrated that among adults, gorilla subspecies vary in maxillomandibular proportions. For example, adult *G. g. beringei* has wider mandibular corpora and symphyses relative to both mandibular and basicranial lengths when compared to either *G. g. gorilla* or *G. g. graueri*. *Gorilla g. beringei* also has a relatively wider mandibular condyle, higher temporomandibular joint and higher ramus, similar to Groves's (1970*a*) earlier findings. Relative to basicranial length, both *G. g. graueri* and *G. g. beringei* have longer mandibular and palatal lengths. Adult gorilla subspecies also differ in bizygomatic breadth relative to mandibular length but not relative to basicranial length. A summary of these differences is provided in the Appendix.

More recently, Taylor (1998*a,b*, 2002; Taylor and Ravosa, 1999) showed that the African apes, ontogenetically scaled for a number of craniomandibular features (Shea, 1982, 1983*a,b*, 1985*a*; Taylor, 2002), depart from ontogenetic scaling in some aspects of the masticatory complex that are consistent with a trend towards extreme folivory. Features that consistently differentiate *G. g. beringei*, a specialized herbivore, from *G. g. gorilla* and *Pan* include the presence of a relatively thicker mandibular corpus and symphysis, a relatively taller mandibular ramus and temporomandibular joint elevated higher above the occlusal plane of the mandible, and a relatively larger insertion site for the masseter muscle. These structural differences are suggestive of morphological adaptations to intensification of folivory in *G. g. beringei*. However, the patterning of craniomandibular variation among the African apes did not always conform to predictions based on comparative studies of primate masticatory form and function, data from *in vivo* experimental studies of stress and strain in prosimian and anthropoid primates, and theoretical biomechanical expectations (DuBrul, 1977; Hylander, 1979*a,b*, 1985; Ravosa, 1990, 1996*a,b*; Cole, 1992; Daegling, 1992; Antón, 1996*a,b*). In a number of instances, the morphological results were either inconsistent with, or ran contrary to, predictions based on dietary differences (Taylor, 2002).

The observed differences among the African apes provide justification for testing a similar set of predictions among gorilla subspecies, a claim explicitly proposed earlier by Groves (1970*a*), who states: "The differences I have described between mountain and lowland gorillas can be said to be an intensification of those between the gorilla as a whole and the chimpanzee. The ecological and especially dietary data from the two species support this view." The timeliness of this investigation is reinforced by more thorough and refined studies of gorilla behavioral ecology, which substantiate a greater degree of frugivory for *G. g. gorilla* as compared to other gorilla subspecies (Remis, 1997; Tutin *et al.*, 1997; Goldsmith, this volume). A detailed understanding of the ontogenetic covariance patterns of the craniomandibular complex in gorillas should also help to clarify some of the patterning that characterizes the African apes (Taylor, 2002).

While previous work has revealed proportion differences among gorilla subspecies adults (Taylor, 1998*b*, 1999), in this study, I use ontogenetic allometry to examine developmental and structural variation in masticatory morphology in gorillas. While few would argue that body weight is a mechanical influence on the masticatory system (Hylander, 1985; Bouvier, 1986*a*), there are arguably sound nonmechanical reasons for hypothesizing that factors influencing changes in body size, such as hormones and metabolism, will similarly effect changes in skull size (Smith, 1993). Moreover, while body size (i.e., body weight) differences among gorilla subspecies may be negligible (Groves, 1970*b*; Smith and Jungers, 1997; cf. Taylor, 1997*a* for skeletal weight), this issue remains moot given that weight estimates are based on samples too small to yield reliable data. Body weight differences aside, adult gorillas differ regionally in skull and tooth dimensions (Coolidge, 1929; Groves 1979*a*; Uchida, 1996; Taylor, 1998*b*, 1999), and though there has been no definitive conclusion regarding the significance of this size variation, even subtle regional differences in size may be biomechanically and evolutionarily important. Ontogenetic allometry provides a test of whether the ontogenetic covariance patterns for masticatory features differ among gorilla subspecies, independent of any growth-allometrically induced changes, providing a robust assessment of whether divergent (i.e., novel, size-corrected) proportions in mountain gorillas fit the predictions of biomechanical models based upon established dietary differences. Apart from any growth or size effects, ontogenetic allometry also conveys important information about scaling relationships of jaw dimensions during growth that may be mechanically advantageous. While actual stress levels during growth can only be demonstrated experimentally, ontogenetic data can at least address whether the ability to resist masticatory forces remains the same, or alters, during growth.

### *Morphological correlates of folivory/herbivory*

Experimental work on several prosimian and anthropoid primates has revealed a series of masticatory stresses routinely encountered along the corpus and at the symphysis during mastication (Hylander, 1979*a*,*b*, 1981, 1984, 1985; Hylander *et al.*, 1987). As some of the expected morphological differences among gorilla subspecies are linked to increased forces associated with these masticatory stresses, these loading regimes are reviewed briefly below. It should be stated clearly at the outset that in assessing the functional consequences of changes in masticatory form, I assume that differences between the more folivorous and more frugivorous gorillas are associated with improving resistance to relatively greater masticatory loads in the more folivorous group in order to maintain similar levels of stress and strain (Hylander, 1985). There is evidence of dynamic similarity in peak strain magnitudes in the postcranium across diversity of taxa (Biewener, 1982; Rubin and Lanyon, 1984), and of preservation of functional equivalence in jaw stress and strain levels ontogenetically and interspecifically in primates (Vinyard and Ravosa, 1998).

Experimental data demonstrate that during the power stroke of mastication, the balancing-side mandibular corpus is bent in the sagittal plane

(Hylander, 1979*a,b*) (Fig. 6.2A). The amount of bending is directly proportional to the level of balancing-side jaw muscle force. Elevated bending loads are most efficiently resisted by an increase in the depth of the mandibular corpus.

Sagittal bending of the mandibular corpus causes dorsoventral shear stress at the mandibular symphysis (Hylander, 1979*a,b*) (Fig. 6.2B). Dorsoventral shear stress derives from the increase in balancing-side muscle force coupled with the bite point reaction forces that occur during unilateral mastication. Increasing the depth of the mandibular symphysis, or overall area of cortical bone, will theoretically serve to best resist elevated loads associated with dorsoventral shear.

During unilateral mastication, the mandibular corpus and symphysis experience lateral transverse bending, or "wishboning" stress (Hylander, 1984, 1985) (Fig. 6.2C). In anthropoid primates, wishboning stress derives from high levels of activity of the balancing-side deep masseter muscle during the late phase of the power stroke of mastication, coupled with the decline in activity of both the balancing-side superficial masseter muscle and working-side deep and superficial masseter muscles (Hylander, 1986; Hylander *et al.*, 1987, 1996, 2000; Hylander and Johnson, 1994). Wishboning bends the mandible in its plane of curvature, causing the mandible to function as a curved beam. This results in particularly high levels of tensile stress along the lingual surface of the symphysis (Fig. 6.2C). In primates with fused mandibular symphyses, increasing the labiolingual thickness or orienting the long axis of the symphysis more horizontally have been hypothesized to best counter elevated wishboning loads, while lateral transverse bending of the mandibular corpus may be resisted by increasing the transverse thickness of the mandible (Hylander, 1985). Irrespective of dietary variation, wishboning stress intensifies with an increase in symphyseal curvature simply because the symphysis functions as a curved beam. Because symphyseal curvature has been shown to increase with ontogenetic and interspecific size increase in some primates (Hylander, 1985; Ravosa, 1996*a,b*; Vinyard and Ravosa, 1998), larger primates with fused symphyses may experience relatively greater wishboning forces as a function of scale.

Lastly, experimental studies demonstrate that the mandibular corpora experience twisting about their long axes during unilateral mastication and incision (Hylander, 1979*a,b*, 1981) (Fig. 6.2D). Axial torsion is produced primarily by the masseter muscle force resultants, which are positioned lateral to the mandible, and by the bite force associated with the working-side mandibular corpus, and is arguably the most important loading regime in the molar region of the working-side mandibular corpus (Hylander, 1979*a,b*, 1984). The forces that combine to produce axial torsion evert the basilar corpus and invert the alveolar corpus with the net effect that the mandibular symphysis is bent in the vertical plane (Hylander, 1979*a,b*, 1981). Muscle force magnitudes, orientation of the superficial and deep masseter muscles, and timing of muscle recruitment all influence the degree of axial torsion (Hylander, 1985; Hylander and Johnson, 1994). Torsional loads are best resisted by redistributing cortical bone more evenly about the neutral axis of the mandible to increase buccolingual thickness of the premolar and molar corpus (Hylander, 1979a,b, 1985). The optimal biomechanical response to symphyseal bending is to increase the depth of the symphysis.

Apart from these masticatory loading patterns, a variety of structural features have been hypothesized to confer mechanical advantages to primates characterized by highly fibrous diets. For example, comparative data reveal that folivores exhibit larger masseter muscles (Turnbull, 1970; Herring and Herring, 1974; Ravosa, 1990; Antón, 1996a). Investigators have argued that more anteriorly positioned masticatory muscles along with anteroposterior shortening and vertical deepening of the face should confer a mechanical advantage by positioning the masticatory musculature closer to the bite points (i.e., improving the load-to-lever arm ratio) and reducing the bending moments in the face (DuBrul, 1977; Hylander, 1977, 1979a; Ravosa, 1990; Spencer and Demes, 1993; Antón, 1996a). As bite force is inversely proportional to jaw length (Hylander, 1985; Spencer and Demes, 1993), relative shortening of the jaw also provides for an increase in the amount of muscle force that may be converted into usable bite force, particularly at $M_1$ (Hylander, 1979a; Spencer, 1998). A temporomandibular joint elevated higher above the occlusal plane may reduce the fatigue failure associated with frequent chewing cycles by distributing occlusal loads more evenly along the postcanine teeth (Herring and Herring, 1974; Ward and Molnar, 1980). Finally, folivores have been shown to exhibit wider mandibular condyles (expanded mediolaterally), a feature that has been associated with enhanced use of the postcanine dentition (Smith *et al.*, 1983; Bouvier, 1986a; Takahashi and Pan, 1994) and the distribution of loads on the lateral condyle during unilateral mastication (Hylander and Bays, 1979). By contrast, anteroposteriorly longer condyles have been associated with anterior tooth function (Smith *et al.*, 1983; Bouvier, 1986a).

### Morphological predictions in gorilla subspecies

In this study, I evaluate the following: First, do gorilla subspecies differ in skull size? Second, if gorillas differ in skull size, what are the ontogenetic bases for any differences in masticatory morphology between subspecies? In other words, which shape differences can be accounted for by ontogenetic scaling, or shared patterns of ancestral growth covariance (Shea, 1995), and which are nonallometric shape differences that require explanation beyond the correlated

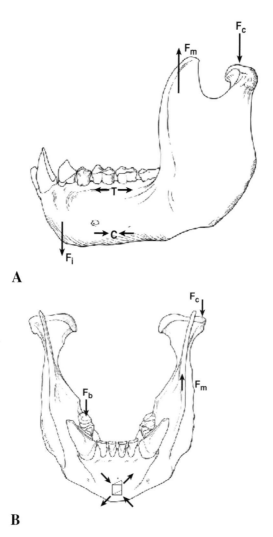

Fig. 6.2. Adult male *Gorilla gorilla gorilla* mandible. (A) Lateral view, left side. $F_c$ represents the condylar reaction force, and $F_m$ the resultant muscle force, each on the balancing side mandible. $F_i$ represents the force transmitted through the symphysis from the balancing side to the working side. During the power stroke of mastication, the balancing-side mandibular corpus is bent in the sagittal plane. This results in tension ($T$) along the alveolar corpus and compression ($C$) along the basal corpus. Sagittal bending loads are efficiently resisted by an increase in corpus depth. (B) Frontal view. Sagittal bending of the balancing-side mandibular corpus produces dorsoventral shear stress at the symphysis. $F_m$ and $F_c$ represent the vertical components of the balancing-side muscle force and bite force, respectively. Dorsoventral shear stress results from the transmission across the symphysis of the vertical component of jaw muscle force from the balancing to working side of the mandible. In primates with fused symphyses, dorsoventral shear is efficiently countered by an increase in symphyseal depth.

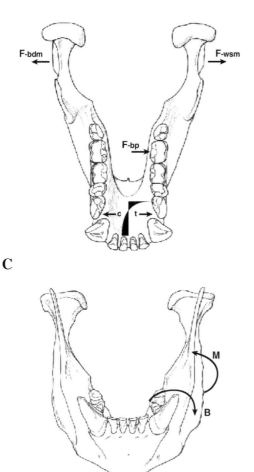

**C**

**D**

Fig. 6.2. *(cont.)*. (C) Superior view. During mastication, the transverse component of bite force on the working-side corpus ($F$-wsm), transverse component of the adductor muscle force on the balancing-side corpus ($F$-bdm), together result in lateral transverse bending, or "wishboning," at the symphysis. During wishboning, the mandible functions as a curved beam, such that the labial aspect of the symphysis experiences relatively high tensile ($t$) stresses and relatively low compressive ($c$) stresses. Wishboning loads are best resisted by increasing the labiolingual thickness of the mandibular symphysis. (D) Frontal view. During the power stroke of mastication, the working-side mandibular corpus is twisted about the long axis. The muscle force ($M$) on the working side everts the basal corpus and inverts the alveolar corpus, while the bite force ($B$) on the working side produces the opposite effect. Resistance to axial torsion is most effectively achieved by increasing the transverse thickness of the postcanine mandibular corpus. (Adapted from Hylander, 1988; Hylander and Johnson, 1994; Ravosa 1996*b*.)

effects of size increase? Finally, do nonallometric shape differences fit the predictions of intensification of folivory?

In testing predictions regarding how the masticatory apparatus should be expected to vary as a function of dietary preference, certain assumptions are implicit. One assumption is that the dietary demands of a specialized folivore like *G. g. beringei* require cyclical loads of greater duration and frequency as compared to either *G. g. gorilla* or *G. g. graueri*. Thus, I expect *G. g. beringei* to exhibit structural alterations of the masticatory complex that improve resistance to fatigue failure from repetitive loading of the jaws (Hylander, 1979*a*). It is also possible that the jaws of *G. g. beringei* are subjected to relatively greater bending moments during the mastication of such items as bark and bamboo, which are woody plants and extremely tough to process mechanically (Lucas *et al.*, 2000). Therefore, *G. g. beringei* should also exhibit features that improve the generation and dissipation of greater occlusal loads (DuBrul, 1977; Hylander, 1979*a*; Antón, 1996*a*). Second, some of the morphological predictions tested in this study derive from experimental studies of masticatory stresses conducted on a limited set of prosimian and anthropoid primates (Hylander, 1979*a*,*b*, 1984). These loading regimes are assumed to hold for gorillas, but have never been demonstrated directly for these taxa. Finally, I rely on several simplifying biomechanical assumptions: First, that linear measurements provide reasonable (if only first) approximations of the actual biomechanical properties of the mandibular corpus and symphysis, and second, that skeletal measures provide reasonable estimates of relative muscle force. These assumptions, while necessary, are not without problems (Daegling, 1990, 1992; Antón, 1999, 2000; Daegling and Hylander, 2000).

Based on the hypothesis that a specialized folivore/herbivore is expected to exhibit structural features that both generate and dissipate higher bite forces and reduce the risk of fatigue failure associated with repetitive loading of the jaws, *G. g. beringei* is predicted to exhibit a relatively deeper mandibular corpus to resist elevated loads associated with parasagittal bending; a relatively deeper mandibular symphysis to resist elevated loads associated with dorsoventral shear; a relatively wider mandibular corpus and symphysis to resist elevated loads linked to torsion and wishboning, respectively; relatively wider mandibular condyles and temporomandibular joints elevated higher above the occlusal plane to resist fatigue failure and more evenly distribute occlusal loads; relatively taller faces vertically, and shorter faces anteroposteriorly, for generation and dissipation of greater occlusal loads; and a more anteriorly positioned masseter muscle to improve mechanical leverage.

*Gorilla g. graueri*, like *G. g. gorilla*, consumes fruit when seasonally available, but comminutes greater quantities of leaves, pith, and bark than its western lowland congener (Watts, 1984; Yamagiwa *et al.*, 1996); there is no evidence to indicate that *G. g. graueri* masticates bamboo, even at higher elevations. Whether leaf-eating requires larger average bite force than fruit may be open to debate (Beecher, 1979; Hylander, 1979*a*), but assuming that *G. g. graueri* engages in repetitive loading of the jaws to a greater extent than *G. g. gorilla*, *G. g. graueri* at a minimum is expected to converge on the masticatory pattern for *G. g. beringei* in ways that facilitate efficient load distribution and reduce the risk of fatigue failure from repetitive chewing cycles. Thus, compared to *G. g. gorilla*, *G. g. graueri* is predicted to exhibit a relatively deeper mandibular corpus, relatively wider mandibular condyles and a temporomandibular joint elevated relatively higher above the occlusal surface.

### Gorillas as a comparative model

Gorillas provide a unique opportunity to test mechanical hypotheses about jaw form and function for several reasons. First, as all gorillas consume terrestrial herbaceous vegetation, the primary difference in diet among these taxa lies in the degree to which gorillas consume leaves and other folivorous material as compared to fruit. In essence, these apes vary along a dietary continuum that allows us to examine the extent to which the mastication of structurally more resistant foods impacts the masticatory complex.

Second, gorillas are sufficiently closely related to be distinguished only at the level of the subspecies (cf. Sarmiento and Oates, 2000; Groves, this volume; Oates *et al.*, this volume). Therefore, comparisons among these closely related taxa are afforded a measure of phylogenetic control. This is important because controlling for phylogeny improves the chances that differences in the masticatory apparatus reflect novel responses to dietary specialization rather than phylogenetic history (Lauder, 1982). The relatively minor differences in size among subspecies make gorillas an excellent case study for allometric size control.

Finally, patterns of morphological variation among subspecies provide an important framework for evaluating and interpreting the functional, adaptive, and evolutionary significance of such patterns at higher taxonomic levels and in the fossil record. The implications of this last point are nicely illustrated by Inouye (this volume). Inouye demonstrates that some of the patterns of morphological variation shown to differentiate between *Pan* and *Gorilla* also differentiate among subspecies of *Gorilla* and species of *Pan*, and this is highlighted against the backdrop of ontogenetic scaling for the majority of comparisons for both *Pan* and *Gorilla* as well as gorilla subspecies. The overall patterning of variation has been taken as contradictory evidence by some that *Pan* and *Gorilla* share a unique, derived morphological/behavioral complex and by others that knuckle-walking has evolved in parallel. Inouye argues that whatever the causal basis for the observed patterns of variation, parsimony dictates that the explanation be congruent whether one is comparing genera, species, or subspecies. Thus, subspecies comparisons can compel us to formulate, and adhere to, logically consistent interpretations of morphology across all taxonomic levels.

I note there has been some debate regarding the importance of subspecific variation from a phylogenetic perspective, or even whether subspecies should be treated as biologically meaningful units from an evolutionary standpoint (Tattersall, 1986; Kimbel, 1991). The mere fact that the taxonomic status of gorillas is currently being debated, whether defined as subspecies or as populations (*sensu* Mayr, 1982), argues for the importance of investigating variation below the level of the species. Beyond issues of gorilla taxonomy, cogent arguments have been made (including by some contributors to this volume, e.g., Albrecht *et al.*, Leigh *et al.*) for the importance of establishing both intraspecific and interspecific patterns of variation, in order to better understand and recognize incipient speciation, ecophenotypic versus microevolutionary adaptation, and paleontological species (Mayr, 1982; Jolly, 1993; Shea *et al.*, 1993). It has also been argued (Leigh *et al.*, this volume) that variation below the species level may provide an important metric for prioritizing conservation efforts.

## Materials and methods

### Samples

I use a mixed-sex, cross-sectional sample of 210 specimens of *Gorilla* (116 *G. g. gorilla*, 58 *G. g. graueri*, and 36 *G. g. beringei*). Developmental ages are assigned using a combination of dental eruption pattern, dental wear, and status of basilar suture fusion (Cheverud, 1981; Shea, 1983*a*) resulting in seven discrete age classes (Taylor, 2002) (Table 6.1). Whenever possible, only adult specimens with little to moderate dental wear are used, as corpus and symphysis measurements can be affected by local bony remodeling and, in particular, resorption of the bony alveoli associated with aging and disease (Lovell, 1990). *Gorilla g. beringei* is an exception, based upon the extremely limited number of specimens available. However, any alveolar bone loss associated with aging in *G. g. beringei* provides a more conservative test of the hypothesis that *G. g. beringei* exhibits structural features that enhance their ability to comminute a tougher diet. All specimens are wild-caught and of known locality.

Table 6.1. *Numbers of specimens comprising the material studied*

| | Gorilla gorilla gorilla[a] | | | Gorilla gorilla graueri[b] | | | Gorilla gorilla beringei[c] | | |
|---|---|---|---|---|---|---|---|---|---|
| Dental stage | M | F | ? | M | F | ? | M | F | ? |
| 1 Incompletely erupted deciduous dentition | 1 | 4 | 0 | 0 | 0 | 3 | 2 | 0 | 1 |
| 2 All deciduous teeth fully erupted | 10 | 2 | 3 | 0 | 0 | 3 | 1 | 0 | 0 |
| 3 Deciduous dentition fully erupted plus $M^1$ partially or fully erupted | 10 | 12 | 2 | 1 | 7 | 3 | 0 | 1 | 2 |
| 4 $M^2$ partially or fully erupted | 5 | 10 | 0 | 0 | 3 | 0 | 0 | 2 | 0 |
| 5 C, $M^3$ erupting | 5 | 4 | 0 | 1 | 1 | 0 | 0 | 1 | 0 |
| 6 C, $M^3$ fully erupted, full permanent dentition with little wear, basilar suture unfused | 12 | 9 | 0 | 6 | 2 | 1 | 2 | 0 | 0 |
| 7 Full permanent dentition with moderate wear and basilar suture fused | 14 | 13 | 0 | 14 | 13 | 0 | 13 | 11 | 0 |
| Total | 57 | 54 | 5 | 22 | 26 | 10 | 18 | 15 | 3 |

[a] *Gorilla g. gorilla* from Powell-Cotton Museum, Kent, U.K.; Field Museum of Natural History, Chicago, IL; Cleveland Museum of Natural History, Cleveland, OH.
[b] *Gorilla g. graueri* from Central African Museum, Tervuren, Belgium; Laboratoire d' Anthropologie, UCL, Louvain-la-Neuve, Belgium.
[c] *Gorilla g. beringei* from Central African Museum, Tervuren, Belgium; National Museum of Natural History, Washington, DC.

## Measurements

Measurement definitions and sources are presented in Table 6.2. Not all measurements could be obtained on each specimen due to damage or wear; therefore, not all dimensions are incorporated in every analysis. The selection of measurements is based on their association in primates and other mammals with dissipating large occlusal loads and improving mechanical advantage. Variables include mandibular corpus depth and width (orthogonal to depth) obtained at $M_1$ and $M_2$, symphyseal depth and width, condylar width, mandibular ramus height, height of the temporomandibular joint above the occlusal plane of the mandible, and length of the masseter muscle lever arm (maximum linear distance between the posterior mandibular condyle and anterior zygoma) (Figs. 6.3A–C). Bizygomatic and bicondylar breadths are used as measures of craniofacial breadth (Figs. 6.3B, D). Bizygomatic breadth relative to bicondylar

Table 6.2. *Measurement definitions and sources*

| Variable | Definition |
| --- | --- |
| Corpus depth | Maximum depth of mandibular corpus taken perpendicular to the occlusal plane from the buccal alveolus to the inferior border of the mandible at the midpoints of $M_1$ and $M_2$ (Hylander, 1979a) |
| Corpus width | Maximum buccolingual width of the mandibular corpus taken perpendicular to corpus depth through a transverse plane at the midpoints of $M_1$ and $M_2$ (Hylander, 1979a) |
| Symphyseal depth | Maximum depth of the mandibular symphysis from gnathion to infradentale perpendicular to the occlusal plane (Hylander, 1985) |
| Symphyseal width | Maximum labiolingual width of the mandibular symphysis taken perpendicular to symphysis depth (Hylander, 1985) |
| Condylar width | Maximum mediolateral dimension of the articular surface in the coronal plane (Bouvier, 1986b) |
| Ramal height | Distance from gonion to uppermost condyle (Antón 1996a) |
| Temporomandibular joint height | Maximum distance from the occlusal plane to the uppermost condyle (Ravosa, 1990) |
| Masseter lever arm length | Distance from posterior condyle to maxillary root of the zygoma (Ravosa, 1990) |
| Bizygomatic breadth | Distance between right and left zygion |
| Bicondylar breadth | Distance between right and left condylion laterale (Hylander, 1985) |
| Facial height | Maximum linear distance between gonion and zygion with the mandible in occlusion (Antón, 1996a) |
| Basicranial length | Distance from nasion to basion (Coolidge, 1929) |
| Mandibular length | Distance between posterior condyle and incision (Bouvier, 1986a) |
| Palatal length | Distance between staphylion and orale |

breadth is used as an indirect measure of the horizontal component of the masseter muscle, which has important implications for torsion. The distance between gonion and zygion serves as a measure of facial height (Fig. 6.3D). Additional measures of skull and facial size include basicranial, mandibular, and palatal lengths. Mandibular length also serves as a biomechanical lever arm (Fig. 6.3B). All measurements were obtained with a Mitutoyo digital calipers accurate to 0.01 mm.

### Methodological approach

I use ontogenetic allometry to investigate variation in masticatory form among gorilla subspecies. In the vein of Huxley (1932) and Thompson (1942), I follow the hierarchical approach to the study of size and its consequences advocated by a long line of investigators (Gould, 1966, 1971, 1975; Shea, 1983*a,b*, 1985*a*, 1995; Martin, 1989; Lauder and Reilly, 1990; Reilly and Lauder, 1990; Harvey and Pagel, 1991; Emerson and Bramble, 1993) to identify and differentiate between structural differences that result from ontogenetic scaling, or the differential extension or truncation of common patterns of relative growth, and those which result independent of the effects of size (Gould, 1966, 1975; Shea, 1984, 1985*a,b*). Even if size differences are minimal, as they appear to be in the case of gorilla subspecies, when a pattern of divergence emerges it is indicative of derived discordant allometries that suggest novel transformations. It is well understood by those familiar with allometry that ontogenetic scaling in and of itself is not an ultimate explanation for why we observe correlated shape change with increase or decrease in size. However, ontogenetic scaling serves as an important starting-point for separating size from other possible factors that might serve as proximate explanations of the observed patterns of differentiation. This approach provides an important perspective for identifying novel shape changes that suggest additional selection pressures beyond those linked to size change which may be associated with the maintenance of functional equivalence (Jungers and Fleagle, 1980; Shea, 1983*a,b*, 1986, 1995; Cheverud, 1984; Atchley, 1987; Müller, 1990; Atchley and Hall, 1991).

Apart from providing an appropriate "criterion of subtraction" (Shea, 1995), ontogeny can have an important influence on adult form. Departures from isometry have been linked to important functional shifts ontogenetically and interspecifically by numerous investigators (Cock, 1963; Dodson, 1975; Cochard, 1985; Shea, 1985*b*, 1986; Jungers and Cole, 1992; Taylor, 1995; Velhagen and Roth, 1997). Conversely, the importance of preservation of geometric similarity as "adaptive" has also been addressed (Gould, 1966, 1971; Dodson, 1975; Shea, 1982). Thus, observed ontogenetic trends and allometric relationships during ontogeny may provide important insights into development, historical constraint, and functional shifts during growth (cf. Godfrey *et al.*, 1998). There is ample evidence as well to indicate that variation in adult morphology may be the result of selection acting earlier in development (Müller, 1990).

### Statistical approach

Within each ontogenetic series, I performed regression analysis on log-transformed data to explore patterns of relative growth among gorilla subspecies. Following Bouvier (1986*a*), I used mandibular length as the independent variable. I relied on ordinary least-squares (OLS) regression analysis because analysis of covariance (ANCOVA) provides a robust parametric test of differences in trajectories that is not available for Model II regression. OLS is justified because correlation coefficients in most cases are high (i.e., $>0.90$).

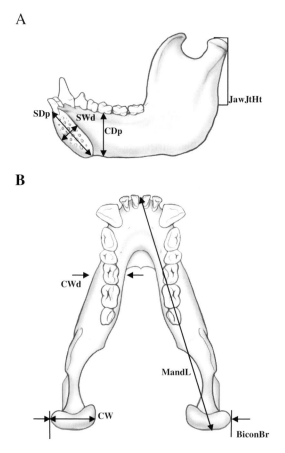

Fig. 6.3. Craniomandibular measurements. (A) SDp, symphyseal depth; SWd, symphyseal width; CDp, corpus depth; JawJtHt, temporomandibular joint height. (B) CWd, corpus width; MandL, mandibular length; CW, condylar width; BiconBr, bicondylar breadth. (C) RHt, mandibular ramus height; MLA, masseter lever arm length. (D) Bizygom, bizygomatic breadth; GonZyg, gonion–zygion distance.

However, in instances of low correlation coefficients (i.e., <0.90), I also used Model II regression, and in such cases I treated differences as statistically significant only if both methods yielded concordant results. To distinguish allometric shape differences from those unrelated to size change, I used ANCOVA to test for significant elevations/transpositions in ontogenetic allometries. I also examined bivariate plots to empirically assess the patterning and degree of variation between taxa.

Two questions were always under consideration when interpreting results: First, do gorillas systematically diverge in the predicted direction and in ways

**C**

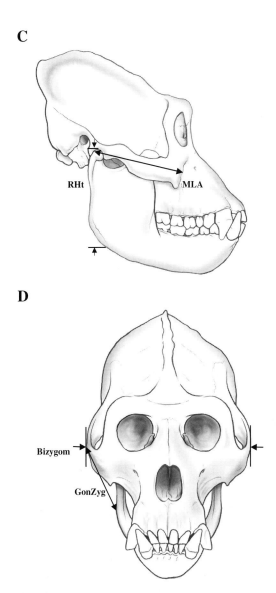

**D**

Fig. 6.3. *(cont.)*.

consistent with established dietary differences, biomechanical theory, and empirically demonstrated loading regimes? And second, to what extent do statistically significant findings result in marked separations (i.e., elevations, transpositions) between taxa?

I also used principal components analysis (PCA) to provide multivariate summaries of patterns of size and shape variation (Jolicoeur, 1963; Shea, 1985*a*; cf. Klingenberg *et al.*, 1996; Ackermann and Cheverud, 2000). PCA is based on the pooled, within-group covariance matrix of log-transformed values. The first principal component may be interpreted as summarizing the variance due to ontogenetic scaling. The second and subsequent components reflect slope and position differences that represent nonallometric (but not isometric) shifts in form. Thus, PCA is an effective means of distinguishing between the variance in adult jaw form that results from ontogenetic allometry, and that which departs from a common vector of allometric growth and reflects proportion differences presumably related to diet. Only the first two factors are presented, as subsequent factors appeared to represent residual individual variation within each subspecies and did not effect any significant separation among taxa.

### Results

Descriptive data on adults of each subspecies are presented in the Appendix. Regression statistics for 16 bivariate comparisons are shown in Table 6.3. For the most part, correlations between linear dimensions are high and significant. The correlation coefficient for width of the mandibular corpus at $M_1$ is low but significant in all taxa; width of the mandibular corpus at $M_2$ is also low and significantly correlated with mandibular length only in *G. g. graueri*. Standard errors are low except in those regressions with low correlation coefficients. Possibly due to relatively small samples, the highest standard errors are almost always associated with *G. g. beringei*.

Ontogenetic trajectories exhibit a mix of allometry and isometry but only mandibular corpus depth at $M_1$ (and $M_2$ based on Model II regressions), and basicranial length versus mandibular length, and basicranial length versus palatal length, are consistently positively allometric across subspecies. A number of ontogenetic trajectories are negatively allometric across all subspecies, including mandibular corpus width at $M_1$, symphyseal width, temporomandibular joint height, and bizygomatic and bicondylar breadths. Isometry of OLS slopes, tested against a slope of 1.0, holds for masseter lever arm length in all taxa, and for mandibular corpus depth at $M_2$, symphyseal depth, condylar width, mandibular ramus height, gonion–zygion distance and bizygomatic versus bicondylar breadths in one or more taxa. Isometry of bivariate slopes occurs with the greatest frequency in *G. g. beringei*. Consistent with Model II regression, RMA slopes were higher than those derived from OLS, converting

Table 6.3. *Regression statistics*

| Variable | Gorilla gorilla gorilla | | | | Gorilla gorilla graueri | | | | Gorilla gorilla beringei | | | |
|---|---|---|---|---|---|---|---|---|---|---|---|---|
| | $k^a$ | $y$ | $r$ | 95% CI | $k^a$ | $y$ | $r$ | 95% CI | $k^a$ | $y$ | $r$ | 95% CI |
| *Versus mandibular length* | | | | | | | | | | | | |
| $M_1$ corpus depth | 1.16 | −0.97 | 0.95 | ±0.09 | 1.19 | −1.04 | 0.97 | ±0.08 | 1.14 | −0.92 | 0.96 | ±0.14 |
| $M_2$ corpus depth[b] | **1.00** | −0.64 | 0.77 | ±0.24 | 1.21 | −1.11 | 0.92 | ±0.18 | **1.03** | −0.69 | 0.87 | ±0.25 |
| | 1.30 | −1.30 | | ±0.26 | 1.32 | −1.35 | | ±0.18 | 1.21 | −1.07 | | ±0.28 |
| $M_1$ corpus width[b] | 0.17 | 0.91 | 0.41 | ±0.09 | 0.15 | 0.96 | 0.45 | ±0.09 | 0.21 | 0.86 | 0.47 | ±0.17 |
| | 0.70 | −0.24 | | ±0.22 | 0.64 | −0.13 | | ±0.20 | 0.63 | −0.07 | | ±0.27 |
| $M_2$ corpus width | 0.13 | 1.04 | 0.17 | ±0.46 | 0.22 | 0.85 | 0.48 | ±0.13 | 0.06 | 1.21 | 0.09 | ±0.42 |
| Symphyseal depth | **1.01** | −0.42 | 0.98 | ±0.05 | 0.97 | −0.35 | 0.99 | ±0.04 | **1.03** | −0.47 | 0.98 | ±0.09 |
| Symphyseal width | 0.78 | −0.30 | 0.93 | ±0.06 | 0.75 | −0.24 | 0.94 | ±0.08 | 0.77 | −0.28 | 0.96 | ±0.09 |
| Condylar width | 0.91 | −0.48 | 0.95 | ±0.06 | 0.90 | −0.47 | 0.96 | ±0.08 | **1.07** | −0.81 | 0.96 | ±0.13 |
| Temporomandibular joint height | 0.90 | −0.09 | 0.94 | ±0.07 | 0.77 | 0.18 | 0.95 | ±0.08 | 0.86 | 0.05 | 0.91 | ±0.17 |
| Ramal height | 1.07 | −0.27 | 0.98 | ±0.05 | **0.99** | −0.13 | 0.98 | ±0.06 | 1.11 | −0.35 | 0.98 | ±0.08 |
| Masseter lever arm length | **1.03** | −0.32 | 0.98 | ±0.04 | **1.00** | −0.26 | 0.98 | ±0.06 | **1.00** | −0.23 | 0.98 | ±0.09 |
| Bizygomatic breadth | 0.76 | 0.53 | 0.98 | ±0.04 | 0.81 | 0.41 | 0.99 | ±0.03 | 0.76 | 0.53 | 0.99 | ±0.05 |
| Bicondylar breadth | 0.64 | 0.72 | 0.97 | ±0.04 | 0.67 | 0.64 | 0.97 | ±0.04 | 0.76 | 0.46 | 0.98 | ±0.06 |
| Gonion–zygion distance | 1.12 | −0.42 | 0.97 | ±0.05 | **1.02** | −0.20 | 0.99 | ±0.05 | **1.00** | −0.14 | 0.98 | ±0.09 |
| Basicranial length | 1.44 | −0.83 | 0.98 | ±0.06 | 1.33 | −0.59 | 0.98 | ±0.07 | 1.37 | −0.68 | 0.98 | ±0.14 |
| Bizygomatic breadth versus bicondylar breadth | 1.21 | −0.35 | 0.97 | ±0.07 | 1.16 | −0.27 | 0.97 | ±0.08 | **1.03** | 0.01 | 0.97 | ±0.12 |
| Basicranial length versus palatal length | 1.53 | −1.21 | 0.96 | ±0.09 | 1.43 | −0.96 | 0.98 | ±0.09 | 1.47 | −1.05 | 0.96 | ±0.17 |

[a] Boldfaced *k*-value signifies that the OLS slope does not depart significantly from a slope of 1.0 (isometry).

[b] For $M_2$ corpus depth and $M_1$ corpus width, OLS regression statistics presented on the line, Model II regression statistics below. $M_2$ corpus width is uncorrelated with mandibular length in *G. g. gorilla* and *G. g. beringei*; therefore, only OLS regressions are presented.

Table 6.4. *Results of tests of statistical differences in slopes and y-intercepts among* Gorilla *subspecies*[a]

| Variable | *Gorilla gorilla gorilla* versus *Gorilla gorilla graueri* | | *Gorilla gorilla graueri* versus *Gorilla gorilla beringei* | | *Gorilla gorilla gorilla* versus *Gorilla gorilla beringei* | |
|---|---|---|---|---|---|---|
| | *k* | *y* | *k* | *y* | *k* | *y* |
| *Versus mandibular length* | NS | NS | NS | NS | NS | NS |
| $M_1$ corpus depth | | | | | | |
| $M_2$ corpus depth | NS | NS | NS | **0.003** | NS | NS |
| $M_1$ corpus width | NS | **0.000** | NS | **0.002** | NS | **0.000** |
| Symphyseal depth | NS | NS | NS | NS | NS | NS |
| Symphyseal width | NS | 0.017 | NS | **0.000** | NS | **0.000** |
| Condylar width | 0.011 | — | NS | **0.000** | NS | **0.000** |
| Temporomandibular joint height | NS | NS | NS | **0.000** | NS | **0.000** |
| Ramal height | NS | 0.011 | 0.011 | — | NS | **0.000** |
| Masseter lever arm length | NS | **0.001** | NS | **0.000** | NS | 0.025 |
| Bizygomatic breadth | NS | **0.000** | NS | **0.000** | NS | NS |
| Bicondylar breadth | NS | 0.024 | NS | **0.000** | NS | 0.013 |
| Gonion–zygion distance | 0.011 | — | NS | **0.000** | NS | **0.005** |
| Basicranial length | 0.024 | — | NS | **0.001** | NS | NS |
| *Bizygomatic breadth versus bicondylar breadth* | NS | NS | NS | NS | NS | NS |
| *Basicranial length versus palatal length* | NS | **0.000** | NS | NS | NS | **0.000** |

[a] Results based on analysis of covariance of log-transformed values. Boldface $p$-values indicate a significant pairwise difference in slopes or $y$-intercepts ($\alpha < 0.05$) after protecting against Type I error using the sequential Bonferroni correction method; values not in bold indicate $p < 0.05$ prior to the Bonferroni adjustment; NS indicates $p > 0.05$.

some slopes from isometry to positive allometry. Where correlation coefficients are high ($r > 0.90$), differences between regression statistics were negligible.

ANCOVA results are presented in Table 6.4. Based on the Bonferroni correction, there are no slope differences among taxa, but few differences achieve statistical significance even without the adjustment. Numerous differences in $y$-intercepts signify dissociations from a common ontogenetic trajectory near the onset of growth. The fewest differences in scaling patterns occur in comparisons between western and eastern lowland gorillas, the greatest number between eastern lowland and eastern mountain subspecies.

It is biologically instructive to combine statistical analyses with empirical observation of bivariate plots to gain insight into the patterning and degree of differentiation among taxa. Because of disproportionate sampling among subspecies and across age classes, data are averaged by dental stage for ease of interpretation; however, bivariate regression analyses and results of statistical tests are based on all data points. Results that consistently fit the predicted pattern of differentiation based on diet are illustrated in Figs. 6.4A–E. There is a trend towards increasing thickness of the mandibular corpus with intensification of folivory (Fig. 6.4A; Tables 6.3 and 6.4). For several comparisons, *G. g. graueri* commences ontogeny elevated above the other two taxa, possibly a reflection of disproportionate sampling at early dental stages, but *G. g. beringei* transposes above both *G. g. gorilla* and *G. g. graueri* during growth. Thus, by dental stage 3 *G. g. beringei* has a relatively thicker mandibular symphysis, elevated temporomandibular joint and higher mandibular ramus (Figs. 6.4B–D). Facial height (gonion–zygion distance) is also significantly greater in *G. g. beringei* (Fig. 6.4E).

Apart from these dissociations of ontogenetic allometries, no other morphological differences consistently fit the predictions of increasing dietary resistance and some actually run counter to expectations based on intensification of folivory. For example, an inverse relationship was predicted between degree of folivory and mandibular length, yet at comparable skull size, mandibular length is relatively greatest in *G. g. graueri* and effects no separation between *G. g. beringei* and *G. g. gorilla* (Tables 6.3 and 6.4). Depths of the mandibular corpus and symphysis were predicted to increase with specialization towards folivory but gorilla subspecies are ontogenetically scaled for these bivariate relationships (Figs. 6.5A,B; Tables 6.3 and 6.4).

I used a combination of cranial and mandibular variables to evaluate differences in skull size between subspecies. These variables included basicranial, mandibular, occlusal, and palatal lengths, bizygomatic and bicondylar breadths, condylar width, symphyseal depth and width, mandibular ramus height, gonion–zygion distance, and condylion posterior – anterior zygoma distance. Figure 6.6A shows a plot of the factor I scores of the PCA for each subspecies separately by dental stage. Size differentiation is visually apparent by dental stage 6 but samples of *G. g. beringei* are unacceptably small at this dental stage to allow for significance testing. At dental stage 7, results of a Bonferroni-adjusted analysis of variance indicate *G. g. beringei* is significantly (df = 2,55; $F = 5.81$; $p = 0.004$) larger in overall craniomandibular size as compared to both *G. g. gorilla* and *G. g. graueri*; the latter two subspecies do not differ in craniomandibular size as adults.

Results of a grouped PCA on the variance–covariance matrix for all three subspecies are shown in Table 6.5. Variables were entered into this analysis if ontogenetic trajectories were found to be statistically different regardless of the degree of separation. Additionally, to ensure maximization of ontogenetic specimens, I excluded width and depth of the mandibular corpus at $M_2$ since

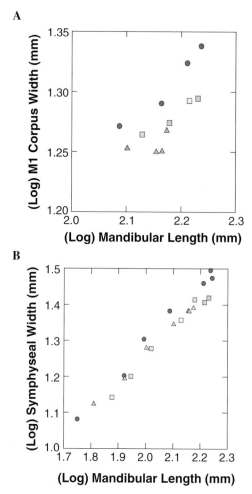

Fig. 6.4. Bivariate plots of log-transformed values of craniomandibular dimensions that resulted in significant elevations or transpositions of *G. g. beringei* in the direction predicted by intensification of folivory (see also Tables 6.3 and 6.4). Plots are of mean values for each age group. All dimensions are scaled against mandibular length. (A) *Gorilla g. beringei* has a significantly wider corpus than *G. g. gorilla* and *G. g. graueri*. (B–D) During early ontogeny gorilla subspecies have similar proportions, but *G. g. beringei* transposes above *G. g. gorilla* and *G. g. graueri* during growth resulting in a significantly wider mandibular symphysis, temporomandibular joint elevated higher above the occlusal plane, and higher mandibular ramus. (E) *Gorilla g. beringei* has a significantly higher face, reflected in gonion–zygion distance, as compared to *G. g. gorilla* and *G. g. graueri*.

**C**

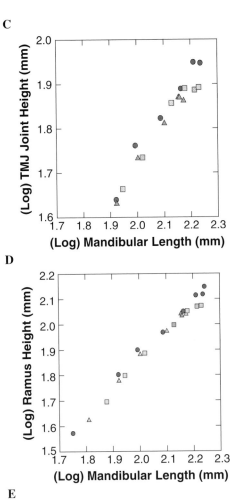

**D**

**E**

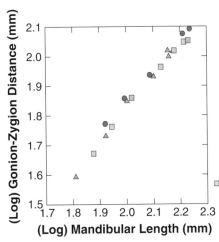

Fig. 6.4. *(cont.)*.

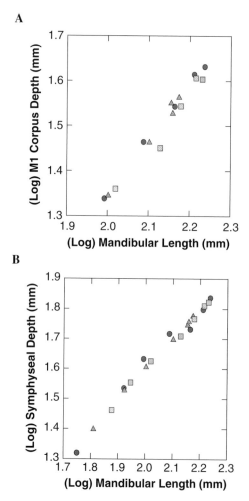

Fig. 6.5. Bivariate plots of examples of craniomandibular proportions that do not fit the predictions of structural modifications of the masticatory complex associated with intensification of folivory. ●, *Gorilla g. beringei*; ▲, *G. g. gorilla*; ■, *G. g. graueri*. (A) *Gorilla g. beringei* was predicted to exhibit a relatively deeper mandibular corpus but there are no differences amongst gorilla subspecies for this bivariate comparison. (B) *Gorilla g. beringei* was predicted to exhibit a relatively deeper mandibular symphysis, but gorillas do not differ in this bivariate comparison.

incorporation of these variables would restrict the analysis to individuals between dental stages 4 and 7. In the grouped PCA, the first principal component accounts for 94% of the variation. The divergent loadings on the first principal component reflect shape variation related to allometric growth (Jolicoeur, 1963; Shea, 1985), consistent with the various departures from isometry of bivariate slopes. Individuals are distributed according to general craniomandibular size,

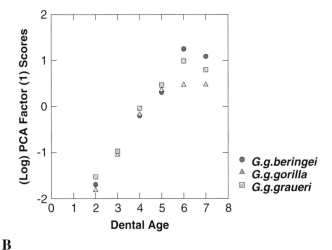

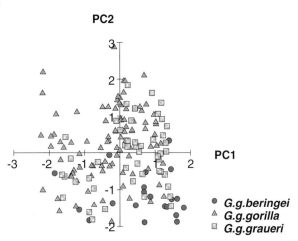

Fig. 6.6. (A) Plot of the Factor I scores from a multigrouped PCA on the covariance matrix of log-transformed values illustrating the multivariate pattern of size differentiation among adults. Gorilla subspecies diverge in overall skull size by dental stage 6 and by dental stage 7, *G. g. beringei* has a significantly ($p < 0.01$) larger skull. (B–E) Plots of factor scores for PCAs (see also Tables 6.5–6.8) on the covariance matrix of log-transformed values for craniomandibular dimensions for: (B) All three subspecies. The first axis distributes the smaller/younger gorillas to the left and larger/older gorillas to the right. There is overlap among all three subspecies, but note the distribution on or below the first axis of *G. g. beringei*. (C) *G. g. beringei* and *G. g. gorilla*. As in Fig. 6.6 B, there is overlap among subspecies. However, whereas *G. g. gorilla* occupies much of the multivariate space, almost all specimens of *G. g. beringei* distribute below PC I. (D) *Gorilla g. beringei* and *G. g. graueri*. There is a clear trend for *G. g. beringei* and *G. g. graueri* to separate along the second axis with only one specimen of *G. g. beringei* falling above PC1. (E) *Gorilla g. gorilla* and *G. g. graueri*. There is considerable overlap between these two subspecies with no discernable degree of separation along PC2.

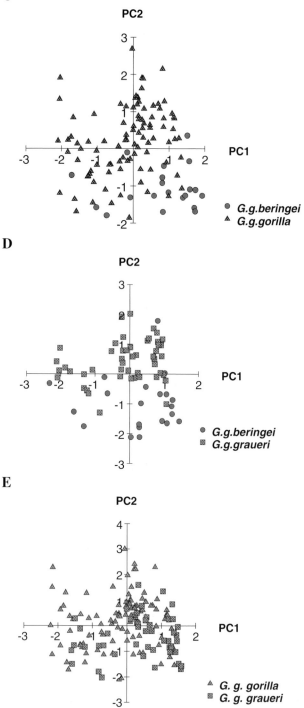

Fig. 6.6. *(cont.)*.

Table 6.5. *Principal component loadings for a PCA of log-transformed dimensions for ontogenetic series of the three subspecies of* Gorilla

| Variable | PCA axis 1 | PCA axis $2^a$ |
|---|---|---|
| $M_1$ corpus width | 0.027 | **−0.030** |
| Symphyseal width | 0.081 | **−0.021** |
| Masseter lever arm length | 0.104 | 0.008 |
| Gonion–zygion distance | 0.106 | 0.007 |
| Basicranial length | 0.072 | 0.006 |
| Mandibular length | 0.101 | 0.006 |
| Condylar width | 0.088 | −0.005 |
| Bizygomatic breadth | 0.079 | 0.005 |
| Palatal length | 0.113 | −0.003 |
| Ramal height | 0.101 | 0.003 |
| Temporomandibular joint height | 0.092 | 0.001 |
| Bicondylar breadth | 0.066 | −0.000 |
| Percentage of variance | 94.16 | 1.55 |

[a] Boldfaced values represent the variables that loaded most strongly in the PCA.

with the youngest/smallest scattered to the left of axis 1 and the oldest/largest to the right (Fig. 6.6B). The second principal component accounts for 1.6% of the variance and produces a shift of *G. g. beringei* below the other two gorilla subspecies. Thus, the second principal axis, orthogonal to the first, summarizes variation related to shape. Results of ANCOVA reveal that the second principal component effects a significant ($p < 0.001$) separation of *G. g. beringei* from the other two taxa, but no significant ($p > 0.05$) separation between *G. g. gorilla* and *G. g. graueri*. Widths of the mandibular corpus and symphysis load most strongly in this analysis.

To assess more clearly the patterning of differentiation among the three subspecies, I performed a PCA on all pairwise subspecies comparisons (Tables 6.6–6.8). A plot of component scores for the log-transformed values for *G. g. gorilla* and *G. g. beringei* is presented in Fig. 6.6C. There is considerable overlap between subspecies, but *G. g. beringei* clusters below the first axis, whereas *G. g. gorilla* distributes across the entire multivariate morphospace. Width of the mandibular corpus and symphysis are primarily responsible for the group separation between *G. g. gorilla* and *G. g. beringei* (Table 6.6). The component scores for the PCA on the two eastern gorilla subspecies indicate that these same two variables differentiate *G. g. graueri* from *G. g. beringei* (Fig. 6.6D; Table 6.7). Additionally, condylar width, mandibular and palatal lengths, and

Table 6.6. *Principal component loadings for a PCA of log-transformed dimensions for ontogenetic series of* G. g. gorilla *and* G. g. beringei

| Variable | PCA axis 1 | PCA axis 2[a] |
|---|---|---|
| $M_1$ corpus width | 0.029 | **−0.031** |
| Symphyseal width | 0.083 | **−0.021** |
| Gonion–zygion distance | 0.107 | 0.009 |
| Masseter lever arm length | 0.104 | 0.008 |
| Basicranial length | 0.070 | 0.006 |
| Condylar width | 0.089 | −0.006 |
| Mandibular length | 0.098 | 0.005 |
| Bizygomatic breadth | 0.078 | 0.005 |
| Palatal length | 0.111 | −0.004 |
| Ramal height | 0.103 | 0.003 |
| Temporomandibular joint height | 0.096 | 0.001 |
| Bicondylar breadth | 0.066 | 0.000 |
| Percentage of variance | 94.17 | 1.69 |

[a]Boldfaced values represent the variables that loaded most strongly in the PCA.

Table 6.7. *Principal component loadings for a PCA of log-transformed dimensions for ontogenetic series of* G. g. graueri *and* G. g. beringei

| Variable | PCA axis 1 | PCA axis 2[a] |
|---|---|---|
| $M_1$ corpus width | 0.025 | **−0.021** |
| Symphyseal width | 0.071 | **−0.018** |
| Condylar width | 0.089 | **−0.018** |
| Mandibular length | 0.098 | **0.018** |
| Palatal length | 0.106 | **0.015** |
| Temporomandibular joint height | 0.086 | **−0.014** |
| Basicranial length | 0.074 | 0.008 |
| Bizygomatic breadth | 0.081 | 0.006 |
| Ramal height | 0.096 | −0.005 |
| Masseter lever arm length | 0.101 | 0.005 |
| Gonion–zygion distance | 0.098 | 0.004 |
| Bicondylar breadth | 0.069 | −0.000 |
| Percentage of variance | 94.22 | 2.11 |

[a]Boldfaced values represent the variables that loaded most strongly in the PCA.

Table 6.8. *Principal component loadings for a PCA of log-transformed dimensions for ontogenetic series of* G. g. gorilla *and* G. g. graueri

| Variable | PCA axis 1 | PCA axis 2[a] |
|---|---|---|
| $M_1$ corpus width | 0.023 | **−0.027** |
| Symphyseal width | 0.075 | **−0.019** |
| Temporomandibular joint height | 0.086 | 0.010 |
| Palatal length | 0.109 | −0.010 |
| Gonion–zygion distance | 0.103 | 0.007 |
| Ramal height | 0.097 | 0.007 |
| Masseter lever arm length | 0.101 | 0.006 |
| Basicranial length | 0.071 | 0.004 |
| Bizygomatic breadth | 0.078 | 0.002 |
| Mandibular length | 0.100 | −0.002 |
| Bicondylar breadth | 0.064 | −0.001 |
| Condylar width | 0.082 | 0.001 |
| Percentage of variance | 94.13 | 1.57 |

[a] Boldfaced values represent the variables that loaded most strongly in the PCA.

temporomandibular joint height load strongly, suggesting these variables may also be important in effecting the separation between these two subspecies. The combination of positive and negative loadings on the second component indicates that whereas *G. g. beringei* has a relatively wider mandibular corpus, symphysis, and condyle, and a relatively higher temporomandibular joint, *G. g. graueri* has a relatively longer mandible and palate. In the final PCA (Table 6.8), the first component accounts for 94% of the variance and the second component for 1.6% of the variance. Interestingly, the same two variables, width of the mandibular corpus at $M_1$ and symphyseal width, exhibit the strongest loadings, but there is extensive overlap between *G. g. gorilla* and *G. g. graueri* with no appreciable multivariate separation (Fig. 6.6E). There is notably more overlap in comparisons involving *G. g. gorilla* than between *G. g. beringei* and *G. g. graueri*, possibly due to relatively smaller samples of both eastern subspecies.

**Discussion**

Ontogenetic allometric analysis of the masticatory complex in gorillas reveals a mix of proportion differences that may be explained by both ontogenetic scaling and ontogenetic dissociations (Table 6.4). *Gorilla g. beringei* does depart from the other two gorilla subspecies in ways that fit some of the predictions of intensification of folivory, confirming to some extent Groves's (1970a) earlier observations of a distinctive jaw morphology in *G. g. beringei*. Features such as width of the mandibular corpus and symphysis not only distinguish *G. g. beringei* from the other gorilla subspecies, but separate *Gorilla* from *Pan* as well, the latter dedicated frugivores (Taylor and Ravosa, 1999; Taylor, 2002; cf. Daegling, 1989). The most notable separations occur between *G. g. beringei* and *Pan*; differences between *G. g. gorilla* and *Pan*, and among gorilla subspecies, are less marked. Furthermore, if *P. paniscus* comminutes a more resistant diet as compared to *P. troglodytes*, this dietary differences is not evident in their masticatory morphology (Taylor, 2002), though it must be stated that the evidence in favor of a tougher diet in *P. paniscus* is far from clear (Badrian and Malenky, 1984; Kano and Mulavwa, 1984; Wrangham, 1986; Malenky and Stiles, 1991; Chapman *et al.*, 1994; Malenky and Wrangham, 1994; Malenky *et al.*, 1994; cf. Wrangham *et al.*, 1996). Because the mandibular features delineated above systematically differentiate among the African apes along a dietary axis of increasing folivory, they provide the strongest link between dietary resistance and masticatory morphology in these taxa. This interpretation accords with similar findings of increasing mandibular robusticity in larger extant and fossil apes (Ravosa, 2000). Nevertheless, the pattern of differentiation as a whole as it relates to diet is not as strong as might be expected based on hypothesized differences in dietary consistency among taxa and theoretical predictions of masticatory stress and load resistance. The functional, evolutionary, and taxonomic implications of these results are addressed below.

### *Do gorilla subspecies differ in craniomandibular size?*

Results of this study demonstrate an unambiguous pattern of craniomandibular size differentiation among gorilla subspecies, particularly in the facial region (Fig. 6.6A), substantiating Groves's (1970*b*) earlier observation of a gradient in skull size among adults from west to east. As shown in the bivariate plot of the first component scores from a multigroup PCA by dental stage (Fig. 6.6A), size differentiation occurs by about dental stage 6, which comprises dentally mature adults with unfused basilar sutures; in other words, individuals for whom cranial growth is near completion, but has not yet ceased. At dental stage 7, *G. g. beringei* is significantly larger in overall size as compared to either *G. g. graueri* or *G. g. gorilla*. Much of the size difference can be attributed to measures that reflect facial size, such as mandibular, occlusal, and palatal lengths. These results concur with those of Stumpf *et al.* (this volume), who find that a similar set of facial dimensions effects a separation of eastern from western gorillas along the first axis in a canonical-variates analysis of cranial dimensions. Thus, regardless of whether any single dimension of the skull differs significantly between subspecies (cf. Uchida, 1996), *G. g. beringei* clearly exhibits the largest adult face. The issue of whether size differentiation occurs prior to adulthood is likely to remain unresolved given that additional data at earlier stages of growth for *G. g. beringei* are unavailable.

Size differences among adults have functional implications, as larger jaws are capable of producing absolutely higher bite force per unit muscle force, and resisting absolutely greater masticatory loads, simply as a function of size increase. Thus, *G. g. beringei* may be characterized by differences in functional capabilities independent of any allometric shifts during growth or nonallometric proportion differences. If this is the case, however, then similar advantages must accrue to larger males as compared to females, though sex differences are not likely the results of targeted selection for improved masticatory efficiency, but rather, the secondary effects of selection for differential size increase (Spencer and Hogard, 2001).

Heterochrony may account for the differences in adult skull size among gorilla subspecies. Indeed, Taylor (2002) has suggested that heterochrony likely accounts for differences in mandibular size and shape between *P. troglodytes* and *P. paniscus*, the latter exhibiting a strongly paedomorphic mandible relative to various estimates of body size. Evaluating heterochrony in gorillas is difficult because one of the key parameters of a heterochronic analysis, body size (e.g., weight), is not well established in any of the gorilla subspecies. Body weight data for all three taxa are scant but the most recent estimates suggest that among adult males, *G. g. beringei* (162.5 kg) are the smallest and *G. g. gorilla* (170.4 kg) and *G. g. graueri* (175.2 kg) differ by only 5 kg (Smith and Jungers, 1997). By contrast, among adult females *G. g. beringei* (97.5 kg) are the largest while *G. g. gorilla* (71.5 kg) and *G. g. graueri* (71.0 kg) do not appear to differ. The greatest difference in weight for males amounts to only 7%. While the 27% difference in adult weights between females is more marked, this figure is based on only a single *G. g. beringei* female. Body size estimates derived from skeletal weights indicate that adult males of *G. g. beringei* are larger than *G. g. gorilla* (Taylor, 1997*a*), which differs from results based on body mass estimates, but there are no significant differences between subspecies females and no skeletal weight data for *G. g. graueri*.

Setting body weight aside, results of this study clearly demonstrate that *G. g. beringei* is significantly larger in facial size as compared to the other two gorilla subspecies. Furthermore, *G. g. gorilla* and *G. g. beringei* are ontogenetically scaled for roughly half of the bivariate relationships evaluated in this study. While a heterochronic analysis is beyond the scope of this paper, and is being explored more fully elsewhere (A.B. Taylor, unpublished data), it is possible to hypothesize about the heterochronic processes that may have produced differentiation in skull size among gorilla subspecies. On morphological and behavioral grounds, mountain gorillas have been argued as representing the derived condition relative to *G. g. gorilla*, though the phylogenetic affinities of *G. g. graueri* remain controversial. Assuming *G. g. beringei* is derived relative to *G. g. gorilla*, selection may have favored larger facial size in the eastern mountain gorillas, possibly related to diet, resulting in facial peramorphosis. While only an explicit heterochronic analysis can address the underlying processes responsible for differences in facial and/or body size between *G. g. gorilla* and *G. g. beringei*, such analyses will necessarily be limited without firm data on both body size and ancestral polarity. Heterochrony cannot, of course, explain the pattern of variation observed between *G. g. graueri* and *G. g. beringei*, as differences in facial or skull size have not been established for these taxa and they exhibit numerous departures from common ontogenetic allometries. And while *G. g. gorilla* and *G. g. graueri* are ontogenetically scaled for most bivariate relationships, they do not differ significantly in craniomandibular size.

It is also possible that differences in jaw size among subspecies reflect ecophenotypic responses to a poorer quality diet and/or seasonal variation. The onset of size differentiation late in ontogeny is certainly suggestive of a local remodeling process, consistent with the near-completion of growth, coupled with repetitive cyclical loading of the jaws. The morphological correlates of dietary consistency have been well established in primates and other mammals (Beecher and Corruccini, 1981; Corruccini and Beecher, 1982, 1984; Beecher *et al.*, 1983). In particular, experimental studies of the effects of dietary consistency on the mandible have demonstrated that groups fed harder diets manifest external morphological changes along with high levels of secondary Haversion remodeling (Bouvier and Hylander, 1981). These data clearly support the possibility that differentially high dynamic stresses incurred during the lifetime of an individual can result in morphological responses to resist relatively greater masticatory loads and fatigue failure. As local populations of *G. g. graueri* parallel the same pattern of altitudinal-dependent variation in dietary composition that characterizes the entire genus (Yamagiwa *et al.*, 1996), a comparative study of masticatory morphology within *G. g. graueri* could provide an indirect test of ecophenotypic plasticity.

### Ontogeny and allometry of jaw form in gorillas

All subspecies exhibit a mix of allometry and isometry of craniomandibular proportions during growth. However, few scaling trends reflect strong positive allometry relative to mandibular length. By contrast, as shown here and elsewhere (Taylor, 2002), strong positive allometry characterizes scaling of linear dimensions against basicranial length. These different scaling patterns clearly reflect differential rates of growth of the skull and mandible. Thus, conclusions drawn from allometric analyses will always be relative to the chosen independent variable, and scaling relationships based on one measure of size should not be assumed to characterize other bivariate relationships.

Slopes that deviate from isometry suggest that the ability to resist masticatory forces alters during growth. Relative to mandibular length, for example, depth of the mandibular corpus scales with moderate positive allometry, suggesting that the ability to resist parasagittal bending loads along the mandibular corpus improves during growth, similar to what Cole (1992) found for *Cebus*. Apart from corpus depth, the overall trend is for scaling trajectories to exhibit moderate to strong negative allometry or to hover around isometry (Table 6.3). The negative allometry of symphyseal width versus mandibular length suggests that the ability of gorillas to resist dorsoventral shear and wishboning loads at the mandibular symphysis decreases during growth. Width of the mandibular corpus also scales with negative allometry relative to mandibular length, indicating that resistance to torsional loads about the long axis of the mandibular corpus similarly decreases during ontogeny. Comparable results have been demonstrated for *Cebus* (Cole, 1992) and *Pan* (Taylor, 2002). Cole (1992) suggested that the developing permanent molars likely account for the relatively thicker corpus in younger as compared to older individuals. Taylor (2002) demonstrated that in the African apes, the slopes for mandibular corpus width are highest in the deciduous corpus and decrease considerably following the eruption of $M_1$. This results in a relatively less thick mandibular corpus in older individuals, which Taylor also attributed to the developing dentition.

Height of the temporomandibular joint above the occlusal plane scales with strong negative allometry in all taxa, and condylar width scales with strong negative allometry in *G. g. gorilla* and *G. g. graueri*; the latter scales isometrically in *G. g. beringei*. Thus, efficiency of load distribution along the postcanine tooth row and mandibular condyle generally decreases as gorillas age. In *G. g. beringei*, isometry of condylar width relative to mandibular length suggests that load distribution along the condyle does not alter during growth.

Isometry reflects preservation of shape with size increase during ontogeny. Thus, isometry of symphyseal depth versus mandibular length, for example, may indicate that the ability of *G. g. gorilla* and *G. g. beringei* to resist vertical bending forces remains constant during growth; the slope is only slightly negatively allometric in *G. g. graueri* and it is reasonable to question whether the value of this slope represents a meaningful difference in ability to resist vertical bending. Isometry of masseter muscle lever arm length in all gorilla subspecies suggests that levels of masseter muscle force production are also maintained during growth.

By contrast, masseter lever arm length does not scale isometrically in chimpanzees. In chimpanzees, presumably characterized by a less mechanically resistant diet than even the most frugivorous of gorillas, masseter lever arm length departs significantly (df $= 1,281$; $F = 34.19$; $p = 0.000$) from isometry during growth and scales with positive allometry relative to mandibular length ($y = 1.10$, $r = 0.97$). This is a particularly interesting result. First, evidence indicates that juvenile primates may not be as successful as adults in their foraging efforts, and that juveniles bypass tougher foods routinely exploited by adult conspecifics (Terborgh, 1983; van Schaik and van Noordwijk, 1986). From these data it can be inferred that dietary toughness increases during growth. Second, adult gorillas, being significantly larger than juveniles, require more food intake per day, and so can be expected to engage in absolutely more chewing cycles than juveniles; although juveniles spend a greater percentage of their time foraging than would be expected based on their nutritional demands (Janson and van Schaik, 1993), this is because they appear to be relatively less successful in their foraging efforts and not because they ingest more food. While the precise effects of scale on masticatory muscle mass and muscle force during ontogeny are unknown, it has been argued that maximum masticatory muscle force scales with negative allometry relative to body mass interspecifically across primates, suggesting that larger taxa may be allometrically constrained to produce relatively less masticatory muscle force (Hylander, 1985). Thus, if masseter muscle lever arm length serves as a valid proxy for masseter muscle force (cf. Antón, 1999), these results suggest that force production capabilities are maintained during growth despite the potential for a more mechanically resistant diet with ontogenetic size increase.

Far from being "trivial" (Godfrey *et al.*, 1998), isometry may be interpreted as revealing the maintenance of functional equivalence throughout growth and/or interspecifically across taxa. Gould (1966, 1971) has argued that in cases of strong positive allometry, ontogenetic scaling or upward transpositions tend to be disadvantageous. Downward transpositions of positive allometries in wing lengths in birds and tooth proportions in hyaenas and cats (Kurtén, 1954 as cited in Gould, 1966), are examples of downward shifts that have been argued as necessary to preclude maladaptive proportions associated with size increase. Others have argued that the absence of a change in shape, where a change would be predicted based on phenotypic and/or genetic correlations, should be taken as evidence that selection has acted to preserve shape; in other words, that no change in shape is an adaptation (Price *et al.*, 1984). It is of interest, therefore, that relative to mandibular length, few ontogenetic trajectories exhibit positive allometry and none can be said to exhibit strong positive allometry (Table 6.3). Moreover, isometry during growth characterizes *G. g. beringei* with greater frequency than either of the other two subspecies (Table 6.3). In light of the increase in facial size shown to characterize *G. g. beringei*, there is at least some evidence to suggest that ontogenetic scaling of positive allometries – change in shape with increase in size – has been minimized in the mountain gorilla. Stronger negative allometry characterizes some of the ontogenetic trajectories, including symphyseal and condylar widths, temporomandibular joint height, and bizygomatic and bicondylar breadths; $M_1$ corpus width is also strongly negatively allometric but low correlation coefficients make this relationship more suspect (Table 6.3).

### Adaptive differences in masticatory form

There are some ontogenetic dissociations of masticatory proportions among gorilla subspecies that fit the predictions of intensification of folivory. Some of the structural differences suggest the jaws of *G. g. beringei* may be better able to resist fatigue failure associated with repetitive loading. For example, both regression analysis and PCA indicate that *G. g. beringei* has a relatively wider mandibular corpus, indicating that mountain gorillas are able to resist elevated torsional loads about the long axis of the mandibular corpus. The suggestion that axial loads increase with pronounced folivory concurs to some extent with Ravosa's (2000) findings that both extant and fossil apes tend to exhibit relatively wider mandibular corpora as compared to cercopithecines. Elevated torsional loads would be expected if *G. g. beringei* recruits greater amounts of the horizontal component of the masticatory muscle, and produces relatively more bite force along the postcanine corpus (Hylander, 1985; Antón, 1996a). *Gorilla g. beringei* does have a relatively higher mandibular ramus, which may reflect a larger attachment site for the masseter muscle and, by inference, a relatively larger muscle mass. However, based on the morphological results, the horizontal component of the masseter muscle does not appear to be a source of relatively larger torsional loads in *G. g. beringei*, as there is little evidence of a relatively wider skull or greater zygomatic flaring that would be indicative of a more horizontally directed masseter muscle (Tables 6.3 and 6.4). There is no systematic pattern of differentiation between gorilla subspecies for measures of skull breadth that serve as proxies of zygomatic flaring, such as bizygomatic and bicondylar breadths, and gorillas are ontogenetically scaled for bizygomatic versus bicondylar breadths, indicating no differences in relative orientation of the masseter muscle from its origin along the zygomatic arch to its insertion in the ascending ramus of the mandible and coronoid process (Raven, 1950). Recruitment of more of the existing horizontal component, regardless of degree of orientation, along with other factors not accounted for in this study, such as the medial pterygoid moment arm, may well contribute to increases in axial torsion in *G. g. beringei*. In this regard, however, it must be noted that chimpanzees exhibit significantly thicker mandibular corpora as compared to bonobos (Taylor, 2002), though there are no dietary data to indicate that chimpanzees masticate a more resistant diet. In light of these mixed morphological results, the relatively wider mandibular corpus in *G. g. beringei* is best interpreted cautiously as an adaptation to resist elevated torsional loads.

Interestingly, Uchida (1996) observed that *G. g. beringei* adults are characterized by a uniformly deep corpus, whereas the mandibular corpus becomes distally more shallow in both *G. g. gorilla* and *G. g. graueri*. This amounts to a smaller robusticity index in *G. g. beringei* (breadth versus height of the mandibular corpus) (Daegling, 1989), though the biomechanical significance of such corpus proportions is suspect without an appropriate lever arm (Bouvier, 1986*a*; Daegling, 1989; Daegling and Hylander, 1998, 2000). In this study, allometrically controlled shape comparisons reveal that relative to mandibular length, gorillas are ontogenetically scaled for depth of the mandibular corpus, signifying that any shape differences between adults can be attributed to the shared, correlated effects of size change. Sexual dimorphism in canine size has been shown to impact the robusticity index, at least in some apes (Daegling and Grine, 1991; Taylor, 2000; Daegling, 2001). The fact that sex differences may account for variation in robusticity indices among gorillas does not contradict the theoretical implications of increased robusticity as a means of improving resistance to bending loads. It does, however, suggest that at least in some circumstances, the presence of a deeper corpus need not be explained as the direct result of selection for improved resistance (Gould and Lewontin, 1979; Gould and Vrba, 1982).

Both regression analysis and PCA also substantiate the importance of a relatively thicker mandibular symphysis in *G. g. beringei*. A relative increase in symphyseal width would effectively counter wishboning loads at the symphysis (Hylander, 1985). Experimental evidence from macaques has shown that wishboning occurs from the working-side deep masseter muscle during the early phase of the power stroke, coupled with the balancing-side deep masseter muscle which peaks during the late phase of the power stroke of mastication (Hylander and Johnson, 1994), and that wishboning is associated with lateral transverse bending at the corpus. There is morphological evidence to suggest hominoids experience wishboning at the symphysis, though wishboning may not be as important a component of the hominoid masticatory regime as it appears to be in other primates (Daegling, 2001). Wishboning is also influenced by the architecture of the symphysis. During wishboning, the symphysis functions mechanically as a curved beam; the more curved the symphysis, the greater the wishboning stresses along the lingual aspect of the symphysis. Symphyseal curvature has also been shown to increase with both ontogenetic and interspecific increase in size in papionins (Vinyard and Ravosa, 1998).

Assessment of symphyseal curvature using measures of palatal length and breadth as well as mandibular length and bicondylar breadth provide no evidence to indicate that *G. g. beringei* has a relatively more curved symphysis than either *G. g. gorilla* or *G. g. graueri* (Taylor, 1998*a*, 1999, 2002). These results concur with Ravosa's (2000) finding that in general, larger-bodied apes exhibit relatively less symphyseal curvature than smaller-bodied apes and cerco-pithecines. However, in comparisons between *G. g. gorilla* and *P. t. troglodytes*, adult male gorillas exhibited a significant ($p < 0.01$) increase in symphyseal curvature, as measured by mandibular arch width (at $M_1$) and mandibular length (Taylor, 2002). *Gorilla g. beringei* was not included in this last analysis because wear and disease did not permit reliable measurements of mandibular arch width, so the possibility that *G. g. beringei* has a relatively more curved symphysis than *Pan* or gorilla subspecies, as measured by mandibular arch width and mandibular length, cannot be ruled out. Daegling (2001) finds that orientation of the symphysis, rather than alterations in biomechanical shape, provides evidence that both small- and large-bodied hominoids experience lateral transverse bending, though these results do not preclude the possibility that resistance to relatively greater wishboning loads may be accomplished by alterations in shape. In sum, the combination of a relatively thicker mandibular symphysis and corpus in *G. g. beringei* provides some support for a biomechanical model linking improved resistance to lateral transverse bending at the corpus and symphysis with pronounced folivory in gorillas (Hylander, 1985).

Other features, such as the relatively higher mandibular ramus and temporomandibular joint, suggest that *G. g. beringei* is able to generate relatively higher muscle and bite forces than other gorillas. The advantage of a relatively higher mandibular ramus is the ability to recruit relatively greater amounts of muscle force afforded by a potentially larger cross-sectional area of the adductor musculature (Freeman, 1988). The absence of differences in corpus depth between subspecies, however, means that higher bite forces would of necessity be dissipated through a relatively thicker mandibular corpus. The advantages of a temporomandibular joint elevated relatively higher above the occlusal surface of the mandible include improved mechanical leverage associated with relative elongation of the masseter muscle lever arm (DuBrul, 1977), a more uniform distribution of bite forces along the postcanine tooth row (Ward and Molnar, 1980), and a more efficient cutting and shearing mechanism (Rensberger, 1973). This morphological complex comprises an adaptive pattern that has been linked to the generation and dissipation of large masticatory forces associated with hard-object feeding in macaques (Antón, 1996*a*) and extreme herbivory in the robust australopithecines and giant panda (DuBrul, 1977).

### Discrepancies between theoretical predictions and morphology

Apart from the allometric dissociations detailed above, gorillas deviate from the predicted pattern of morphological differentiation in numerous instances. First, despite differences in degree of folivory, *G. g. gorilla* and *G. g. graueri* share similar masticatory morphologies. *Gorilla g. graueri* exhibits none of the predicted morphologies based upon the expectation of a more herbaceous diet as compared to *G. g. gorilla*. *Gorilla g. graueri* does not converge on any aspect of the morphological pattern that characterizes *G. g. beringei*, and perhaps more importantly, *G. g. gorilla* and *G. g. graueri* cannot be distinguished based on multivariate patterns of craniomandibular variation.

In addition, *G. g. beringei* does not exhibit all of the predicted morphologies. For example, *G. g. beringei* was predicted to exhibit a relatively shorter face. Reduced prognathism was hypothesized to confer a mechanical advantage by positioning the masticatory musculature closer to the molar bite points (i.e., improving the load-to-lever arm ratio) and reducing the bending moments in the face (DuBrul, 1977; Hylander, 1977, 1979a; Ravosa, 1990; Spencer and Demes, 1993; Antón, 1996a). As bite force is inversely proportional to jaw length (Hylander, 1985; Spencer and Demes, 1993), relative shortening of the jaw should provide for an increase in the amount of muscle force that may be converted into usable bite force (Hylander, 1979a; Spencer, 1998).

It is already well established that *G. g. beringei* has an absolutely longer mandible and palate as compared to *G. g. gorilla* (Coolidge, 1929; Groves, 1970a; Uchida, 1996), though there is some question as to whether palatal length is significantly greater in *G. g. beringei* as compared to *G. g. graueri* (cf. Groves, 1970a; Uchida, 1996). Comparison of ontogenetic allometries reveals that the two eastern subspecies exhibit longer than expected palates when scaled against basicranial length as compared to *G. g. gorilla* (Fig. 6.7); static adult interspecific allometries show a similar pattern when palatal length is scaled against midfacial breadth (Uchida, 1996). Mandibular length is also longer than expected when scaled against basicranial length in *G. g. graueri* (Tables 6.3 and 6.4); *G. g. beringei* has slightly longer jaws than predicted, but does not differ significantly ($p = 0.071$) from *G. g. gorilla*.

Increased prognathism in *G. g. graueri* and *G. g. beringei*, as reflected in a longer mandible and/or palate, would be expected to result in relatively higher bending moments along the face and might decrease the amount of muscle force that could be converted into usable bite force. It is worth emphasizing that Hylander (1979a) suggests folivores are probably not subjected to excessively high bending moments. However, *G. g. beringei* is the only subspecies to comminute both bark and bamboo, which are woody plants and arguably sufficiently tough to require relatively higher force production capabilities beyond what is required for the mastication of leaves, yet *G. g. beringei* is relatively more prognathic than *G. g. gorilla* based on palatal and possibly mandibular length. Deposition of bone posteriorly characterizes palatal morphogenesis in some nonhuman primates (McCollum, 1997), suggesting that a relatively elongated palate may not necessarily result in increased facial prognathism.

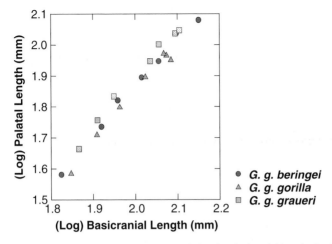

Fig. 6.7. Bivariate plot of palatal length scaled against basicranial length. *Gorilla g. beringei* and *G. g. graueri* have significantly longer palates as compared to *G. g. gorilla*, indicating increased prognathism in the two eastern subspecies. These results confirm Groves's (1970*a*) earlier observations based on adults. There are no differences in relative length of the palate between the two eastern subspecies.

Nevertheless, assuming palatal length reflects length of the anterior face, and that diet is an important predictor of facial length, relative shortening of the face is expected to correlate with increase in dietary resistance. Conversely, under the same assumptions, if bending moments do not differ among subspecies as a consequence of diet (i.e., there are no differences in diet that result in relatively elevated facial bending moments), then all three subspecies should exhibit the same degree of facial prognathism. In gorillas, neither pattern holds.

Extensive overlap among gorillas and *Pan* within the 95% confidence limits for mandibular length relative to basicranial length (Fig. 6.8) suggests the African apes are ontogenetically scaled for this measure of facial prognathism as well (similar results obtain for palatal versus basicranial length; not shown). In other words, mandibular and palatal lengths in gorillas are comparable to those of chimpanzees and bonobos scaled to the same skull size despite even greater differences in degree of folivory. Thus, results indicate that facial shortening, at least as reflected in mandibular and palatal lengths, cannot always be interpreted as an adaptation to dietary resistance. Aside from reducing facial length, there are alternative means of producing greater bite force, including increasing the absolute cross-sectional area of the jaw musculature and modifying muscle recruitment patterns or muscle architecture. The relative enlargement of the postcanine dentition in the eastern lowland gorillas may explain the relative

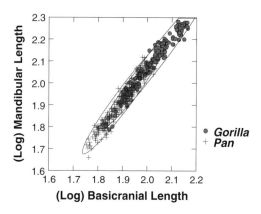

Fig. 6.8. Bivariate plot of mandibular length versus basicranial length for *Gorilla* and *Pan*. The 95% confidence ellipse for *Pan* fits well within the 95% confidence ellipse for *Gorilla*, indicating that the African apes are ontogenetically scaled for this bivariate comparison. Thus, mandibular length in gorillas is comparable to *Pan* extrapolated to larger skull size.

increase in palatal and mandibular lengths in these taxa (Uchida, 1996). Increased prognathism in *G. g. graueri* may also be the correlated effect of increase in skull size.

*Gorilla g. beringei*, as previously mentioned, also does not exhibit relatively broader, more flaring zygomatic arches, one of the hallmark features of primates and other mammals known to be (or presumed to have been, as in the case of robust australopithecines: Hylander, 1988), characterized by resistant diets. DuBrul (1977) has argued, for example, that the lateral expansion of the zygomatic arches as seen in both the dietarily specialized *Ailuropoda* (giant panda) and *Paranthropus* reflects a pattern of structural convergence that facilitates transverse motion of the coronoid process during mastication, and provides for an increase in both mass and the horizontal component of the adductor musculature. Similar arguments have been advanced for macaques (Antón, 1996a). Rak (1983) additionally suggested that the maxillary zygomatic process functions to buttress the palate as the medial pterygoid muscle contracts during unilateral mastication. Similarities in zygomatic structure, already noted, raise questions about the extent to which gorillas differ in transverse motion and horizontal muscle forces associated with the masseter muscle, and may be an indication that large vertical shearing forces (Hylander, 1979a) are equally important in all gorillas despite varying degrees of folivory (and differences in dental morphology). Even if *G. g. beringei* recruits greater amounts of the horizontal component of the masseter muscle, the African apes demonstrate that a flared zygomatic is not consistently associated with increase in degree of folivory and so cannot be argued as requisite for maintaining a more resistant diet (Taylor, 2002).

The most notable deviation from the expected pattern of morphological differentiation is the similarity in mandibular corpus depth once gorillas have been scaled to common mandibular lengths (Fig. 6.5A; Tables 6.3 and 6.4). These results suggest gorillas have similar resistance capabilities to parasagittal bending loads despite the extreme folivory/herbivory that characterizes *G. g. beringei*. Gorillas thus contrast with previous findings of relatively deeper corpora in leaf-eating primates, including folivorous colobines, platyrrhines, and some prosimians (Hylander, 1979*a*; Bouvier, 1986*b*; Ravosa, 1991). Surprisingly, the African apes as a group all appear to be equally effective at resisting parasagittal bending loads (Daegling, 1990; Taylor, 2002). Thus, these apes do not conform to theoretical predictions linking a relatively deep mandibular corpus with repetitive cycling of the jaws associated with intensification of folivory. In light of these findings, it is possible that the African apes maintain a critical threshold for corpus strength that permits the mastication of leaves – a biological function that requires relatively small forces regardless of the frequency of this behavior – resulting in an average biomechanical situation (Oxnard, 1972, 1979), whereas the significant difference in jaw form and function among the African apes may lie in the production of large forces associated with the comminution of woody plants or in kinematic behaviors.

One additional point made earlier that bears re-emphasizing is that masticatory stresses and loading regimes drawn from experimental data on prosimian and anthropoid primates are assumed to hold for the African apes, but the validity of this assumption can only be evaluated with *in vivo* experimental data on these apes, which seems logistically impractical. It is possible that even subtle differences in the ways in which the mandible is loaded during unilateral mastication and incision could influence theoretical and mechanical expectations of resistance capabilities to internal forces, and these could account for some of the contradictory results. Data on the mechanical properties of foods masticated by gorilla subspecies would also go a long way towards addressing the hypothesis that foods consumed by some gorilla subspecies are actually tougher or more resistant than foods consumed by other gorilla taxa.

### *Summary of morphological findings*

Several conclusions may be drawn from these results. First, gorillas differ significantly in facial size, with *G. g. beringei* exhibiting a larger face as compared to both *G. g. gorilla* and *G. g. graueri*. It seems plausible that ontogenetic scaling of positive allometries has been minimized in *G. g. beringei*, possibly as a means of maintaining functional equivalence without entering size ranges that may be maladaptive.

Gorillas are also characterized by some allometric dissociations of ontogenetic trajectories. Compared to both *G. g. gorilla* and *G. g. graueri*, *G. g. beringei* exhibits a relatively wider mandibular corpus and symphysis, elevated temporomandibular joint, higher mandibular ramus and greater facial height. In these instances, *G. g. beringei* does diverge from the presumed ancestral condition in ways that suggest improved resistance to elevated masticatory loads associated with increasing folivory. Some of these differences, such as the relatively thicker mandibular corpus and symphysis, distinguish *G. g. beringei* from the other two subspecies, and differentiate more generally between gorillas and *Pan* (Taylor, 2002). These shape transformations require additional explanation beyond what may be accounted for by the effects of size increase and correlated shape change (Shea, 1995), and thus may be related to degree of dietary resistance.

Not all predictions are borne out by the data, however, and some results run contrary to expectations based upon degree of folivory. *Gorilla g. graueri*, for example, does not fit any of the morphological predictions based on diet; for the most part, the two lowland gorilla subspecies are indistinguishable based on craniomandibular measurements incorporated in this study. And whereas gorillas do not differ in some of the predicted morphologies deemed particularly crucial for maintaining a highly resistant diet, such as a relatively deeper mandibular corpus, the finding of a relatively longer palate in both *G. g. graueri* and *G. g. beringei* runs counter to predictions of reduced facial prognathism with increasing dietary resistance. As gorillas do not always diverge in the expected direction based upon known dietary variation, some of the predicted morphologies cannot be logically interpreted as adaptations to diet.

Finally, while there is a subtle but consistent trend for *G. g. beringei* to depart from the other subspecies in ways predicted by a diet of pronounced folivory, there are varying degrees of overlap among subspecies and few distinct separations among taxa. As judged from a combination of the 95% confidence intervals and bivariate and multivariate plots, statistically significant findings are often associated with relatively modest degrees of separation among gorilla subspecies. Thus, although *G. g. beringei* differs from the other two subspecies in some of the predicted morphologies, differences amongst gorillas are not as great as might be anticipated based upon dietary variation, and when compared to patterns of morphological differentiation seen in other closely related primates that vary in degree of folivory or dietary resistance (e.g., Ravosa, 1990; Cole, 1992; Daegling, 1992; Antón, 1996*a*). It is worth noting in this regard that while significant findings do not always or consistently differentiate subspecies along a dietary axis of intensification of folivory, relatively small ontogenetic and adult samples of the most dietarily specialized subspecies make it difficult to assess just how different *G. g. beringei* is from the other two taxa. The best-studied and most behaviorally specialized gorilla subspecies is also the most poorly represented morphologically, and anyone engaged in comparative morphology involving *G. g. beringei* is faced with this unfortunate paradox. Nevertheless, while diet may be one factor responsible for the differences in craniomandibular morphology shown here, diet alone does not appear sufficient to explain the patterns of morphological differentiation among gorilla subspecies. Neither can morphological variation between *Gorilla* and *Pan* be attributed principally to dietary specialization (Taylor, 2002). Other factors, such as developmental mechanisms, historical constraint, and drift, may have influenced the pattern of structural variation shown here (Smith, 1983).

### Evolutionary implications

One of the interesting results to emerge from this analysis is the multivariate patterning of shape variation among gorilla subspecies. Based on the PCAs, width of the mandibular corpus and symphysis are important in consistently separating *G. g. beringei* from the other gorilla subspecies, but differences in condylar width, mandibular length, palatal length, and temporomandibular joint height are additionally important in distinguishing *G. g. beringei* from *G. g. graueri*. In other words, the multivariate patterns of divergence between *G. g. beringei* and the two lowland subspecies are not quite identical. As already noted, adequate samples with which to interpret differences between *G. g. beringei* and other taxa are always problematic. Nevertheless, given an a priori expectation that morphological differences should be greatest between *G. g. gorilla* and *G.g. beringei*, and the incorporation of relatively large samples of both *G. g. gorilla* and *G. g. graueri*, attributing the observed patterns of morphological differentiation to sampling issues seems oversimplified.

*Gorilla g. graueri*, more frugivorous than *G. g. beringei* but less so than *G. g. gorilla* (Yamagiwa *et al.*, 1996), does appear to be "intermediate" in jaw form based upon comparisons of absolute jaw dimensions among adults (Taylor, 1998*b*, 1999). However, in allometrically controlled comparisons, *G. g. graueri* converges on the morphological pattern of its western lowland congener; these two subspecies are largely indistinguishable based upon the craniomandibular measurements used here (e.g., Figs. 6.4A–B). Bivariate analyses show that *G. g. gorilla* and *G. g. graueri* depart from a common allometric trend in only a few instances, and not in ways consistent with theoretical predictions of repetitive cycling of the jaws with increasing folivory. Moreover, based upon PCA, there is no discernable differentiation between these two lowland subspecies (Fig. 6.6E). In the absence of a genuine shared masticatory pattern between

*G. g. graueri* and *G. g. beringei*, variation in any single feature may plausibly be attributed to the random effects of genetic drift. In contrast to absolute jaw dimensions, *G. g. graueri* is significantly larger than *G. g. beringei* in some dental proportions (Uchida, 1998), while converging on the masticatory pattern of *G. g. gorilla* in allometrically controlled comparisons, and so cannot be viewed as "intermediate" in this case (cf. Albrecht and Miller, 1993).

It is of interest that the shared masticatory pattern shown in this investigation between *G. g. gorilla* and *G. g. graueri* contrasts with the tendency for *G. g. graueri* to converge on the dental morphology of *G. g. beringei* (Uchida, 1998). In other words, dental proportions are either larger in *G. g. graueri* as compared to *G. g. beringei*, or the eastern subspecies converge to the exclusion of *G. g. gorilla* (Uchida, 1998). The absence of differences in jaw proportions in allometrically controlled comparisons between the two lowland congeners, coupled with fairly marked differences in the dentition, together provide an interesting morphological mix. On the one hand, these data indicate a lack of congruence between the functional models of masticatory form and diet. Alternatively, they may signify a progressive series of structural modifications of the teeth and jaws, possibly reflecting both microevolutionary and ecophenotypic processes, to an ever-increasing specialized diet from *G. g. gorilla* to *G. g. beringei*.

Groves and Stott (1979) earlier proposed that the morphology of *G. g. graueri* was influenced both by a lengthy period of uninterrupted gene flow between lowland and highland populations and selection for adaptations to high altitude, while *G. g. beringei* remained reproductively isolated and subject to strong selection for adaptations to high altitude and dietary specialization. Leigh *et al.* (this volume) speculate even further that whereas the habitat of the Virunga gorillas has contracted, restricting these gorillas to a more specialized environment, populations of *G. g. graueri* may have expanded relatively recently and quickly into multi-level habitats. If these dispersal patterns are historically accurate, they may offer a partial explanation for the patterning of craniodental morphology in gorillas. Specifically, *G. g. graueri* would not have been under the same intensity of selection as *G. g. beringei*, and as in the case of chimpanzees (Shea and Coolidge, 1988), their expanded habitat may have resulted in a more generalized pattern of craniomandibular morphology. *Gorilla g. graueri* would also have had less time to respond to the selection pressures of more localized environments beyond any dental adaptations they may have evolved.

A more detailed analysis of craniomandibular morphology among local populations of *G. g. graueri* that range from low to high altitudes may reveal what, if any, features are consistently associated with intensification of folivory. For example, craniomandibular adaptations linked to specialized folivory would predict greater similarities between the Virunga mountain and Kahuzi gorillas, than between the former and gorillas of the Itebero region, based upon the relative proportions of leaves, pith, stems, and bark consumed by each (Yamagiwa *et al.*, 1996). It is difficult to make predictions as to where the Bwindi gorillas should be expected to fall morphologically. While their geographic proximity to the Virunga gorillas might suggest a pattern of dietary and thus morphological convergence with *G. g. beringei*, Bwindi gorillas have a rather extensive altitudinal range extending between 1400 m and 2300 m. In this range, Bwindi gorillas do not quite overlap with either the extreme high end-range for western lowland gorillas or with the lower end-range for *G. g. beringei* (Watts, 1998; Goldsmith, this volume; Remis, this volume). Preliminary craniodental data are suggestive of a Bwindi gorilla morphology that reflects a more frugivorous diet as compared to that of *G. g. beringei* (Sarmiento *et al.*, 1996), but there has as yet been no comprehensive investigation, allometric or otherwise, of masticatory morphology in these gorillas. Finally, important data on the mechanical properties of foods masticated by these taxa (Lucas *et al.*, 2000), and not simply on the distribution and overlap of food species consumed, are crucial for establishing robust biomechanical links between diet and morphology.

### Gorilla taxonomy

The patterns of variation in jaw morphology shown here are reasonably consistent with findings from previous studies of gorilla morphology. Varying degrees of differentiation among gorilla subspecies have been observed in comparisons of dental (Uchida, 1996, 1998), skull (Groves, 1970*b*; Casimir, 1975; Albrecht *et al.*, this volume) and postcranial (Schultz, 1930, 1934; Groves, 1979; Inouye, 1992, this volume; Taylor, 1997*a,b*) morphology. Despite earlier suggestions of a deep molecular divergence between eastern and western gorillas, prompting the preliminary suggestion that gorilla subspecies be elevated to species status (Ruvolo *et al.*, 1994), additional molecular studies have tended to uphold the current taxonomy (Jensen-Seaman *et al.*, this volume). Furthermore, from a qualitative standpoint, the molecular evidence largely mirrors the pattern of morphological differentiation, supporting a deeper genetic divergence between *G. g. gorilla* and *G. g. beringei*, regardless of the basis of the divergence estimates (mitochondrial versus nuclear DNA, for example) whereas genetic differences between *G. g. graueri* and *G. g. beringei* are relatively less marked.

As noted earlier, however, variation in mandibular morphology may reflect the inherent plasticity of living bone and its response to mechanical stress during the life of an individual (Wolff, 1870). Variation due to plasticity certainly has functional implications, but selection need not be invoked to explain the divergent morphological patterns. In addition, there is scant information regarding the genetic basis of jaw morphogenesis in primates. We know from experimental studies that formation and regulation of the mandible varies among vertebrates. One of the key differences relates to the timing of migration of neural crest cells relative to neural fold development, a difference that has the potential to impact the occurrence of inductive interactions prior to or during cellular migration (Macdonald and Hall, 2001). The mammalian pattern, as reflected in both rats and mice (Vermeij-Keers and Poelmann, 1980; Tan and Morriss-Kay, 1985; Nichols, 1986), is characterized by a relatively early migration of neural crest cells, which suggests a lower chance of inductive interactions pre- or mid-cellular migration in mammals as compared to birds and amphibians. Nevertheless, comparisons among inbred strains of mice have revealed differences in the timing of initiation of mandibular skeletogenesis (Macdonald and Hall, 2001) and notable differences in adult morphology have been linked to relatively minor changes in the timing of earlier development (Brylski and Hall, 1988; Raff *et al.*, 1990; Miyake *et al.*, 1997).

Thus, advocating the use of comparative data on jaw form as a means of addressing issues of taxonomic affinity in *Gorilla* would be ill advised. I note only that results of this investigation do not contradict the current single-species, multi-subspecies classification. In allometrically controlled comparisons, degrees of divergence in both craniomandibular size and shape are greatest in comparisons between *G. g. beringei* and *G. g. gorilla*, despite findings of statistically significant differences between *G. g. beringei* and both lowland gorilla subspecies. Furthermore, though not presented here, preliminary results of multivariate discriminant function analyses on the craniomandibular dimensions used in this study demonstrate essentially the same pattern of differentiation: whether on raw or size-corrected data, gorillas cluster (with some overlap) into three distinct groups that correspond to the three currently recognized subspecies, and Mahalanobis distances are greatest in comparisons between *G. g. gorilla* and the Virunga mountain gorilla (A.B. Taylor, unpublished data). In this sense, these results do not depart radically from findings of divergence based upon other morphologies (Uchida, 1996) and genetics (Garner and Ryder, 1996).

While the Nigerian gorillas are only now being evaluated morphologically, data indicate they are significantly smaller than other western lowland gorillas in measures of skull size and in molar surface area (Sarmiento and Oates, 2000). Preliminary ecological data for these gorillas indicate a more seasonally variable diet comprised of less fruit and more herbs, bark, and leaves (Oates *et al.*, this volume), which potentially contrasts with the decrease in molar tooth size, although the allometry of tooth and skull size in these gorillas is unknown. Additional evidence, though quite preliminary, suggests the Nigerian gorillas may be genetically distinct as well (Sarmiento and Oates, 2000; Oates *et al.*, this volume). More extensive analyses of morphology, genetics, and behavior are needed to resolve the taxonomic status of these Cross River gorillas.

## Conclusions

Ontogenetic, allometric analysis of the masticatory complex in *Gorilla* reveals a mix of features that are ontogenetically scaled among gorilla subspecies, and some nonallometric features that may be related to intensification of folivory/herbivory in *G. g. beringei. Gorilla g. beringei* is characterized by a significantly larger face, which is currently the most robustly established size difference among gorilla subspecies and justifies the use of an ontogenetic allometric approach in this case. In general, dimensions scaled against mandibular length either exhibit weak to moderate allometry or isometry. The relatively high frequency of isometry and weak positive allometry during growth in *G. g. beringei* suggests that ontogenetic scaling of positive allometries has been minimized in this subspecies, possibly to prevent ontogenetic scaling of positive allometries into maladaptive size ranges. Though there are no other established differences in size among gorilla subspecies, should body or other regional size differences prove to be significant, this would suggest that gorillas are characterized by a mosaic pattern of size change, with the larger face in *G. g. beringei* likely related to diet, and possibly reflecting phenotypic plasticity.

Some of the novel shape differences between *G. g. beringei* and the other two subspecies are consistent with the hypothesis that *G. g. beringei* is characterized by adaptations to extreme folivory/herbivory. These features include a relatively wider mandibular corpus and symphysis, wider mandibular condyles, higher mandibular rami, temporomandibular joint elevated higher above the occlusal plane, and a vertically taller face. Some of these same features have been shown to distinguish gorillas from *Pan*, bolstering arguments that such structural modifications are adaptations to increase muscle and bite force, resist higher external and internal forces, and reduce the risk of fatigue failure associated with repetitive loading of the jaws.

However, while some features differentiate *G. g. beringei* from the other two subspecies in the expected direction based on diet, other results are either inconsistent with, or contradict, biomechanical models of masticatory form and function. Among the most notable discrepancies is the fact that *G. g. graueri* does not differ from *G. g. gorilla* in any of the predicted features despite ecological data supporting a more folivorous diet for *G. g. graueri*. The African apes as a group are plagued by similar inconsistencies. Therefore, increase in degree of folivory cannot always be associated with predictable morphological adaptations of the cranium and mandible. Contrasting these findings with results obtained from other comparative studies, it must be concluded that variation in craniomandibular form should be interpreted cautiously, particularly when variation cannot be substantiated by data on the mechanical properties of foods, when force production capabilities are based solely on skeletal indicators, and without *in vivo* validation of the stress and strains encountered during masticatory behavior.

## Acknowledgments

A portion of this manuscript was originally presented at the 1999 American Association of Physical Anthropologists symposium entitled "Revisions of the genus *Gorilla*: 70 years after Coolidge". I thank Michele Goldsmith for co-organizing and co-chairing the symposium and for her work as co-editor of this volume, a role she was willing to assume even after both of us had a reasonble idea of what we were getting ourselves into. I sincerely thank Drs. Susan Antón, David Daegling, Matthew Ravosa, Brian Shea, Dennis Slice and Chris Vinyard for their ongoing discussions of ontogeny, allometry, adaptation, and masticatory biomechanics. Thank also to Drs. Gene Albrecht, Susan Antón, John Fleagle, Colin Groves, Sandra Inouye, Steven Leigh, Matthew Ravosa, Brian Shea, Michael Siegel, Russel Tuttle, Chris Vinyard, and two anonymous reviewers for reading and critiquing various versions of this manuscript. Dr. James Patton, Museum of Vertebrate Zoology, University of California, Berkeley, kindly provided access to African ape skulls during initial stages of this work. I am also grateful to the following individuals who provided access to the material used in this study and assistance during data collection: Dr. Wim Van Neer, Central African Museum, Tervuren, Belgium; Dr. Bruce Latimer and Lyman Jellema, Cleveland Museum of Natural History, Ohio; John Harrison and Malcolm Harman, The Powell-Cotton Museum and Quex House and Gardens, Birchington, Kent, U.K.; Dr. Richard Thorington and Linda Gordon, Department of Mammalogy, U.S. National Museum of Natural History, Smithsonian Institution, Washington, D.C.; Drs. Bruce Patterson and Matt Ravosa, Field Museum of Natural History, Chicago. This study was supported in part by funds from the LSB Leakey Foundation.

## References

Ackermann, R.R. and Cheverud J.M. (2000). Phenotypic covariance structure in tamarins (genus *Saguinus*): A comparison of variation patterns using matrix correlation and common principal component analysis. *American Journal of Physical Anthropology*, **111**, 489–501.

Albrecht, G.H. and Miller, J.M.A. (1993). Geographic variation in primates: A review with implications for interpreting fossils. In *Species, Species Concepts, and Primate Evolution*, eds. W.H. Kimbel and L.B. Martin, pp. 123–162. New York: Plenum Press.

Antón, S.C. (1996a). Cranial adaptation to a high attrition diet in Japanese macaques. *International Journal of Primatology*, **17**, 401–427.

Antón, S.C. (1996b). Tendon-associated bone features of the masticatory system in Neandertals. *Journal of Human Evolution*, **31**, 391–408.

Antón, S.C. (1999). Macaque masseter muscle: Internal architecture, fiber length and cross-sectional area. *International Journal of Primatology*, **20**, 441–462.

Antón, S.C. (2000). Macaque pterygoid muscles: Internal architecture, fiber length, and cross-sectional area. *International Journal of Primatology*, **21**, 131–156.

Atchley, W.R. (1987). Developmental quantitative genetics and the evolution of onto-genies. *Evolution*, **41**, 316–330.

Atchley, W.R. and Hall, B.K. (1991). A model for development and evolution of complex morphological structures. *Biological Review,* **66**, 101–157.

Badrian, N.L. and Malenky, R.K. (1984). Feeding ecology of *Pan paniscus* in the Lomako Forest, Zaire. In *The Pygmy Chimpanzee: Evolutionary Biology and Behavior*, ed. R.L. Susman, pp. 275–299. New York: Plenum Press.

Beecher, R.M. (1979). Functional significance of the mandibular symphysis. *Journal of Morphology*, **159**, 117–130.

Beecher, R.M. and Corruccini, R.S. (1981). Effects of dietary consistency on craniofacial and occlusal development in the rat. *Angle Orthodontics*, **51**, 61–69.

Beecher, R.M., Corruccini, R.S., and Freeman, M. (1983). Craniofacial correlates of dietary consistency in a nonhuman primate. *Journal of Craniofacial Genetics and Developmental Biology*, **3**, 193–202.

Biewener, A.A. (1982). Bone strength in small mammals and bipedal birds: Do safety factors change with body size? *Journal of Experimental Biology*, **98**, 289–301.

Bouvier, M. (1986a). A biomechanical analysis of mandibular scaling in Old World monkeys. *American Journal of Physical Anthropology*, **69**, 473–482.

Bouvier, M. (1986b). Biomechanical scaling of mandibular dimensions in New World Monkeys. *International Journal of Primatology*, **7**, 551–567.

Bouvier, M. and Hylander, W.L. (1981). Effect of bone strain on cortical bone structure in macaques (*Macaca mulatta*). *Journal of Morphology*, **167**, 1–12.

Brylski, P. and Hall, B.K. (1988). Ontogeny of a macroevolutionary phenotype: The external cheek pouches of geomyoid rodents. *Evolution*, **42**, 391–395.

Casimir, M.J. (1975). Some data on the systematic position of the Eastern gorilla population of the Mt. Kahuzi region (République du Zaïre). *Zeitschrift für Morphologie und Anthropologie*, **66**, 188–201.

Chapman, C.A., White, F.J., and Wrangham, R.W. (1994). Party size in chimpanzees and bonobos: A re-evaluation of theory based on two similarly forested sites. In *Chimpanzee Cultures*, eds. R.W. Wrangham, W.C. McGrew, F.B.M. de Waal, and P.G. Heltne, pp. 41–57. Cambridge, MA: Harvard University Press.

Cheverud, J.M. (1981). Epiphyseal union and dental eruption in *Macaca mulatta*. *American Journal of Physical Anthropology*, **56**, 157–169.

Cheverud, J.M. (1984). Quantitative genetics and developmental constraints on evolution by selection. *Journal of Theoretical Biology*, **101**, 155–171.

Cochard, L.R. (1985). Ontogenetic allometry of the skull and dentition of the rhesus monkey (*Macaca mulatta*). In *Size and Scaling in Primate Biology*, ed. W.L. Jungers, pp. 231–256. New York: Plenum Press.

Cock, A.G. (1963). Genetical aspects of metrical growth and form in animals. *Quarterly Review of Biology*, **41**, 131–190.

Cole, T.M. III (1992). Postnatal heterochrony of the masticatory apparatus in *Cebus apella* and *Cebus albifrons*. *Journal of Human Evolution*, **23**, 253–282.

Coolidge, H.J. (1929). A revision of the genus *Gorilla*. *Memoirs of the Museum of Comparative Zoology, Harvard*, **50**, 291–381.

Corruccini, R.S. and Beecher, R.M. (1982). Occlusal variation related to soft diet in a nonhuman primate. *Science*, **218**, 74–76.

Corruccini, R.S. and Beecher, R.M. (1984). Occlusofacial morphological integration lowered in baboons raised on soft diet. *Journal of Craniofacial and Genetic Developmental Biology*, **4**, 135–142.

Daegling, D.J. (1989). Biomechanics of cross-sectional size and shape in the hominoid mandibular corpus. *American Journal of Physical Anthropology*, **80**, 91–106.

Daegling, D.J. (1990). Geometry and biomechanics of hominoid mandibles. PhD thesis, State University of New York at Stony Brook, Stony Brook, NY.

Daegling, D.J. (1992). Mandibular morphology and diet in the genus *Cebus*. *International Journal of Primatology*, **13**, 545–570.

Daegling, D.J. (2001). Biomechanical scaling of the hominoid mandibular symphysis. *Journal of Morphology*, **250**, 12–23.

Daegling, D.J. and Grine, F.E. (1991). Compact bone distribution and biomechanics of early hominid mandibles. *American Journal of Physical Anthropology*, **86**, 321–339.

Daegling, D.J. and Hylander, W.L. (1998). Biomechanics of torsion in the human mandible. *American Journal of Physical Anthropology*, **105**, 73–87.

Daegling, D.J. and Hylander, W.L. (2000). Experimental observation, theoretical models, and biomechanical inference in the study of mandibular form. *American Journal of Physical Anthropology*, **112**, 541–551.

Dodson, P. (1975). Functional and ecological significance of relative growth in *Alligator*. *Journal of Zoology, London*, **1975**, 315–355.

DuBrul, L.E. (1977). Early hominid feeding mechanisms. *American Journal of Physical Anthropology*, **47**, 305–320.

Emerson, S.B. and Bramble, D.M. (1993). Scaling, allometry, and skull design. In *The Skull*, vol. 3, *Functional and Evolutionary Mechanisms*, eds. J. Hanken and B.K. Hall, pp. 384–421. Chicago, IL: University of Chicago Press.

Freeman, P.W. (1988). Frugivorous and animalivorous bats (Microchiroptera): Dental and cranial adaptations. *Biological Journal of the Linnean Society*, **33**, 249–272.

Garner, K.J. and Ryder, O.A. (1996). Mitochondrial DNA diversity in gorillas. *Molecular Phylogenetics and Evolution*, **6**, 39–48.

Godfrey, L.R., King, S.J., and Sutherland, M.R. (1998). Heterochronic approaches to the study of locomotion. In *Primate Locomotion: Recent Advances*, eds. E. Strasser, J. Fleagle, A. Rosenberger, and H. McHenry, pp. 277–307. New York: Plenum Press.

Gould, S.J. (1966). Allometry and size in ontogeny and phylogeny. *Biological Review*, **41**, 587–640.

Gould, S.J. (1971). Geometric similarity in allometric growth: A contribution to the problem of scaling in the evolution of size. *American Naturalist*, **105**, 113–136.

Gould, S.J. (1975). Allometry in primates, with emphasis on scaling and the evolution of the brain. In *Approaches to Primate Paleobiology*, ed. F. Szalay, pp. 244–292. Basel, Switzerland: S. Karger.

Gould, S.J. and Lewontin, R.C. (1979). The spandrels of San Marco and the Panglossian paradigm: A critique of the adaptationist programme. In *The Evolution of Adaptation by Natural Selection*, eds. J. Maynard Smith and R. Holliday, pp. 147–164. London: Royal Society.

Gould, S.J. and Vrba, E.S. (1982). Exaptation: A missing term in the science of form. *Paleobiology*, **8**, 4–15.

Groves, C.P. (1967). Ecology and taxonomy of the gorilla. *Nature*, **213**, 890–893.

Groves, C.P. (1970*a*). *Gigantopithecus* and the mountain gorilla. *Nature*, **226**, 973–974.

Groves, C.P. (1970*b*). Population systematics of the gorilla. *Journal of the Zoological Society of London*, **161**, 287–300.

Groves, C.P. (1986). Systematics of the Great Apes. In *Evolution, Ecology, Behavior and Captive Maintenance*, ed. D.M. Rumbaugh, pp. 2–102. Basel, Switzerland: S. Karger.

Groves, C.P. (2001). *Primate Taxonomy*. Washington, D.C.: Smithsonian Institution Press.

Groves, C.P. and Stott, K.W., Jr. (1979). Systematic relationships of gorillas from Kahuzi, Tshiaberimu and Kayonza. *Folia Primatologica*, **32**, 161–179.

Harvey, P.H. and Pagel, M.D. (1991). *The Comparative Method in Evolutionary Biology*. Oxford, U.K.: Oxford University Press.

Herring, S.W. and Herring, S.E. (1974). The superficial masseter and gape in mammals. *American Naturalist*, **108**, 561–576.

Huxley, J.S. (1932). *Problems of Relative Growth*. London: MacVeagh.

Hylander, W.L. (1975). Incisor size and diet in anthropoids with special reference to Cercopithecidae. *Science*, **26**, 1095–1098.

Hylander, W.L. (1977). The adaptive significance of Eskimo craniofacial morphology. In *Orofacial Growth and Development*, eds. A.A. Dahlberg and T.M. Graber, pp. 129–170. Paris: Mouton.

Hylander, W.L. (1979*a*). The functional significance of primate mandibular form. *Journal of Morphology*, **160**, 223–240.

Hylander, W.L. (1979*b*). Mandibular function in *Galago crassicaudatus* and *Macaca fascicularis*: An *in vivo* approach to stress analysis of the mandible. *Journal of Morphology*, **159**, 253–296.

Hylander, W.L. (1981). Patterns of stress and strain in the macaque mandible. In *Craniofacial Biology*, ed. C.S. Carlson, pp. 1–37. Ann Arbor, MI: University of Michigan.

Hylander, W.L. (1984). Stress and strain in the mandibular symphysis of Primates: A test of competing hypotheses. *American Journal of Physical Anthropology*, **64**, 1–46.

Hylander, W.L. (1985). Mandibular function and biomechanical stress and scaling. *American Zoologist*, **25**, 315–330.

Hylander, W.L. (1986). *In vivo* bone strain as an indicator of masticatory bite force in *Macaca fascicularis*. *Archives of Oral Biology*, **31**, 149–157.

Hylander, W.L. (1988). Implications of *in vivo* experiments for interpreting the functional significance of "robust" australopithecine jaws. In *Evolutionary History of the "Robust" Australopithecines*, ed. F.E. Grine, pp. 55–83. New York: Aldine de Gruyter.

Hylander, W.L. and Bays, R. (1979). An *in vivo* strain gage analysis of squamosal dentary joint reaction force during mastication and incision in *Macaca mulatta* and *Macaca fascicularis*. *Archives of Oral Biology*, **24**, 689–697.

Hylander, W.L. and Johnson, K. (1994). Jaw muscle function and wishboning of the mandible during mastication in macaques and baboons. *American Journal of Physical Anthropology*, **94**, 523–548.

Hylander, W.L, Johnson, K.R., and Crompton, A.W. (1987). Loading patterns and jaw movements during mastication in *Macaca fascicularis*: A bone strain, electromyographic, and cineradiographic analysis. *American Journal of Physical Anthropology*, **72**, 287–314.

Hylander, W.L., Johnson, K.R., Ravosa, M.J., and Ross, C.F. (1996). Mandibular bone strain and jaw-muscle recruitment patterns during mastication in anthropoids and prosimians. *American Journal of Physical Anthropology, Supplement* **22**, 128–129.

Hylander, W.L., Ravosa, M.J., Ross, C.F., Wall, C.E., and Johnson, K.R. (2000). Symphyseal fusion and jaw-adductor muscle force: An EMG study. *American Journal of Physical Anthropology*, **112**, 469–492.

Inouye, S.E. (1992). Ontogeny and allometry of African ape manual rays. *Journal of Human Evolution*, **23**, 107–138.

Janson, C. and van Schaik, C.P. (1993). Ecological risk aversion in juvenile primates: Slow and steady wins the race. In *Juvenile Primates, Life History, Development and Behavior*, eds. M.E. Pereira and L.A. Fairbanks, pp. 57–74. Oxford, U.K.: Oxford University Press.

Jolicoeur, P. (1963). The multivariate generalization of the allometry equation. *Biometrics*, **19**, 497–499.

Jolly, C.J. (1993). The seed-eaters: A new model of hominid differentiation based on a baboon analogy. *Man*, **5**, 5–26.

Jungers, W.L. and Cole, M.S. (1992). Relative growth and shape of the locomotor skeleton in lesser apes. *Journal of Human Evolution*, **23**, 93–106.

Jungers, W.L. and Fleagle, J.G. (1980). Postnatal growth allometry of the extremities in *Cebus albifrons* and *Cebus apella*: A longitudinal and comparative study. *American Journal of Physical Anthropology*, **53**, 471–478.

Kano, T. and Mulavwa, M. (1984). Feeding ecology of the pygmy chimpanzees (*Pan paniscus*) of Wamba. In *The Pygmy Chimpanzee: Evolutionary Biology and Behavior*, ed. R.L. Susman, pp. 233–274. New York: Plenum Press.

Kimbel, W. (1991). Species, species concepts and hominid evolution. *Journal of Human Evolution*, **20**, 355–372.

Klingenberg, C.P., Neuenschwander, B.E., and Flury, B.D. (1996). Ontogeny and individual variation: Analysis of patterned covariance matrices with common principal components. *Systematic Biology*, **45**, 135–150.

Lauder, G. (1982). Historical biology and the problem of design. *Journal of Theoretical Biology*, **97**, 57–67.

Lauder, G.V. and Reilly, S.M. (1990). Metamorphosis of the feeding mechanism in tiger salamanders (*Ambystoma tigrinum*): The ontogeny of cranial muscle mass. *Journal of Zoology, London*, **222**, 59–74.

Lovell, N.C. (1990). Skeletal dental pathology of free-ranging Mountain Gorillas. *American Journal of Physical Anthropology*, **81**, 399–412.

Lucas, P.W., Turner, I.M., Dominy, N.J., and Yamashita, N. (2000). Mechanical defenses to herbivory. *Annals of Botany*, **86**, 913–920.

Macdonald, M.E. and Hall, B.K. (2001). Altered timing of the extracellular-matrix-mediated epithelial–mesenchymal interaction that initiates mandibular skeletogenesis in three inbred strains of mice: Development, heterochrony, and evolutionary change in morphology. *Journal of Experimental Zoology*, **291**, 258–273.

Malenkey, R.K. and Stiles, E.W. (1991). Distribution of terrestrial herbaceous vegetation and its consumption by *Pan paniscus* in the Lomako Forest, Zaire. *American Journal of Primatology*, **23**, 153–169.

Malenky, R.K. and Wrangham, R.W. (1994). A quantitative comparison of terrestrial herbaceous food consumption by *Pan paniscus* in the Lomako Forest, Zaire, and *Pan troglodytes* in the Kibale Forest, Uganda. *American Journal of Primatology*, **32**, 1–12.

Malenky, R.K., Kuroda, S., Vineberg, E.O., and Wrangham, R.W. (1994). The significance of terrestrial herbaceous foods for bonobos, chimpanzees, and gorillas. In *Chimpanzee Cultures*, eds. R.W. Wrangham, W.C. McGrew, F.B.M. de Waal, and P.G. Heltne, pp. 59–75. Cambridge, MA: Harvard University Press.

Martin, R.D. (1989). Size, shape and evolution. In *Evolutionary Studies: A Centenary Celebration of the Life of Julian Huxley*, eds. M. Keynes and G.A. Harrison, pp. 96–141. Basingstoke, U.K.: Macmillan.

Mayr, E. (1982). Of what use are subspecies? *Auk*, **99**, 593–595.

McCollum, M.A. (1997). Mechanical and spatial determinants of *Paranthropus* facial form. *American Journal of Physical Anthropology*, **93**, 259–273.

McCrossin, M.L. and Benefit, B.R. (1993). Recently recovered *Kenyapithecus* mandible and its implications for great ape and human origins. *Proceedings of the National Academy of Sciences U.S.A.*, **90**, 1962–1966.

Miyake, T., Cameron, A.M., and Hall, B.K. (1997). Variability of embryonic development among three inbred strains of mice. *Growth, Development and Aging*, **61**, 141–155.

Morell, V. (1999). Ecology returns to speciation studies. *Science*, **284**, 2106.

Müller, G.B. (1990). Developmental mechanisms: A side-effect hypothesis. In *Evolutionary Innovations*, ed. M.H. Nitecki, pp. 99–130. Chicago, IL: University of Chicago Press.

Nichols, D.H. (1986). Formation and distribution of neural crest mesenchyme to the first pharyngeal arch region of the mouse embryo. *American Journal of Anatomy*, **176**, 221–231.

Oxnard, C.E. (1972). Functional morphology of primates: Some mathetmatical and physical methods. In *The Functional and Evolutionary Biology of Primates*, ed. R.H. Tuttle, pp. 305–336. Chicago, IL: University of Chicago Press.

Oxnard, C.E. (1979). Some methodological factors in studying the morphological–behavioral interface. In *Environment, Behavior, and Morphology: Dynamic*

*Interactions in Primates*, eds. M.E. Morbeck, H. Preuschoft, and N. Gomberg, pp. 209–227. New York: Gustav Fischer.

Price, T.D., Grant, P.R., and Boag, P.T. (1984). Genetic changes in the morphological differentiation of Darwin's ground finches. In *Population Biology and Evolution*, eds. K. Wohrmann and V. Loeschke, pp. 49–66. Berlin: Springer-Verlag.

Raff, R.A., Parr, B., Parks, A., and Wray, G. (1990). Heterochrony and other mechanisms of radical evolutionary change in early development. In *Evolutionary Innovations*, ed. M.H. Nitecki, pp. 71–98. Chicago, IL: University of Chicago Press.

Rak, Y. (1983). *The Australopithecine Face*. New York: Academic Press.

Raven, H.C. (1950). *The Anatomy of the Gorilla*. New York: Columbia University Press.

Ravosa, M.J. (1990). Functional assessment of subfamily variation in maxillomandibular morphology among Old World monkeys. *American Journal of Physical Anthropology*, **82**, 199–212.

Ravosa, M.J. (1991). Structural allometry of the prosimian mandibular corpus and symphysis. *Journal of Human Evolution*, **20**, 3–20.

Ravosa, M.J. (1996a). Jaw morphology and function in living and fossil Old World Monkeys. *International Journal of Primatology*, **17**, 909–932.

Ravosa, M.J. (1996b). Mandibular form and function in North American and European Adapidae and Omomyidae. *Journal of Morphology*, **229**, 171–190.

Ravosa, M.J. (2000). Size and scaling in the mandible of living and extinct apes. *Folia Primatologica*, **71**, 305–322.

Reilly, S.M., and Lauder, G.V. (1990). Metamorphosis of cranial design in tiger salamanders (*Ambystoma tigrinum*): A morphometric analysis of ontogenetic change. *Journal of Morphology*, **104**, 121–137.

Remis, M.J. (1997). Western lowland gorillas (*Gorilla gorilla gorilla*) as seasonal frugivores: Use of variable resources. *American Journal of Primatology*, **43**, 87–109.

Rensberger, J.M. (1973). *Sanctimus* (Mammalia, Rodentia) and the phyletic relationships of the large arikareean geomyoids. *Journal of Paleontology*, **47**, 835–853.

Rubin, C.T. and Lanyon, L.E. (1984). Dynamic strain similarity in vertebrates: An alternative to allometric limb bone scaling. *Journal of Theoretical Biology*, **107**, 321–327.

Ruvolo, M., Pan, D., Zehr, S., Goldberg, T., Disotell T.R., and von Dornum, M. (1994). Gene trees and hominoid phylogeny. *Proceedings of the National Academy of Sciences U.S.A.* **91**, 8900–8904.

Sarmiento, E.E., & Oates, J. (2000). The Cross River gorillas: A distinct subspecies *Gorilla gorilla diehli* Matschie 1904. *American Museum Novitates*, **3304**, 1–55.

Sarmiento, E.E., Butynski, T.M., and Kalina, J. (1996). Gorillas of the Bwindi-Impenetrable Forest and Virunga Volcanoes: Taxonomic implications of morphological and ecological differences. *American Journal of Primatology*, **40**, 1–21.

Schultz, A. (1930). The skeleton of the trunk and limbs of higher primates. *Human Biology*, **2**, 303–435.

Schultz, A.H. (1934). Some distinguishing characters of the mountain gorilla. *Journal of Mammalogy*, **15**, 51–61.

Seaman, M.L., Deinard, A.S., & Kidd, K.K. (1999). Incongruence between mitochondrial and nuclear DNA estimated of divergence between gorilla subspecies. *American Journal of Physical Anthropology, Supplement* **28**, 259.

Shea, B.T. (1982). Growth and size allometry in the African Pongidae: Cranial and postcranial analyses. PhD dissertation, Duke University, Chapel Hill, NC.

Shea, B.T. (1983*a*). Size and diet in the evolution of African ape craniodental form. *Folia Primatologica*, **40**, 32–68.

Shea, B.T. (1983*b*). Allometry and heterochrony in the African apes. *American Journal of Physical Anthropology*, **62**, 275–289.

Shea, B.T. (1984). An allometric perspective on the morphological and evolutionary relationships between pygmy (*Pan paniscus*) and common (*Pan troglodytes*) chimpanzees. In *The Pygmy Chimpanzee: Evolutionary Biology and Behavior*, ed. R.L. Susman, pp. 89–130. New York: Plenum Press.

Shea, B.T. (1985*a*). Ontogenetic allometry and scaling: A discussion based on growth and form of the skull in African apes. In *Size and Scaling in Primate Biology*, ed. W.L. Jungers, pp. 175–206. New York: Plenum Press.

Shea, B.T. (1985*b*). Bivariate and multivariate growth allometry: Statistical and biological considerations. *Journal of Zoology, London*, **206**, 367–390.

Shea, B.T. (1986). Scapula form and locomotion in chimpanzee evolution. *American Journal of Physical Anthropology*, **70**, 475–488.

Shea, B.T. (1995). Ontogenetic scaling and size correction in the comparative study of primate adaptations. *Anthropologie*, **33**, 1–16.

Shea, B.T. and Coolidge, H.T. (1988). Craniometric differentiation and systematics in the genus *Pan*. *Journal of Human Evolution*, **17**, 671–685.

Shea, B.T., Leigh, S.R., and Groves, C.P. (1993). Multivariate craniometric variation in chimpanzees: Implications for species identification in paleoanthropology. In *Species, Species Concepts, and Primate Evolution*, eds. W.H. Kimbel and L.B. Martin, pp. 265–296. New York: Plenum Press.

Smith, R.J. (1983). The mandibular corpus of female primates: Taxonomic, dietary, and allometric correlates of interspecific variations in size and shape. *American Journal of Physical Anthropology*, **61**, 315–330.

Smith, R.J. (1993). Categories of allometry: Body size versus biomechanics. *Journal of Human Evolution*, **24**, 173–182.

Smith, R.J. and Jungers, W.L. (1997). Body mass in comparative primatology. *Journal of Human Evolution*, **32**, 523–559.

Smith, R.J., Petersen, C.E., and Gipe, D.P. (1983). Size and shape of the mandibular condyle in primates. *Journal of Morphology*, **177**, 59–68.

Spencer, M.A. (1998). Force production in the primate masticatory system: Electromyographic tests of biomechanical hypotheses. *Journal of Human Evolution*, **34**, 25–54.

Spencer, M.A. and Demes, B. (1993). Biomechanical analysis of masticatory function–configuration in Neandertals and Inuits. *American Journal of Physical Anthropology*, **91**, 1–20.

Spencer, M.A. and Hogard, R. (2001). Biomechanics of sexual dimorphism in the anthropoid masticatory system. *American Journal of Physical Anthropology, Supplement* **32**, 141.

Takahashi, L.K. and Pan, R. (1994). Mandibular morphometrics among macaques: The case of *Macaca thibetana*. *International Journal of Primatology*, **15**, 597–621.

Tan, S.S. and Morriss-Kay, G.M. (1985). The development and distribution of the cranial neural crest in the rat embryo. *Cell and Tissue Research*, **240**, 403–416.

Tattersall, I. (1986). Species recognition in human paleontology. *Journal of Human Evolution*, **15**, 165–175.

Taylor, A.B. (1995). Effects of ontogeny and sexual dimorphism on scapula morphology in the mountain gorilla (*Gorilla gorilla beringei*). *American Journal of Physical Anthropology*, **98**, 431–445.

Taylor, A.B. (1997*a*). Relative growth, ontogeny, and sexual dimorphism in *Gorilla* (*Gorilla gorilla gorilla* and *G. g. beringei*): Evolutionary and ecological considerations. *American Journal of Primatology*, **43**, 1–33.

Taylor, A.B. (1997*b*). Scapula form and biomechanics in gorillas. *Journal of Human Evolution*, **34**, 529–533.

Taylor, A.B. (1998*a*). Ontogeny and function of maxillomandibular form in *Gorilla*. *Abstracts of Contributions to the Dual Congress* **1998**, 65.

Taylor, A.B. (1998*b*). Masticatory form and function in gorillas (*G. g. gorilla* and *G. g. beringei*). *American Journal of Physical Anthropology, Supplement* **26**, 216–217.

Taylor, A.B. (1999). Variation in masticatory form in lowland gorillas (*Gorilla gorilla gorilla* and *G. g. graueri*): An ontogenetic approach. *American Journal of Physical Anthropology, Supplement* **28**, 262.

Taylor, A.B. (2000). Tooth or consequences: Is jaw robusticity the correlated effect of increased tooth size? *American Journal of Physical Anthropology, Supplement* **30**, 300.

Taylor, A.B. (2002). Masticatory form and function in the African apes. *American Journal of Physical Anthropology*, **117**, 133–156.

Taylor, A.B. and Ravosa, M.J. (1999). Ontogeny and function in the evolution of African ape masticatory form. *American Journal of Primatology*, **49**, 183.

Terborgh, J.W. (1983). *Five New World Primates: A Study in Comparative Ecology*. Princeton, NJ: Princeton University Press.

Thompson, D.W. (1942). *On Growth and Form*. London: Cambridge University Press.

Turnbull, W.D. (1970). Mammalian masticatory apparatus. *Field Museum of Natural History Fieldiana: Geology*, **18**, 149–356.

Tutin, C.E.G., Ham, R.M., White, L.J.T., and Harrison, M.J.S. (1997). The primate community of the Lopé Reserve, Gabon: Diets, responses to fruit scarcity, and effects on biomass. *American Journal of Primatology*, **42**, 1–24.

Uchida, A. (1996). *Craniodental Variation among the Great Apes*, Peabody Museum Bulletin no. 4. Cambridge, MA: Harvard University Press.

Uchida A. (1998). Variation in tooth morphology of *Gorilla gorilla*. *Journal of Human Evolution*, **34**, 55–70.

van Schaik, C.P. and van Noordwijk, M.A. (1986). The hidden costs of sociality: Intragroup variation in feeding strategies in Sumatran long-tailed macaques (*Macaca fascicularis*). *Behaviour*, **99**, 296–315.

Velhagen, W.A. and Roth, V.L. (1997). Scaling of the mandible in squirrels. *Journal of Morphology*, **232**, 107–132.

Vermeij-Keers, C. and Poelmann, R.E. (1980). The neural crest: A study on cell degeneration and the improbability of cell migration in mouse embryos. *Netherlands Journal of Zoology*, **30**, 74–81.

Vinyard, C.J. and Ravosa, M.J. (1998). Ontogeny, function, and scaling of the mandibular symphysis in papionin primates. *Journal of Morphology*, **235**, 157–175.

Vogel, V.C. (1961). Zur systematischen Untergliederung der Gattung *Gorilla* anhand van Untersuchungen der Mandibel. *Zeitschrift für Säugetierkunde*, **26**, 1–12.

Ward, S.C. and Molnar, S. (1980). Experimental stress analysis of topographic diversity in early hominid gnathic morphology. *American Journal of Physical Anthropology*, **53**, 383–395.

Ward, S., Brown, B., Hill, A., Kelley, J., and Downs, W. (1999). *Equatorius*: A new hominoid genus from the Middle Miocene of Kenya. *Science*, **285**, 1382–1386.

Watts, D.P. (1984). Composition and variability of mountain gorilla diets in the central Virungas. *American Journal of Primatology*, **7**, 323–356.

Watts, D.P. (1998). Seasonality in the ecology and life histories of mountain gorillas (*Gorilla gorilla beringei*). *International Journal of Primatology*, **19**, 929–948.

Wolff, J. (1870). Ueber die innere Architectur der Knochen und ihre Bedeutung für die Frage vom Knochenwachsthum. *Archives für pathologische Anatomie und Physiologic und für kulinische Medizin (Virchows Archiv)*, **50**, 389–453.

Wrangham, R.W. (1986). Ecology and social evolution in two species of chimpanzee. In *Ecology and Social Evolution: Birds and Mammals*, eds. D.L. Rubenstein and R.W. Wrangham, pp. 352–378. Princeton, NJ: Princeton University Press.

Wrangham, R.W., Chapman, C.A., Clark-Arcadia, A.P., and Isabirye-Basuta, G. (1996). Social ecology of Kanyawara chimpanzees: Implications for understanding the costs of great ape groups. In *Great Ape Societies*, eds. W.C. McGrew, L.F. Marchant, and T. Nishida, pp. 45–57. Cambridge, U.K.: Cambridge University Press.

Yamagiwa, J., Maruhashi, T., Yumoto, T., and Mwanza, N. (1996). Dietary and ranging overlap in sympatric gorillas and chimpanzees in Kahuzi-Biega National Park, Zaire. In *Great Ape Societies*, eds. W.C. McGrew, L.F. Marchant, and T. Nishida, pp. 82–98. Cambridge, U.K.: Cambridge University Press.

# Appendix

Table A6.1. *Descriptive statistics for adult linear dimensions for each subspecies separately by sex*

| | Gorilla gorilla gorilla | | | | | | Gorilla gorilla graueri | | | | | | Gorilla gorilla beringei | | | | | |
|---|---|---|---|---|---|---|---|---|---|---|---|---|---|---|---|---|---|---|
| | M | | | F | | | M | | | F | | | M | | | F | | |
| Dimension | $x$ | SD | $n$ | $x$ | SD | $n$ | $x$ | SD | $n$ | $x$ | SD | $n$ | $x$ | SD | $n$ | $x$ | SD | $n$ |
| Basicranial length | 131.13 | 9.2 | 12 | 114.16 | 3.6 | 12 | 136.11 | 6.8 | 13 | 114.00 | 2.8 | 13 | 134.34 | 10.7 | 10 | 115.72 | 3.1 | 9 |
| Palatal length | 103.92 | 8.3 | 13 | 84.42 | 5.2 | 13 | 120.58 | 7.6 | 14 | 97.96 | 8.1 | 13 | 120.11 | 10.3 | 12 | 95.78 | 6.1 | 9 |
| Bicondylar breadth | 140.42 | 9.6 | 12 | 123.03 | 4.2 | 13 | 143.32 | 5.3 | 14 | 126.38 | 6.0 | 13 | 148.70 | 5.3 | 11 | 129.63 | 7.8 | 9 |
| Bizygomatic breadth | 174.38 | 6.4 | 14 | 144.84 | 6.6 | 12 | 176.12 | 8.6 | 14 | 145.27 | 5.2 | 13 | 175.41 | 9.8 | 11 | 148.30 | 7.1 | 8 |
| Mandibular length | 169.84 | 11.7 | 13 | 142.78 | 13.6 | 13 | 202.11 | 10.0 | 14 | 165.69 | 7.4 | 13 | 204.70 | 8.9 | 11 | 164.15 | 7.2 | 9 |
| $M_1$ corpus depth | 41.44 | 3.5 | 14 | 35.15 | 2.9 | 13 | 45.45 | 3.3 | 14 | 35.78 | 3.0 | 13 | 44.71 | 3.6 | 11 | 37.78 | 3.2 | 10 |
| $M_2$ corpus depth | 38.73 | 3.6 | 14 | 33.08 | 3.1 | 13 | 42.22 | 3.6 | 14 | 33.77 | 2.6 | 13 | 42.31 | 4.5 | 11 | 35.96 | 2.6 | 10 |
| $M_1$ corpus width | 19.21 | 1.8 | 14 | 17.87 | 1.4 | 13 | 20.33 | 1.4 | 14 | 18.96 | 1.1 | 13 | 21.83 | 1.7 | 11 | 20.38 | 1.2 | 10 |
| $M_2$ corpus width | 20.66 | 1.9 | 14 | 19.56 | 1.7 | 13 | 21.38 | 1.1 | 13 | 20.54 | 1.0 | 13 | 22.88 | 2.2 | 11 | 22.14 | 1.8 | 10 |
| Symphyseal depth | 63.87 | 7.2 | 14 | 52.55 | 3.3 | 13 | 71.81 | 3.8 | 14 | 57.75 | 3.1 | 13 | 71.17 | 5.2 | 11 | 54.96 | 2.9 | 10 |
| Symphyseal width | 27.31 | 2.0 | 14 | 22.88 | 1.6 | 13 | 27.83 | 2.5 | 14 | 23.38 | 1.8 | 13 | 31.03 | 2.7 | 11 | 26.75 | 1.8 | 10 |
| Condylar width | 34.83 | 3.6 | 13 | 30.06 | 1.8 | 13 | 35.87 | 3.1 | 14 | 29.84 | 2.1 | 13 | 39.16 | 3.6 | 11 | 34.64 | 1.8 | 10 |
| Gonion–zygion distance | 115.41 | 6.9 | 13 | 97.11 | 6.0 | 13 | 122.53 | 8.4 | 14 | 101.19 | 4.5 | 13 | 127.33 | 4.4 | 10 | 108.25 | 4.2 | 7 |
| Ramal height | 120.20 | 6.4 | 14 | 104.14 | 5.1 | 13 | 128.42 | 6.2 | 13 | 107.53 | 4.6 | 13 | 139.55 | 3.0 | 11 | 120.79 | 4.6 | 10 |

Table A6.2. *Descriptive statistics for craniomandibular shape ratios for gorilla subspecies adults*

| Ratio | Gorilla gorilla gorilla | | | Gorilla gorilla graueri | | | Gorilla gorilla beringei | | |
|---|---|---|---|---|---|---|---|---|---|
| | $x$ | SE | $n$ | $x$ | SE | $n$ | $x$ | SE | $n$ |
| *Versus mandibular length* | | | | | | | | | |
| Corpus depth ($M_1$) | 25.67 | 0.35 | 26 | 24.74 | 0.35 | 26 | 25.73 | 0.43 | 17 |
| Corpus depth ($M_2$) | 24.05 | 0.33 | 26 | 23.10 | 0.33 | 26 | 24.33 | 0.41 | 17 |
| Corpus width ($M_1$) | 12.51 | 0.23 | 26 | 12.05 | 0.23 | 26 | 12.99 | 0.28 | 17 |
| Corpus width ($M_2$) | 13.56 | 0.29 | 26 | 12.88 | 0.29 | 25 | 13.93 | 0.36 | 17 |
| Symphyseal depth | 38.92 | 0.40 | 26 | 39.57 | 0.40 | 26 | 39.67 | 0.50 | 17 |
| Symphyseal width | 16.87 | 0.22 | 26 | 15.67 | 0.22 | 26 | 17.80 | 0.28 | 17 |
| Condylar width | 21.02 | 0.62 | 26 | 20.08 | 0.62 | 26 | 23.09 | 0.76 | 17 |
| Temporomandibular joint height | 49.81 | 0.75 | 26 | 47.49 | 0.77 | 25 | 55.65 | 0.93 | 17 |
| Ramal height | 75.48 | 0.72 | 26 | 72.25 | 0.73 | 25 | 81.38 | 0.89 | 17 |
| Masseter lever arm length | 57.16 | 0.65 | 24 | 54.42 | 0.64 | 25 | 58.63 | 0.80 | 16 |
| Bizygomatic breadth | 107.26 | 0.82 | 24 | 98.35 | 0.80 | 26 | 102.09 | 1.05 | 15 |
| Bicondylar breadth | 84.96 | 2.40 | 25 | 82.74 | 2.35 | 26 | 87.01 | 2.91 | 17 |
| Gonion–zygion distance | 68.41 | 1.83 | 26 | 68.34 | 1.83 | 26 | 73.22 | 2.41 | 15 |
| Basicranial length | 121.33 | 1.15 | 23 | 130.84 | 1.11 | 25 | 127.99 | 1.48 | 14 |
| *Bizygomatic breadth versus bicondylar breadth* | 121.11 | 1.15 | 24 | 119.13 | 1.08 | 27 | 118.86 | 1.40 | 16 |
| *Palatal length versus basicranial length* | 76.22 | 1.22 | 24 | 87.22 | 1.16 | 26 | 86.23 | 1.37 | 19 |

Table A6.3.  *Results of one-way ANOVAs among adult subspecies of* Gorilla
*for craniomandibular dimensions*

| | *Gorilla gorilla gorilla* versus *Gorilla gorilla beringei* | *Gorilla gorilla gorilla* versus *gorilla Gorilla graueri* | *Gorilla gorilla beringei* versus *gorilla Gorilla graueri* |
|---|---|---|---|
| *Versus mandibular length* | | | |
| Corpus depth ($M_1$) | NS | NS | NS |
| Corpus depth ($M_2$) | NS | NS | * |
| Corpus width ($M_1$) | * | * | NS |
| Corpus width ($M_2$) | * | * | * |
| Symphyseal depth | NS | NS | NS |
| Symphyseal width | ** | ** | ** |
| Condylar width | * | NS | ** |
| Temporomandibular joint height | ** | * | ** |
| Ramal height | ** | ** | ** |
| Masseter lever arm length | NS | ** | ** |
| Bizygomatic breadth | * | ** | NS |
| Bicondylar breadth | NS | NS | NS |
| Gonion–zygion distance | NS | NS | NS |
| Basicranial length | ** | ** | NS |
| *Bizygomatic breadth versus bicondylar breadth* | NS | NS | NS |
| *Palatal length versus basicranial length* | ** | ** | NS |

*, $p < 0.05$; **; $p < 0.01$; NS, not significant.

Table A6.4. *Descriptive statistics for shape ratios derived from linear dimensions versus basicranial length for adult subspecies of* Gorilla

| Versus basicranial length | Gorilla gorilla gorilla | | | Gorilla gorilla graueri | | | Gorilla gorilla beringei | | |
|---|---|---|---|---|---|---|---|---|---|
| | $x$ | SE | $n$ | $x$ | SE | $n$ | $x$ | SE | $n$ |
| Corpus depth ($M_1$) | 31.42 | 0.53 | 24 | 32.47 | 0.51 | 26 | 32.30 | 0.63 | 17 |
| Corpus depth ($M_2$) | 29.38 | 0.54 | 24 | 30.40 | 0.52 | 26 | 30.83 | 0.65 | 17 |
| Corpus width ($M_1$) | 15.12 | 0.29 | 24 | 15.79 | 0.28 | 26 | 16.99 | 0.35 | 17 |
| Corpus width ($M_2$) | 16.44 | 0.33 | 24 | 16.91 | 0.32 | 25 | 18.17 | 0.39 | 17 |
| Symphyseal depth | 47.52 | 0.71 | 24 | 51.73 | 0.68 | 26 | 50.08 | 0.84 | 17 |
| Symphyseal width | 20.41 | 0.32 | 24 | 20.50 | 0.31 | 26 | 23.02 | 0.38 | 17 |
| Condylar width | 24.95 | 0.75 | 24 | 26.17 | 0.72 | 26 | 29.52 | 0.89 | 17 |
| Temporomandibular joint height | 60.38 | 1.02 | 24 | 61.93 | 0.98 | 26 | 71.49 | 1.21 | 17 |
| Ramal height | 91.80 | 1.08 | 24 | 94.41 | 1.04 | 26 | 103.78 | 1.29 | 17 |
| Masseter lever arm length | 69.66 | 0.94 | 24 | 71.06 | 0.91 | 26 | 74.59 | 1.06 | 19 |
| Bizygomatic breadth | 130.03 | 1.23 | 23 | 128.41 | 1.16 | 26 | 129.00 | 1.43 | 17 |
| Bicondylar breadth | 102.43 | 3.00 | 23 | 108.15 | 2.82 | 26 | 110.41 | 3.59 | 16 |
| Gonion–zygion distance | 82.75 | 2.44 | 24 | 89.69 | 2.35 | 26 | 93.27 | 3.09 | 15 |

Table A6.5. *Results of one-way ANOVAs among adult subspecies of* Gorilla *for craniomandibular dimensions versus basicranial length*

| Versus basicranial length | Gorilla gorilla gorilla versus Gorilla gorilla beringei | Gorilla gorilla gorilla versus Gorilla gorilla graueri | Gorilla gorilla beringei versus Gorilla gorilla graueri |
|---|---|---|---|
| Corpus depth ($M_1$) | NS | NS | NS |
| Corpus depth ($M_2$) | NS | NS | * |
| Corpus width ($M_1$) | ** | NS | * |
| Corpus width ($M_2$) | ** | NS | NS |
| Symphyseal depth | NS | NS | NS |
| Symphyseal width | ** | NS | ** |
| Condylar width | ** | NS | * |
| Temporomandibular joint height | ** | NS | ** |
| Ramal height | ** | NS | ** |
| Masseter lever arm length | ** | NS | * |
| Bizygomatic breadth | NS | NS | NS |
| Bicondylar breadth | NS | NS | NS |
| Gonion–zygion distance | * | NS | NS |

*, $p < 0.05$; **, $p < 0.01$; NS, not significant.

# 7 Intraspecific and ontogenetic variation in the forelimb morphology of Gorilla

SANDRA E. INOUYE

## Introduction

Recent field, morphological, and molecular studies have increased our knowledge of variation in behavior, morphology, and genetics among the different subspecies of *Gorilla*. The behavioral and morphological variation between the eastern mountain gorilla (*Gorilla gorilla beringei*) and western lowland gorilla (*Gorilla gorilla gorilla*) is more extensive than previously thought, and mountain gorillas are viewed as more terrestrial and less arboreal than western lowland gorillas. This behavioral difference between mountain and western lowland gorillas is purportedly associated with different habitats, social organization, and a suite of morphological features.

Work that examines and evaluates the extent of morphological and behavioral variation among the subspecies of *Gorilla* is important for a number of reasons. Western lowland gorillas are more arboreal than previously thought (e.g., Fay, 1989; Fay *et al.*, 1989; Remis, 1995, 1998), while mountain gorillas are more terrestrial and climb trees less (Akeley, 1929; Schaller, 1963; Fossey, 1983; Watts, 1984; Tuttle and Watts, 1985; Doran, 1996). Morphological differences in the postcranium presumably relate to differences between mountain and western lowland gorillas in postural and locomotor behavior, associated with greater arboreality in the latter. Although some studies have examined the subspecific morphological variation of the postcranium in adult *Gorilla* (e.g., Schultz, 1927, 1930, 1934; Groves, 1972; Jungers and Susman, 1984; Larson, 1995; Sarmiento and Marcus, 2000) we need to supplement these studies of variation in mountain and lowland gorilla postcrania from an ontogenetic perspective (Inouye, 1992, 1994*a,b*; Taylor, 1995, 1997*a,b*; Inouye and Shea, 1997). This approach is advocated because it allows us to see when differences emerge and how morphologies develop; furthermore, selection acts not only on adults, but throughout ontogeny. Whenever local or overall body size differences exist, we can identify those characters that are special morphologies or adaptations independent of size from those features that are size-related or size-required. Thus, we can more precisely explicate the functional link between morphological variation and purported behavioral differences between the sub-species of *Gorilla*. Moreover, a good knowledge of the functional relevance of morphological features in *Gorilla* can help us make informed contributions to the current debate about the evolution of knuckle-walking (Shea and Inouye, 1993; Inouye, 1994*a*; Dainton and Macho, 1999; Richmond and Strait, 2000).

Another reason why a closer examination of subspecific variation in *Gorilla* postcrania is important is that recent discussions about the morphological and behavioral variation among the subspecies of *Gorilla* factor into debates regarding the taxonomic status of these taxa. Based on the extent of variation and conservation concerns, some favor the taxonomic elevation of the subspecies designations of *Gorilla* to the species level. Some proponents of this view suggest that the morphological differences between *G. g. gorilla* and *G. g. beringei* warrant a species level distinction (Groves, this volume; Stumpf *et al.*, this volume), while the molecular differences between *G. g. gorilla* and *G. g. beringei* may be as great as those between *P. troglodytes* and *P. paniscus* (Ruvolo *et al.*, 1994). Therefore, to contribute to this taxonomic issue, it is important to examine carefully ontogenetic and allometric variation in postcranial features purportedly associated with terrestrial/arboreal differences between mountain and western lowland gorillas to determine how many and which features truly differ between them once we properly account for any size differences that may exist.

This study adds to this body of work on variation in *Gorilla*, and addresses the ontogenetic and allometric variation in the skeleton of the forelimb of mountain and western lowland gorillas. This study specifically examines the ontogenetic variation in morphological features that purportedly correlate with functional differences in the locomotor repertoire of mountain and western lowland gorillas, in order to determine which features are underlain by concordant growth allometries and which features are products of restructured covariance that may be linked to new functions. This ontogenetic, allometric study helps to elucidate the underlying patterns of postcranial variation between *G. g. gorilla* and *G. g. beringei* and its relationship to positional (locomotor and postural) behavior, with hopes of clarifying some of these important issues.

### *Subspecies of* Gorilla

Gorillas are the largest of the extant apes and occupy West Africa, northeastern central Africa, and southwestern Uganda. There are two (Coolidge, 1929) or three (Groves, 1967, 1970*a,b*; Leigh *et al.*, this volume) well-recognized sub-species of *Gorilla gorilla*, and perhaps more (Groves, this volume; Albrecht *et al.*, this volume). Of the three well-recognized subspecies, the western lowland gorilla, *Gorilla gorilla gorilla*, lives in the forests of Cameroon, Equatorial Guinea, Gabon, Congo, Cabinda, the western part of Central African Republic, and north of the Zaïre river (Coolidge, 1936; Groves, 1971; Jones and Sabater Pi, 1971; Fossey, 1982; Emmons *et al.*, 1983; Tutin and Fernandez, 1984*a*,*b*). *Gorilla gorilla beringei*, the mountain gorilla, lives on the Virunga mountains, which lie along the international borders of Zaïre, Rwanda, and Uganda. Lastly, *G. g. graueri* inhabits the eastern lowland and highland forests of northeastern Zaïre, east of the Lualaba River and west of the section of the Western Rift (Tuttle, 1986).

Gorillas are very sexually dimorphic and adult females are about half the weight of adult male *Gorilla*, resulting predominantly from bimaturism, or a difference in age of sexual maturity (Leigh, 1992; Leigh and Shea, 1995, 1996). The actual size differences between the three subspecies remain unclear, partly due to small sample sizes for the mountain and eastern lowland gorillas. The mean adult weights of the three subspecies according to Smith and Jungers (1997) are as follows: *G. g. beringei*, 162.5 kg for males ($n = 5$) and 97.5 kg for females ($n = 1$), *G. g. gorilla*, 170.4 kg ($n = 10$) for males and 71.5 kg ($n = 3$) for females, and *G. g. graueri*, 175.2 kg ($n = 4$) for males and 71.0 kg ($n = 2$) for females. However, this is in contrast to Groves (1970*b*), who indicates that the mean adult male weights are: western lowland, 139 kg ($n = 32$), mountain, 156 kg ($n = 6$), and eastern lowland, 163 kg ($n = 4$). Female mountain gorillas are somewhat larger than their western lowland counterparts, and the ranges for male mountain and lowland gorillas clearly overlap, but Groves (1970*b*) indicates that male mountain gorillas are larger than western lowland gorillas.

### Taxonomic issues

The debate about the number of well-defined subspecies of *Gorilla* has a long history, and has yet to be resolved (Groves, 1967, 1970*a*,*b*, this volume; Albrecht *et al.*, this volume; Leigh *et al.*, this volume; Stumpf *et al.*, this volume). Based on an early study of crania, Coolidge (1929) argued for a single species of *Gorilla* and two subspecies – the western lowland gorilla, *G. g. gorilla*, and the eastern or mountain gorilla, *G. g. beringei*. Coolidge and Schultz noted many differences in nearly all parts of the body between eastern and western gorillas, enough to convince Schultz (1934) and others that the "short-armed" *G. g. beringei* should be classified as a single species. However, Groves (1970*b*) challenged this assertion, arguing that the eastern group should be divided into the eastern lowland gorilla, *G. g. graueri*, and the mountain gorilla, *G. g. beringei*, thus creating three distinct subspecies of *Gorilla*. Some subsequent craniometric studies support the argument for three subspecies of *Gorilla* (Albrecht and Miller, 1993; Gelvin *et al.*, 1997; Leigh *et al.*, this volume), while other studies argue for different numbers of species and/or subspecies (see chapters in this volume by Albrecht *et al.*, Groves, and Stumpf *et al.*).

### *Variation in positional and locomotor behavior in* Gorilla

Many studies demonstrate that mountain gorillas are quite distinct behaviorally from lowland gorillas. Field studies show the mountain gorilla to be terrestrial and herbivorous (Akeley, 1929; Schaller, 1963; Fossey, 1983; Watts, 1984; Tuttle and Watts, 1985; Doran, 1996) in contrast to the western lowland gorillas, which are more arboreal and frugivorous (e.g., Fay, 1989; Fay *et al.*, 1989; Nishihara, 1995; Remis, 1995, 1998; Carroll, 1996). Remis shows that variation in body size influences positional behavior in some predictable ways, but there is also a considerable amount of variation in positional behavior that is influenced by other factors, such as habitat and social context (Remis, 1995, 1998, 1999). For example, Remis (1995, 1998) predicts and shows that the large, male lowland gorillas, compared to female lowland gorillas, face difficulties with balance, substrate size and the energetics of climbing, and these affect their frequency of climbing, size of substrates used and modes of positional behavior. These differences probably occur because female gorillas, compared to males, are constrained less by their body size and therefore can enter trees of different shapes and varying fruit availability (Remis, 1999). This is in contradistinction to mountain gorillas, which exhibit fewer sex differences in feeding and positional behavior (Tuttle and Watts, 1985).

Remis (1998) finds that lowland gorillas, compared to mountain gorillas, use trees at a much higher frequency and use small substrates found at great heights in the canopy. In fact, lowland gorillas do not appear to engage in significantly less arboreal suspensory locomotion than do chimpanzees (Doran and Hunt, 1994; Remis, 1995, 1998). Mountain gorillas, on the other hand, use large, low-lying boughs and fallen trunks, and have a less varied locomotor repertoire compared to their lowland counterparts. These marked differences in behavior between the subspecies of *Gorilla* are interesting because there is considerable overlap in adult body size between lowland gorillas and mountain gorillas, yet there is not a similar overlap in behavior (Remis, 1998). Remis (1998) explains that these behavioral differences between lowland and mountain gorillas are probably a reflection of the reliance on ground foods and the scarcity of arboreal substrates and foods in the habitat of the latter. Remis (1998) emphasizes that besides body size, other factors such as seasonality, social context, and patch size influence variation in substrate use and positional behavior.

Table 7.1. *Summary of cranial and postcranial features of mountain gorillas, in comparison to lowland gorillas*

*According to Coolidge (1929)*
- Longer palate
- Generally narrower skull
- Thicker pelage
- Large amount of black hair
- Fleshy callosity on the crest of the head

*According to Schultz (1927, 1930, 1934)*
- Higher face
- Narrower width between eyes
- Greater length of the trunk
- Higher-situated nipples
- Narrower hips
- Greater length of the neck
- Shorter lower limbs in relation to height of the trunk
- Longer lower limbs in relation to length of the upper limbs
- Much shorter upper limbs
- Broader and shorter hand
- Great toe reaching farther distally
- Great toe branching from sole more distally
- Less convex joint at base of hallux
- Relatively shorter outer lateral toes
- Usually webbed toes
- Smaller average number of thoracolumbar vertebrae
- Absolutely and relatively shorter humerus
- Absolutely and relatively longer clavicle
- Absolutely and relatively longer ilium
- Relatively longer radius
- Concavely curved vertebral border of the scapula
- Shorter scapulae
- Lower scapular indices
- Higher ratio of infraspinous fossa versus scapular spine length
- Higher ratio of spine length versus scapular length

## *Morphological variation in* Gorilla

Coolidge (1929) and Schultz (1927, 1934) document a number of cranial differences between *G. g. gorilla* and *G. g. beringei* (Table 7.1), and a number of subsequent studies discuss craniometric and dental features that distinguish the subspecies of *Gorilla* from one another (Coolidge, 1929; Groves, 1967, 1970*a,b*, 1972; Albrecht and Miller, 1993; Gelvin *et al.*, 1997). Elsewhere in this volume, others review and discuss these differences between mountain and western lowland gorillas in detail as well as present new data on cranial and mandibular variation in *Gorilla* (Albrecht *et al.*, Groves, Leigh *et al.*, Stumpf *et al.*, Taylor, this volume).

In addition to craniometric variation, a number of studies document postcranial differences among the subspecies of *Gorilla* (e.g., Schultz, 1927, 1930, 1934; Groves, 1972; Jungers and Susman, 1984; Sarmiento and Marcus, 2000; Taylor, 1995, 1997*a*,*b*; Inouye, 1992, 1994*a*,*b*). Schultz's seminal works on *Gorilla* (1927, 1930, 1934) document significant morphological differences between *G. g. beringei* and *G. g. gorilla*, which are summarized in Table 7.1. Schultz (1927) finds that although *G. g. beringei* and *G. g. gorilla* overlap in range of variation for intermembral indices, *G. g. beringei* has a slightly lower value than both *G. g. gorilla* and *G. g. graueri*. This proportional difference is mainly due to the absolutely and relatively shorter arms and relatively longer legs of *G. g. beringei* compared to *G. g. gorilla*. Schultz (1934) also shows that the number of thoracolumbar vertebrae averages 16.7 in *G. g. gorilla*, though only 16.0 in *G. g. beringei*.

*Gorilla g. gorilla* also has longer scapulae, higher scapular indices, and lower ratios of infraspinous fossa versus scapular length and spine length versus scapular length than *G. g. beringei* (Schultz, 1930, 1934). Schultz also notes that mountain gorillas have scapulae with concave vertebral borders. He cautions, however, that these averages are based on a small sample of *G. g. beringei* ($n = 8$) in comparison to that of *G. g. gorilla* ($n = 96$). A subsequent study by Larson (1995) shows that mountain and lowland gorillas are similar in most features of the scapula and proximal humerus that relate to the line of action and attachments of the rotator cuff muscles. Orientation of the scapular spine, relative width of the infraspinous fossa at the neck of the scapula, angle between the infraspinatus and supraspinatus insertion facets, and orientation of the subscapularis insertion facet are similar in *G. g. gorilla* and *G. g. beringei* (Larson, 1995). However, there is less overlap between the subspecies for the angle between the infraspinatus insertion facet and the long axis of the humerus (Larson, 1995).

Schultz (1927, 1934) further examined fetal, infant, and adult gorillas, and shows in a series of ratios that mountain gorillas differ from lowland gorillas in the following: the hips (between greater trochanters) are narrower and longer in relation to the height of the trunk, the nipples are situated higher on the trunk, the neck is longer in relation to the trunk, the lower limb is shorter in relation to the trunk, the upper limb and upper arm are very much shorter in relation to the trunk, the hand is broader in relation to the length of the hand with a proportionately strong and long thumb, the free portion of the great toe is shorter, the cleft between the great toe and second toe is shorter, the lateral toes are shorter, and the great toe reaches more distally along the total length of the foot. As Schultz notes (1934), some of these proportional differences may be explained in part by differences between mountain and lowland gorillas in certain dimensions, such as the longer trunk length and shorter hand length in the mountain gorilla.

Schultz (1927) and others conclude from comparative morphological studies of African apes that while gorillas in general have become less adapted for arboreal life compared to other apes, this condition has been exacerbated in mountain gorillas. For example, gorillas, especially mountain gorillas, have relatively shorter pedal digits than does *Pan*, with less plantar curvature (Morton, 1924; Keith, 1926; Schultz, 1927, 1968; Tuttle, 1970, 1972; Fossey, 1983) and an increased development along the inner border, especially of the hallux (Morton, 1927). These proportions are similar to those of the hand (Schultz, 1927, 1934); the thumb is longer and the hand is relatively shorter and broader in mountain gorillas than in the lowland forms (Schultz, 1927, 1934). Also, in mountain gorillas, the lateral toes are often webbed in between the digits up to the first interphalangeal joint, making these toes even shorter functionally (Schaller, 1963; Tuttle, 1970, 1972; Fleagle, 1988). In describing the foot of mountain gorillas, Schultz (1968:132) says that "The grasping ability of the first toe is decidedly more limited than in typical western gorillas and the other toes appear to be remarkably short and better suited for terrestrial than arboreal life." Schultz (1927, 1934) and others (Schaller, 1963; Tuttle, 1970, 1972; Taylor, 1995, 1997*a*) argue that the suite of morphological differences between *G. g. gorilla* and *G. g. beringei* in the scapula, limb proportions, hands, and feet reflect the greater adaptation of *G. g. beringei* to terrestrial life.

### Allometric variation in Gorilla

Many of the documented differences between mountain and lowland gorillas are expressed in the form of proportions (e.g., Morton, 1924; Keith, 1926; Schultz, 1927, 1930, 1934, 1968; Tuttle, 1970, 1972; Fossey, 1983), which unfortunately makes it difficult to determine which dimensions comprising the ratio are responsible for the proportional differences. Furthermore, these studies do not tell us how these proportions change throughout ontogeny. Some have built upon these comparative, static studies of *Gorilla* morphology, using ontogenetic and allometric approaches (Jungers and Susman, 1984; Inouye, 1992, 1994*a,b*; Shea and Inouye, 1993; Taylor, 1995, 1997*a*) to examine the postcranial variation among the subspecies of *Gorilla*.

#### Scapula

Taylor (1995, 1997*a*) expands upon Schultz's (1930, 1934) earlier scapular studies utilizing a significantly larger, ontogenetic, data set. Taylor finds that mountain gorillas have significantly longer scapular spines and shorter scapulae than lowland gorillas of comparable superior border lengths. However, at comparable body weight estimates (based on dry skeletal weights), mountain gorillas exhibit significantly shorter spines and superior borders than do lowland gorillas. Taylor

Table 7.2. *Means and standard deviations for bar-glenoid angle in adults*

| Taxonomic group | *n* | Bar-glenoid angle (SD) |
|---|---|---|
| *Gorilla gorilla gorilla* | 44 | 128.7 (5.3) |
| *Gorilla gorilla beringei* | 22 | 127.0 (6.5) |
| *Pan paniscus* | 8 | 126.1 (9.0) |
| *Pan troglodytes* | 22 | 127.0 (5.4) |

(1995, 1997*b*) also demonstrates that differences in scapular dimensions between male and female mountain gorillas are attributable to ontogenetic scaling. In other words, males and females share a common growth trajectory, and males continue to grow along this trajectory after cessation of female growth.

Inouye and Shea (1997) show that the bar-glenoid angle of the scapula varies a fair amount in great apes and humans. However, there is no significant difference between mountain and lowland gorillas in this angle (Table 7.2), and the adult ranges of all Great Apes and humans overlap extensively. Furthermore, they find no significant allometric relationship between bar-glenoid angle and glenoid cavity length throughout ontogeny for either subspecies of *Gorilla*. In other words, this angle does not change with size throughout ontogeny in either *Gorilla* or *Pan* (or *Pongo*). This contrasts with humans, which exhibit an allometric pattern in this angle wherein the range of variation overlaps with the African apes at the smaller size ranges. Inouye and Shea (1997) conclude that the bar-glenoid angle does not tightly correlate with function in humans and, as such, cannot be a good morphological indicator of degree of arboreality among African apes and humans (hominids) at smaller size ranges. The lack of correspondence of this angle to behavioral differences between mountain and lowland gorillas further strengthens their conclusions.

### Hands and feet

Inouye (1992, 1994*a,b*) explores the relationship of hand dimensions to body weight surrogates during ontogeny, and shows that most dimensions of the third and fifth rays, which include the third metacarpal and the proximal and middle phalanges, are ontogenetically concordant between mountain and lowland gorillas (Table 7.3). However, there is a clear difference in the relative length of the metacarpals; mountain gorillas have relatively shorter metacarpals compared to their western lowland counterparts (Inouye, 1992, 1994*a,b*).

The difference between mountain and lowland gorillas in metacarpal length parallels a difference between *Gorilla* and *Pan*; *Gorilla* has shorter metacarpals than *Pan* at comparable body sizes (Inouye, 1992, 1994*a*). This trend is similar for all of the metacarpals (Inouye, 1994*a*). This pattern of shorter manual digits in mountain gorillas compared to lowland gorillas is identical for the feet; mountain gorillas have relatively shorter lateral metatarsals compared to lowland

Table 7.3. *ANCOVAs between* G. g. beringei *and* G. g. gorilla *for regressions of dimensions of fifth metacarpal and fifth proximal phalanx against size surrogates*

| Variable | *Gorilla gorilla gorilla* versus *Gorilla gorilla beringei* |
|---|---|
| *Fifth metacarpal* | |
| Metacarpal length | NS/* |
| Mediolateral midshaft width | NS/NS |
| Dorsopalmar midshaft depth | NS/NS |
| Mediolateral head width | NS/NS |
| Dorsopalmar head width | NS/NS |
| Biepicondylar width | NS/NS |
| *Fifth proximal phalanx* | |
| Proximal phalanx length | NS/* |
| Mediolateral midshaft width | NS/NS |
| Dorsopalmar midshaft depth | NS/NS |
| Mediolateral basal width | NS/NS |
| Dorsopalmar basal width | NS/NS |
| Mediolateral trochlear width | NS/NS |
| Dorsopalmar trochlear width | NS/NS |
| Biepicondylar width | NS/NS |

NS, not significant.
*, significant at $p < 0.001$.

gorillas (Inouye, 1994*a*). The relatively shorter metacarpals and metatarsals seem to support Schultz's claims of adaptations for greater terrestriality in the mountain gorillas.

Previous work shows the possible size-related component to expression of knuckle-walking features and locomotion (Tuttle, 1967, 1969*a*; Tuttle and Basmajian, 1974; Susman, 1979; Susman and Creel, 1979). The dorsal metacarpal ridge, a purported knuckle-walking adaptation (Tuttle, 1967, 1969*a*,*b*,*c*), is variable in its presence in all the African ape taxa (Inouye, 1994*a*). Inouye shows that it is not only absent in some juveniles, as Susman (1979) pointed out, but it is also absent in many African ape adults (Table 7.4). Inouye also shows that there is no difference between adult mountain and lowland gorillas or between adult bonobos and chimpanzees in metacarpal ridge height for each of the second to fifth rays. However, there is a significant difference between *Pan* and *Gorilla*; on average, *Gorilla* has absolutely larger metacarpal ridges (Table 7.5), almost twice the size, than the ridges on the corresponding rays of *Pan* (Inouye, 1994*a*). Further allometric analyses, however, clarify that the metacarpal ridges are ontogenetically scaled between *Pan* and *Gorilla* (Shea and Inouye, 1993; Inouye, 1994*a*).

Table 7.4. *Percentage and number of specimens that exhibit metacarpal ridges based on the measured angle and ridge height*

| Species | Metacarpal II | Metacarpal III | Metacarpal IV | Metacarpal V |
|---|---|---|---|---|
| Subadult and adult | | | | |
| *Gorilla gorilla gorilla* | 73 (63/86) | 91 (80/88) | 85 (73/86) | 76 (62/82) |
| *Gorilla gorilla beringei* | 56 (09/46) | 80 (12/15) | 56 (09/16) | 56 (09/16) |
| *Pan paniscus* | 65 (13/20) | 81 (17/21) | 67 (14/21) | 63 (12/19) |
| *Pan troglodytes* | 38 (33/88) | 71 (60/84) | 62 (56/91) | 33 (28/84) |
| Adults only | | | | |
| *Gorilla gorilla gorilla* | 85 (39/46) | 100 (49/49) | 94 (44/47) | 82 (37/45) |
| *Gorilla gorilla beringei* | 56 (05/09) | 88 (07/08) | 56 (05/09) | 33 (03/09) |
| *Pan paniscus* | 67 (06/09) | 80 (08/10) | 60 (06/10) | 56 (05/09) |
| *Pan troglodytes* | 52 (25/48) | 79 (37/47) | 76 (38/50) | 47 (22/47) |

Table 7.5. *Average heights and standard deviations of metacarpal ridges, adults only*

| | Metacarpal II | | Metacarpal III | | Metacarpal IV | | Metacarpal V | |
|---|---|---|---|---|---|---|---|---|
| Species | Height (mm) (SD) | $n$ | Height (mm) (SD) | $n$ | Height (mm) (SD) | $n$ | Height (mm) (SD) | $n$ |
| *Gorilla gorilla gorilla* | 1.41(1.01) | 39 | 2.41(1.77) | 49 | 3.04(1.52) | 44 | 1.77(1.47) | 37 |
| *Gorilla gorilla beringei* | 1.06(0.76) | 5 | 1.84(1.53) | 7 | 1.77(1.14) | 5 | 1.03(0.45) | 3 |
| *Pan paniscus* | 0.72(0.44) | 6 | 0.78(0.42) | 8 | 0.93(0.67) | 6 | 0.58(0.32) | 5 |
| *Pan troglodytes* | 0.87(0.65) | 25 | 1.42(0.96) | 37 | 1.50(0.95) | 38 | 0.62(0.47) | 22 |

There are several possible explanations for the variation in the metacarpal ridges (Inouye, 1994*a*). Comparing ridge height to hand posture during knuckle-walking, Inouye (1994*a*) concludes that development of the metacarpal ridge is not related to frequency of overextension (versus other joint postures during knuckle-walking) at the metacarpophalangeal joint. Metacarpal ridge height scales ontogenetically between *Pan* and *Gorilla*, while frequency of overextension does not (Inouye, 1994*a*). Variation in metacarpal ridge height may relate to the degree of overextension at the metacarpophalangeal joint, but kinematic data are needed to evaluate this hypothesis. Other functional explanations of the metacarpal ridges include a relationship between ridge height and overall frequency of knuckle-walking, compared to other modes of locomotion, and a relationship between metacarpal ridge height and amount of body weight borne on the hands (Inouye, 1994*a*).

Table 7.6. *Percentage and number of specimens that exhibit metatarsal ridges based on the measured angle and ridge height*

| Species | Metatarsal II | Metatarsal III | Metatarsal IV | Metatarsal V |
|---|---|---|---|---|
| Subadults and adults | | | | |
| Gorilla gorilla gorilla | 48 (38/80) | 62 (49/79) | 54 (44/81) | 47 (33/70) |
| Gorilla gorilla beringei | 21 (03/14) | 62 (08/13) | 53 (08/15) | 38 (05/13) |
| Pan paniscus | 60 (09/15) | 28 (05/18) | 11 (02/19) | 12 (02/17) |
| Pan troglodytes | 12 (09/78) | 19 (15/78) | 12 (09/75) | 15 (11/71) |
| Adults only | | | | |
| Gorilla gorilla gorilla | 40 (19/47) | 58 (28/48) | 65 (32/49) | 65 (26/40) |
| Gorilla gorilla beringei | 29 (02/07) | 71 (05/07) | 63 (05/08) | 43 (03/07) |
| Pan paniscus | 43 (03/07) | 20 (02/10) | 10 (01/10) | 11 (01/09) |
| Pan troglodytes | 17 (08/46) | 30 (14/47) | 19 (09/47) | 23 (11/47) |

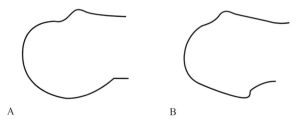

A                                       B

Fig. 7.1. Metacarpal and metatarsal tori. (A) The right third metacarpal and (B) the right third metatarsal from a male adult *Gorilla*. Note the development of a torus on the metatarsal head.

An important point made from that study (Inouye, 1994*a*) is that even though metacarpal ridge height covaries with body size, knuckle-walking does not biomechanically require the presence of metacarpal ridges since ridges are variably present in both smaller and larger *Gorilla* and *Pan*. Furthermore, whether or not the ridges relate to the frequency or ability to knuckle-walk, it is difficult to argue that the metacarpal ridges are products of knuckle-walking, *per se*, since the metatarsals of African apes also exhibit ridges on the dorsum (Fig. 7.1; Table 7.6). These metatarsal ridges follow a pattern of variation similar to the development and expression of metacarpal ridges (Inouye, 1994*a*).

*Other postcranial comparisons*

Jungers and Susman (1984) examined differences between the lengths of various extremities in the same sexes of *G. g. gorilla* and *G. g. beringei*. They found no significant differences in their comparisons and therefore concluded that several of the characters Schultz described as distinguishing the mountain from lowland gorilla are highly questionable. For example, they found that the intermembral indices are almost identical, and mountain gorillas may have a somewhat greater intermembral index than lowland gorillas (Jungers and Susman, 1984). Also, they found that the humerus of mountain gorillas was not significantly different in length from that of lowland gorillas. Other parts of the postcranial skeleton, such as scapular length, scapular breadth, humeral length, femoral length, femoral circumference, and iliac length were found to exhibit patterns of ontogenetic scaling between trajectories of *G. g. beringei* and *G. g. gorilla* (Taylor, 1997*b*).

### Allometry and ontogeny

Most of the early work on *Gorilla* postcrania (e.g., Schultz, 1927, 1930, 1934; Groves, 1972) and functional interpretations about morphological variation were based on adult forms. However, differences in overall body size in *Gorilla*, especially in such a sexually dimorphic genus, may have a significant effect on postcranial proportions and thus influence our functional interpretations of the morphological variation (e.g., Jungers and Susman, 1984; Inouye, 1992, 1994*a*,*b*; Shea and Inouye, 1993; Taylor, 1995, 1997*a*,*b*). Shea (e.g., 1981, 1983*a*, 1985*a*,*b*, 1986) and others have shown that this is the case for many cranial proportions that are distinct in adult *Pan* and *Gorilla* but are similar once body size is properly considered. In other words, it is necessary to evaluate which differences between mountain and lowland gorilla adults are attributed to differences in overall size, and which may be attributed to discordant ontogenetic trajectories, which signify restructured patterns of covariance or proportional changes reflecting a correlative functional shift.

Whether or not mountain and lowland gorillas differ in body size is debatable (Groves, 1970*b*; Smith and Jungers, 1997), and therefore one may question why one should use an ontogenetic approach to examining variation between these groups. An ontogenetic and allometric approach is advocated because there may be body size differences between the taxa. Of course, it is necessary in order to discern which differences between male and female adults are the result of overall size differences and which are due to sex effects. Also, some studies have shown that in spite of much overlap in body size, there are localized size differences between the mountain and lowland gorillas – in the face, for example (see Taylor, this volume). Lastly, even though they may not differ in size, examination of growth allometries can show us *how* these adult differences between mountain and lowland gorillas are produced and knowledge of their postnatal development can aid in our functional interpretation of their morphologies.

*Ontogenetic approaches to form–function relationships*

Many comparative studies examine variation in shape and function and then make correlative interpretations as to the relationship between form and function. Hypotheses derived from these interpretations may subsequently be used to infer the behavior of an extinct organism from its fossilized morphology. However, many shape differences are related to body size; thus, careful and complete comparative analysis requires recognition of the effects of allometry (Huxley, 1932; Gould, 1966; McMahon and Bonner, 1983; Schmidt-Nielsen, 1984). Furthermore, behavioral or ecological characters (Clutton-Brock and Harvey, 1983; Fleagle, 1985), such as the proportion of time spent in different diurnal activities (Clutton-Brock and Harvey, 1977*a*), home range size (Clutton-Brock and Harvey, 1977*b*), reproductive strategies (Shea, 1987), diet (Kay, 1981; Fleagle, 1985), and other aspects of life history (Shea, 1990) of primates are also in part size-related phenomena. These confounding influences of size can pose a problem to comparative studies.

While it is important that we should question size itself because it is a significant factor in an animal's life history and requirements, Shea (1995) points out a view that he (Shea, 1984, 1985*a,b*, 1986, 1995) and others (Pilbeam and Gould, 1974; Martin, 1989) share, namely that there is a consensus in biology that one should attempt to distinguish allometric shape differences from those unrelated to size when making comparisons between groups (Gould, 1966, 1975; Harvey and Pagel, 1991), and this is no less important in the realm of biological anthropology. Allometry, the study of changes in shape with changes in size, is one approach that we may use to elucidate the relationship between two characters.

*Ontogenetic "criterion-of-subtraction"*

There are three different types of allometry that one can use in comparative studies, and each involves different samples and has different foci. Ontogenetic allometry focuses on the changing proportions during growth of an individual (longitudinal) or species (cross-sectional). Static adult intraspecific allometry focuses on size-related proportional changes among adults of a group, such as male–female proportional differences within a species. Lastly, static interspecific allometry focuses on size-related proportional changes among adults of different species. These types of allometry are very different, and attempts to infer growth patterns from static adult intraspecific allometry or static interspecific allometry have been problematic (e.g., Cock, 1966; Lande, 1979; Atchley *et al.*, 1981; Shea, 1981, 1983*b*, 1995; Cheverud, 1982). There are also different approaches or uses of allometry (Gould, 1966, 1975; Shea, 1981, 1984, 1995). Allometry can be used as a "criterion-of-subtraction" by elucidating the size–shape trends between two variables during growth and examining the residuals or departures from these patterns across species of differing size (Gould, 1975; Smith, 1984*a,b*; Shea, 1985*b*, 1995; Pagel and Harvey, 1988;

Shea and Bailey, 1996). The line, the phenotypic covariance during ontogeny, used as the criterion-of-subtraction, is assumed to represent the genetic covariance between one feature and another and overall size. This is interpreted as similar to the line that quantitative geneticists predict based on empirically derived knowledge of genetic covariance of certain features with each other and overall size (Lande, 1979). Residuals are taken to represent departures from the baseline of (ancestral) covariance, perhaps indicating specific selection for proportions that depart from the baseline allometric relationship.

Multidisciplinary studies that incorporate laboratory and field studies of morphology, performance, and fitness are the best way to elucidate adaptations (Bock, 1980; Arnold, 1981), and certainly comparative studies are limited and cannot directly elucidate adaptations in the same manner. However, what ontogenetic comparative studies can do is demonstrate when ancestral patterns of size–shape or other covariance are restructured or repatterned in the presence of putative selective agents, or altered functions and/or new environments. Allometric shape differences between groups which can be explained via ontogenetic scaling (the differential extension or truncation of a shared common growth trajectory between two groups) of the inherited ancestral pattern of covariance with local or global size are viewed as nonadaptive (Shea, 1995), *sensu* Huxley (1932) and Gould and Lewontin (1979). Conversely, discordant growth trajectories that reflect restructured patterns of genetic covariance presumably selected in response to external factors are classified as adaptations (Shea, 1995).

A recent misconception about ontogenetic scaling is the claim that the shared ontogenetic trajectory is assumed to be inherently functionally neutral by utilizing this approach (Godfrey *et al.*, 1998). In fact, a shared ontogenetic trajectory can represent the relationship between changing proportions and correlative functional change; those studying allometry have traditionally been very interested in the functional basis of shape transformation during growth (e.g., Schultz, 1927, 1934; Huxley, 1932; Gould, 1966; Jungers and Fleagle, 1980; Carrier, 1983; Grand, 1983; Peters, 1983; Pounds *et al.*, 1983 ). Obviously, a shared ontogenetic trajectory that is positively or negatively allometric does signal a change in proportions along the line, and it is not assumed that a feature that is ontogenetically scaled necessarily has the same function at different ends of the size spectrum. However, at any point along the shared growth trajectory, one assumes similar function unless there is demonstrated evidence to the contrary. For example, human hindlimbs grow with positive allometry during ontogeny, and as the hindlimbs grow proportionately longer there is also a correlative biomechanical change in bipedalism. However, one would assume functional equivalence for different humans that share the same point on the line.

A large part of the confusion about the interpretation of ontogenetic scaling relates to the distinction of *function* versus *adaptation* and primitive versus derived functions. An ontogenetic trajectory can certainly represent a trajectory of changing function along different points of the line. However, functional and proportional differences between groups that may be explained by a shared ontogenetic trajectory is not equivalent to an adaptation. Shape differences between groups that can be accounted for by ontogenetic scaling of the ancestral pattern of covariance with size are assumed to be *nonadaptive* (Shea, 1995). Also, dissociation of ontogenetic trajectories may be linked to functional equivalence at different sizes, so there is no necessary relationship mapping coincidence and departure of ontogenetic trajectories onto functional neutrality and functional change.

## Materials and methods

### *Samples and measurements*

The overall composition of the skeletal samples used in this study is described in Table 7.7. The samples of *Gorilla gorilla gorilla* (western lowland gorilla) and *Gorilla gorilla beringei* (mountain gorilla) include roughly equal numbers of males and females within each taxon, and both include adult and subadult specimens. Skeletal measurements used in this study are presented in Table 7.8.

#### *Shoulder*
I examined a set of eight variables relating to weight carriage and stability of the shoulder region during knuckle-walking (Figs. 7.2 and 7.3). Digital calipers were used to measure glenoid cavity width and length, acromial length, coracoid length, humeral head widths, and clavicular length. Humeral torsion, as described by Evans and Krahl (1945), as the counterclockwise angle between the bicondylar axis and the long axis of the humeral head (Fig. 7.4) was also measured.

Table 7.7. *Skeletal sample for bivariate and multivariate analyses*

| Species | *n* | Sex | Museum |
|---|---|---|---|
| *Gorilla gorilla gorilla* | 60 | M | Powell-Cotton Museum and Quex House |
| | 66 | F | |
| | 2 | ? | |
| *Gorilla gorilla beringei* | 18 | M | Musée Royale de l'Afrique Centrale, |
| | 16 | F | U.S. National Museum (Smithsonian |
| | 7 | ? | Institution) |

Table 7.8. *Linear measurements of the forelimb*

| Bone | Dimension |
| --- | --- |
| *Shoulder region* | |
| Scapula | Coracoid process length |
| | Acromial length |
| | Glenoid cavity length |
| | Glenoid cavity width |
| | Glenoid cavity depth |
| Clavicle | Clavicular length |
| Humerus | Humeral anteroposterior head diameter |
| | Humeral mediolateral head diameter |
| *Elbow region* | |
| Humerus | Humeral mediolateral articular surface width |
| | Distal trochlear width |
| | Distal capitular width |
| | Capitular height |
| | Olecranon fossa width |
| | Olecranon fossa length |
| | Olecranon fossa depth |
| | Anteroposterior trochlear length |
| | Posterior articular surface width |
| Ulna | Olecranon anteroposterior proximal depth |
| | Olecranon anteroposterior distal depth |
| | Proximal trochlear notch width |
| | Middle trochlear notch width |
| | Distal trochlear notch width |
| | Trochlear/radial notch combined width |
| | Olecranon length |
| | Coronoid process projection |
| | Medial height of the trochlear notch |
| | Midline height of the trochlear notch |
| Radius | Radial anteroposterior head diameter |
| | Radial mediolateral head diameter |
| *Wrist region* | |
| Ulna | Ulnar styloid process length |
| | Ulnar mediolateral head diameter |
| | Ulnar anteroposterior head diameter |
| Radius | Scaphoid articular surface width of radius |
| | Lunate articular surface width of radius |
| | Radial styloid process length |
| Scaphoid | Mediolateral radial articular facet width |
| | Anteroposterior radial articular facet width |
| Lunate | Mediolateral radial articular facet width |
| | Anteroposterior radial articular facet width |

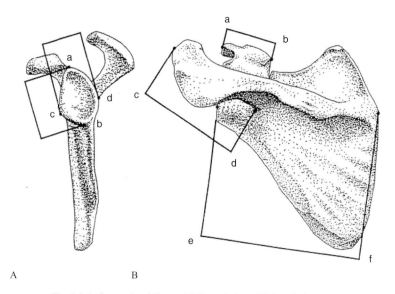

Fig. 7.2. Left scapula of *Homo*. (A) Lateral view; (B) dorsal view.

*Elbow*

I examined a suite of 21 variables relating to stability of the elbow joint region during knuckle-walking (Figs. 7.3 and 7.5). This data set primarily describes articular surface and olecranon process and fossa dimensions in the region of the distal humerus and proximal ulna and radius. Capitular and trochlear articular surface dimensions and olecranon fossa dimensions on the humerus, trochlear notch dimensions of the ulna, and articular head widths of the radius were measured.

*Wrist*

I examined a set of ten variables relating to stability in the wrist region (Fig. 7.6). This data set is composed entirely of articular dimensions of the proximal wrist joint, which include the distal radius and ulna and the proximal articular dimensions of the scaphoid and lunate bones.

### Statistical methods of analysis

*Multivariate regression analyses*

A principal components analysis (PCA) was used to analyze the data because of some of its advantages over bivariate techniques. For example, a PCA can simultaneously consider many linear variables, summarizes large numbers of

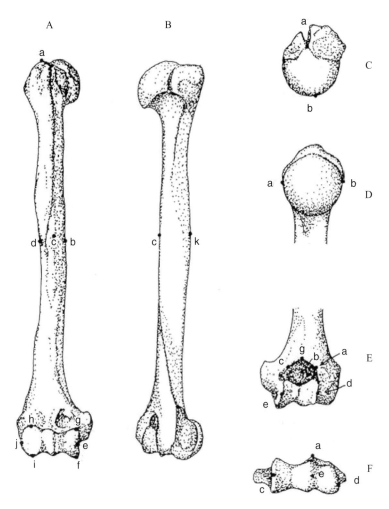

Fig. 7.3. Right humerus of *Pan*. (A) Anterior view; (B) lateral view; (C) superior view; (D) medial view; (E) posterior view; (F) inferior view.

variables in a few dimensions, and has the potential to reveal biologically meaningful patterns of covariation among variables that are sometimes over-looked when utilizing bivariate analytical techniques (Davies and Brown, 1972; Oxnard, 1978; Neff and Marcus, 1980; Shea, 1985a). Thus, I used PCAs to search initially for patterns of variation in the data sets, and when there ap-peared to be a difference between groups in the variable component loadings, an analysis of variance (ANOVA) or analysis of covariance (ANCOVA) was used to test for the difference.

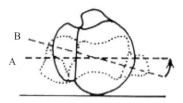

Fig. 7.4. Humeral torsion measurement. Humeral torsion is determined by positioning the (A) bicondylar axis parallel to a horizontal surface and measuring the angle from the (B) long axis of the humeral head counterclockwise to the bicondylar axis. This angle is subtracted from 90 degrees to allow comparison to previous measurements of torsion.

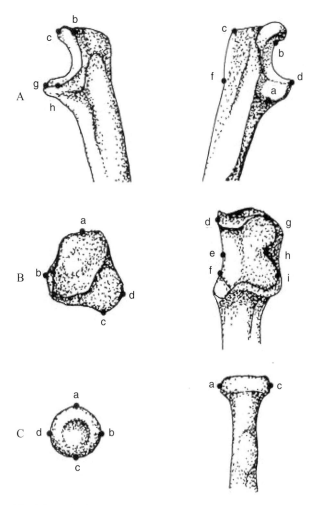

Fig. 7.5. Proximal ulnar and proximal radial measurements. (A) (left) Medial and (right) lateral view of the proximal ulna; (B) (left) cranial and (right) anterior view of the proximal ulna; (C) (left) cranial and (right) anterior view of the proximal radius.

212

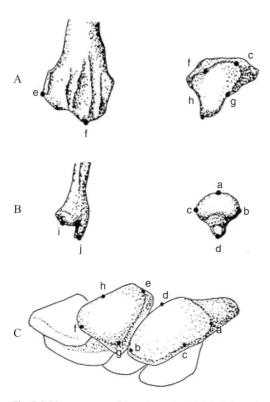

Fig. 7.6. Measurements of the wrist region. (A) (left) Posterior and (right) inferior
view of the distal radius; (B) (left) medial and (right) inferior view of the distal ulna;
(C) cranial view of the (left) lunate and (right) scaphoid bones.

*Bivariate regression analyses*

After initial PCAs determined differences between groups, bivariate analyses
were subsequently run to confirm the results of the PCA analyses. Bivariate ordinary least-squares (OLS) regression was used to describe scaling trajectories
of each group, and ANCOVA of OLS regressions was used to detect the differences in the slopes and positions (*y*-intercepts) of the ontogenetic trajectories of
different groups. Skeletal variables were used as surrogates of size in all regression analyses since actual body weight and age information are not available for
most of the specimens in this study. To reduce the potential bias of the size variable on scaling patterns, the variable of interest was regressed against three
size variables – humeral diaphyseal length, pubis length, and a composite size
variable. Subsequent ANCOVAs of these regressions were compared to one
another; only those ANCOVAs which yield overlapping and significant differences between groups using all three size surrogates were believed to actually

portray true significant differences in the scaling patterns and not artifacts of
the size variable.

## Results

### *Shoulder*

Male and female gorillas do not generally differ significantly from one another
in shoulder morphology, once allometric influences are taken into account.
PCAs and subsequent bivariate analyses of the shoulder variables within *G. g.
gorilla* (Fig. 7.7) and within *G. g. beringei* reveal no apparent difference between
males and females within each group. In other words, at comparable sizes, male
and female lowland gorillas and male and female mountain gorillas are similar
to one another for these features of the shoulder. Furthermore, *Pan* exhibits
the same results as *Gorilla* for comparable analyses; there are no differences
between male and female *Pan paniscus* and *Pan troglodytes* in these shoulder
dimensions (Inouye, 1994*a*), which concurs with Taylor (1995).

Overall, there is little difference in the shoulder morphology between mountain and lowland gorillas (Fig. 7.8; Table 7.9). Bivariate and multivariate analyses reveal that there are no significant differences between *G. g. gorilla* and
*G. g. beringei* at comparable sizes for features of the shoulder, which again
concurs with Taylor (1997*a*).

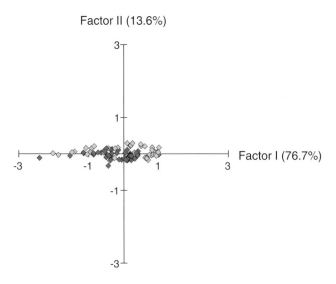

Fig. 7.7. Plot of factor scores for a PCA on the shoulder variables for *Gorilla gorilla
gorilla*: ◆, females; ◇, males.

Table 7.9. *ANCOVAs between African ape groups for shoulder dimensions regressed against different size surrogates*

| Variable | Gorilla gorilla gorilla versus Gorilla gorilla beringei | Pan paniscus versus Pan troglodytes | Gorilla versus Pan |
|---|---|---|---|
| Coracoid process length | NS/NS | NS/NS | */— |
| Acromial length | NS/NS | NS/NS | NS/* |
| Glenoid cavity length | NS/NS | NS/NS | NS/* |
| Glenoid cavity width | NS/NS | NS/NS | NS/NS |
| Glenoid cavity depth | NS/NS | NS/NS | NS/NS |
| Clavicular length | NS/NS | NS/* | NS/* |
| Humeral anteroposterior head diameter | NS/NS | NS/NS | NS/NS |
| Humeral mediolateral head diameter | NS/NS | NS/NS | NS/NS |

NS, no significant difference at $p < 0.001$.

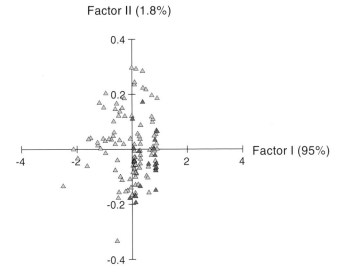

Fig. 7.8. Plot of factor scores for a PCA on the shoulder variables for *Gorilla*: ▲, *Gorilla gorilla beringei*; △, *Gorilla gorilla gorilla*.

Similarly, comparisons between bonobos and chimpanzees yielded only a significant difference in clavicular length out of all the shoulder variables (Table 7.9), showing that chimpanzees have longer clavicles than bonobos at comparable sizes. However, when comparing *Pan* to *Gorilla*, some significant differences emerge. Multivariate analyses, confirmed by subsequent bivariate analyses, reveal that gorillas have longer glenoid cavity lengths and shorter acromion processes, clavicles, and coracoid processes than comparably sized chimpanzees and bonobos (Table 7.9).

Table 7.10. *Means and standard deviations for medial humeral torsion in adult apes*

| Taxonomic group | *n* | Medial humeral torsion angle (SD) |
| --- | --- | --- |
| *Gorilla gorilla gorilla* | 52 | 62.1(6.6) |
| *Gorilla gorilla beringei* | 17 | 74.2(7.5) |
| *Pan paniscus* | 10 | 62.1(14.2) |
| *Pan troglodytes* | 50 | 58.4(7.7) |

Table 7.11. *ANOVA of means of medial humeral torsion for adult apes*

| Taxonomic groups | *p*-value |
| --- | --- |
| *Gorilla gorilla gorilla* versus *Gorilla gorilla beringei* | 0.000 |
| *Pan paniscus* versus *Pan troglodytes* | NS |
| *Pan* versus *Gorilla* | 0.000 |

A salient difference that does emerge between mountain and lowland gorilla adults is the degree of humeral torsion (Table 7.10). *G. g. beringei* has a significantly higher average degree of humeral medial torsion (74.2°) than does *G. g. gorilla* (62.1°) (Table 7.11). This is in contrast to bonobo and chimpanzee adults, which do not differ significantly from one another in humeral torsion (averages of 62.1° and 58.4°, respectively); this may be attributed to the relatively small sample size of bonobos accompanied by the larger error variance in the measurement. ANOVA indicates a significant difference between adults of *Pan* and *Gorilla* for humeral torsion, with greater medial torsion in *Gorilla*; this is predominantly due to the higher degree of medial humeral torsion in mountain gorillas compared to lowland gorillas, chimpanzees, and bonobos. Unfortunately, samples of *G. g. beringei* yielded insignificant negative correlations in regressions, likely due to small sample sizes, and thereby prevented further subspecific ontogenetic allometric comparisons of humeral torsion in *Gorilla*. Thus, only ontogenetic allometric comparisons between *Pan* and *Gorilla* were made. Although ANCOVAs indicate that *Pan* and *Gorilla* are ontogenetically scaled, this may be an artifact due to the relatively weak but significant correlations of the regressions for each group.

### Elbow

Bivariate and multivariate analyses of elbow dimensions show that there is no significant difference between male and female gorillas of both subspecies

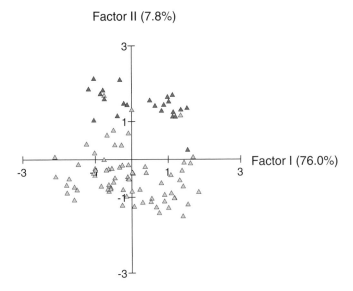

Fig. 7.9. Plot of factor scores for a PCA on the elbow variables for *Gorilla*: ▲, *Gorilla gorilla beringei*; △, *Gorilla gorilla gorilla*.

once allometric factors are properly considered. However, comparisons between mountain and western lowland gorillas reveal some differences in elbow morphology (Fig. 7.9). Multivariate analyses reveal that there is a significant separation between mountain and lowland gorillas on PC2, and the variables that load most strongly on PC2 are mostly dimensions of the trochlear notch and olecranon fossa. Further bivariate analyses reveal that at comparable body size, mountain gorillas have a relatively shorter height of the olecranon fossa, shorter trochlear notch midline height, narrower distal trochlear notch, and narrower capitular width than their western lowland counterparts at comparable body sizes (Table 7.12). A comparison of these results to identical analyses for *Pan* reveals that three of the five variables that affect the separation of chimpanzees from bonobos (Inouye, 1994*a*) are the same variables that separate mountain from western lowland gorillas.

### Wrist

The lunate and scaphoid anteroposterior widths had large error variances and insignificant correlations with size surrogates; therefore, these variables were omitted from the multivariate analyses. Multivariate analyses and subsequent

Table 7.12. *ANCOVAs between* Gorilla gorilla gorilla *and* Gorilla gorilla beringei *and* Pan paniscus *and* Pan troglodytes *for elbow and wrist dimensions regressed against all size surrogates (p <0.001)*

| Variable | Gorilla | Pan |
|---|---|---|
| Humeral mediolateral articular surface width | NS/NS | **NS/\*** |
| Distal trochlear width | NS/NS | NS/NS |
| Distal capitular width | **NS/\*** | **NS/\*** |
| Capitular height | NS/NS | NS/NS |
| Olecranon fossa width | NS/NS | **NS/\*** |
| Olecranon fossa length | **NS/\*** | **NS/\*** |
| Olecranon fossa depth | NS/NS | NS/NS |
| Anteroposterior trochlear length | NS/NS | NS/NS |
| Posterior articular surface width | NS/NS | NS/NS |
| Olecranon anteroposterior proximal depth | NS/NS | NS/NS |
| Olecranon anteroposterior distal depth | NS/NS | NS/NS |
| Proximal trochlear notch width | NS/NS | NS/NS |
| Middle trochlear notch width | NS/NS | NS/NS |
| Distal trochlear notch width | **NS/\*** | **NS/\*** |
| Trochlear/radial notch combined width | NS/NS | **NS/\*** |
| Radial anteroposterior head diameter | NS/NS | NS/NS |
| Radial mediolateral head diameter | NS/NS | NS/NS |
| Medial height of the trochlear notch | NS/NS | **NS/\*** |
| Midline height of the trochlear notch | **NS/\*** | NS/NS |
| Olecranon length | NS/NS | NS/NS |
| Coronoid process projection | NS/NS | NS/NS |
| Ulnar styloid process length | NS/NS | —[a] |
| Ulnar mediolateral head diameter | NS/NS | NS/NS |
| Scaphoid articular surface width of radius | NS/NS | NS/NS |
| Lunate articular surface width of radius | NS/NS | NS/NS |
| Radial styloid process length | NS/NS | NS/NS |
| Mediolateral radial articular facet width of scaphoid | NS/NS | NS/NS |
| Ulnar anteroposterior head diameter | NS/NS | NS/NS |
| Mediolateral radial articular facet width of lunate | **NS/\*** | —[a] |

[a] Nonsignificant regression for *Pan paniscus* at $p <0.05$.
Bold text indicates those results that are significant.

ANOVAs indicate that there are no sex differences in either mountain gorillas or western lowland gorillas for the remaining eight wrist features. For comparisons between the subspecies of *Gorilla*, only one feature of the wrist significantly separates the two groups (Table 7.12). Mediolateral width of the lunate bone loads most strongly on PC2 (Fig. 7.10), and separates the mountain from western lowland gorillas. Bivariate analyses confirm this difference and show that the

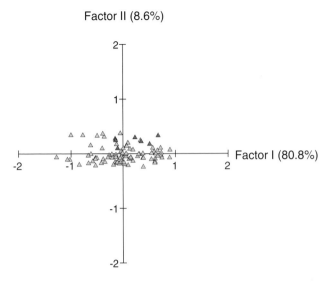

Fig. 7.10. Plot of factor scores for a PCA on the wrist variables for *Gorilla*: ▲, *Gorilla gorilla beringei*; △, *Gorilla gorilla gorilla*.

lowland gorillas have wider lunate bones than mountain gorillas at comparable sizes. In comparable multivariate and bivariate analyses of *Pan*, mediolateral width of the lunate bone is also the only feature that separates chimpanzees from bonobos (Inouye, 1994*a*).

### Summary of results

The results of ontogenetic, allometric comparisons between mountain gorillas and western lowland gorillas may be summarized as follows:

- Male and female mountain gorillas and male and female lowland gorillas are ontogenetically scaled for all of the features examined.
- Mountain and lowland gorillas are ontogenetically concordant for the shoulder dimensions examined. However, adult mountain gorillas have a higher average degree of humeral torsion than do adult lowland gorillas.
- Some features of the elbow depart from ontogenetic scaling. Mountain gorillas have a relatively shorter olecranon fossa height, shorter trochlear notch midline height, narrower distal trochlear notch, and narrower capitular width than do comparably sized western lowland gorillas.

- Mountain and lowland gorillas are ontogenetically concordant for most of the wrist dimensions examined. However, lowland gorillas have significantly wider lunates than do comparably sized mountain gorillas.
- Overall, the results from the analyses within *Gorilla* are similar to comparable analyses within *Pan*. Some of the features of the elbow and wrist regions that separate mountain from western lowland gorillas also separate chimpanzees from bonobos in comparable multivariate and bivariate analyses.

### Discussion

This ontogenetic study of variation in the forelimb of *Gorilla* touches on at least three important issues in anthropology. Information from this work contributes to discussions about the variation in locomotor profiles between mountain and western lowland gorillas and the associated morphological differences between them. The results also have implications for the current debate on the evolution of knuckle-walking in *Pan* and *Gorilla*, and contributes to discussions concerning the taxonomic status of the subspecies of *Gorilla*.

### *Arboreality versus terrestriality: the relationship between form and function*

Traditional morphometric studies of the postcrania of mountain and western lowland gorillas examine adult specimens. In contrast, this study examines the ontogeny of postcranial features, which provides a different perspective on the differences between adult mountain and western lowland gorillas. There is considerable allometry in the shoulder, elbow, and wrist regions of *Gorilla* during ontogeny; ontogenetic scaling describes much of the variation between males and females within each subspecies in all of the features examined. Futhermore, ontogenetic concordance describes much of the variation between *G. g. gorilla* and *G. g. beringei* in dimensions of the shoulder, elbow, wrist, and hand regions. Thus, mountain and western lowland gorillas do not differ in these dimensions at comparable body sizes. However, against this backdrop of ontogenetic concordance, the growth allometries of some features are discordant between mountain and lowland gorillas, which may relate to greater terrestriality in the former and/or greater arboreality in the latter.

In the shoulder region, mountain and western lowland gorillas are ontogenetically concordant with one another for most of the features examined; these results are similar to other work by Taylor (1995, 1997*a,b*) on *Gorilla*. Only one dimension of the shoulder region, humeral torsion, clearly distinguishes mountain from western lowland gorillas. In this case, mountain gorillas have a much higher degree of medial humeral torsion than their western lowland counterparts.

Some suggest that pronation of the hands in *Gorilla*, compared to *Pan*, may be influenced by variation in shoulder morphology, specifically medial torsion of the humeral shaft (Larson, 1988; Inouye, 1994*a*). Gorillas use pronated hand postures more frequently than *Pan* throughout ontogeny and, as adults, gorillas have a higher degree of humeral torsion (Inouye, 1994*a*). In fact, Inouye (1994*a*) shows that there is an increasing trend in the frequency of pronated hand postures during knuckle-walking going from *Pan paniscus, Pan troglodytes*, to *Gorilla gorilla gorilla*. Since we do not have kinematic data on knuckle-walking for mountain gorillas, we can only hypothesize that perhaps the increase in medial humeral torsion in mountain gorillas correlates with more frequent use of pronated forelimb postures during knuckle-walking. In this case, the greater humeral torsion in the mountain gorillas compared to lowland gorillas seems to support the claims of the greater degree of terrestriality in the former.

Most dimensions of the hand are also ontogenetically concordant with one another in comparisons of mountain and lowland gorillas. However, at comparable body sizes, mountain gorillas have shorter metacarpals than do lowland gorillas; this trend is also seen in proportions of the foot (Inouye, 1994*a*). The relatively shorter metacarpals and metatarsals of mountain gorillas seems to support Schultz's argument that they are ill-equipped for life in the trees in comparison to lowland gorillas.

The differences between mountain and lowland gorillas in the elbow and wrist due to discordant growth allometries are somewhat more difficult to interpret. Features of the elbow and wrist region (e.g., shorter height of the olecranon fossa, shorter trochlear notch midline height, narrower distal trochlear notch, narrower capitular width, and narrower lunate bones) seem to reflect less stability at the elbow and wrist in mountain gorillas compared to lowland gorillas. By logical extension, this lesser elbow stability does not reflect a greater usage of mountain gorilla forelimbs in terrestrial, knuckle-walking postures. In any case, the variation in the elbow and wrist cannot be simplistically interpreted *vis-à-vis* arboreal versus terrestrial trends polarizing mountain and lowland gorillas. The morphology of the wrist and elbow regions, like other features, likely represent compromises between different types of behaviors (e.g., climbing, feeding, and knuckle-walking) of *Gorilla*.

## Evolution of knuckle-walking in the African apes

The variation between mountain and lowland gorillas is interesting in light of recent studies on knuckle-walking and its link to the origin of human bipedalism (Shea and Inouye, 1993; Inouye, 1994*a,b*; Dainton and Macho, 1999; Richmond and Strait, 2000). The evolution of human bipedalism has been studied for about a century, with still no consensus as to the mode of locomotion that preceded bipedalism. Washburn suggested back in 1967 that African apes and humans shared a terrestrial, troglodytian ancestor, but this idea was later challenged by the suggestion that we evolved from a predominantly arboreal ancestor (Stern and Susman, 1983; Susman *et al.*, 1984). Very recently, some researchers, building upon the work of Tuttle (1969*a*) and Washburn (1967), attempted to clarify whether or not our ancestors passed through a knuckle-walking stage by examining the wrist bones of great apes (Dainton and Macho, 1999) and some hominids (Richmond and Strait, 2000). These studies led to divergent views, with one concluding that knuckle-walking evolved in parallel in *Pan* and *Gorilla* (Dainton and Macho, 1999), while the other argued that knuckle-walking is a derived feature of the African ape and human clade (Richmond and Strait, 2000). While Dainton and Macho's study concords with a model of arboreal ancestry for humans, Richmond and Strait's study favors the view of an ancestor for the African apes and humans that was adapted for knuckle-walking and arboreal climbing.

The most parsimonious choice between these scenarios, using only morphological data, depends upon several lines of evidence, including demonstrated biomechanical explication of knuckle-walking adaptations, evidence (or lack thereof) of knuckle-walking adaptations in fossils, and the extent of similarity or difference between *Pan* and *Gorilla* in the knuckle-walking complex. The extent of variation among African apes in the knuckle-walking complex, and the presence or absence of these features in the fossil record, are critical factors in determining whether one views the knuckle-walking complex as a shared–derived feature of the African ape and human clade or as a complex that has evolved in parallel lineages leading to *Pan* and *Gorilla*. Deciding between these two scenarios hinges partly upon a key question that one must ask – how similar or different do the African apes need to be, *vis-à-vis* the knuckle-walking complex, before we can conclude that knuckle-walking evolved only once or in parallel in *Pan* and *Gorilla*?

There have been some studies on the variation in knuckle-walking features between *Pan* and *Gorilla* (Tuttle, 1967, 1969*a,b,c*, 1970, 1972, 1975; Susman, 1979; Inouye, 1992, 1994 *a,b*; Dainton and Macho, 1999; Richmond and Strait, 2000). However, there has been less work on ontogenetic, allometric variation in knuckle-walking features at the subspecies level in *Gorilla* (Inouye, 1992,

1994*b*). This closer examination of knuckle-walking features at the subspecies level in *Gorilla* may shed some light on this longstanding debate.

Dainton and Macho's (1999) analyses of 27 features of the wrist in subadult *Pan* and *Gorilla* resulted in 15 (56%) of these exhibiting shared growth trajectories (ontogenetic scaling) and 12 exhibiting transpositions of *Pan* above *Gorilla*. Of the dimensions that differ, Dainton and Macho (1999) attribute a small number of differences to behavioral variation between genera, but they report that the majority of differences indicate heterochronic modifications of development during evolution. This pattern is similar to that found by Inouye (1992, 1994*a,b*) in her analyses of the hand. Overall, she found that there was pervasive ontogenetic scaling in features of the hand, but some features, such as metacarpal and proximal phalangeal lengths, clearly had different growth trajectories in *Gorilla* and in *Pan*. In addition, behavioral data (Inouye, 1994*b*) suggest that there are some subtle differences in hand positions and the number of digits used by *Pan* and *Gorilla* while knuckle-walking. Inouye suggests that these differences between *Pan* and *Gorilla* may relate to kinematic differences in knuckle-walking behavior, increased efficiency of knuckle-walking in the latter, or perhaps to other locomotor behaviors. Dainton and Macho (1999) suggest that the heterochronic differences in wrist morphology between *Pan* and *Gorilla* and the documented differences in hand morphology (Inouye, 1992, 1994*a,b*) may relate to a higher frequency of arboreal behavior in *Pan* (which somehow restricts efficient quadrupedal weight bearing on the fifth digit), or perhaps the larger body size of *Gorilla* may require unique modification of wrist and hand morphology. In other words, these heterochronic differences purportedly reflect the independent acquisition of solutions to the same problem of walking quadrupedally on the ground. Based on this evidence, they conclude that the compendium of data suggests that it is most parsimonious to view knuckle-walking as independently evolved in the two lineages.

This may be a logical conclusion if one affords greater weight to the departures from ontogenetic scaling than to the features that follow a pattern of ontogenetic scaling. But what if one weights the number of similiarities (i.e., ontogenetically concordant features) more heavily? Most of the features in the wrist and hand are ontogenetically concordant; in other words, the majority of the features do not differ between *Pan* and *Gorilla*, once body size is properly considered. Moreover, *Pan* and *Gorilla* are very close phylogenetically and are the only taxa that use this unique form of quadrupedal locomotion. Although *Pongo* uses its hands during quadrupedal terrestrial locomotion in a variety of fist-walking and knuckling postures (Susman, 1974; Tuttle, 1967, 1975, 1976), it does not knuckle-walk. One might question why *Pan* and *Gorilla* would independently converge on knuckle-walking when *Pongo* does not.

There are some differences between *Pan* and *Gorilla* that are not allometric and require another explanation, but with regard to resolving issues of phylogeny and character evolution, one must (1) assess the evolutionary significance of variation in the whole morphological complex in question (similarities as well as differences), and (2) ask whether or not the amount and degree of difference is surprising, given 5 million years of evolutionary divergence between *Pan* and *Gorilla*. Certainly, the argument for parallel development of knuckle-walking in *Pan* and *Gorilla* would be much more compelling if the majority of features were not ontogenetically scaled, were markedly divergent, and these could be linked to fundamental functional differences during knuckle-walking. In this case, departures from ontogenetic scaling would provide strong evidence of adaptations for kinematically distinct types of knuckle-walking. However, evaluating the evolution of this unique morphological and behavioral complex is much more difficult when two closely related species exhibit a mosaic of many similarities and some differences. For example, there are many different species of bipedal hominids, but given the evolutionary novelty of bipedalism, we would conclude that the similarity in adaptations for bipedalism far outweigh the small and relatively subtle differences between the different species, and thus bipedalism is not likely homoplastic.

At this juncture, one may wonder how examination of ontogenetic variation in mountain and lowland gorillas would contribute to the evolution of knuckle-walking debate. Previous ontogenetic studies of *Gorilla* postcrania (Inouye, 1992, 1994*a*; Taylor, 1997*a,b*) show a predominant pattern of ontogenetic scaling between mountain and western lowland gorillas for many features. However, they also depart from ontogenetic scaling in features of the hand, shoulder, elbow, and wrist. Mountain gorillas have shorter fingers, larger ridges on the third rather than fourth metacarpal, more humeral torsion, shorter scapular spines, and shorter superior scapular borders than western lowland gorillas at comparable body size. In fact, there is also quite a bit of variation between *Pan paniscus* and *Pan troglodytes* in the knuckle-walking complex (Inouye, 1992, 1994*a*); in other words, within both *Pan* and *Gorilla* there are departures from ontogenetic scaling in postcranial features likely related to knuckle-walking.

Departures from ontogenetic scaling in comparisons of *Pan* and *Gorilla* wrist and hand morphology have been used as evidence for the independent evolution of knuckle-walking (Dainton and Macho, 1999). Following a consistent and logical argument based on the minor differences in hand and wrist morphology by Dainton and Macho (1999), one would have to conclude that if knuckle-walking evolved in parallel in *Pan* and *Gorilla*, then it also requires that knuckle-walking evolved independently for the subspecies of *Gorilla* (and species of *Pan*). Based on the known variation in the knuckle-walking complex of the African apes and assuming that knuckle-walking evolved in parallel for

*Pan* and *Gorilla*, one is required to conclude that knuckle-walking evolved independently not twice, but at least four times (for *G. g. gorilla*, *G. g. beringei*, *P. paniscus*, and *P. troglodytes*) and perhaps more (for *G. g. graueri*).

My intention here is not to attempt to settle the debate about the evolution of knuckle-walking, but rather, I wish to emphasize that the ontogenetic and allometric variation in African ape postcrania is merely one line of evidence to be used in this debate, and one should be very cautious when making assumptions and interpretations regarding evolutionary transformations based only on ontogenetic allometry. Simple minor departures from or sharing of ontogenetic allometries for a single or few characters, in isolation, do not necessarily and simply signal when a complex morphological system is shared–derived or homoplastic. In the extreme, a clear case could be made for one or the other when the whole morphological complex is ontogenetically scaled or when it fundamentally departs from ontogenetic scaling in its entirety. Unfortunately, this is not usually the case, and the interpretations are less clear.

### *Taxonomic status of* Gorilla

The variation in *Gorilla* postcrania, examined in this study and elsewhere, does not provide, by itself, a sound basis for supporting or rejecting the establishment of multiple species of *Gorilla*. There is some variation between mountain and lowland gorillas in postcranial morphology, but there is also much ontogenetic scaling of postcranial dimensions. In this study, analyses of growth data revealed that 1 of 9 shoulder dimensions, 6 of 21 elbow features, 1 of 8 wrist features, and 2 of 14 features of the hand differed between mountain and lowland gorillas. Only 17% of all of the features considered in this study were determined to be nonallometric differences between mountain and western lowland gorillas. However, many of the differences may relate to the greater terrestriality of mountain gorillas, which is congruent with findings of previous studies (Coolidge, 1929; Schultz, 1934) and therefore may be viewed as compelling evidence to argue for a singles species of *Gorilla*.

Morphometric comparisons made elsewhere (Inouye, 1994*a*) between bonobos and chimpanzees show that for the identical set of dimensions, there are approximately the same number of nonallometric differences distinguishing *Pan paniscus* from *Pan troglodytes*. Moreover, the nonallometric features that distinguish bonobos from chimpanzees in each area of the postcranial skeleton are mostly the same features that distinguish mountain from western lowland gorillas in comparable multivariate and bivariate analyses. These data, along with some past studies of *Gorilla* postcrania (Inouye, 1992, 1994*a*), show that the differences between *P. paniscus* and *P. troglodytes* are generally of the same type and extent as those separating *G. g. gorilla* from *G. g. beringei*. In this case, one may argue that because the differences between mountain and western lowland gorillas parallel the differences between bonobos and chimpanzees, it is ample evidence for a species-level distinction between the mountain and western lowland gorillas.

It is apparent that these data are inherently problematic when used in isolation to evaluate the taxonomic status of *Gorilla*. Furthermore, different vectors of comparison may yield somewhat different answers to this question. For example, although mountain gorillas may be as morphologically distinct from lowland gorillas as chimpanzees are from bonobos, how different are they behaviorally? Some of the most compelling evidence for specific separation of *P. troglodytes* from *P. paniscus* comes not only from morphological studies, but also from behavioral studies (e.g., Badrian and Badrian, 1977; Savage *et al.*, 1977; MacKinnon, 1978; Savage-Rumbaugh and Wilkerson, 1978; Kuroda, 1979, 1980; Kano, 1980).

Furthermore, this study, like all simple observations of characters that are similar or different between groups, does not answer the question of *how* these taxa might have speciated (or not speciated). For example, behavioral interactions among primates are complex, and this needs to be examined in detail in local populations when trying to determine whether animals are or are not species (Albrecht and Miller, 1993). Speciation is a process, and the best we can do with morphological variation is to make inferences about genetic divergence, reproductive isolation, or other issues of interest; morphological variation can provide only indirect evidence about the dynamic evolutionary processes related to speciation (Shea *et al.*, 1993).

Another problem is that there is no theoretically predictable or necessary relationship between genetic and morphological divergence (Shea *et al.*, 1993). What is still true today is that "We still do not know how genotypes produce phenotypes, which is a huge gap in our understanding of the evolutionary process" (Futuyma, 1979). We need to understand how genes relate to adaptation before we can determine what is indicative of speciation. For example, contrary to the notion that big mutations are harmful, some studies have shown that perhaps only a few major genes can lead to big, adaptive effects during ecological speciation (Schemske and Bradshaw, 1999). In other words, big mutations may lead the way in adaptive events.

Furthermore, simple observation of morphological variation does not distinguish between genetic variation and phenotypic plasticity, which represent alternative and mutually exclusive means of dealing with environmental variability (Marshall and Jain, 1968; Jain, 1979). We need to understand the developmental reaction norm – the set of multivariate ontogenetic trajectories produced by a single genotype in response to naturally occurring environmental variation (Schlichting and Pigliucci, 1998). In other words, we need to understand the genetic control of plastic responses before we can distinguish between morphologies produced by genetic divergence and those that are plastic responses to the environment.

The relevance of the features in question must be explicated at least in terms of function, genetic heritability, fitness, geographic variation, and a plausible selective scenario, to name a few. Although the results from this study are relatively congruent with previous morphological studies on the mountain and lowland gorillas, this study, like other isolated morphological studies, does not provide any simple answers to the questions regarding taxonomic status of the subspecies of *Gorilla*. That answer lies in the assessment of data from the total picture presented by morphological, ecological, behavioral, geographic, interbreeding, and genetic studies and subsequent inferences about gene exchange, fitness, and selection.

## Conclusions

Ontogenetic, allometric analyses of the forelimb within *Gorilla* reveal a mix of features that are ontogenetically concordant between *G. g. gorilla* and *G. g. beringei*, and some nonallometric features that may relate to the greater terrestriality of the mountain gorilla. Most of the growth allometries of shoulder and wrist dimensions are concordant between mountain and western lowland gorillas. However, the ontogenetic allometries of humeral torsion and many features of the elbow are discordant between mountain and lowland gorillas. Of these distinctions, the greater degree of humeral torsion supports the notion that mountain gorillas are adapted more for terrestriality than are western lowland gorillas. Other features of the elbow region that seemingly do not follow this locomotor polarization between *G. g. beringei* and *G. g. gorilla* may reflect a compromise in various locomotor demands, or perhaps illustrate the complex inputs into morphological shape.

The implications of ontogenetic, allometric variation in the forelimb of African apes for the evolution of knuckle-walking can be interpreted in various and sometimes conflicting ways. This can be the case for any morphological complex in question, and gorillas provide a particularly good illustration of some of the complex issues to consider when using ontogenetic, allometric analyses. Ontogenetic analyses on a suite of characters that reveal mixed patterns of ontogenetic scaling and nonallometric differences can be difficult to distinguish as homoplastic or shared–derived features, especially when there are only two groups that share the features in question. As in any comparative study, there are a variety of inputs to consider. We need to decide how similar or different two groups must be in order to consider the complex as shared–derived or homoplastic; take into account the variation in the total morphological pattern; consider behavioral, ecological, and life-history information about the complex in question; and finally, consider various phylogenetic issues and selection scenarios. When morphological patterns of a complex in question are difficult to interpret, and the most parsimonious evolutionary paths for this complex are unclear, a more reserved and conservative stance should be adopted on interpretations of evolutionary, morphological transformations of the complex.

This study does not resolve debates concerning the taxonomic status of the subspecies of *Gorilla*. An evaluation of morphological patterns, by itself, does not answer how these groups might have speciated; we can only make inferences about the dynamic evolutionary processes that relate to speciation, such as genetic divergence, reproductive isolation, and phenotypic expression of genotypes (Shea *et al.*, 1993). However, it does increase our grasp of the morphological variation in *Gorilla* which perhaps can provide more fuel for this taxonomic debate and a better platform for speculation about the evolutionary processes which resulted in the current groups.

## Acknowledgments

The research reported on here is part of my PhD (1994) thesis from Northwestern University. I would like to thank Drs. B.T. Shea, M.D. Dagosto, L.R. Cochard and R.H. Tuttle for their help in that endeavor. I also thank Mr. Derek Howlett, Dr. W. Van Neer, Dr. Richard Thorington, and the helpful museum staff at the Powell-Cotton Museum and Quex House, Museé Royale de l'Afrique Centrale, and the National Museum of Natural History, for facilitating my research. I gratefully acknowledge the financial support of the National Science Foundation (NSF BNS 9000-964) and Northwestern University (0100-510-152Y) for this research. I would also like to thank A.B. Taylor and M.L. Goldsmith for inviting me to participate in the 1999 Columbus, Ohio AAPA symposium on "A revision of the genus *Gorilla*: 70 Years after Coolidge", and to contribute to this volume. I give a special thanks to B.T. Shea for his helpful comments, input, and support during the production of this chapter.

## References

Akeley, C.E. (1929). *In Brightest Africa*. New York: Doubleday.
Albrecht, G.H. and Miller, J.M.A. (1993). Geographic variation in primates: A review with implications for interpreting fossils. In *Species, Species Concepts, and Primate Evolution*, eds. W.H. Kimbel and L.B. Martin, pp. 123–161. New York: Plenum Press.

Arnold, S.J. (1981). Behavioral variation in natural populations. I: Phenotypic, genetic and environmental correlations among chemoreceptive responses to prey in the garter snake, *Thamnophis elegans*. *Evolution*, **35**, 489–509.

Atchley, W.R., Rutledge, J.J., and Cowley, D.E. (1981). Genetic components of size and shape. II: Multivariate covariance patterns in the rat and mouse skull. *Evolution*, **35**, 1037–1055.

Badrian, A. and Badrian, N. (1977). Pygmy chimpanzees. *Oryx*, **14**, 463–472.

Bock, W.J. (1980). The definition and recognition of biological adaptation. *American Zoologist*, **20**, 217–227.

Carrier, D.R. (1983). Postnatal ontogeny of the musculo-skeletal system in the Black-tailed jack rabbit (*Lepus californicus*). *Journal of Zoology, London*, **201**, 27–55.

Carroll, R.W. (1996). Feeding ecology of lowland gorillas (*Gorilla gorilla gorilla*) in the Dzanga-Sangha Dense Forest Reserve of the Central African Republic. *Mammalia*, **52**, 311–323.

Cheverud, J.M. (1982). Relationships among ontogenetic, static, and evolutionary allometry. *American Journal of Physical Anthropology*, **59**, 139–149.

Clutton-Brock, T.H. and Harvey, P.H. (1977*a*). Species differences in feeding and ranging behavior in primates. In *Primate Ecology: Studies of Feeding and Ranging Behavior in Lemurs, Monkeys and Apes*, ed. T.H. Clutton-Brock, pp. 556–584. London: Academic Press.

Clutton-Brock, T.H. and Harvey, P.H. (1977*b*). Primate ecology and social organization. *Journal of Zoology, London*, **183**, 1–39.

Clutton-Brock, T.H. and Harvey, P.H. (1983). The functional significance of variation in body size among mammals. In *Advances in the Study of Mammalian Behavior*, eds. J. F. Eisenberg and D.G. Kleiman, pp. 632–665. New York: American Society of Mammalogy.

Cock, A.G. (1966). Genetical aspects of metrical growth and form in animals. *Quarterly Review of Biology*, **41**, 131–190.

Coolidge, H.J. (1929). A revision of the genus *Gorilla*. *Memoirs of the Museum of Comparative Zoology, Harvard*, **50**, 291–383.

Coolidge, H.J. (1936). Zoological results of the George Vanderbilt African expedition of 1934. IV: Notes on four gorillas from the Sanga River Region. *Proceedings of the Academy of Natural Sciences, Philadephia*, **88**, 479–501.

Dainton, M. and Macho, G.A. (1999). Did knuckle walking evolve twice? *Journal of Human Evolution*, **36**, 171–194.

Davies, R.G. and Brown, V. (1972). A multivariate analysis of postembryonic growth in two species of *Ectobius* (Dictyoptera: Blattidae). *Journal of Zoology, London*, **168**, 51–79.

Doran, D.M. (1996). The comparative and positional behavior of the African apes. In *Great Ape Societies*, eds. W.C. McGrew and T. Nishida, pp. 213–224. Cambridge, U.K.: Cambridge University Press.

Doran, D.M. and Hunt, K.D. (1994). Comparative locomotor behavior of chimpanzees and bonobos. In *Chimpanzee Cultures*, eds. R.W. Wrangham, W.C. McGrew, F.B.M. DeWaal, and P.G. Heltne, pp. 93–108. Cambridge, MA: Harvard University Press.

Emmons, L.H., Gautier-Hion, A., and Dubost, G. (1983). Community structure of the frugivorous–folivorous forest mammals of Gabon. *Journal of the Zoological Society of London*, **199**, 209–222.

Evans, F.G. and Krahl, V.E. (1945). The torsion of the humerus: A phylogenetic survey from fish to man. *American Journal of Anatomy*, **76**, 303–337.

Fay, J.M. (1989). Partial completion of a census of the western lowland gorilla (*Gorilla g. gorilla* (Savage and Wyman)) in southwestern Central African Republic. *Mammalia*, **53**, 203–215.

Fay, J.M., Agnagna, M., Moore, J., and Oko, R. (1989). Gorillas (*Gorilla gorilla gorilla*) in the Likouala Swamp forests of north central Congo: Preliminary data on populations and ecology. *International Journal of Primatology*, **10**, 477–486.

Fleagle, J.G. (1985). Size and adaptation in primates. In *Size and Scaling in Primate Biology*, ed. W.L. Jungers, pp. 1–19. New York: Plenum Press.

Fleagle, J.G. (1988). *Primate Adaptation and Evolution*. San Diego, CA: Academic Press.

Fossey, D. (1982). Reproduction among free-living mountain gorillas. *American Journal of Primatology, Supplement* **1**, 97–104.

Fossey, D. (1983). *Gorillas in the Mist*. Boston, MA: Houghton Mifflin.

Futuyma, D.J. (1979). *Evolutionary Biology*. Sunderland, MA: Sinauer Associates.

Gelvin, B.R., Albrecht, G.H., and Miller, J.M.A. (1997). The hierarchy of craniometric variation among gorillas. *American Journal of Physical Anthropology, Supplement* **24**, 117.

Godfrey, L.R., King, S.J., and Sutherland, M.R. (1998). Heterochronic approaches to the study of locomotion. In *Primate Locomotion*, eds. E. Strasser, J. Fleagle, A. Rosenberger, and H. McHenry, pp. 277–308. New York: Plenum Press.

Gould, S.J. (1966). Allometry and size in ontogeny and phylogeny. *Biological Reviews*, **41**, 587–640.

Gould, S.J. (1975). Allometry in primates, with emphasis on scaling and the evolution of the brain. *Contributions to Primatology*, **5**, 244–292.

Gould, S.J. and Lewontin, R.C. (1979). The spandrels of San Marco and the Panglossian paradigm: A critique of the adaptationist programme. *Proceedings of the Royal Society of London, Series B*, **205**, 581–598.

Grand, T.I. (1983). The anatomy of growth and its relation to locomotor capacity in Macaca. In *Advances in the Study of Mammalian Behavior*, eds. J. F Eisenberg and D.G. Kleiman, pp. 5–24. New York: American Society of Mammalogy.

Groves, C.P. (1967). Ecology and taxonomy of the gorilla. *Nature*, **213**, 890–893.

Groves, C.P. (1970*a*). *Gorillas*. New York: Arco Publishing Co.

Groves, C.P. (1970*b*). Population systematics of the gorilla. *Journal of Zoology, London*, **161**, 287–300.

Groves, C.P. (1971). Distribution and place of origin of the gorilla. *Man*, **6**, 44–51.

Groves, C.P. (1972). Phylogeny and classification of primates. In *Pathology of Simian Primates*, vol. 1, ed. R.N. Twistleton-Wichham-Fiennes, pp. 11–57. Basel Switzerland: S. Karger.

Harvey, P.H. and Pagel, M.D. (1991). *The Comparative Method in Evolutionary Biology*. Oxford, U.K.: Oxford University Press.

Huxley, J.S. (1932). *Problems of Relative Growth*. London: MacVeagh.

Inouye, S.E. (1992). Ontogeny and allometry of African apes manual rays. *Journal of Human Evolution*, **23**, 107–138.

Inouye, S.E. (1994*a*). The ontogeny of knuckle-walking behavior and associated morphology in the African apes. PhD thesis, Northwestern University, Evanston, IL.

Inouye, S.E. (1994*b*). Ontogeny of knuckle-walking hand postures in African apes. *Journal of Human Evolution*, **26**, 459–486.

Inouye, S.E. and Shea, B.T. (1997). What's your angle? Size-correction and bar-glenoid orientation in "Lucy" (A. L. 288–1). *International Journal of Primatology*, **18**, 629–650.

Jain, S.K. (1979). Adaptive strategies: Polymorphism, plasticity, and homeostasis. In *Topics in Plant Population Biology*, eds. O.T. Solbrig, S. Jain, G.B. Johnson, and P.H. Raven, pp. 160–187. New York: Columbia University Press.

Jones, C. and Sabater Pi, J. (1971). Comparative ecology of *Gorilla gorilla* (Savage and Wyman) and *Pan troglodytes* (Blumenbach) in Rio Muni, West Africa. *Biblioteca Primatologia*, **13**, 1–96.

Jungers, W.L. and Fleagle, J.G. (1980). Postnatal growth allometry of the extremities in *Cebus albifrons* and *Cebus apella*: A longitudinal and comparative study. *American Journal of Physical Anthropology*, **53**, 471–478.

Jungers, W.L. and Susman, R.L. (1984). Body size and skeletal allometry in African apes. In *The Pygmy Chimpazee: Evolutionary Biology and Behavior*, ed. R.L. Susman, pp. 131–177. New York: Plenum Press.

Kano, T. (1980). Social behavior of the wild pygmy chimpanzees (*Pan paniscus*) of Wamba: A preliminary report. *Journal of Human Evolution*, **9**, 243–260.

Kay, R.F. (1981). The nut-crackers: A new theory of the adaptations of the Ramapithecinae. *American Journal of Physical Anthropology*, **55**, 141–151.

Keith, A. (1926). The gorilla and man as contrasted forms. *Lancet*, **210**, 490–492.

Kuroda, S. (1979). Grouping of the pygmy chimpanzees. *Primates*, **20**, 161–183.

Kuroda, S. (1980). Social behavior of the pygmy chimpanzees. *Primates*, **21**, 181–197.

Lande, R. (1979). Quantitative genetic analysis of multivariate evolution, applied to brain:body allometry. *Evolution*, **33**, 402–416.

Larson, S.G. (1988). Subscapularis function in gibbons and chimpanzees: Implications for interpretation of humeral head torsion in hominoids. *American Journal of Physical Anthropology*, **76**, 449–462.

Larson, S.G. (1995). New characters for the functional interpretation of primate scapulae and proximal humeri. *American Journal of Physical Anthropology*, **98**, 13–35.

Leigh, S.R. (1992). Ontogeny and body size dimorphism in anthropoid primates. PhD thesis, Northwestern University.

Leigh, S.R. and Shea, B.T. (1995). Ontogeny and the evolution of adult body size dimorphism in apes. *American Journal of Primatology*, **36**, 37–60.

Leigh, S.R. and Shea, B.T. (1996). Ontogeny of body size variation in African apes. *American Journal of Physical Anthropology*, **99**, 43–65.

MacKinnon, J. (1978). *The Ape Within Us*. New York: Holt, Rinehart, & Winston.

Marshall, D.R. and Jain, S.K. (1968). Phenotypic plasticity of *Avena fatua* and *A. barbata*. *American Naturalist*, **102**, 457–467.

Martin, R.D. (1989). Size, shape and evolution. In *Evolutionary Studies: A Centenary Celebration of the Life of Julian Huxley*, eds. M. Keynes and G.A. Harrison, pp. 96–141. Basingstoke, U.K.: Macmillan.

McMahon, T.A. and Bonner, J.T. (1983). *On Size and Life*. New York: Scientific American Library.

Morton, D.J. (1924). Evolution of the human foot. II. *American Journal of Physical Anthropology*, **5**, 305–336.

Morton, D.J. (1927). Human origin: Correlation of previous studies of primate feet and posture with other morphological evidence. *American Journal of Physical Anthropology*, **10**, 173–203.

Neff, N.A. and Marcus, L.F. (1980). *A Survey of Multivariate Methods for Systematics*. New York: privately published.

Nishihara, T. (1995). Feeding ecology of western lowland gorillas in the Nouabale-Ndoki National Park, Congo. *Primates*, **36**, 151–168.

Oxnard, C.E. (1978). One biologist's view of morphometrics. *Annual Review of Ecology and Systematics*, **9**, 219–241.

Pagel, M.D. and Harvey, P.H. (1988). Recent developments in the analysis of comparative data. *Quarterly Review of Biology*, **63**, 413–440.

Peters, S.E. (1983). Postnatal development of gait behavior and functional allometry in the domestic cat (*Felis catus*). *Journal of Zoology, London*, **199**, 461–486.

Pilbeam, D. and Gould, S.J. (1974). Size and scaling in human evolution. *Science*, **186**, 892–901.

Pounds, J.A., Jackson, J.F., and Shively, S.H. (1983). Allometric growth of the hind limbs of some terrestrial iguanid lizards. *American Naturalist*, **110**, 201–207.

Remis, M. (1995). Effects of body size and social context on the arboreal activities of lowland gorillas in the Central African Republic. *American Journal of Physical Anthropology*, **97**, 413–433.

Remis, M. (1998). The gorilla paradox: The effects of body size and habitat on the positional behavior of lowland and mountain gorillas. In *Primate Locomotion*, eds. E. Strasser, J. Fleagle, A. Rosenberger, and H. McHenry, pp. 95–106. New York: Plenum Press.

Remis, M. (1999). Tree structure and sex differences in arboreality among western lowland gorillas (*Gorilla gorilla gorilla*) at Bai Hokou, Central African Republic. *Primates*, **40**, 383–396.

Richmond, B.G. and Strait, D.S. (2000). Evidence that humans evolved from a knuckle-walking ancestor. *Nature*, **404**, 382–385.

Ruvolo, M., Pan, D., Zehr, S., Goldberg, T., Disotell, T.R., and von Dornum, M. (1994). Gene trees and hominoid phylogeny. *Proceedings of the National Academy of Sciences U.S.A.*, **91**, 8900–8904.

Sarmiento, E.E. and Marcus, L.F. (2000). The os navicular of humans, great apes, OH 8, Hadar, and *Oreopithecus*: Function, phylogeny, and multivariate analyses. *American Museum Novitates*, **3288**.

Savage, E.S., Wilkerson, B.J., and Bakeman, R. (1977). A spontaneous gestural communication among conspecifics in the pygmy chimpanzee (*Pan paniscus*). In *Progress in Ape Research*, ed. O. H. Bourne, pp. 97–116. New York: Academic Press.

Savage-Rumbaugh, E.S. and Wilkerson, B.J. (1978). Socio-sexual behavior in *Pan paniscus* and *Pan troglodytes*: A comparative study. *Journal of Human Evolution*, **7**, 327–344.

Schaller, G. (1963). *The Mountain Gorilla: Ecology and Behavior*. Chicago, IL: University of Chicago Press.

Schemske, D.W. and Bradshaw, H.D., Jr. (1999). Pollinator preference and the evolution of floral traits in monkeyflowers (*Mimulus*). *Proceedings of the National Academy of Sciences U.S.A.*, **96**, 11910–11915.

Schlichting, C.D. and Pigliucci, M. (1998). *Phenotypic Evolution: A Reaction Norm Perspective*. Sunderland, MA: Sinauer Associates.

Schmidt-Nielsen, K. (1984) *Scaling: Why Is Animal Size So Important?* Cambridge, U.K.: Cambridge University Press.

Schultz, A.H. (1927). Studies on the growth of the gorilla and of other higher primates, with special reference to a fetus of gorilla, preserved in the Carnegie Museum. *Memoirs of the Carnegie Museum*, **11**, 1–87.

Schultz, A.H. (1930). The skeleton of the trunk and limbs of higher primates. *Human Biology*, **2**, 303–438.

Schultz, A.H. (1934). Some distinguishing characters of the mountain gorilla. *Journal of Mammalogy*, **12**, 15–61.

Schultz, A.H. (1968). The recent hominoid primates. In *Perspectives on Human Evolution*, vol. 1, eds. S.L. Washburn and P. Jay, pp. 122–195. New York: Holt, Rinehart, & Winston.

Shea, B.T. (1981). Relative growth of the limbs and trunk of the African apes. *American Journal of Physical Anthropology*, **56**, 179–202.

Shea, B.T. (1983*a*). Size and diet in the evolution of African ape craniodental form. *Folia Primatologia*, **40**, 32–68.

Shea, B.T. (1983*b*). Phyletic size change and brain/body scaling: A consideration based on the African pongids and other primates. *International Journal of Primatology*, **4**, 33–62.

Shea, B.T. (1984). An allometric perspective on the morphological and evolutionary relationships between pygmy (*Pan paniscus*) and common (*Pan troglodytes*) chimpanzees. In *The Pygmy Chimpazee: Evolutionary Biology and Behavior*, ed. R.L. Susman, pp. 89–130. New York: Plenum Press.

Shea, B.T. (1985*a*). Bivariate and multivariate growth allometry: Statistical and biological considerations. *Journal of Zoology, London*, **206**, 367–390.

Shea, B.T. (1985*b*). Ontogenetic allometry and scaling: A discussion based on growth and form of the skull in African apes. In *Size and Scaling in Primate Biology*, ed. W.L. Jungers, pp. 175–206. New York: Plenum Press.

Shea, B.T. (1986). Scapula form and locomotion in chimpanzee evolution. *American Journal of Physical Anthropology*, **70**, 475–488.

Shea, B.T. (1987). Reproductive strategies, body size, and encephalization in primate evolution. *International Journal of Primatology*, **8**, 139–156.

Shea, B.T. (1990). Dynamic morphology: Growth, life history, and ecology in primate evolution. In *Primate Life History and Evolution*, ed. C.J. DeRousseau, pp. 325–352. New York: Wiley-Liss.

Shea, B.T. (1995). Ontogenetic scaling and size correction in the comparative study of primate adaptations. *Anthropologie*, **33**, 1–2, 1–16.

Shea, B.T. and Bailey, R.C. (1996). Allometry and adaptation of body proportions and stature in African pygmy chimpanzees: Size correction in the comparative study of primate adaptations. *American Journal of Physical Anthropology*, **100**, 311–340.

Shea, B.T. and Inouye, S.E. (1993). Knuckle-walking ancestors. *Science*, **259**, 293–294.

Shea, B.T., Leigh, S.R., and Groves, C.P. (1993). Multivariate craniometric variation in chimpanzees. In *Species, Species Concepts, and Primate Evolution*, eds. W.H. Kimbel and L.B. Martin, pp. 265–296. New York: Plenum Press.

Smith, R.J. (1984*a*). Determination of relative size: The "criterion of subtraction" problem in allometry. *Journal of Theoretical Biology*, **108**, 131–142.

Smith, R.J. (1984*b*). Allometric scaling in comparative biology: Problems of concept and method. *American Journal of Physiology*, **246**, R152–R160.

Smith, R.J. and Jungers, W.L. (1997). Body mass in comparative primatology. *Journal of Human Evolution*, **32**, 523–559.

Stern, J.T., Jr. and Susman, R.L. (1983). The locomotor anatomy of *Australopithecus afarensis*. *American Journal of Physical Anthropology*, **60**, 279–317.

Susman, R.L. (1974). Facultative terrestrial hand postures in an orangutan (*Pongo pygmaeus*) and pongid evolution. *American Journal of Physical Anthropology*, **40**, 27–38.

Susman, R.L. (1979). Comparative and functional morphology of hominoid fingers. *American Journal of Physical Anthropology*, **50**, 215–236.

Susman, R.L. and Creel, N. (1979). Functional and morphological affinities of the subadult hand (O.H. 7) from Olduvai Gorge. *American Journal of Physical Anthropology*, **51**, 311–332.

Susman, R.L., Stern, J.T., Jr., and Jungers, W.L. (1984). Arboreality and bipedality in Hadar hominids. *Folia Primatologia*, **43**, 113–156.

Taylor, A.B. (1995). Scapula morphology in the mountain gorilla. *American Journal of Physical Anthropology*, **98**, 431–445.

Taylor, A.B. (1997*a*). Scapula form and biomechanics in gorillas. *Journal of Human Evolution*, **33**, 529–553.

Taylor, A.B. (1997*b*). Relative growth, ontogeny, and sexual dimorphism in gorilla (*Gorilla gorilla gorilla* and *G.g. beringei*): Evolutionary and ecological considerations. *American Journal of Primatology*, **43**, 1–31.

Tutin, C.E.G. and Fernandez, M. (1984*a*). Nationwide census of gorilla (*Gorilla g. gorilla*) and chimpanzee (*Pan t. troglodytes*) populations in Gabon. *American Journal of Primatology*, **6**, 313–336.

Tutin, C.E.G. and Fernandez, M. (1984*b*). Ape census in Gabon. *Primate Eye*, **21**, 16–17.

Tuttle, R.H. (1967). Knuckle-walking and the evolution of hominoid hands. *American Journal of Physical Anthropology*, **26**, 171–206.

Tuttle, R.H. (1969*a*). Knuckle-walking and the problem of human origins. *Science*, **166**, 953–961.

Tuttle, R.H. (1969*b*). Quantitative and functional studies on the hands of the Anthropoidea and the Hominoidea. *Journal of Morphology*, **128**, 309–364.

Tuttle, R.H. (1969c). Terrestrial trends in the hands of the Anthropoidea. In *Proceedings of the 2nd International Congress of Primatology*, Atlanta, 1968, vol. 2, pp. 192–200. Basel, Switzerland: S. Karger.

Tuttle, R.H. (1970). Postural, propulsive and prehensile capabilities in the cheiridia of chimpanzees and other great apes. In *The Chimpanzee*, vol. 2, ed. G.H. Bourne, pp. 167–253. Basel, Switzerland: S. Karger.

Tuttle, R.H. (1972). Relative mass of cheiridial muscles in catarrhine primates. In *The Functional and Evolutionary Biology of Primates*, ed. R.H. Tuttle, pp. 262–291. The Hague: Mouton.

Tuttle, R.H. (1975). Knuckle-walking and knuckle-walkers: A commentary on some recent perspectives on hominoid evolution. In *Primate Functional Morphology and Evolution*, ed. R.H. Tuttle, pp. 203–212. The Hague: Mouton.

Tuttle, R.H. (1976). Knuckling behavior in captive orangutans and a wounded baboon. *American Journal of Physical Anthropology*, **45**, 123–134.

Tuttle, R.H. (1986). *Apes of the World: Studies on the Lives of Great Apes and Gibbons, 1929–1985*. Park Ridge, NJ: Noyes.

Tuttle, R.H. and Basmajian, J.V. (1974). Electromyography of forearm musculature in *Gorilla* and problems related to knuckle-walking. In *Primate Locomotion*, ed. F.A. Jenkins, Jr., pp. 293–347. New York: Academic Press.

Tuttle, R.H. and Watts, D.P. (1985). The positional behavior and adaptive complexes of *Pan* and *Gorilla*. In *Primate Morphophysiology, Locomotor Analyses and Human Bipedalism*, ed. S. Kondo, pp. 261–288. Tokyo: University of Tokyo Press.

Washburn, S.L. (1967). Behavior and the origin of man. *Proceedings of the Royal Anthropological Institute of Great Britain and Ireland*, **3**, 21–27.

Watts, D.P. (1984). Composition and variation of mountain gorilla diets in the Central Virungas. *American Journal of Primatology*, **7**, 323–3556.

# Part 2
## *Molecular genetics*

# 8 An introductory perspective: Gorilla *systematics, taxonomy, and conservation in the era of genomics*

OLIVER A. RYDER

The papers in this section shed new insights into the evolutionary divergence of gorilla populations and conservation status of wild populations of gorillas. In *Gorilla Biology: A Multidisciplinary Perspective*, the chapters on genetic studies recount findings that reflect queries coming from many perspectives. Medical science, anthropology, cognitive sciences, behavioral ecology, population genetics, and conservation biology all incorporate areas of interest for which genetic studies involving other species have provided new information. Of course, it is anticipated that insights into gorilla biology and conservation will occur as a result of genetic investigations involving this endangered species.

But how many species and subspecies do gorillas comprise, and how has the evolution of distinct lineages resulted in the morphological, ecological behavioral, and familial variety that is still in the process of being described?

Insofar as conservation efforts for gorillas and other endangered species are focused on maintaining viable populations in their habitats, the importance of understanding the distribution of heritable variation and the evolutionary patterns that result in its geographic distribution across landscapes suggests that genetic methods will continue to receive attention and yield useful insights. This would seem to be especially true as the technology for measuring genetic variation becomes more sophisticated and widely applied in many biological disciplines. Methods of investigation of genome expression now allow identification of genes whose activity and/or regulation serve as a source of adaptation for survival and reproduction. This ushers in a new era of investigation in biology and provides the possibility for evaluating evolution of life history strategies in new ways (Ideker *et al.*, 2001).

Papers in this section on genetics, as well as the discussion of the gorillas of the Cross River region in the section on gorilla conservation, contribute new information to the discussion that, we can anticipate, will continue and likely intensify in the context of consideration of conservation management planning and action.

Genetic studies of gorillas only became possible in the relatively recent past. Initially, only captive animals or post-mortem specimens could provide research materials, although this circumstance allowed for some of the first studies on genetic make-up of gorillas, such as their diploid chromosome number (Hamerton *et al.*, 1961) and genetic variation within the species (Moor-Jankowski and Socha, 1979; Socha and Moor-Jankowski, 1986). It was a decade later, with the advent of the polymerase chain reaction (PCR), that wildlife studies could be undertaken with noninvasive samples and among the first of these were studies of wild gorilla populations (Garner and Ryder, 1989, 1992, 1996; Field *et al.*, 1998). Previous studies of other ape species, notably orangutans, had demonstrated that useful information about evolution and genetic isolation of great ape populations could be derived from genetic analyses (Seaunez, 1982; Chemnick and Ryder, 1994; Zhi *et al.*, 1996; Zhang *et al.*, 2001). Nonetheless, the first genetic studies of gorillas were brought forward in an arena of controversy regarding the systematics of the populations, as has been well detailed in chapters in this volume by Jensen-Seaman *et al.*, Clifford *et al.*, and Oates *et al.*

Mitochondrial DNA was first utilized for phylogenetic studies of great apes at a time when a considerably larger controversy existed regarding hominoid phylogeny (Ferris *et al.*, 1981). While utility of mitochondrial DNA has continually been challenged as will be discussed further below, it remains a crucially useful tool in studies of molecular evolution and systematics and in the context of conservation (Avise, 1989, 1995). The prospect of gathering mitochondrial DNA sequence data from samples obtained by noninvasive means has contributed to the interest in this approach and the productivity of its application (e.g., Gagneux *et al.*, 1999). However, because of its smaller effective population size, matrilineal inheritance, and potential for sex bias in dispersal, concordance with findings from nuclear-encoded loci is often called for. Additionally, problems associated with small sample sizes, limited numbers of informative characters, and the propensity for homoplasy due to multiple mutations at a single position have not always been sufficiently addressed in light of current understanding.

Yet, incorporation of significant and sufficient amounts of nuclear-encoded information is currently difficult to accomplish because of the lower mutation rate of nuclear DNA. The effort currently required to obtain sufficient phylogentically informative data from nuclear loci requires collection of larger amounts of sequence data and has thereby restrained investigations of this kind. Many new contributions are accumulating in the area of comparative genetics and now, comparative genomics (Gagneux and Varki, 2001; Hacia, 2001). The impetus for these emerging studies is the Human Genome Project, an effort that now incorporates the necessity to place our understanding of human genetic variation in an evolutionary context. It is clear, though, that there are numerous conservation implications as well as fundamental insights into fields of anthropology and cognitive sciences that may also be derived.

At this time it is important to recognize that the intraspecific phylogeography of gorillas and other great apes is largely unknown. As new methods of investigation and analysis supported largely by investment in understanding the human genome are applied, insights into prehistoric patterns of the distribution of human genetic variation are appearing regularly (Weiss, 1998; Hammer *et al.*, 2001). We find ourselves at a point where basic information about the species to which humans are most closely related – the chimpanzees and gorillas – such as sex bias in dispersal, gene flow over geographic distances, effective population size, the breeding structure of populations, although receiving increased focus and, attention (e.g., Ruvolo, 1997) is still far from being fully understood.

In addition, vulnerabilities of small populations include the risk of genetic disease (Ryder, 1988). Epidemiological assessments and application of healthcare intervention including the use of new diagnostic methods, demonstrate that a broadened understanding of gorilla biology is crucial for the effort to undertake the variety of measures required to provide the best outlook for survival of gorilla populations. Lack of information concerning disease exposure of gorillas that have not been in contact with humans confounds efforts to evaluate the potential disease impacts of human presence on habituated groups of gorillas. Dr. William Karesh of the Wildlife Conservation Society has collected samples from solitary male gorillas in Odzala National Park in the Democratic Republic of Congo prior to habituation of gorilla groups for approved tourism viewing in the National Park (W. Karesh, personal communication). This prospective assessment requires high-quality samples from wild populations.

In this volume, studies of mitochondrial DNA (Clifford *et al.*, Jensen-Seaman *et al.*, Oates *et al.*) and nuclear DNA including coding loci, introns, pseudogenes, and a 3′ UTR (Jensen-Seaman *et al.*) and microsatellites (Clifford *et al.*) form the basis for these new studies. Different geographic regions encompassing the distribution of gorillas have been examined. Jensen-Seaman *et al.* concentrate their efforts on evaluating the genetic differentiation between the eastern and western populations of gorillas using Kahuzi and Bwindi for the eastern populations (although relatively few individuals from eastern populations were included in their analyses for all but the DRD4 locus). Clifford *et al.* obtained the most thorough set of samples yet assembled for scientific analysis, a portion of these having been utilized for studies of microsatellite variation and mitochondrial sequencing analysis. Their samples included extensive sampling of *Gorilla gorilla graueri* and regions from the western portion of the range of (lowland) gorillas including the Cross River area. Oates *et al.* focus on the Cross River area on the border of Nigeria and Cameroon.

All studies found a deep divergence between mitochondrial sequences of gorillas in the east as compared with those of the western portions of the range, as has been previously suggested (Garner and Ryder, 1996). However, the implications of previous studies with regard to the validity of major phylogenetic lineages (and their interpretation in a systematics context, e.g., as subspecies and species) has been challenged (Sarmiento *et al.*, 1996) based on ecological and behavioral adaptations not reflected in the phylogenetic analysis incorporating portions of the mitochondrial DNA.

Furthermore, misleading information regarding phylogeny of other taxa, the result of paralogous sequences of mitochondrial DNA inserted into the nuclear genome (Numts), suggested that additional analysis would be crucial for understanding the true phylogeny of gorillas (Jensen-Seaman, 2000; Bensasson *et al.*, 2001). In addition, Jensen-Seaman (2000) identified multiple sequences that were notably divergent within individual gorillas that amplified with mitochondrial primers. Other evidence presented in his dissertation from the Kidd Laboratory at Yale (Jensen-Seaman, 2000) was consistent with the hypothesis that the deepest divergence between western lowland gorilla (*Gorilla gorilla gorilla*) mitochondrial lineages, identified earlier by Garner and Ryder, was an artifact of transfer of DNA sequences between cytoplasmic mitochondria and the nucleus. The studies of Garner and Ryder were conducted on the control region or D-loop of the mitochondrial DNA and, in this volume, Jensen-Seaman *et al.* (utilizing the mitochondrial NADH5 locus) and larger population sampling of individuals of known geographic provenance, contribute to the consensus for the deep phylogenetic separation of eastern and western gorillas. Additionally, unpublished results of Zhang *et al.* have supplemented the initial study by Garner and Ryder (1996) utilizing mitochondrial ND5 data derived from sampling the same individuals studied by Garner and Ryder (1996). Their findings are consistent with those of Jensen-Seaman (2000) regarding overestimation of divergence of mitochondrial sequences of western lowland gorillas based on the D-loop analysis, most likely as a result of Numts. Nonetheless, the intraspecific phylogeny of gorillas suggested originally by Garner and Ryder (1996) remains upheld (Zhang *et. al.*, 2001; Jensen-Seaman *et al.*, this volume) Thus, it is important to emphasize that all current studies of gorilla mitochondrial variation support reciprocal monophyly of eastern and western gorillas and, within the east, between *Gorilla gorilla graueri* and the gorillas of the Virunga mountains and Bwindi forests (*Gorilla gorilla beringei).*

The position of the Cross River gorillas within the intraspecific phylogeny of gorillas is not resolved by the studies presented here. Identified Numts and PCR amplifications from samples obtained in the Cross River area for mitochondrial D-loop region obscure the results (Oates *et al.*, this volume). However, identification of additional informative loci, e.g., COII (Ruvolo *et al.*, 1994), NADH5

(Jenson-Seaman *et al.*, this volume) and ND5 (Zhang *et al.*, 2001), suggest that mitochondrial DNA sequences (which are more readily obtainable from hair than nuclear loci) may yet contribute to clarification of the degree of phylogenetic uniqueness of the highly endangered gorillas of Nigeria and Cameroon. The limitations of poor-quality samples impact the analysis of the Cross River gorilla population. Hair samples may provide inaccurate results if Numts are present (Greenwood and Pääbo, 1999). The difficulties entailed in amplifying longer stretches of mtDNA for analysis are still less daunting than the hurdles to obtaining nuclear sequences, which are more difficult to amplify and analyze because of their lower relative concentration. These obstacles stand in the way of the production of data relevant to assessment of the phylogenetic status of the Cross River gorillas, inferences about gene flow from adjacent populations, breeding structure of this population, and other comparative aspects of their population structure.

Mitochondrial DNA analysis has elicited a revolution in molecular evolutionary studies. Additionally, single nuclear loci can be under differing selection pressures, altering estimations of rates of molecular evolution (Jensen-Seaman *et al.*, this volume). Thus, while comparative studies of nuclear and mitochondrial DNA sequences are still relatively few in number, available examples often show concordance between nuclear and mitochondrial results, though typically, as expected, mitochondrial DNA provides greater indications of population substructuring. Recent studies of nuclear (Eggert *et al.*, 2000; Roca *et al.*, 2001) and mitochondrial DNA (Eggert, 2001) sequences in African elephants appear to confirm the existence of defined evolutionary lineages of African elephants corresponding to adaptation to forest and savanna habitats. Studies of nuclear genetic variation encoded on the X-chromosome of chimpanzees has been compared to intraspecific mitochondrial divergence (Kaessmann *et al.*, 1999). Their findings provide a picture of intraspecific variation in which recognized subspecies, corresponding to reciprocally monophyletic mitochondrial DNA clades, also partition X-chromosome variation by distance analysis.

Jensen-Seaman *et al.* urge continuing caution in interpretation of mitochondrial DNA sequence divergence within gorillas and offer possible alternative explanations for discordance in mitochondrial and nuclear data which their contribution details. Selection may have altered the pattern of nucleotide replacement in different lineages. Or, mitochondrial DNA evolution may not reflect the true species phylogeny. Different patterns of male versus female migration in *Gorilla* and *Pan* might produce misleading inferences, or differences in geographic structure of mitochondrial variation in *Gorilla* and *Pan* alike could lead to apparent discordance between data derived from nuclear and mitochondrial sequences. Clarification of these alternatives clearly indicates the need for additional studies. Jensen-Seaman *et al.* (this volume) draw strongly on the DRD4 locus for which the greatest number of independent chromosomes was analyzed in their study. In humans, DRD4 appears to be under selection (Ding *et al.*, 2002). If species of *Pan* and humans are different from gorillas with respect to mutation rate of the DRD4 locus, then different selection pressures may have arisen since divergence of the common ancestor of the *Homo–Pan* clade with that of the clade leading to extant gorillas. The evaluation of this hypothesis clearly requires a larger investigation of DRD4 sequences in orangutans.

These chapters depict new findings and work in progress in this fast-moving field. Technological developments enabled by the Human Genome Project can be expected to provide new data at hitherto unanticipated rates. What is unclear is whether the samples required for such highly technically sophisticated analyses will be available. Crucial will be progress in obtaining data from non-invasive samples. Gorillas may disappear from some areas which they currently occupy, or populations in these areas may be severely reduced and historic patterns of gene flow disrupted. The only samples for analysis may be currently in hand and their experimental use now may eliminate possible future studies. Considering the importance of genetic studies for phylogeography, behavior, and conservation, this aspect of resources for genetic study needs more critical evaluation. The incentives for undertaking current investigations, such as publications and funding opportunities, may compromise future studies involving superior technology.

## References

Avise, J.C. (1989). A role for molecular genetics in the recognition and conservation of endangered species. *Trends in Ecology and Evolution*, **4**, 279–281.

Avise, J.C. (1995). Mitochondrial DNA polymorphism and a connection between genetics and demography of relevance to conservation. *Conservation Biology*, **9**, 686–690.

Bensasson, D., Zhang, D.X., Hartl, D.L., and Hewitt, G.M. (2001). Mitochondrial pseudogenes: Evolution's misplaced witnesses. *Trends in Ecology and Evolution*, **16**, 314–321.

Chemnick, L. and Ryder, O.A. (1994). Cytological and molecular divergence of orangutan subspecies. In *Proceedings of the International Conference on "Orangutans: The Neglected Ape"*, eds. J.J. Ogden, L.A. Perkins, and L. Sheeran, pp. 74–79. San Diego, CA: Zoological Society of San Diego.

Ding, Y.C., Chi, H.C., Grady, D.L., Morishima, A., Kidd, J.R., Kidd, K.K., Flodman, P., Spence, M.A., Schuck, S., Swanson, J.M., Zhang, Y.P., and Moyzis, R.K. (2002). Evidence of positive selection acting at the human dopamine receptor D4 gene locus. *Proceedings of the National Academy of Sciences U.S.A.*, **99**, 309–314.

Eggert, L.S. (2001). The evolution and conservation of the African forest elephant. PhD thesis, University of California, San Diego, CA.

Eggert, L.S., Ramakrishnan, U., Mundy, N.I., and Woodruff, D.S. (2000). Polymorphic microsatellite DNA markers in the African elephant (*Loxodonta africana*) and their use in the Asian elephant (*Elephas maximus*). *Molecular Ecology*, **9**, 2223–2225.

Ferris, S.D., Brown, W.M., Davidson, W.S., and Wilson, A.C. (1981). Extensive polymorphism in the mitochondrial DNA of apes. *Proceedings of the National Academy of Sciences U.S.A.*, **78**, 6319–6323.

Field, D., Chemnick, L., Robbins, M., Garner, K., and Ryder, O.A. (1998). Paternity determination in captive lowland gorillas and orang-utans and wild mountain gorillas by microsatellite analysis. *Primates*, **39**, 199–209.

Gagneux, P. and Varki, A. (2001). Genetic differences between humans and great apes. *Molecular Phylogenetics and Evolution*, **18**, 2–13.

Gagneux, P., Wills, C., Gerloff, U., Tautz, D., Morin, P.A., Boesch, C., Fruth, B., Hohmann, G., Ryder, O.A., and Woodruff, D.S. (1999). Mitochondrial sequences show diverse evolutionary histories of African hominoids. *Proceedings of the National Academy of Sciences U.S.A.*, **96**, 5077–5082.

Garner, K.J. and Ryder, O.A. (1989). Assessment of genetic diversity in gorillas by amplification of mitochondrial D-loop DNA. *Journal of Cellular Biochemistry, Supplement* **13E**, 285.

Garner, K.J. and Ryder, O.A. (1992). Some applications of PCR to studies in wildlife genetics. In *Biotechnology and the Conservation of Genetic Diversity*, eds. H.D.M. Moore, W.V. Holt, and G.M. Mace, pp. 167–181. Oxford: Clarendon Press.

Garner, K.J. and Ryder, O.A. (1996). Mitochondrial DNA diversity in gorillas. *Molecular Phylogenetics and Evolution*, **6**, 39–48.

Greenwood, A.D. and Pääbo, S. (1999). Nuclear insertion sequences of mitochondrial DNA predominate in hair but not in blood of elephants. *Molecular Ecology*, **8**, 133–137.

Hacia, J.G. (2001). Genome of the apes. *Trends in Genetics*, **17**, 637–645.

Hamerton, J.L., Fraccaro, M., De Carli, L., Nuzzo, F., Klinger, H.P., Hulliger, L., Taylor, A., and Lang, E.M. (1961). Somatic chromosomes of the gorilla. *Nature*, **192**, 225.

Hammer, M.F., Karafet, T.M., Redd, A.J., Jarjanazi, H., Santachiara-Benerecetti, S., Soodyall, H., and Zegura, S.L. (2001). Hierarchical patterns of global human Y-chromosome diversity. *Molecular Biology and Evolution*, **18**, 1189–1203.

Ideker, T., Galitski, T., and Hood, L. (2001). A new approach to decoding life: Systems biology. *Annual Review of Genomics and Human Genetics*, **2**, 343–372.

Jensen-Seaman, M.I. (2000). Evolutionary genetics of gorillas. PhD thesis, Yale University, New Haven, CT.

Kaessmann, H., Wiebe, V., and Pääbo, S. (1999). Extensive nuclear DNA sequence diversity among chimpanzees. *Science*, **286**, 1159–1162.

Moor-Jankowski, J. and Socha, W.W. (1979). Blood groups of Old World monkeys: Evolutionary and taxonomic implications. *Journal of Human Evolution*, **8**, 445–451.

Roca, A.L., Georgiadis, N., Pecon-Slattery, J., and O'Brien, S.J. (2001). Genetic evidence for two species of elephant in Africa. *Science*, **293**, 1473–7.

Ruvolo, M. (1997). Genetic diversity in hominoid primates. *Annual Review of Anthropology*, **26**, 515–540.

Ruvolo, M., Pan, D., Zehr, S., Goldberg, T., Disotell, T.R., and von Dornum, M. (1994). Gene trees and hominoid phylogeny. *Proceedings of the National Academy of Sciences U.S.A.*, **91**, 8900–8904.

Ryder, O.A. (1988). Founder effects and endangered species. *Nature*, **331**, 396.

Sarmiento, E.E., Butynski, T.M., and Kalina, J. (1996). Gorillas of Bwindi-Impenetrable forest and the Virunga volcanoes: Taxonomic implications of morphological and ecological differences. *American Journal of Primatology*, **40**, 1–21.

Seaunez, H.N. (1982). Chromosome studies in the orang-utan (*Pongo pygmaeus*): Practical applications for breeding and conservation. *Cytogenetics and Cell Genetics*, **23**, 137–140.

Socha, W.W. and Moor-Jankowski, J. (1986). Blood groups of apes and monkeys. In *Primates: The Road to Self-Sustaining Populations*, ed. K. Benirschke, pp. 921–932. New York: Springer-Verlag.

Weiss, K.M. (1998). Coming to terms with human variation. *Annual Review of Anthropology*, **27**, 273–300.

Zhang, Y.W., Ryder, O.A., and Zhang, Y.P. (2001). Genetic divergence of orang-utan subspecies (*Pongo pygmaeus*). *Journal of Molecular Evolution*, **52**, 516–526.

Zhi, L., Karesh, W.B., Janczewski, D.N., Frazier-Taylor, H., Sajuthi, D., Gombek, F., Andau, M., Martenson, J.S., and O'Brien, S.J. (1996). Genomic differentiation among natural populations of orang-utan (*Pongo pygmaeus*). *Current Biology*, **6**, 1326–1336.

# 9 Mitochondrial and nuclear DNA estimates of divergence between western and eastern gorillas

MICHAEL I. JENSEN-SEAMAN, AMOS S. DEINARD,
AND KENNETH K. KIDD

## Introduction

### Gorilla distribution and taxonomy

Gorillas are found discontinuously in the tropical forests of equatorial Africa (Fig. 9.1) (for more detailed discussion of the distribution of gorilla populations see Groves, 1970*b*, 1971; Hall *et al.*, 1998; Omari *et al.*, 1999; Plumptre *et al.*, this volume). The largest discontinuity in gorillas' distribution is between the gorillas from West Africa (Nigeria, Cameroon, Gabon, Equatorial Guinea, Republic of Congo, and Central African Republic) and those from East/central Africa (eastern Democratic Republic of Congo, Rwanda, and Uganda).

The taxonomy of gorillas (as well as the other great apes) is currently under debate. During the past two decades most authors have used a "one-species–three-subspecies" taxonomy (e.g., Groves, 1986; Fleagle, 1988; Uchida, 1996). Recently, however, not only are gorillas increasingly being considered as two separate species (i.e., *Gorilla gorilla* and *Gorilla beringei*: e.g., Groves, 1996, 2001, this volume; Sarmiento and Butynski, 1996), but the exact number of recognized subspecies is also undergoing revision (Sarmiento and Butynski, 1996; Sarmiento *et al.*, 1996; Oates *et al.*, 1999, this volume; Groves, 2001, this volume; Stumpf *et al.*, this volume). For the purposes of this discussion (and to avoid confusion), gorillas from West Africa (*G. g. gorilla* and *G. g. diehli*) and from East/central Africa (*G. g. graueri* and *G. g. beringei*, or *G. b. graueri* and *G. b. beringei* if two species of gorillas are recognized) will be referred to here simply as "western gorillas" and "eastern gorillas", respectively.

### Previous genetic results

A very early study of the ABO blood group variation in gorillas reported that all western lowland gorillas ($n = 13$) were type B, while mountain gorillas

247

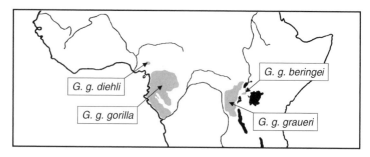

Fig. 9.1. Geographic distribution of extant gorilla populations. A traditional conservative taxonomy is shown here for clarity where all gorillas are lumped into a single species. Western gorillas include the western lowland gorilla (*G. g. gorilla*) and the Cross River gorilla (*G. g. diehli*); eastern gorillas include the eastern lowland gorilla (*G. g. graueri*) and the mountain gorilla (*G. g. beringei*). Distributions are shown as continuous in this figure, but in reality gorilla populations only exist in small isolated pockets of forest in both West and East Africa.

($n = 2$) were type A (Candela *et al.*, 1940).[1] Later studies using improved methodology and additional individuals demonstrated that all gorillas (both eastern and western) were in fact type B (Socha *et al.*, 1973; Wiener *et al.*, 1976). Studies of additional blood group systems (MN, He, Rh, VAB, CEF, G, and H) further supported the similarity between western and eastern gorillas (Socha *et al.*, 1973; Wiener *et al.*, 1976). With respect to taxonomy, these authors concluded that:

> the blood grouping results give no indication of any noteworthy differences which could distinguish between mountain and lowland gorillas. This contrasts with our findings for chimpanzees, where sharp differences in the results of blood grouping tests support the concept of *Pan troglodytes* and *Pan paniscus* as separate species.
>
> (Wiener *et al.*, 1976:320)

Protein electrophoretic distance studies conducted in the 1970s gave a different story – that eastern and western gorillas had diverged as far back in time as had chimpanzees and bonobos (Sarich and Cronin, 1976; Sarich, 1977). These studies estimated that eastern and western gorillas separated about half as far back as the *Homo–Pan–Gorilla* divergence. Sarich (1977) and Sarich and Cronin (1976) estimated that this latter divergence occurred 4 million years

---

[1] In this and later blood group studies (e.g., Socha *et al.*, 1973, 1995; Wiener *et al.*, 1976) the authors use the term "mountain gorilla" to refer to all eastern gorillas, many of which were actually eastern lowland gorillas (subspecies *G. g. graueri*). This is probably because zoo records were slow to change following the demonstration by Groves (1967) of a third subspecies (see discussion in Groves, 1970*a*).

ago, which gives a date of 2 million years ago for the eastern–western gorilla split, as well as for the chimpanzee–bonobo split.

In the 1990s two studies using mitochondrial DNA (mtDNA) examined the genetic divergence between western and eastern gorillas. Ruvolo *et al.* (1994) sequenced 684 bp of the mtDNA COII locus in four western gorillas and four eastern gorillas. Somewhat unexpectedly, they found a divergence between eastern and western gorillas as great or greater than that between chimpanzees and bonobos. This led them to state that the "establishment of additional gorilla and orangutan species designations may even be warranted" (Ruvolo *et al.*, 1994: 8903; see also discussion in Morell, 1994). Similar results were reported by Garner and Ryder (1996), using a 250-bp segment of the mtDNA control region (also known as the displacement loop, or D-loop), who estimated the divergence between eastern and western gorillas to be about as large as that between chimpanzees and bonobos, and the deepest split within western gorilla mtDNA haplotypes to be about so large as well. Clifford *et al.* (this volume) demonstrate in a phylogeographic analysis that much of the mtDNA diversity in western gorillas is found between major groups of geographically distinct gorilla populations.

In contrast, recent studies utilizing DNA sequence data from the Y-chromosome suggested a smaller degree of divergence between western and eastern gorillas than between chimpanzees and bonobos (Burrows and Ryder, 1997; Altheide and Hammer, 1999, 2000). These Y-chromosome studies found no nucleotide substitutions between eastern and western gorilla haplotypes, while observing differences between chimpanzees and bonobos. The study by Burrows and Ryder (1997) relied on small sample sizes and very few observed mutations (one mutation between chimpanzees and bonobos; zero between eastern and western gorillas). However, the studies by Altheide and Hammer (1999, 2000) used many individuals and found multiple fixed differences between the *Pan* species, but none between western and eastern gorillas.

### Goals of the study

Previous estimates using DNA sequence data are insufficient to estimate accurately the degree of genetic divergence between gorilla populations for several reasons. First, mtDNA is inherited only maternally and therefore may not reflect the genetic history of the entire species. For the same reason, the paternally inherited Y-chromosome may not be representative of the whole species. It has been shown that primate species with drastically different male/female demographic patterns can have very different patterns of their mtDNA, autosomal

DNA, and Y-chromosomal DNA (Melnick and Hoelzer, 1992). Although both male and female gorillas may transfer from their natal group, substantial differences between the sexes do occur (Harcourt, 1978). In addition, the existing mtDNA data come from relatively short regions, and therefore may be sampling too few sites. Furthermore, the mtDNA D-loop may be evolving too fast (and therefore possess too much homoplasy) to provide reasonable estimates of divergences as deep as that between western and eastern gorillas. Finally, the observation of a discrepancy between the mtDNA and Y-chromosome DNA suggests that perhaps the early estimates of divergence between western and eastern gorillas based on mtDNA may not be truly representative of the entire genome.

In order to address these concerns we sequenced one additional mitochondrial locus, as well as eight independent noncoding loci from the nuclear genome in at least one western gorilla, one eastern gorilla, one common chimpanzee, and one bonobo. The nuclear DNA (nucDNA) data total over 10 000 bp in each taxon. This comprises the largest data set to date to address this issue. We also compare these results to the previously published data mentioned above.

## Materials and methods

### Individuals examined

The mitochondrial NADH5 locus was amplified and sequenced in two eastern gorillas: one *graueri* (captive; Houston Zoological Gardens) and one *beringei* (wild-living; Bwindi-Impenetrable National Park, Uganda). The nuclear loci were amplified and sequenced from captive gorillas, chimpanzees, and bonobos, with a few wild-living gorillas (*graueri* and *beringei*) included at the DRD4 locus (see Table 9.1 for details). Additional sequences were taken from previously published studies (Miyamoto *et al.*, 1987; Deinard and Kidd, 1995, 1998, 1999, 2000, Arnason *et al.*, 1996; Xu and Arnason, 1996*a*; Iyengar *et al.*, 1998).

### Loci examined

The region of the mtDNA NADH5 gene examined is homologous to nucleotides 12 697–13 932 (1236 bp) of the human reference sequence (Anderson *et al.*, 1981), and includes codons 121–533 of the gene. This region was chosen because it appeared to have an intermediate mutation rate in hominoids based on the comparison of complete mitochondrial genomes of Bornean and Sumatran

Table 9.1. *Nuclear loci and numbers of individuals included in this study*

| Locus | Western gorillas[a]<br>G. g. gorilla | Eastern gorillas<br>G. g. graueri | <br>G. g. beringei | Chimpanzees<br>P. troglodytes | Bonobos<br>P. paniscus | Class | Chromosome | Length (bp) |
|---|---|---|---|---|---|---|---|---|
| *ADH* | | | | | | | 4q | 1682–1699 |
| ADH1 | 10 | 4 | 0 | 11 | 5 | intronic | | 587–606 |
| intron 2 | | | | | | | | |
| ADH2 | 1 | 1 | 0 | 1 | 1 | intronic | | 518–519 |
| intron 2 | | | | | | | | |
| ADH3 | 1 | 1 | 0 | 1 | 1 | intronic | | 575–576 |
| intron 2 | | | | | | | | |
| *DRD2* | | | | | | | 11q | 1693 |
| intron 1 | 1 | 1 | 0 | 1 | 1 | intronic | | 248 |
| intron 2 | 1 | 1 | 0 | 1 | 1 | intronic | | 821 |
| intron 3 | 15 | 1 | 0 | 45 | 18 | intronic | | 301 |
| 3' UTR | 1 | 1 | 0 | 1 | 1 | UTR | | 323 |
| *DRD4*[b] | 18 | 7 | 3 | 16 | 6 | intergenic | 11p | 1130–1135 |
| Ψ-ηγ-globin[b,c] | 1 | 1 | 0 | 1 | 1 | pseudogene | 11p | 2256–2264 |
| *HOXB7/B6*[d] | 15 | 3 | 0 | 45 | 18 | intergenic | 17q | 1034 |
| *PAH* | | | | | | | 12q | 670 |
| intron 1 | 1 | 1 | 0 | 1 | 1 | intronic | | 288 |
| intron 7 | 1 | 1 | 0 | 1 | 1 | intronic | | 382 |
| *PABX* | 21 | 1 | 0 | 45 | 18 | intergenic | Xp | 725 |
| *RBP3* | | | | | | | 10q | 814 |
| intron 1 | 1 | 1 | 0 | 1 | 1 | intronic | | 451 |
| intron 2 | 1 | 1 | 0 | 1 | 1 | intronic | | 363 |

[a] All western gorillas are from captive individuals. According to zoo records they are all western lowland gorillas (*G. g. gorilla*), but it is possible that some may be Cross River gorillas (*G. g. diehli*) since this distinction has only recently been recognized.

[b] The exact number of As in polyA stretches at DRD4 and ψ-ηγ-globin were not determined and were not included in the analyses.

[c] The western gorilla, chimpanzee, and bonobo ψ-ηγ-globin sequences were taken from GenBank. The *G. g. graueri* sequence was obtained by us.

[d] A polymorphic (CA)$_n$ short tandem repeat was excluded from the HOXB7/B6 analyses.

orangutans (Xu and Arnason, 1996*b*). The nuclear loci are a combination of several noncoding regions from eight loci (Table 9.1). Some of the "loci" are actually two or three noncontiguous regions from the same genomic region, and are combined as if they were one locus. The loci are from the following regions (with the abbreviations used here in parentheses): each second intron of ADH1, ADH2, and ADH3 (ADH); parts of intron 1, intron 2, intron 3, and the 3′ untranslated region of DRD2 (DRD2); an intergenic region upstream of the DRD4 gene (DRD4); part of the η-globin pseudogene (ψ-η-globin); an intergenic region between the HOXB7 and HOXB6 genes (HOX); parts of intron 1 and intron 7 of PAH (PAH); an intergenic region of the pseudoautosomal boundary of the X-chromosome (PABX); and parts of intron 1 and intron 2 of RBP3 (RBP3). These nuclear regions were chosen because they had previously been well characterized for other studies (Miyamoto *et al.*, 1987; Deinard and Kidd, 1995, 1998, 1999, 2000, Deinard, 1997; Iyengar *et al.*, 1998; Osier *et al.*, 1999).

### Laboratory methods

Genomic DNA was extracted from blood or cells as previously described (Maniatis *et al.*, 1982), or using purification kits (Qiagen, Valencia, CA). Genomic DNA was extracted from shed or plucked hairs using a modification of a Chelex resin-based protocol (Walsh *et al.*, 1991), with the addition of proteinase K (Jensen-Seaman, 2000; T. Goldberg, personal communication). Loci were amplified with PCR using standard conditions and protocols. Primer sequences and PCR conditions for all reactions used in this study are given in Table 9.2. PCR products were purified using minicolumns (Qiagen), and sequenced with the same primers used for PCR and with additional internal sequencing primers using fluorescent dye-terminator chemistry and run on either an ABI373 or ABI377 automated sequencer (Applied Biosystems, Foster City, CA). All loci were sequenced in both directions in all taxa. Gametic phase of all polymorphisms at DRD4 was determined using molecular haplotyping methods, including allele-specific PCR, allele-specific restriction enzyme digestion, and TA-cloning of PCR products (Jensen-Seaman, 2000).

### Data analyses

DNA sequences were edited and aligned using the Lasergene software (DNA-Star, Madison, WI). Nuclear DNA sequences were recoded prior to analysis in two ways. First, insertion/deletion mutations (indels) were recoded to count as observed substitutions. Although indels are often ignored for simplicity in estimating genetic divergence, they do contain information (Giribet and Wheeler, 1999; Simmons and Ochoterena, 2000). It is important to include all possible information when the number of observed substitutions is small, as is expected when comparing closely related taxa. Second, nucleotide sites found to be polymorphic within any of the relevant taxa (western gorillas, eastern gorillas, chimpanzees, or bonobos) were recoded as ambiguous nucleotides (e.g., Y = C or T) within the taxon. In this way all of the data from each locus for each taxon could be represented by a single sequence. This was done instead of using multiple haplotypes for each taxon since the gametic phases of multiply heterozygous individuals were determined in only some individuals for a few loci.

DNA sequences were analyzed using the DAMBE software (Xia, 2000), which accommodates ambiguous nucleotides in distance estimation by counting such nucleotides as one-half of a mutation. Genetic distances between taxa were estimated using a Kimura two-parameter model (Kimura, 1980) or a Jukes–Cantor one-parameter model (Jukes and Cantor, 1969) of nucleotide substitution for the mtDNA locus and the nuclear loci, respectively. The absolute numbers of observed differences at the mitochondrial loci, including transversions and transitions, were calculated using the MEGA software (Kumar *et al.*, 1993). Novel sequences have been deposited in GenBank (accession numbers AF240446–AF240447 and AY017012–AY017042).

## Results

### *MtDNA*

Results from the mitochondrial NADH5 locus are shown in Table 9.3, along with results from analyses of previously published mtDNA sequences. The amount of nucleotide sequence divergence between eastern and western gorillas at the NADH5 locus is very nearly the same as that between the two species of *Pan*. In this respect the NADH5 results are consistent with the COII and D-loop data (Ruvolo *et al.*, 1994; Garner and Ryder, 1996). The only minor inconsistency is seen in the number of transversions, which accumulate more slowly and are assumed to be less subject to homoplasy. More than twice as many transversions are seen between chimpanzees and bonobos than are seen between western and eastern gorillas at NADH5, whereas the opposite is the case for the D-loop. Three transitions and no transversions were observed between the *graueri* and *beringei* NADH5 sequences, yielding an estimated genetic distance of 0.24.

Table 9.2. *PCR primers and conditions*[a]

| Locus[b] | Forward primer[c] | Reverse primer[c] | [Mg++] (mM) | Annealing temperature (°C) |
|---|---|---|---|---|
| *NADH5* | CCCAACATCAACCAATTCTTCAAATAC | CGTGGGTGCAGATGTGTAGAAATG | 1.5 | 58 |
|  | GCATCGGCCAGCCACACC | TGATGTTGGGGTAAAATCCGAGTATG | 1.5 | 58 |
| *ADH* |  |  |  |  |
| ADH1 intron 2 | TTGCACCTCCTAAGGCCC | AACCACGTGGTCATCTGTG | 1.5 | 57 |
| ADH2 intron 2 | CTAAGGCTTATGAAGTTCGC | CAATACATGAGTGCCTGAATGC | 1.5 | 58 |
| ADH3 intron 2 | TTGCACCTCCTAAGGCTC | CTAACCACATGCTCATCTGAA | 1.5 | 58 |
| *DRD2* |  |  |  |  |
| intron 1 | GATACCCACTTCAGGAAGTC | GATGTGTAGGAATTAGCCAGG | 1.5 | 60 |
| intron 2 | CCTCTGAGGCTTACTGTCTG | AAAACTAGGGAGGGTCAGAG | 1.5 | 60 |
| intron 3 | ACTTGTGTGCCATCAGCATC | GCGTATTGTACAGCATGGGC | 2.25 | 59 |
| 3' UTR | AGTGTTCGCTTGGCTCCA | TGGTATTTACATGGGTCTCCC | 1.5 | 55 |
| *DRD4* |  |  |  |  |
| entire region | GGGAAAAGTGGTCAAACATGAGTAGC | GAAGGAGCAGGCACBGTGAGC | 1.5 | 58 |
| fragment 1 | ATCAACGAGCGTCCTACTACACTAAAATGG | CCCCCAGCCTMYACTAAACTTC | 1.5 | 58 |
| fragment 2 | GAGCCATRATTGCAGTRAGATGTGAT | GATCCAGAAACAATGAGAAAAGACAGACAAC | 1.5 | 57 |
| fragment 3 | GGGGTTGTCTGTCTTTTCTCATTGTTTC | GAAGGAGCAGGCACBGTGAGC | 1.5 | 57 |
| Ψ-η-*globin* | AAGATGGGGCAACTGTTCACTGGTA | GAAACAAACTAGGGAAATAAGGGGAAAACT | 1.5 | 58 |
|  | GAGCAGCACACTTCTCCCCCA | GCCGCTAATAACCTGAATGCTGAC | 1.5 | 58 |
|  | TCCTATTTGTCAGCATTCAGGTTATTAGC | CTCACCAGGAAGTTCTCAGGGTCC | 1.5 | 58 |

| | | | |
|---|---|---|---|
| *HOXB7/B6* | CAGCGTTTGAATATTGGCCCCTGTC | CCTAAAATCGGCCCCTCCAC | 1.5 | 59 |
| | GAAGTCTTTGTACTTTGATGTGGAGGG | ATACACAGCGGGCTCAGCCATC | 1.5 | 57 |
| *PAH* | | | | |
| intron 1 | GCAGGAAACTATCTGACTTTGG | TGGCAGTTCTGGAGGCCAGA | 1.5 | 60 |
| intron 7 | GAAGCATATTGTATCTGCCC | CACATGTCCCAACAGCTCAT | 2.25 | 58 |
| *PABX* | CTGCAGAAACAAGCTCATCAG | AATTCTTAACAGGACCCATTTAGG | 2.25 | 58.5 |
| *RBP3* | | | | |
| intron 1 | CTGTAAGAGTTGGAGGT | GTGCTTCAGTTTTTCCATGG | 1.5 | 58 |
| intron 2 | ACCCTGTGACACAGCAGAGG | ACCAGGGTTGTCATGTGTTG | 1.5 | 60 |

[a]PCR reactions contained standard buffer (Applied Biosystems), with either 1.5 mM or 2.25 mM MgCl$_2$ as indicated. Dimethyl sulfoxide (DMSO) was added to a final concentration of 5% to the DRD4 entire region, DRD4 fragment 2, DRD4 fragment 3, and the first HOXB7/B6 fragment. A three-step PCR cycle was used with a denaturing temperature of 94 °C, the annealing temperature shown in the table, and an extension temperature of 72 °C. The length of each step varied according to the size of the amplification product; 30–40 cycles were used depending on the amount of template added.

[b]Loci were amplified with the primer pair shown as a single fragment with the following exceptions: NADH5, ψ-ηglobin, and HOXB7/B6 were amplified as overlapping fragments. DRD4 was either amplified as a single fragment, or alternatively as three overlapping fragments when the template was DNA extracted from shed hair.

[c]PCR primers are shown 5′ to 3′. Ambiguous bases are shown according to standard code (Y = C or T; R = A or G; M = A or C; and B = C, G, or T).

Table 9.3. *Results from mitochondrial loci*

| Locus | Length (bp) | Transitions[a] E/W[c] | Transitions[a] C/B[d] | Transversions[a] E/W | Transversions[a] C/B | Total substitutions[a] E/W | Total substitutions[a] C/B | Distance[b] E/W | Distance[b] C/B |
|---|---|---|---|---|---|---|---|---|---|
| COII[e] | 684 | 19.5 | 16.8 | 1.83 | 1.83 | 21.3 | 18.6 | 3.21 | 2.74 |
| D-loop[f] | 250 | 27.3 | 26.4 | 12.8 | 6.0 | 40.0 | 32.4 | 16.02 | 12.95 |
| NADH5[g] | 1236 | 59.5 | 58 | 3 | 7 | 62.5 | 65 | 5.32 | 5.53 |

[a] Absolute numbers of observed substitutions, calculated as the average of all pairwise comparisons between relevant taxa.
[b] Kimura two-parameter distances (Kimura, 1980), given as a percentage.
[c] E/W refers to the comparison between eastern and western gorillas.
[d] C/B refers to the comparison between chimpanzees and bonobos.
[e] Sequence data from Ruvolo *et al.* (1994). Distances given here differ slightly from that study (Ruvolo *et al.*, 1994:8902, Table 2) because of different methods of computing genetic distance.
[f] Sequence data from previous studies (Morin *et al.*, 1994; Garner and Ryder, 1996; Deinard, 1997; Goldberg and Ruvolo, 1997; Saltonstall *et al.*, 1998; Gagneux *et al.*, 1999; Deinard and Kidd, 2000; Jensen-Seaman, 2000). Numbers given here differ slightly from Garner and Ryder (1996:44, Table 1) because additional sequences were included, a slightly longer region of sequence was analyzed, and different methods for computing distances were used.
[g] Sequence data from both the present study and previous studies (Arnason *et al.*, 1996; Xu and Arnason, 1996*a*). NADH5 sequences from eastern gorillas have been deposited in GenBank (accession numbers AF240446 and AF240447).

## Nuclear DNA

Estimates of the sequence divergence between western and eastern gorillas, and between chimpanzees and bonobos, from the eight nuclear loci are shown in Table 9.4. More nucleotide substitutions were observed between chimpanzees and bonobos than between western and eastern gorillas at seven of the eight loci. The total number of observed substitutions between chimpanzees and bonobos is 52, while only 33.5 were seen between western and eastern gorillas, which is significantly different from an expected equal number of substitutions ($p < 0.05$, $\chi^2$ test). The single locus that shows a greater divergence between western and eastern gorillas is the $\psi$-$\eta$-globin pseudogene, which accounts for over half of all observed substitutions between these gorilla taxa.

At one of the loci (DRD4), the gametic phase of all haplotypes was completely determined with molecular methods (for details, see Jensen-Seaman, 2000) for multiple individuals of all taxa ($n = 37, 20, 32$, and 12 independent chromosomes of western gorillas, eastern gorillas, chimpanzees, and bonobos, respectively). The network of all observed haplotypes is shown in Fig. 9.2. Chimpanzees and bonobos are reciprocally monophyletic, and do not share

Table 9.4. *Results from nuclear loci*

| Locus[a] | Length | Substitutions[b] | | Distance[b,c] | |
| --- | --- | --- | --- | --- | --- |
| | | E/W | C/B | E/W | C/B |
| ADH | 1682–1699 | 3 | 5 | 0.1913 | 0.3191 |
| DRD2 | 1693 | 2 | 7.5 | 0.1246 | 0.4685 |
| DRD4 | 1130–1135 | 3 | 7.5 | 0.2910 | 0.7296 |
| Ψ-η-globin | 2256–2264 | 18 | 16 | 0.8333 | 0.7403 |
| HOXB7/B6 | 1034 | 4.5 | 6 | 0.4767 | 0.6363 |
| PAH | 670 | 1 | 1.5 | 0.1703 | 0.2555 |
| PABX | 725 | 1.5 | 4.5 | 0.2347 | 0.7064 |
| RBP3 | 814 | 0.5 | 4 | 0.0705 | 0.5655 |
| TOTAL | 10 004–10 034 | 33.5 | 52 | | |
| AVERAGE | | | | 0.2991 | 0.5527 |

[a] See text for explanation of loci abbreviations.
[b] Substitutions and distances were estimated with polymorphic sites (variable within a taxon) coded as ambiguous, in order to count as half of one mutation (see text).
[c] Jukes–Cantor (1969) distances.

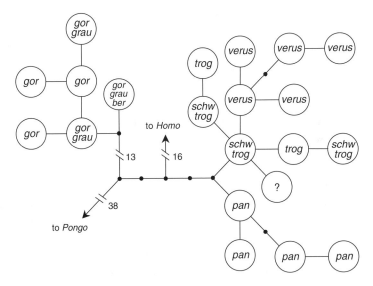

Fig. 9.2. Haplotype network of DRD4 hominoid haplotypes (nuclear DNA). Large circles represent observed haplotypes. Lines connecting circles represent one mutation. Dots along lines represent haplotypes inferred to have existed. The abbreviations *gor, grau, ber, trog, schw, verus,* and *pan* within circles identify those haplotypes found within *G. g. gorilla, G. g. graueri, G. g. beringei, P. t. troglodytes, P. t. schweinfurthii, P. t. verus,* and *P. paniscus*, respectively. No haplotypes were found to be unique to eastern gorillas (*G. g. graueri* and *G. g. beringei*).

any haplotypes. In contrast, western and eastern gorillas share several haplotypes. In fact, eastern gorillas do not possess any unique haplotypes. Rather, all DRD4 haplotypes observed in eastern gorillas were also seen in western gorillas.

## Discussion

### *Sequencing results*

The difference between the mitochondrial and nuclear DNA estimates of divergence between western and eastern gorillas – relative to that between chimpanzees and bonobos – seems to be "real" in the sense that the estimates given here are truly reflective of the evolutionary histories of each of the genomes. In other words, neither the mtDNA nor the nucDNA estimates are likely to be aberrant due to sampling artifacts or other reasons. The combined mtDNA data total over 2 kb, and the nucDNA over 10 kb, of sequence in all four relevant taxa.

All three mtDNA loci give very similar results, as should be expected from completely linked loci which must share identical evolutionary histories. The protein-coding COII and NADH5 loci differ slightly in that COII shows the east–west gorilla split to be a little larger than the chimpanzee–bonobo split, while NADH5 shows the opposite. Nonetheless, given the imperfect nature of the molecular clock and the difficulties in divergence estimation from mtDNA loci due to large amounts of homoplasy, such small differences are inconsequential.

Estimating DNA sequence divergence between closely related taxa using nucDNA is somewhat problematic. The low mutation rate of nucDNA means there will be few informative sites at any given locus. Indeed, far fewer varying sites were observed in the combined nucDNA data set of over 10 kb than in the mtDNA NADH5 locus, which is about eight times shorter. In addition, the larger effective population size of nucDNA results in haplotypes being shared between populations long after they have split. Two populations that have separated will remain paraphyletic at most nuclear loci long after they have achieved reciprocal monophyly at a mitochondrial locus. One consequence is that multiple individuals of all taxa should ideally be sampled for each nuclear locus – which was only done for the DRD4 locus in this study. However, at DRD4 where many individuals were examined, the difference between the patterns seen in *Gorilla* versus *Pan* was most striking. Several DRD4 haplotypes were found to be shared between western and eastern gorillas, whereas chimpanzees and bonobos were completely reciprocally monophyletic with no haplotype-sharing (Fig. 9.2).

The results from the nucDNA loci in this study are congruent with those from studies utilizing Y-chromosomal sequence data (Burrows and Ryder, 1997; Altheide and Hammer, 1999, 2000). These studies found multiple fixed differences between chimpanzees and bonobos, while western and eastern gorillas possessed identical haplotypes. As different portions of the genomes of the African apes seem to be evolving semi-independently, it appears that the paternally inherited Y-chromosome is behaving more like the rest of the nuclear genome than the maternally inherited mitochondrial genome.

### Explaining the difference between Pan and Gorilla

Considering both the amount of sequence divergence and the pattern of haplotype-sharing, the most likely conclusion is that the nuclear genomes of eastern and western gorillas indeed have a more recent common ancestry than those of chimpanzees and bonobos – either because of a later divergence date or because of subsequent migration – while the mitochondrial genomes of western and eastern gorillas diverged at about the same time as those of chimpanzees and bonobos. Four nonexclusive scenarios are envisioned which may explain this apparent inconsistency: (1) selection has acted in such a way as to confuse the evolutionary history of one or more of the genomes; (2) mtDNA is a single locus, and like any single locus, may not accurately reflect the organismal history; (3) greater male versus female migration occurs in *Gorilla* but not *Pan*; and (4) more geographic structure to mtDNA than nucDNA is found in the ancestral *Gorilla* population, but not *Pan*.

If selection is operating on either genome to produce this pattern, then it is more likely to be on the mtDNA since the mitochondrial genome acts as a single linkage unit comprising mostly coding sequence while the nuclear genome is represented here by eight independent loci. Nonetheless, it is difficult to imagine a scenario under which any type of selection would generate this type of discrepancy.

Different loci are expected to vary in their evolutionary history because of the random segregation of ancestral polymorphism. For this reason, multiple independent loci are required to estimate accurately phylogenies or divergence times (Pamilo and Nei, 1988; Wu, 1991; Ruvolo, 1996). Because of the ancestral polymorphism and the stochastic nature of mutation, estimates of sequence divergence between taxa from different loci should not be identical, but rather should form a distribution around the mean. MtDNA is only one of the nine loci studied here. Perhaps it just happens to be one of the loci that falls farther from the mean, even though because of its smaller effective population size and higher mutation rate it should track population changes more closely than nucDNA (Moore, 1995).

If, following the initial divergence between western and eastern gorillas, substantial male migration occurred between the daughter populations in the absence of female migration, then the nuclear genome would show a more recent divergence than mtDNA, with extensive haplotype-sharing. This may be possible. While not all male gorillas emigrate from their natal group, those that do may travel much farther than emigrating females. Males may travel for several years before establishing their own group, while females transfer to a neighboring group in a process that may take only minutes (Harcourt, 1978). Perhaps there was a major biogeographic barrier that males were capable of breaching that females could not. The results from the Y-chromosome are consistent with a sex-biased migration model (Burrows and Ryder, 1997; Altheide and Hammer, 1999). Of course, some degree of female-mediated gene flow could have occurred under this scenario, with the immigrant mtDNA haplotypes lost to genetic drift.

Under most circumstances mtDNA should show greater geographic subdivision than nucDNA (Birky *et al.*, 1983). This is because of its smaller effective population size and because a single immigrant female can only add one mtDNA haplotype to the population whereas she could add two different nuclear haplotypes if she were heterozygous (Birky *et al.*, 1983, 1989). For whatever reason, this pattern of a greater degree of geographic subdivision in mtDNA versus nucDNA may have been more exaggerated in the ancestral *Gorilla* population than the ancestral *Pan*. If so, following the initial split creating the daughter populations of gorillas the divergence between their mitochondrial genomes would be greater than that in their nuclear genomes, to a greater extent than in *Pan*.

### *Implications for* Gorilla *taxonomy and biogeography*

Previously, the suggestion had been made that since the genetic divergence between western and eastern gorillas as ascertained from mitochondrial loci was as large or larger than that between chimpanzees and bonobos, and since there is nearly universal agreement that these latter apes are indeed good species, then perhaps we should recognize two species of gorillas – *G. gorilla* and *G. beringei* (Morell, 1994; Ruvolo *et al.*, 1994; Garner and Ryder, 1996; Ryder *et al.*, 1999). If, based on the evidence presented here, western and eastern gorillas have not been reproductively isolated for as long as the recognized species of *Pan*, then should we definitely consider all gorillas to be of a single species? Of course not. There is no reason to believe that the divergence between chimpanzees and bonobos represents any sort of minimum amount of difference required for speciation. Nor is it clear what, if any, taxa should be held up as a standard for comparison of other taxa (Jolly *et al.*, 1995; Ruvolo, 1997), especially since there is little consistency across taxa in such measures (Johns and Avise, 1998). Furthermore, there is no biological basis for the idea that speciation necessarily occurs in a clock-like fashion.

A consensus remains to be reached on exactly how molecular data should be used in taxonomy. DNA sequence data can test the monophyly of groups, and provide reasonable estimates of the time of the splitting of taxa. But how these time estimates relate to taxonomic rank depends on one's systematic philosophy. For example, Goodman and colleagues (1998) suggest a very strict adherence to the idea that divergence time should be the basis of taxonomic rank. Alternatively, proponents of the phylogenetic species concept would name each distinctive clade a separate species – adherence to this principle would likely create several new species of apes. Although many recent studies have used molecular data to revise primate taxonomy, perhaps the most important aspect missing from such studies is explicit statements as to what species concepts, species definitions, and speciation processes are being used when taxonomic revision is proposed (e.g., Morin *et al.*, 1994; Xu and Arnason, 1996*b*; Zhi *et al.*, 1996; Gonder *et al.*, 1997; but see Wyner *et al.*, 1999).

Gorillas and chimpanzees are sympatric throughout much of equatorial Africa. The comparison of the geographic distribution of genetic variation among sympatric taxa can reveal how these taxa may have responded similarly or differently to biogeographic forces common to the region (Avise *et al.*, 1987; Evans *et al.*, 1997; Avise, 1998; Bermingham and Moritz, 1998). If the dating of the deepest split within *Gorilla* is the same as in *Pan* – as it is from the mtDNA data – then one may speculate that both splits may have been caused by the same biogeographic forces in the African Plio-Pleistocene. However, the observation that western and eastern gorillas may have a more recent common ancestry than do chimpanzees and bonobos suggests that different events have led to the oldest splits within these two genera of African apes.

### Implications for other species

It is often the case that a single genetic locus – usually mitochondrial – is examined in studies of primate phylogeny, taxonomy, and biogeography. Since any single locus may not be representative of the entire genome, caution should be taken when interpreting results from a single locus. Since most primates exhibit male dispersal from the natal group, it may be quite common to see large discrepancies between the patterns of mtDNA diversity and patterns of diversity in the rest of the genome. Recent years have seen several suggestions for taxonomic revision of great apes based on a single study of a single mitochondrial locus (Morin *et al.*, 1994; Ruvolo *et al.*, 1994; Xu and Arnason, 1996*b*; Gonder *et al.*, 1997; Ryder *et al.*, 1999). Although such proposals almost always include a statement to the effect that several independent nuclear loci should also be examined, this is seldom done.

As more species are studied using genetic markers from the mitochondrial genome, the autosomal genome, and the X- and Y-chromosomes, generalities may emerge that help to explain the causes of incongruence between genomes. In turn, this may help us better understand our own evolutionary history. Differences in the patterns of the genetic diversity of modern humans as ascertained from mtDNA, autosomal DNA, and Y-chromosomal DNA have led to different interpretations of our evolutionary past (e.g., Hammer and Zegura, 1996; Harding *et al.*, 1997; Jorde *et al.*, 1998), with conflicting interpretations over what may be the underlying causes of such differences (Hey, 1997; Fay and Wu, 1999; Hey and Harris, 1999). Additional data from the African apes can provide the necessary comparative framework in which to best understand the patterns of human genetic diversity.

## Conclusions

Gorillas from West Africa and from East Africa exhibit pronounced fixed diagnosable differences in morphology (Schultz, 1934; Groves, 1971, 1986, 2001; Uchida, 1996; several chapters in this volume). The genetic divergence of the mitochondrial genomes of western and eastern gorillas has previously been shown to be as large as that between chimpanzees and bonobos (Ruvolo *et al.*, 1994; Garner and Ryder, 1996). Our data from the NADH5 gene of the mitochondrial genome confirms this result. However, our data from eight independent nuclear loci suggest that the nuclear genomes of eastern and western gorillas are not as divergent as those of chimpanzees and bonobos, although it is impossible to provide accurate estimates of the absolute timing of these splits. This apparent discrepancy between the picture from the mitochondrial versus nuclear genomes could be due to selection, differences in male versus female dispersal, differences in the geographic structure of the genetic diversity in the ancestral *Gorilla* and *Pan* populations, or simply because the mitochondrial locus – like any single locus – may be atypical compared to the average locus. Future studies may help refine which of these nonexclusive hypotheses may be more likely, and in doing so shed light on ancient demographic patterns of ancestral populations.

### Acknowledgments

We thank Andrea Taylor and Michele Goldsmith for inviting us to contribute to this volume. We thank the following zoos and centers for kindly providing blood or hair samples of apes: Houston Zoological Gardens, Houston, TX; Royal Zoological Society of Antwerp, Antwerp, Belgium; Zoo Atlanta/Yerkes Regional Primate Center, Atlanta, GA; Wildlife Conservation Park, Bronx, NY; Miami Metrozoo, Miami, FL; Henry Doorly Zoo, Omaha, NE; Arizona Regional Primate Center; Centre International de Recherches Médicales de Franceville, Gabon; Milwaukee County Zoo, Milwaukee, WI; Franklin Park Commonwealth Zoo, Boston, MA; and Lowery Zoo, Tampa, FL. We thank the following individuals who generously gave their time and effort to collecting samples in the wild: Tom Butynski, Michele Goldsmith, Alastair McNeilage, Andrew Plumptre, Martha Robbins, and Esteban Sarmiento. This work was supported by National Science Foundation doctoral dissertation grant SBR 9900100 to MIJ-S and by National Science Foundation doctoral dissertation grant SBR 9315871 to ASD.

### References

Altheide, T.K. and Hammer, M.F. (1999). Y chromosome variation in the Hominoidea. *American Journal of Physical Anthropology, Supplement* **28**, 83.

Altheide, T. and Hammer, M. (2000). Comparing patterns of Y chromosome and mitochondrial DNA variation in the Hominoidea. *American Journal of Physical Anthropology, Supplement* **30**, 95.

Anderson, S., Bankier, A.T., Barrell, B.G., de Bruijn, M.H.L., Coulson, A.R., Drouin, J., Eperon, I.C., Nierlich, D.P., Roe, B.A., Sanger, F., Schreier, P.H., Smith, A.J.H., Staden, R., and Young, I.G. (1981). Sequence and organization of the human mitochondrial genome. *Nature*, **290**, 457–465.

Arnason, U., Xu, X., and Gullberg, A. (1996). Comparison between the complete mitochondrial DNA sequences of *Homo* and the common chimpanzee based on nonchimeric sequences. *Journal of Molecular Evolution*, **42**, 145–152.

Avise, J.C. (1998). The history and purview of phylogeography: A personal reflection. *Molecular Ecology*, **7**, 371–379.

Avise, J.C., Arnold, J., Ball, R.M., Bermingham, E., Lamb, T., Neigel, J.E., Reeb, C.A., and Saunders, N.C. (1987). Intraspecific phylogeography: The mitochondrial DNA bridge between population genetics and systematics. *Annual Review of Ecology and Systematics*, **18**, 489–522.

Bermingham, E. and Moritz, C. (1998). Comparative phylogeography: Concepts and applications. *Molecular Ecology*, **7**, 367–369.

Birky, C.W., Jr., Maruyama, T., and Fuerst, P. (1983). An approach to population and evolutionary genetic theory for genes in mitochondria and chloroplasts, and some results. *Genetics*, **103**, 513–527.

Birky, C.W., Jr., Fuerst, P., and Maruyama, T. (1989). Organelle gene diversity under migration, mutation and drift: Equilibrium expectations, approach to equilibrium, effects of heteroplasmic cells, and comparison to nuclear genes. *Genetics*, **121**, 613–627.

Burrows, W. and Ryder, O.A. (1997). Y-chromosome variation in great apes. *Nature*, **385**, 125–126.

Candela, P.B., Wiener, A.S., and Goss, L.J. (1940). New observations on the blood group factors in Simiidi and Cercopithecidae. *Zoologica*, **25**, 513–521.

Deinard, A.S. (1997). The evolutionary genetics of the chimpanzees. PhD thesis, Yale University, New Haven, CT.

Deinard, A.S. and Kidd, K.K. (1995). Levels of DNA polymorphism in extant and extinct hominoids. In *The Origin and Past of Modern Humans as Viewed from DNA*, ed. S. Brenner and K. Hanihara, pp. 149–170. Singapore: World Scientific Publishing.

Deinard, A.S. and Kidd, K.K. (1998). Evolution of a D2 dopamine receptor intron within the Great Apes and humans. *DNA Sequence*, **8**, 289–301.

Deinard, A. and Kidd, K.K. (1999). Evolution of a HOXB6 intergenic region within the Great Apes and humans. *Journal of Human Evolution*, **36**, 687–703.

Deinard, A. and Kidd, K.K. (2000). Identifying conservation units within captive chimpanzee populations. *American Journal of Physical Anthropology*, **111**, 25–44.

Evans, B.J., Morales, J.C., Picker, M.D., Kelley, D.B., and Melnick, D, J. (1997). Comparative molecular phylogeography of two *Xenopus* species, *X. gilli* and *X. laevis*, in the southwestern Cape Province, South Africa. *Molecular Evolution*, **6**, 333–343.

Fay, J.C. and Wu, C.-I. (1999). A human population bottleneck can account for the discordance between patterns of mitochondrial versus nuclear DNA variation. *Molecular Biology and Evolution*, **16**, 1003–1005.

Fleagle, J.G. (1988). *Primate Adaptation and Evolution*. San Diego, CA: Academic Press.

Gagneux, P., Wills, C., Gerloff, U., Tautz, D., Morin, P.A., Boesch, C., Fruth, B., Hohmann, G., Ryder, O.A., and Woodruff, D. S. (1999). Mitochondrial sequences show diverse evolutionary histories of African hominoids. *Proceedings of the National Academy of Sciences U.S.A.*, **96**, 5077–5082.

Garner, K.J. and Ryder, O.A. (1996). Mitochondrial DNA diversity in gorillas. *Molecular Phylogenetics and Evolution*, **6**, 39–48.

Giribet, G. and Wheeler, W.C. (1999). On gaps. *Molecular Phylogenetics and Evolution*, **13**, 132–143.

Goldberg, T.L. and Ruvolo, M. (1997). Molecular phylogenetics and historical biogeography of East African chimpanzees. *Biological Journal of the Linnean Society*, **61**, 301–324.

Gonder, M.K., Oates, J.F., Disotell, T.R., Forstner, M.R.J., Morales, J.C., and Melnick, D.J. (1997). A new West African chimpanzee subspecies? *Nature*, **388**, 337.

Goodman, M., Porter, C.A., Czelusniak, J., Page, S.L., Schneider, H., Shoshani, J., Gunnell, G., and Groves, C.P. (1998). Toward a phylogenetic classification of primates based on DNA evidence complemented by fossil evidence. *Molecular Phylogenetics and Evolution*, **9**, 585–598.

Groves, C.P. (1967). Ecology and taxonomy of the gorilla. *Nature*, **213**, 890–893.

Groves, C.P. (1970*a*). *Gorillas*. New York: Arco Publishing Co.

Groves, C.P. (1970*b*). Population systematics of the gorilla. *Journal of Zoology, London*, **161**, 287–300.

Groves, C.P. (1971). Distribution and place of origin of the gorilla. *Man*, **6**, 44–51.

Groves, C.P. (1986). Systematics of the Great Apes. In *Comparative Primate Biology*, vol. 1, *Systematics, Evolution and Anatomy*, eds. D.R. Swindler and J. Erwin, pp. 187–217. New York: A.R. Liss.

Groves, C.P. (1996). Do we need to update the taxonomy of gorillas? *Gorilla Journal*, **June**, 3–4.

Groves, C.P. (2001). *Primate Taxonomy*. Washington, D.C.: Smithsonian Institution Press.

Hall, J.S., Saltonstall, K., Inogwabini, B.I., and Omari, I. (1998). Distribution, abundance and conservation status of Grauer's gorilla. *Oryx*, **32**, 122–130.

Hammer, M.F. and Zegura, S.L. (1996). The role of the Y chromosome in human evolutionary studies. *Evolutionary Anthropology*, **5**, 116–133.

Harcourt, A.H. (1978). Strategies of emigration and transfer by primates, with particular reference to gorillas. *Zeitschrift für Tierpsychologie*, **48**, 401–420.

Harding, R.M., Fullerton, S.M., Griffiths, R.C., Bond, J., Cox, M.J., Schneider, J.A., Moulin, D.S., and Clegg, J.B. (1997). Archaic African and Asian lineages in the genetic ancestry of modern humans. *American Journal of Human Genetics*, **60**, 772–789.

Hey, J. (1997). Mitochondrial and nuclear genes present conflicting portraits of human origins. *Molecular Biology and Evolution*, **14**, 166–172.

Hey, J. and Harris, E. (1999). Population bottlenecks and patterns of human evolution. *Molecular Biology and Evolution*, **16**, 1423–1426.

Iyengar, S., Seaman, M., Deinard, A.S., Rosenbaum, H.C., Sirugo, G., Castiglione, C.M., Kidd, J.R., and Kidd, K.K. (1998). Analyses of cross-species polymerase chain reaction products to infer the ancestral state of human polymorphisms. *DNA Sequence*, **8**, 317–327.

Jensen-Seaman, M.I. (2000). Evolutionary genetics of gorillas. PhD thesis, Yale University, New Haven, CT.

Johns, G.C. and Avise, J.C. (1998). A comparative summary of genetic distances in the vertebrates from the mitochondrial cytochrome *b* gene. *Molecular Biology and Evolution*, **15**, 1481–1490.

Jolly, C., Oates, J., and Disotell, T. (1995). Chimpanzee kinship. *Science*, **268**, 185–186.

Jorde, L.B., Bamshad, M., and Rogers, A.R. (1998). Using mitochondrial and nuclear DNA markers to reconstruct human evolution. *BioEssays*, **20**, 126–136.

Jukes, T.H. and Cantor, C.R. (1969). Evolution of protein molecules. In *Mammalian Protein Metabolism*, ed. H. N. Munro, pp. 21–123. New York: Academic Press.

Kimura, M. (1980). A simple method for estimating evolutionary rates of base substitutions through comparative studies of nucleotide sequences. *Journal of Molecular Evolution*, **16**, 111–120.

Kumar, S., Tamura, K., and Nei, M. (1993). *MEGA: Molecular Evolutionary Genetics Analysis*, v 1.01. University Park, PA: Pennsylvania State University.

Maniatis, T., Fritsch, E.F., and Sambrook, J. (1982). *Molecular Cloning: A Laboratory Manual*. Cold Spring Harbor, NY: Cold Spring Harbor Laboratory Press.

Melnick, D.J. and Hoelzer, G.A. (1992). Differences in male and female macaque dispersal lead to contrasting distributions of nuclear and mitochondrial DNA variation. *International Journal of Primatology*, **13**, 379–393.

Miyamoto, M.M., Slightom, J.L., and Goodman, M. (1987). Phylogenetic relations of humans and African apes from DNA sequences in the $\psi$-$\eta$-globin region. *Science*, **238**, 369–373.

Moore, W.S. (1995). Inferring phylogenies from mtDNA variation: Mitochondrial-gene trees versus nuclear-gene trees. *Evolution*, **49**, 718–726.

Morell, V. (1994). Will primate genetics split one gorilla into two? *Science*, **265**, 1661.

Morin, P.A., Moore, J.J., Chakraborty, R., Jin, L., Goodall, J., and Woodruff, D.S. (1994) Kin selection, social structure, gene flow and the evolution of chimpanzees. *Science*, **265**, 1193–1201.

Oates, J.F., McFarland, K.L., Stumpf, R.M., Fleagle, J.G., and Disotell, T.R. (1999). New findings on the distinctive gorillas of the Nigeria–Cameroon border region. *American Journal of Physical Anthropology, Supplement* **28**, 213–214.

Omari, I., Hart, J.A., Butynski, T.M., Birhashirwa, N.R., Upoki, A., M'Keyo, Y., Bengana, F., Bashonga, M., and Bagurubumwe, N. (1999). The Itombwe Massif, Democratic Republic of Congo: Biological surveys and conservation, with an emphasis on Grauer's gorilla and birds endemic to the Albertine Rift. *Oryx*, **33**, 301–322.

Osier, M., Pakstis, A.J., Kidd, J.R., Lee, J.-F., Yin, S.-J., Ko, H.-C., Edenberg, H.J., Lu, R.-B., and Kidd, K.K. (1999). Linkage disequilibrium at the *ADH2* and *ADH3* loci and risk of alcoholism. *American Journal of Human Genetics*, **64**, 1147–1157.

Pamilo, P. and Nei, M. (1988). Relationship between gene trees and species trees. *Molecular Biology and Evolution*, **5**, 568–583.

Ruvolo, M. (1996). A new approach to studying modern human origins: Hypothesis testing with coalescence time distributions. *Molecular Phylogenetics and Evolution*, **5**, 202–219.

Ruvolo, M. (1997). Genetic diversity in hominoid primates. In *Annual Review of Anthropology*, ed. W.H. Durham, pp. 515–540. Palo Alto, CA: Annual Reviews Inc.

Ruvolo, M., Pan, D., Zehr, S., Goldberg, T., Disotell, T.R., and von Dornum, M. (1994). Gene trees and hominoid phylogeny. *Proceedings of the National Academy of Sciences U.S.A.*, **91**, 8900–8904.

Ryder, O.A., Garner, K.J., and Burrows, W. (1999). Non-invasive molecular genetic studies of gorillas: Evolutionary and systematic implications. *American Journal of Physical Anthropology, Supplement* **28**, 238.

Saltonstall, K., Amato, G., and Powell, J. (1998). Mitochondrial DNA variability in Grauer's gorillas of Kahuzi-Biega National Park. *Journal of Heredity*, **89**, 129–135.

Sarich, V.M. (1977). Rates, sample sizes, and the neutrality hypothesis for electrophoresis in evolutionary studies. *Nature*, **265**, 24–28.

Sarich, V.M. and Cronin, J.E. (1976). Molecular systematics of the primates. In *Molecular Anthropology: Genes and Proteins in the Evolutionary Ascent of the Primates*, eds. M. Goodman, R.E. Tashian, and J.H. Tashian, pp. 141–170. New York: Plenum Press.

Sarmiento, E.E. and Butynski, T.M. (1996). Present problems in gorilla taxonomy. *Gorilla Journal*, **June**, 5–7.

Sarmiento, E.E., Butynski, T.M., and Kalina, J. (1996). Gorillas of Bwindi-Impenetrable forest and the Virunga Volcanoes: Taxonomic implications of morphological and ecological differences. *American Journal of Primatology*, **40**, 1–21.

Schultz, A.H. (1934). Some distinguishing characters of the mountain gorilla. *Journal of Mammalogy*, **15**, 51–61.

Simmons, M. and Ochoterena, H. (2000). Gaps as characters in sequence-based phylogenetic analyses. *Systematic Biology*, **49**, 369–381.

Socha, W.W., Wiener, A.S., Moor-Jankowski, J., and Mortelmans, J. (1973). Blood groups of mountain gorillas (*Gorilla gorilla beringei*). *Journal of Medical Primatology*, **2**, 364–368.

Socha, W.W., Blancher, A., and Moor-Jankowski, J. (1995). Red cell polymorphisms in nonhuman primates: A review. *Journal of Medical Primatology*, **24**, 282–304.

Uchida, A. (1996). What we don't know about great ape variation. *Trends in Ecology and Evolution*, **11**, 163–168.

Walsh, P.S., Metzger, D.A., & Higuchi, R. (1991). Chelex 100 as a medium for simple extraction of DNA for PCR-based typing from forensic material. *BioTechniques*, **10**, 506–513.

Wiener, A.S., Socha, W.W., Arons, E.B., Mortelmans, J., and Moor-Jankowski, J. (1976). Blood groups of gorillas: Further observations. *Journal of Medical Primatology*, **5**, 317–320.

Wu, C.-I. (1991). Inferences of species phylogeny in relation to segregation of ancient polymorphisms. *Genetics*, **127**, 429–435.

Wyner, Y., Absher, R., Amato, G., Sterling, E., Stumpf, R., Rumpler, Y., and DeSalle, R. (1999). Species concepts and the determination of historic gene flow patterns in the *Eulemur fulvus* complex. *Biological Journal of the Linnean Society*, **66**, 39–56.

Xia, X. (2000). *Data Analysis in Molecular Biology and Evolution*. Boston, MA: Kluwer.

Xu, X. and Arnason, U. (1996*a*) A complete sequence of the mitochondrial genome of the Western lowland gorilla. *Molecular Biology and Evolution*, **13**, 691–698.

Xu, X. and Arnason, U. (1996*b*). The mitochondrial DNA molecule of Sumatran orangutan and a molecular proposal for two (Bornean and Sumatran) species of orangutan. *Journal of Molecular Evolution*, **43**, 431–437.

Zhi, L., Karesh, W.B., Janczewski, D.N., Frazier-Taylor, H., Sajuthi, D., Gombek, F., Andau, M., Martenson, J.S., and O'Brien, S.J. (1996). Genomic differentiation among natural populations of orang-utan (*Pongo pygmaeus*). *Current Biology*, **6**, 1326–1336.

# 10   *Genetic studies of western gorillas*

STEPHEN L. CLIFFORD, KATE A. ABERNETHY,
LEE J.T. WHITE, CAROLINE E.G. TUTIN, MIKE W. BRUFORD,
AND E. JEAN WICKINGS

## Introduction

### *Classification*

The science of systematics is used to establish evolutionary relationships among species and until the 1960s was based almost entirely on morphological data (Moritz, 1995). The morphological evidence for separating the gorilla into the three subspecies currently recognized, *Gorilla gorilla gorilla* (western lowland gorilla), *Gorilla gorilla graueri* (eastern lowland gorilla), and *Gorilla gorilla beringei* (mountain gorilla), was compiled in the 1960s and recognizes significant differences in cranial and postcranial features (Schaller, 1963; Groves, 1970; Groves and Stott, 1979). At the present time, these three subspecies have distinct geographical distributions, with the western subspecies separated from the eastern and mountain subspecies by more than 800 km. The western lowland gorilla is found in Gabon, Cameroon, Nigeria, Equatorial Guinea, Congo and the Central African Republic (Lee *et al.*, 1988). Groves (1967) proposed four "demes" within the western lowland gorilla range based on clinal variations in skull size. These roughly correspond to: (1) gorillas from the valley of the Sangha River in the Central African Republic, Cameroon, and northern Congo, (2) gorillas from Nigeria, (3) gorillas from southern Cameroon including populations from Dja, and (4) gorillas from coastal and central Gabon and southern Congo. Fig. 10.1 illustrates the current distribution of gorilla populations. More recently Oates *et al.* (1999) have suggested that gorilla populations north of the Sanaga River are morphologically distinct, in accordance with Groves's observations. The eastern lowland gorilla is found in fragmented populations of both lowland and highland habitat in the Democratic Republic of Congo and is almost contiguous in its distribution with the mountain gorilla, which is found in the Virunga Volcanoes of Rwanda and the Bwindi Forest of Uganda. The Bwindi gorillas have also been considered a separate subspecies based on morphology (Groves and Stott, 1979; Sarmiento and Butynski, 1996).

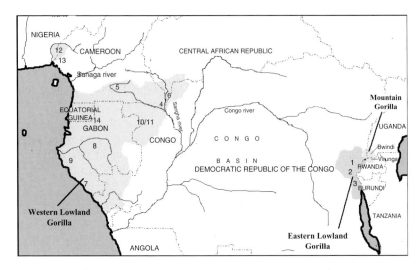

Fig. 10.1. Map of central Africa illustrating the location of sites where gorilla nests were sampled. The numbers correspond to the following: (1) Kasese, (2) Kahuzi-Biega,* (3) Itombwe, (4) Lobéké, (5) Dja, (6) Dzanga,* (7) Petit Loango, (8) Lopé,* (9) Mt. Doudou, (10) Odzala,* (11) Lossi, (12) Afi Mts.,* (13) Mbe Mts., (14) Monte Alen. Gorilla distribution is shaded in grey. *Only those samples marked with an asterisk yielded mitochondrial sequences.

Genetic studies of gorillas are still few (Garner and Ryder, 1996; Field *et al.*, 1998; Gagneux *et al.*, 1999; Oates *et al.*, 1999; Oates *et al.*, this volume) but are becoming increasingly relevant in systematic studies. Genetic data often support existing morphological and ecological data for a given species (Hillis *et al.*, 1996) but not always (Harris and Disotell, 1998). Genetic studies on gorillas, particularly information obtained from mitochondrial DNA (mtDNA) sequences, have revealed large levels of variability within western gorillas (Gagneux *et al.*, 1999), greater than that seen, for example, between bonobos (*Pan paniscus*) and the common chimpanzee (*Pan troglodytes*) (Garner and Ryder, 1996). The geographic basis of this variability in gorillas has not been investigated, since too few samples of known geographic origin have been examined. More recently, in accordance with morphological data, the genetic distinctiveness of populations of gorillas in Nigeria and Cameroon in the Cross River region has been postulated (Oates, *et al.*, 1999, this volume). Additionally, genetic studies have corroborated the morphological and geographical affinity of the eastern lowland and mountain gorillas by confirming their recent divergence from a common ancestor, approximately 100 000 years ago (Ruvolo *et al.*, 1994). The eastern–western gorilla divergence has, however, been dated much earlier, at about 2 million years ago (Ruvolo *et al.*, 1994; Sarich, 1977). Information from gorilla nuclear DNA has revealed less variability than mitochondrial DNA (Seaman *et al.*, 1999). Similarly, nuclear sequences in chimpanzees exhibit much less variation than mitochondrial sequences (Kaessmann *et al.*, 2000).

The Bwindi and Virunga mountain gorillas have been of interest to the geneticist, especially in relation to the classification of Bwindi gorillas as mountain gorillas. Some researchers believe they are mountain gorillas and distinct from the Virunga grouping (Sarmiento and Butynski, 1996). However, genetic studies of mtDNA variation have suggested that both groups are of the same recent genetic origin (Garner and Ryder, 1996; Jensen-Seaman and Kidd, 2001) and that therefore both are taxonomically the same group. In this situation, the morphological differences between the two populations may indicate that ecology and diet are reflected in behavior and morphology in the absence of substantial genetic differentiation. Alternatively, the genetic differentiation that exists has not been found/determined with the genetic marker (mitochondrial DNA) used in this study. In addition, the few eastern lowland gorilla sequences that have been analyzed show low intrasubspecific genetic differentiation (Garner and Ryder, 1996; Seaman *et al.*, 1999).

Doubts about the classical taxonomy of gorillas, based on both morphological and genetic data, have recently led to a proposed reclassification of the gorilla (Groves, 2001). Two species have been proposed: *Gorilla gorilla* (western gorilla) which comprises two subspecies, *G. g. gorilla* (former western lowland gorilla except Nigerian gorillas) and *G. g. diehli* (Nigerian gorillas), and *Gorilla beringei* (eastern gorilla) which consists of three subspecies, *G. b. graueri* (former eastern lowland), *G. b. beringei* (Virunga mountain gorilla) and the as-yet unnamed subspecies for the Bwindi forest (Sarmiento and Butynski, 1996). This chapter outlines the current opinions in genetics and the understanding of gorilla taxonomy, presents new data on genetic variability, and demonstrates the impact these data will have for the future conservation and management of the gorilla species.

### Origins of divergence

The historical events that separated gorilla subspecies may have been driven by numerous periods of forest retraction and expansion, which occurred over the last 2.3 million years (Lanfranchi and Schwartz, 1990; Maley, 1996). The result of rainforest cover retraction has been the creation of forest refuges, usually mountainous regions such as in northwest Cameroon and the Rift Valley of eastern Democratic Republic of Congo, Rwanda, and Uganda, and along major watercourses which did not dry out. Gorillas are restricted to closed canopy forest and populations must have become fragmented, reduced in number and isolated numerous times between periods of forest re-expansion. The necessity of adapting to different habitat types, with associated changes in diet, is thought to have been a driving factor in allopatric divergence. Over long periods of time when different gorilla populations would have been separated, changes in the genetic diversity of different populations would have occurred through selection at certain genes for adaptation and through genetic drift. The longer the populations were separated the greater the degree of genetic differentiation. Differences in ecological niche adaptation may also account for some of the morphological features distinguishing eastern lowland and mountain gorillas, where differences in aboreality and diet may be expressed through skeletal adaptation. Mitochondrial sequences from the hypervariable control region (CR1) have been examined in gorillas and are consistent with the above scenarios (Gagneux *et al.*, 1999).

### *Conservation genetics*

Human pressures currently threaten all gorilla habitats, and populations are often isolated islands surrounded by human activity. Deforestation and the resulting problems associated with it, such as loss of habitat, fragmentation, and increased poaching, all result in intrinsic population pressures. The western gorilla is the least endangered, but may be becoming extinct in parts of its present-day range, especially in Nigeria where fewer than 100 individuals may remain (Harcourt *et al.*, 1989). The mountain gorilla populations of Virunga and Bwindi and the eastern lowland gorilla populations are critically small and are extremely vulnerable to extinction. *Gorilla gorilla graueri* populations are also highly fragmented, but their comparative lack of genetic diversity may indicate that this fragmentation has occurred recently, or that more samples need to be studied. To date only one population from the Kahuzi-Biega region has been studied in depth (Saltonstall *et al.*, 1998). However, whatever the outcome of systematic studies, clearly they indicate that many gorilla populations are under immense pressure and point to the need for strategies to ensure the long-term survival of the gorilla species.

Until recently the greater part of gorilla research has concentrated on morphology, ecology, and behavior but genetic studies are becoming increasingly numerous, and have a potentially important role to play in gorilla conservation. Genetic studies are especially relevant in light of the increasing isolation of gorilla populations as a result of habitat fragmentation. Existing populations harbor genetic variation that has arisen as a result of evolutionary processes. Information from genetic studies can shed light on demographic events such as severe reductions in population size (bottlenecks) and mixing of populations of different genetic origins. Comparison of information obtained from female mediated markers (mitochondrial DNA), male-mediated markers (Y-chromosome) or biparental markers such as autosomal microsatellites can give information on paternity, reveal sex bias in the movements of animals over many generations, and indicate the influence of social structure on the extent of genetic variation. Information from genetic markers is also useful for determining inbreeding, effective population size, and population structure. These processes can lead, over long periods of time, to high levels of genetic differentiation between populations, to reproductive isolation, and eventually to speciation. The information obtained from genetics is increasingly used, in combination with morphological and ecological data, to classify wild populations and define (and redefine) taxonomic units for conservation (Crandall *et al.*, 2000).

The information obtained from molecular genetic data can thus play an important role in the management of species (O'Brien, 1994). When applying these data to conservation management, units of conservation need to be defined. Management units (MUs) can be defined by notional levels of differentiation in allele frequencies between populations (Moritz, 1995). Evolutionarily significant units or ESUs (Ryder, 1986) require information on allele phylogeny (specifically mitochondrial DNA) in relation to their distribution. ESUs have been defined as reciprocally monophyletic for mtDNA loci and also differ significantly in the frequencies of alleles at nuclear loci. An extension of this approach is to define evolutionarily significant areas, within and between which many species show concordant phylogeographic structuring (Avise, 1994). Recently, Crandall *et al.* (2000) have incorporated ecological exchangeability into this approach, although the operational definition of ESUs is still intractable in species such as the gorilla where experiments on exchangeability cannot be carried out.

## Methods

### *Noninvasive sampling strategies*

When studying wild populations, severe limitations are imposed on the type of sample that can be obtained from individuals. The samples obtained need to be composed of material that will yield DNA. Blood and various organs as well as muscle are the best sources of DNA, but these tissues are difficult if not impossible to obtain from wild-living animals, and it is undesirable to sample invasively from any population. The advent of polymerase chain reaction (PCR) amplification of specific regions of DNA means that minute amounts of material (a single cell can suffice) can be subjected to DNA analysis (Saiki *et al.*, 1985; Kocher *et al.*, 1989). The region to be amplified is defined by oligonucleotide primers that recognize specific sequences which flank the sequence of interest. This section of DNA is enzymatically copied thousands of times. The procedure relies on a thermostable DNA polymerase and cycles of heating and cooling of the sample. Hair and feces are two materials that may yield such small amounts of DNA (picogram) and which can be collected noninvasively from wild animals, including gorillas, with little or no disturbance to the individuals. In addition to causing no harm or perturbation to wild individuals, noninvasive sampling has numerous advantages; large numbers of samples can be collected with relatively low effort and samples can be repeatedly collected from the same individuals. Noninvasive sampling facilitates the study of wild populations at the genetic level, studies which otherwise would not be possible (Woodruff, 1993).

### Hair

Higuchi *et al.*, (1988) demonstrated that hair roots were a source of DNA that could be used to examine variability in human genomes. DNA is also present in gorilla hair samples, which can be found in night nests constructed by gorillas each evening (Tutin *et al.*, 1995). Many hairs can be collected from vacated nests up to several days after the gorilla's occupancy. The ability to extract DNA from hair and thus obtain genetic data from wild apes has been previously demonstrated in several studies of chimpanzees *P. troglodytes* (Morin *et al.*, 1992, 1994; Gagneux *et al.*, 1997) and *P. paniscus* (Gerloff *et al.*, 1995).

Only the hair follicle contains sufficient amounts of DNA to make amplification feasible. Chelex was used as the extraction medium (Walsh *et al.*, 1991) and this produces single-stranded DNA in low quantities, which limits the number and type of analyses that can be carried out for each sample. Hairs can be classed as anagen (the normal living phase, when nuclear DNA is present), catagen (hair undergoing apoptosis, or programmed cell death) or telogen (dying phase). DNA can be obtained from anagen hair or telogen if there is germinal tissue attached. To date only 26% of a selection of gorilla hairs examined from Lopé Reserve, Gabon were found to be anagen or telogen with germinal tissue attached and, therefore, potentially able to yield amplifiable DNA (K. Jeffery, unpublished data). Hence, a sample of hairs from any one nest will not necessarily contain DNA.

### Feces

The use of feces and urine as sources of DNA has been examined in other primate studies (Gerloff *et al.*, 1995, 1999; Smith *et al.*, 2000), and more recently its application to other areas of noninvasive research such as the genetics of simian immunodeficiency viruses or SIV (Santiago *et al.*, 2002) has been discussed. Considerable debate has surrounded the potential of fecal material for noninvasive genetic studies of wild populations (e.g., Gerloff *et al.*, 1995; Taberlet *et al.*, 1996; Kohn *et al.*, 1999; Goossens *et al.*, 2000; Waits *et al.*, 2000). Whilst feces are becoming increasingly used as a source suitable for mtDNA analysis, concern has been expressed about the suitability for the amplification of single-copy nuclear DNA markers such as microsatellite loci.

Similar problems are associated with the interpretation of results obtained from both hair and fecal samples (and other noninvasively collected samples). In spite of the great promise of early studies, several years have passed and only a very few comprehensive studies using noninvasive genetic sampling have been published (Morin *et al.*, 1993; Gagneux *et al.*, 1997). Other species, in which it is less easy to identify individuals or where feces are more difficult to find, may prove more difficult to type noninvasively. Problems associated with noninvasively collected samples as tools for molecular genetics will be discussed in the relevant sections below.

## Genetic analysis

### *Mitochondrial DNA*

Mitochondrial DNA is maternally inherited in a single copy, does not undergo recombination, and has been shown on average to evolve five to ten times faster than nuclear DNA. It is particularly useful in population-level studies of DNA variation in mammals including primates (Brown *et al.*, 1979). Mitochondrial DNA found in hair roots is typically degraded and the analysis of long sequences requires the design of multiple PCR primers, each corresponding to smaller fragments which together constitute contiguously the region of interest. Typically, sequences of less than 1 kb (and usually 200–500 bp) may be obtained in a single PCR. The number of mtDNA copies in cells is much higher than nuclear DNA and often amplifies successfully, especially in noninvasively collected samples where nuclear DNA is typically degraded. Unfortunately, mitochondrial DNA sequences are frequently inserted into the nuclear genome (Numt) where they evolve differently (Collura and Stewart, 1995: Zischler *et al.*, 1995) and can lead to erroneous interpretation of results if they are not separated from "true" mitochondrial sequences. The amplification of Numts (because the flanking regions to which the oligonucleotide primers can attach maintain high levels of homology) has been shown to be associated with the use of DNA extracted from hair (Greenwood and Pääbo, 1999).

## *Nuclear DNA*

Repetitive DNA, constituting up to 60% of total DNA, is found dispersed throughout the eukaryotic genome and exists in several forms. One type consists of tandem arrays of repetitive units that can be classed on the basis of size, one class being microsatellites (Tautz, 1993). Microsatellite loci have become the nuclear marker of choice for population studies (Bruford and Wayne, 1993). These loci comprise highly polymorphic di-, tri- or tetranucleotide repeat sequences, which can be repeated 10–30 times (Weber and May, 1989; Edwards *et al.*, 1991). Microsatellies are inherited from generation to generation in a biparental fashion, one allele being passed from the female parent and one from the male parent. The availability of thousands of human microsatellite primers which amplify primate DNA (Coote and Bruford, 1996) means that microsatellite loci can be examined in closely related species such as apes (Field *et al.*, 1998) and more specifically gorillas (Clifford *et al.*, 1999).

Recent studies, whilst confirming the power of this approach of cross-species amplification applied to noninvasive strategies, have shown that amplification from the extremely low amounts of DNA is associated with stochastic problems. These problems include the preferential amplification of only one allele, as well as contamination and artifactual amplification (Gerloff *et al.*, 1995; Taberlet *et al.*, 1996; Gagneux *et al.*, 1997). There are, however, analytical methods that can be used to infer problems of this nature. For example, the presence of excess homozygosity in noninbred populations can point to the presence of null alleles or allelic dropout in the first instance (Bruford and Wayne, 1993; Brookfield, 1996). It then becomes necessary to extract and/or amplify each sample up to seven times to be able to confirm a homozygote result at any given locus (e.g., Taberlet *et al.*, 1996). Such an approach is cumbersome but essential for robust interpretation of results.

### Genetic studies of lowland gorillas throughout central Africa

In a comprehensive study carried out in our laboratories at the Centre International de Recherches Médicales, Franceville (CIRMF), Gabon, and at Cardiff University, Wales, noninvasively collected hair samples were available for 15 gorilla populations from throughout the range of both eastern and western lowland gorillas. These samples were obtained through collaborations with researchers from existing study sites and from ongoing census work. A list of these contributors and sample sites is presented in Table 10.1; site distribution is illustrated in Fig. 10.1. This unique collection of gorilla hair comprises the largest number of samples from wild gorillas of known origin. DNA was

**Table 10.1.** *List of samples collected for this lowland gorilla study*

| Country | Zone[a] | n[b] | Collectors[c] |
|---|---|---|---|
| *Eastern lowland range* | | | |
| Democratic Republic of Congo | 2. Kahuzi-Biega | 204 | D. Bonny, K.P. Kiswele (CNRS); I. Omari, C. Sikubwabo (ICCN); L. White, J. Hall, I. Bila-Isia. H. Simons Morland, E. Williamson, K. Saltonstall, A. Vedder, K. Freeman, B. Curran (WCS); J. Yamagiwa (Kyoto) |
| | 1. Kasese | 41 | Curran (WCS) |
| | 3. Itombwe | 126 | I. Omari, F. Bengana (ICCN); J. Hart (WCS) |
| *Western lowland range* | | | |
| Cameroon | 4. Lobeke | 48 | L. White, L. Usongo (WCS) |
| | 5. Dja | 29 | E. Williamson (ECOFAC); L. Usongo (WCS/ECOFAC) |
| Central African Republic | 6. Dzanga–Sangha | (471)[d] | M. Goldsmith, (SUNY); L. White (WCS) |
| Gabon | 7. Petit Loango | 3 | J. Yamagiwa (Kyoto) |
| | 8. Lopé | (1700)[d] | C. Tutin, K. Abernethy, E. Dimoto, J.T. Dinkagadissi, R. Parnell, P. Peignot, B. Fontaine (CIRMF); M.E. Rogers, L. White, B. Voysey, K. McDonald, (Edinburgh); R. Ham (Stirling); J.G. Emptaz-Collomb |
| | 9. Mt. Doudou | 3 | C. Mbina (WWF); L. White (WCS) |
| Congo | 10. Odzala | 80 | M. Bermejo, G. Illera, F. Maisels (ECOFAC) |
| | 11. Lossi/Odzala | 30 | M. Bermejo, G. Illera (ECOFAC) |
| Nigeria | 12. Afi Mts. | (100)[d] | K. McFarland, J. Oates (CUNY) |
| | 13. Mbe Mts. | 1 | K. McFarland, J. Oates (CUNY) |
| | Confiscations | 3 | P. Hall (ProNatura) |
| Equatorial Guinea | 14. Monte Alen | 9 | M. Bermejo, G. Illera (ECOFAC) |

[a] Zone numbers correspond to locations marked on the maps.
[b] n, total number of individual samples collected per site.
[c] CNRS, Centre National de la Recherche Scientifique; ICCN, Institut Congolais pour la Conservation de la Nature; WCS, Wildlife Conservation Society; ECOFAC, Ecosystèmes Forestiers d'Afrique Centrale; SUNY, State University of New York at Stony Brook; Kyoto, Kyoto University; CIRMF, Centre International de Recherches Médicales Franceville; CUNY, City University of New York; WWF, World-Wide Fund for Nature.
[d] Repeated sampling of same individual.

extracted from the hair samples and subjected to analysis at mitochondrial (control region and COII) and nuclear microsatellite loci. Despite the limitations already described of working with noninvasively collected samples, we anticipated that sufficient genetic data could be obtained from these populations to undertake a detailed study of genetic variability within and between the two classically recognized lowland gorilla subspecies.

### Mitochondrial DNA analysis

Several important substitutions in a 289-bp sequence of the control region of mtDNA have been demonstrated in a subset of 63 gorillas, many of unknown origin (Garner and Ryder, 1996). However, we found it impossible to routinely amplify DNA from hairs using these primers. New primers,

$$(5'\text{-CACCATCAGCACCCAAAGCTAATAT-}3')$$

and

$$(5'\text{-GGGATGGGTTAGTTGGTATTCCGTT-}3')$$

were designed which encompassed smaller sections of the control region (approximately 180 bp) and amplification using these new primers, although still difficult, was more successful. The product yielded enough DNA to allow sequence analysis. A series of up to 30 tandemly repeated cytosine residues exists within the amplified region, which effectively terminates the sequencing reaction thereby limiting the amount of sequence produced. The final region of the sequence chosen for analysis consisted of 104 bp which lies directly left of the run of cytosines but which contains 23 informative substitutions from 67 individuals. Fourteen haplotypes were obtained for the control region which is lower than previously reported (Garner and Ryder, 1996). Haplotypes obtained combined with the equivalent sequences from GenBank are illustrated in Fig. 10.2. Genetic distances and neighbor-joining trees were calculated and constructed using PAUP-4 (Swofford, 1998) and CLUSTAL (Thompson *et al.*, 1994). Bootstrap replicates were repeated 1000 times. The neighbor-joining tree from control region haplotypes is illustrated in Fig. 10.3. The data recapitulate the major evolutionary split between eastern and western gorillas and within the western clade the Nigerian samples are the most distinct.

Other mtDNA regions used in genetic studies of primates have been more conserved sequences, such as cytochrome *b* (Morin *et al.*, 1993) and cytochrome oxidase (COII) (Ruvolo *et al.*, 1991, 1994). Phylogenetically, these regions yield less information than the control region as they are more conserved and

| | |
|---|---|
| **NIGERIA1** | --------------------------G-T------------------------T-------------A--------T-T----A-CA-TAT----A-------- |
| **NIGERIA2** | --------------------------G--------------------------T-------------A--------T-T----A-CA-TAT------------- |
| WLG | --------------------------G--------------------------T-------------A--------T-T----A-C---AT----A-------- |
| WLU | --------------------------G--------------------------T--------------C-------T-C----A-CT--A-------------- |
| **LOPE1** | -----------------T---------------------------G---------------------------------------C------------------ |
| **LOPE2** | -----------------------------------G---------G---------------------------------------C------------------ |
| **COLACDP** | -----------------T---------------------------G-------------------------------------------C-------------- |
| WLE | -----------------T---------------------------G-------------------------------------------C-------------- |
| **KESSELACDP** | -----------------T---------------------------G-------------------------------------------C-------------- |
| WLD | -----------------T-----------------------------------------------------------------------C-------------- |
| WLT | -----------------T-----------------------------------------------------------------------C-------------- |
| WLW | -----------------T-----------------------------------------------------------------------C-------------- |
| WLX | -----------------T-----------------------------------------------------------------------C-------------- |
| **ODZALA** | -----------------T-----------------------------------------------------------------------C-------------- |
| WLZ | -----------------T-----------------------------------------------------G-----------------C-------------- |
| WLA | -----------------T--------G--------------------------C--------------C-------T-C---C-TCT--C-------------- |
| WLY | -------G---------T----------T---------------CC----------------------C-------------T-TC-G-C-------------- |
| WLV | -------G---------T--------------------------TC----------------------C--A-C--------T-TC-G---------------- |
| **CAR1** | --------------------------------------------C--------------------------G------T------C---------A--GC---- |
| **CAR3** | -----------------T-----------------A--------C--------------------------G------T------C---------A--GC---- |
| **CAR2** | -----------------T--------T--------A--------C--------------------------G------T------C---------A--GC---- |
| WLB | -----------------T-----------------A--------C--------------------------G------T------C------------GC---- |
| WLC | -----------------T-----------------A--------C--------------------------G------T------C------------GC---- |
| WLF | -----------------T--------------------------C--------------------------G------T------C------------GC---- |
| WLH | -----------------T-----------------------------------------------------G------T------C--GC-------------- |
| ELJ | ---------T----------------G--------T-----------------C-----------------G------T----A---CC------A---T---- |
| ELK | ---------T----------------G--------T-----------------C-----------------G------T----A---CC------A---T---- |
| ELN | ---------T----------------G--------T-----------------C-----------------G------T----A---CC------A---T---- |
| ELO | ---------T----------------G--------T-----------------C-----------------G------T----A---CC------A---T---- |
| **K-BIEGA1** | ---------T----------------G--------T-----------------C-----------------G------T----A---CC------A---T---- |
| **K-BIEGA2** | ---------T----------------G--------T-----------------C-----------------G------T----A---CC------A---T---- |
| ELI | ---------T----------------G--------T-----------------C------------------------T----A---CC------A---T---- |
| ELM | ---------T----------------G--------T-----------------C------------------------T----A---CC------A---T---- |
| **ITOMBWE** | ---------T----------------G--------------------------C-----------------G------T----A---CC------A---T---- |
| **BWINDI** | ---------T----------------G--------------------------C-----------------------------A---CCC-----A---T---- |
| MP | ---------T----------------G--------------------------C-----------------------------A---CCC-----A---T---- |
| MQ | ---------T----------------G--------------------------C-----------------------------A---CCC-----A---T---- |
| MR | ---------T----------------G--------------------------C-----------------------------A---CCC-----A---T---- |
| MS | ---------T----------------G--------------------------C-----------------------------A---CCC-----A---T---- |
| | *******_*_*******_********_*_******_********__*******_*************__**_*_**_*_***_________****_**__**** |
| CONSENSUS | AAGTATTAGCTAACCCACCAATAATTRTCATGTATGTCGTGCATTACTGCCAGACACCATGAATAATGTACAGTACCACAAACGCTCACTCACCTGTAACACAT |

Fig. 10.2. List of sequences from the mitochondrial control region which were used in sequence analysis. Sequences which are labelled WL, EL and M are from Garner and Ryder (1996) and their GenBank accession numbers are L76749 to L76773. Sequences labeled in bold were derived in this study.

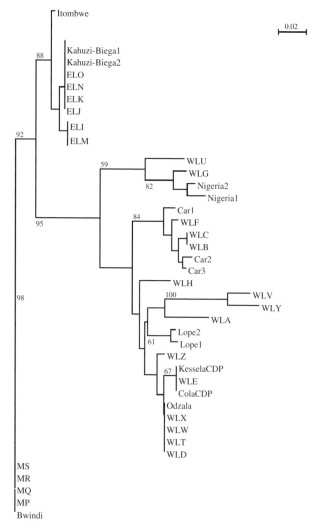

Fig. 10.3. Neighbor-joining tree based on genetic distances derived from control region sequences. Bootstrap replicates were carried out 1000 times. Bootstrap values of less than 50 have not been included. Sequences included comprise those obtained in this study and those derived from Garner and Ryder (1996). Key to sequences are given by location or by GenBank accession numbers and are as follows: *Gorilla gorilla beringei*, Bwindi (Uganda), MP (L76749), MQ (L76750), MR (L76751), MS (L76752); *Gorilla gorilla graueri*, Itombwe (DRC), Kahuzi-Biega (DRC), ELI (L76768), ELJ (L79769), ELK (L76770), ELM (L76771), ELN (L76772), ELO (L76773); *Gorilla gorilla gorilla*; Nigeria (Cross River, Nigeria/Cameroon), Car (Dzanga, Central African Republic), Lopé (Lopé, Gabon), Kessela (captive individual from Centre de Primatologie (CDP), CIRMF, Gabon), Cola (captive individual from Centre de Primatologie (CDP), CIRMF, Gabon), Odzala (Congo), WLA (L76760), WLB (L76761), WLC (L76762), WLD (L76763), WLE (L76764), WLF (L76765), WLG (L76766), WLH (L76767), WLT (L76753), WLU (L76754), WLV (L76755), WLW (L76756), WLX (L76757), WLY (L76758), WLZ (L76759).

show fewer differences. We sequenced a 257-bp sequence of the COII gene which successfully amplified DNA derived from hair in 90% of samples tested (primer sequences are as described by Ruvolo *et al.*, 1991). Genetic distances were obtained using PAUP-4 (Swofford, 1998) and CLUSTAL (Thompson *et al.*, 1994). The eastern populations (*G. g. graueri* and *G. g. beringei*) separate from the other populations with bootstrap values of 100. Within the western grouping there was no significant differentiation, the most distinct populations being those from the Central African Republic (CAR).

A combined analysis of 367 bp of COII and control region sequence revealed nine haplotypes (Fig. 10.4). Genetic distances were calculated and a neighbor-joining tree was obtained (Fig. 10.5). The results correspond with the analysis of control region alone. Mountain and *graueri* separate from each other and from western gorillas. Three groupings emerge within western gorillas: (1) Nigeria, (2) CAR, and (3) all other populations tested from Cameroon, Congo, and Gabon.

### Microsatellite DNA analysis

We have characterized a series of microsatellite markers for gorillas, using primers designed for human loci (Clifford *et al.*, 1999). A marked difference in amplification success rate and accuracy was seen between DNA extracted from blood and from shed hair. Of the 24 polymorphic primer pairs tested, only eight gave consistent amplification when applied to DNA from shed hair (Table 10.2). Further, when these eight loci were used to amplify samples from all populations only five yielded enough genotypes for analyses. Table 10.3 lists the results obtained for these five loci for nine gorilla populations. The population called CDP represents the control captive gorilla population housed at CIRMF, Gabon, where DNA was derived from blood.

Overall, shed hairs were a poor starting material for nuclear DNA analysis, due to telogen root phase, number of hairs collected, and quality of DNA extracted from viable hairs. In total, 504 DNA extractions were carried out, each sample (i.e., individuals from which more than one hair was collected; if sufficient hairs were available, extraction could be carried out in duplicate) being extracted twice and PCRs repeated at least twice. Of the 504 samples, 241 (48%) gave a genotype for at least one locus; however this dropped to 99 (19%) for samples which gave genotypes at three or more loci and less than 2% (10) which gave genotypes at all five loci. All populations except for the CDP group, where DNA was obtained from blood, departed from Hardy–Weinberg equilibrium and heterozygosity was significantly reduced at all loci. The fact that the CDP population did not deviate from Hardy–Weinberg expectations

```
CAR1          --------------------------------------A---------------------------------T-G------T-------------------G------------------
CAR3          --------------------------------------A-----------------------------------G------T-------------------G------------------
CAR2          -----------------------------T--------A-----------------------------------G------T-------------------G------T-----------
LOPE1         --------------------------------------G--------T----------------------------------------T-------------------------------
COLA          --------------------------------------A--------TG---------------------------------------T---C---------------------------
NIGERIA1      ----------G---------------------------G---------C----------------------C----C--------T-T--G-----------------------------
NIGERIA2      ----------G---------------------------G---------C----------------------C----C-------------G-----------------------------
K-BIEGA       --T---------T-------C--------G--------T--------T--------C----------------:G------T----A-T-CC------A---T-----------------
BWINDI        ------------T-------C-----:--G--------G--------T--------C----------------:------------A-T-CCC-----A---T-----------:-----
              **_*******_*_*******_*****_**_********_********___******_**************______****_***____*___*****_**__*****_*****_*****

CONSENSUS     CCCAAGTATTAGCTAACCCATCAATAATTATCATGTATRTCGTGCATCACTGCCAGACACCATGAATAATGTACAGTACCACAAACGCCCAATCACCTGTAACACATAGACGCCCAAGAA

CAR1          ------------------------------------C-----------------------------------------------------------------------------------
CAR3          ------------------------------------C-----------------------------------------------------------------------------------
CAR2          ------------------------------------C---:-----------------------------------------------:-------------------------------
LOPE1         ------------------------------------------------------------------------------------------------------------------------
COLA          ------------------------------------------------------------------------------------------------------------------------
NIGERIA1      -----------------------G------------------------------------------------------------------------------------------------
NIIGERIA2     -----------------------G------------------------------------------------------------------------------------------------
K-BIEGA       ---------G----------------A-------------:------------------C----------------------------:---C--T--------T---------------
BWINDI        ---------G----------------A-------------:------------------C----------------------------:---C-----------T---------------
              *********_*************_**_*********_***_******************_****************************_***_**_********_***************

CONSENSUS     ATAGAGACCATCTGAACCATCCTACCCGCTATTATCTTAGTCCTGATCGCCCTCCCATCTCTACGAATCCTCTATATGACAGATGAAATCAATGACCCCTCCTTCACTATCAAATCAATC

CAR1          --------------------------------------------------------------------------------------------G---------------------------
CAR3          ------------------------------------------------------------------------------------------------------------------------
CAR2          ------------------------------------------------------------------------------------------------------------------------
LOPE1         ------------------------------------------------------------------------------------------------------------------------
COLA          ------------------------------------------------------------------------------------------------------------------------
NIGERIA1      ------------------------------------------------------------------------------------------------------------------------
NIGERIA2      ------------------------------------------------------------------------------------------------------------------------
K-BIEGA       ------------------------------------------------G----------T--C-:------T----------------------------------------:-------
BWINDI        ----------------:------------------------------------------T--C-:------T----------------------------------------:-------
              ****************_*******************************_**********_**_*_******_********************_*******************_*******

CONSENSUS     GGTCACCAATGATACTGAACCTACGAATACACCGACTATGGTGGATTGATCTTTAACTCCTATATACTCCCCCCATTATTCCTAGAACCAGGTGACCTACGACTCCTTGACGTCGACAAC

CAR1          -----
CAR3          -----
CAR2          -----
LOPE1         -----
COLA          -----
NIGERIA1      -----
NIGERIA2      -----
K-BIEGA       -----
BWINDI        ----G
              ****_

CONSENSUS     GCGAA
```

Fig. 10.4. List of Combined 367-bp sequences from the mitochondrial control region and COII which were used in analysis. Sequences CAR1, CAR2, CAR3 are from Dzanga in the Central African Republic, Lope is from Lopé in Gabon, Cola is a captive individual originally from central Gabon, Nigeria1 and Nigeria2 are from the Cross River gorilla populations on the Cameroon–Nigeria border. K-Biega is from Kahuzi-Biega in the

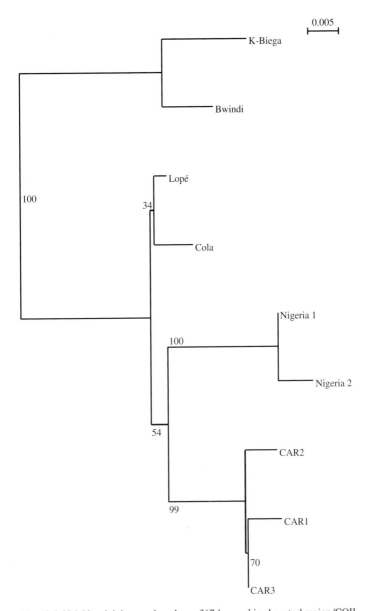

Fig. 10.5. Neighbor-joining tree based on a 367-bp combined control region/COII sequence from nine individuals. Bootstrap replicates were carried out 1000 times. Sequences CAR1, CAR2, CAR3 are from Dzanga in the Central African Republic, Lopé is from Lopé in Gabon, Cola is a captive individual originally from central Gabon, Nigeria 1 and Nigeria 2 are from the Cross River gorilla populations on the Cameroon–Nigeria border, K-Biega is from Kahuzi-Biega in the DRC and Bwindi is from Bwindi in Uganda.

Table 10.2. *Characterization of eight polymorphic microsatellite loci using hair-derived DNA from western lowland gorillas*

| Locus | Variable number tandem repeat | Annealing Temperature (°C) | Number of individuals | Number of alleles | Heterozygosity | | PCR primer sequences 5'–3' |
|---|---|---|---|---|---|---|---|
| | | | | | Observed | Expected | |
| DIS548 | (GATA)$_n$ | 58 | 10 | 4 | 0.400 | 0.737 | GAACTCATTGGCAAAAGGAA GCCTCTTTGTTGCAGTGATT |
| 1S550[a] | (GATA)$_n$ | 58 | 22 | 10 | 0.409 | 0.868 | CCTGTTGCCACCTACAAAAG TAAGTTAGTTCAAATTCATCAGTGC |
| 2S434[a] | (GATA)$_n$ | 58 | 21 | 11 | 0.524 | 0.843 | TAAATCACTAGCCTTTGCCG GCCATCTGTACTGTTCCCAG |
| 2S1326[a] | (GATA)$_n$ | 58 | 21 | 12 | 0.571 | 0.808 | AGACAGTCAAGAATAACTGCCC CTGTGGCTCAAAAGCTGAAT |
| 3S1768[a] | (GATA)$_n$ | 58 | 31 | 11 | 0.523 | 0.601 | GGTTGCTGCCAAAGATTAGA CACTGTGATTTGCTGTTGGA |
| D22S685 | (GATA)$_n$ | 58 | 8 | 5 | 0.500 | 0.833 | TTCTTAGTGGGGAAGGGATC TGAGTTTGATGTTTTTGATAGACA |
| DXS738 | (CA)$_n$ | 54 | 10 | 5 | 0.273 | 0.723 | ATGTGTTGTTGTATTCACCTTGC CCAGCAATAACCATAAGTAAAAC |
| DXS6810[a] | (GATA)$_n$ | 58 | 24 | 7 | 0.063 | 0.678 | ACAGAAAACCTTTTGGGACC CCCAGCCCTGAATATTATCA |

[a]Loci used in large-scale study encompassing DNA derived from noninvasively collected hair samples from all gorilla populations for which samples were available.

Table 10.3. *Heterozygosity values obtained over five microsatellite loci for the gorilla populations*

| Population | Mean sample size per locus | Mean number of alleles | Mean heterozygosity Observed | Mean heterozygosity Hardy–Weinberg expected[a] |
|---|---|---|---|---|
| CDP[b] | 10.0±0.0 | 5.2±0.9 | 0.700±0.148 | 0.747±0.057 |
| Afi Mountains | 3.0±0.4 | 3.4±0.2 | 0.600±0.170 | 0.792±0.034 |
| Dzanga-Sangha | 20.0±2.9 | 8.0±0.6 | 0.617±0.059 | 0.822±0.029 |
| Dja | 5.0±0.8 | 4.4±0.5 | 0.205±0.060 | 0.782±0.020 |
| Itombwe | 8.8±1.7 | 5.8±1.4 | 0.376±0.125 | 0.793±0.040 |
| Kahuzei-Biega | 20.8±1.2 | 8.0±0.8 | 0.512±0.093 | 0.718±0.064 |
| Lobéké | 6.6±0.9 | 4.8±0.5 | 0.289±0.093 | 0.783±0.042 |
| Lopé | 11.6±1.4 | 7.4±0.9 | 0.538±0.101 | 0.842±0.022 |
| Lossi | 11.0±1.8 | 6.4±0.9 | 0.511±0.132 | 0.797±0.034 |

[a] Unbiased estimate.
[b] CDP population represents the captive gorillas housed at CIRMF, Gabon.

or exhibit significant reductions in heterozygosity indicates that the results obtained from the noninvasively sampled populations can be attributed to significant allelic dropout (stochastic loss of alleles due to nonamplification from poor-quality template DNA) and not population structuring. The results presented here are the first from an ongoing study analyzing gorilla populations at microsatellite loci using noninvasively collected materials. They illustrate the immense difficulty encountered when working with nuclear markers and noninvasively collected samples, and provide an indication of the problems that may be encountered when trying to analyze wild populations at microsatellite loci, specifically using DNA extracted from hair.

### Gorilla phylogeography

Our results confirm the divergence between the eastern and western gorillas that has been reported previously (Garner and Ryder 1996) at both mtDNA control region and COII loci. Within the western lowland gorillas our results revealed three highly distinct groupings: populations from Nigeria (Afi Mountains), populations from the Dzanga-Ndoki National Park (Central African Republic), and the remaining populations tested (constituting the largest proportion (75%) of western gorillas). The distinctiveness of the Afi Mountains gorillas has already been previously suggested (Stumpf *et al.*, 1998; Oates *et al.*, 1999) and our genetic results support the presumed isolation of this residual population. We report here the distinctiveness of the most easterly *Gorilla gorilla gorilla* populations from the Dzanga-Ndoki National Park. When control region and COII sequences are combined, the distinctiveness of the Afi Mountains population and the Dzanga population from the other western gorillas is even more evident.

### Nigeria

The samples we had available for analysis comprised a small group – representative of one subpopulation only (Afi Mountains) – located on the Nigerian/Cameroon border, which has been termed the Cross River gorillas. This gorilla population is thought to number only a few hundred individuals at most and is geographically isolated from the nearest western lowland gorilla population in southeast Cameroon by 250 km (Oates *et al.*, this volume). In total, the Cross River gorilla population comprises four to five subpopulations with, in recent history, little or no interchange. The Afi Mountains sample, which is presented here, is from the most northwestern and isolated of all these subpopulations.

This area of Africa in northern Cameroon/Nigeria north of the Sanaga River has been labeled a speciation hotspot (Grubb, 1990). There is evidence for subdivision amongst certain primate species, most notably *Pan t. vellerosus*, the proposed fourth subspecies of chimpanzee (Gonder *et al.*, 1997), and the mandrill–drill separation (Grubb, 1990). It has been proposed that the Sanaga River is a barrier to gene flow between certain species but this does not extend to all species of primates. Smaller monkey species such as *Cercopithecus nictitans* and *C. pogonias* are distributed across this area, but little is known about their genetic variability and a number of races may exist. Gorillas from this region have been previously characterized, based on skeletal morphology, as potentially very distinct (Groves, 1970; Stumpf *et al.*, this volume). The forest around the Cameroon/Nigeria border is of a montane nature with altitudes of up to 1500 m and is different from the type of forest found in the Congo Basin; morphological variation in gorillas may reflect this difference (Stumpf *et al.*, this volume). More recently, preliminary mitochondrial data (Oates *et al.*, 1999, this volume) have suggested that these gorillas are indeed distinct, as distinct from the other western gorillas as the western gorillas are from the eastern groupings. Our results confirm these findings.

### Central African Republic

This is the first study to analyze samples from the Central African Republic (CAR) and results show this group to be a genetically distinct population. Groves proposed one gorilla "deme" from the Sangha River valley, based on skull morphology, to be in this region (Groves, 1967). However the clinal variation reported in skull size for these gorillas was not as large as that reported for Nigerian gorillas. The CAR gorillas included in this study are from a 50-km$^2$ area (Bai Hokou) of the Dzanga-Ndoki National Park. Sequencing of different individuals from this region and comparison with sequences in GenBank confirmed that the mitochondrial haplotype identified was specific to this region. Geographically, the only obvious barriers between these gorillas and the main western gorilla group are the Sangha River and Dja River.

There has been some discussion of a central African lowland refuge around the Congo River which includes all water courses to the north and south of the Congo River basin refuge (Kingdon, 1990). The northern area corresponds at its western edge to the confluence of the Oubangui and the Congo, of which the Sangha is a tributary. Colyn (1991) cites a possible refuge in this area, according to data for butterflies. The Oubangui forms the extreme eastern border of the western gorilla range and it is possible that this population of gorillas may have been subjected to restricted gene flow potential during periods of forest fragmentation.

### Central grouping

The remaining group potentially consists of the largest number of gorillas (more than 90% of the existing populations). However, it must be pointed out that sampling of large numbers of gorilla populations was not possible, and several of the populations that were sampled did not yield any genetic information because the hair samples were too old or too few and because subsequently the DNA was degraded. Future studies may reveal a greater degree of variation than detected in this study in additional extant gorilla populations. For example, gorillas from southern Cameroon south of the Dja River (population 5 in Fig. 10.1) may reveal a distinct genetic lineage, especially in light of Groves's work 30 years ago (Groves, 1967). In addition, the two distinct populations revealed in this study, Cross River and Dzanga, are both found in the extremities of western gorilla distribution and future studies on marginal or peripheral populations may also reveal such diversity.

### Conservation implications

The results obtained in this study reveal distinct population structuring in gorillas. Based on mtDNA sequences, three different groups of gorillas within the western lowland range have emerged. In terms of defining ESUs, the criterion that populations should differ significantly with regard to nuclear allelic frequencies may be overly restrictive given that nuclear genes are expected to retain ancestral polymorphism for longer periods than mtDNA (Moritz, 1995). Often clearly differentiated species that are reciprocally monophyletic for mtDNA retain ancestral polymorphisms at nuclear loci that do not differentiate the species. It has been previously reported in gorillas that estimates of nuclear gene divergence are not as great as divergence derived from mitochondrial sequences (Seaman *et al.*, 1999; Jensen-Seaman *et al.*, this volume). Similarly, nuclear sequences in chimps have exhibited much less variation than mitochondrial sequences (Kaessmann *et al.*, 2000). In the absence of any useful results from our study of nuclear DNA, our mtDNA results challenge the classification of *Gorilla gorilla gorilla* populations. Dzanga gorillas and Cross River gorillas are different from other western lowland gorillas and would merit distinct taxonomic status. The Cross River gorillas are isolated from other gorillas by distance, but Dzanga gorillas appear to be just as isolated and have a limited dispersal pattern. Their level of genetic distinctiveness has not been demonstrated before. In addition, future detailed analysis of other *Gorilla gorilla gorilla* populations may also define new divergent mtDNA haplotypes. These results have implications for the conservation management of these populations and lead to questions concerning the use of specific and subspecific designations of gorillas in conservation. For example, since all gorilla populations are threatened, should conservation efforts be concentrated on all populations regardless of classification, or on certain small isolated populations such as the Cross River gorillas, perhaps to the detriment of others? Our results demonstrate the uniqueness (much of which is probably as yet undefined) of specific western gorilla populations, and strengthen the argument for protection of all such gorilla populations. Similar arguments apply to eastern gorillas. Results of genetic studies provide information on the breakdown of the gorilla species and point to the shortfalls in the use of a rigid classification in the conservation management of gorillas.

### Acknowledgments

This work was funded by the Leverhulme Trust, London. This work was facilitated by the efforts of all the collectors listed in Table 10.1, all of whom we would like to thank, especially Magda Bermejo, Michele Goldsmith and Kelley McFarland. Kathryn Jeffery carried out the analysis of the Lopé hair samples.

### References

Avise, J.C. (1994). *Molecular Markers, Natural History and Evolution*. New York: Chapman & Hall.

Brookfield, J.F.Y. (1996). A simple new method for estimating null allele frequency from heterozygote deficiency. *Molecular Ecology*, **5**, 453–455.

Brown, W.M., George, M., and Wilson, A.C. (1979). Rapid evolution of animal mitochondrial DNA. *Proceedings of the National Academy of Sciences U.S.A.*, **76**, 1967–1971.

Bruford, M.W. and Wayne, R.K. (1993). Microsatellites and their application to population genetics. *Current Opinion in Genetics and Development*, **3**, 939–943.

Clifford, S.L., Jeffery K., Bruford, M.W., and Wickings, E.J. (1999). Identification of polymorphic microsatellite loci in the gorilla (*Gorilla gorilla gorilla*) using human primers: Application to noninvasively collected hair samples. *Molecular Ecology*, **8**, 1556–1558.

Collura, R.V. and Stewart, C.-B. (1995). Insertions and duplications of mtDNA in the nuclear genomes of old world monkeys and hominoids. *Nature*, **378**, 485–489.

Colyn, M. (1991). L'importance zoogéographic du bassin du fleuve Zaïre pour la speciation. *Annales des Sciences Zoologiques*, **264**, 180–185.

Coote, T. and Bruford, M.W. (1996). Human microsatellites applicable for analysis of genetic variation in apes and old world monkeys. *Journal of Heredity*, **87**, 406–410.

Crandall, K.A., Bininda-Emonds, O.R.P., Mace, G.M., and Wayne, R.K. (2000). Considering evolutionary processes in conservation biology. *Trends in Ecology and Evolution*, **15**, 290–295.

Edwards, A., Civitello, A., Hammond, H.A., and Caskey, C.T. (1991). DNA typing and genetic mapping with trimeric and tetrameric tandem repeats. *American Journal of Human Genetics*, **49**, 746–756.

Field, D., Chemick, L., Robbins, M., Garner, K., and Ryder, O. (1998). Paternity determination in captive lowland gorillas and orangutans and wild mountain gorillas by microsatellite analysis. *Primates*, **3**, 199–209.

Gagneux, P., Boesch, C., and Woodruff, D.S. (1997). Microsatellite scoring errors associated with noninvasive genotyping based on nuclear DNA amplified from shed hair. *Molecular Ecology*, **6**, 861–868.

Gagneux, P., Wills, C., Gerloff, U., Tautz, D., Morin, P.A., Boesch C., Fruth, B., Hohmann, G., Ryder, O.A., and Woodruff, D.S. (1999). Mitochondrial sequences show diverse evolutionary histories of African hominoids. *Proceedings of the National Academy of Sciences U.S.A.*, **96**, 5077–5082.

Garner, K.J. and Ryder, O.A. (1996). Mitochondrial DNA diversity in gorillas. *Molecular Phylogenetics and Evolution*, **6**, 39–48.

Gerloff, U., Schlötterer, C., Rassmann, K., Rambold, I., Hohmann, G., Fruth, B., and Tautz, D. (1995). Amplification of hypervariable simple sequence repeats (microsatellites) from excremental DNA of wild bonobos (*Pan paniscus*). *Molecular Ecology*, **4**, 515–518.

Gerloff, U., Hartung, B., Fruth, B., Hohmann, G., and Tautz, D. (1999). Intracommunity relationships, dispersal pattern and paternity success in a wild living community of bonobos (*Pan paniscus*) determined from DNA analysis of faecal samples. *Proceedings of the Royal Society of London, Series B*, **266**, 1189–1195.

Gonder, M.K., Oates, J.F., Disotell, T.R., Forstner, M.R.J., Morales, J.C., and Melnick, D.J. (1997). A new West African chimpanzee subspecies? *Nature*, **388**, 337.

Goossens, B., Latour, S., Vidal, C., Jamart, A., Ancrenaz, M., and Bruford, M.W. (2000). Twenty new microsatellite loci for use with hair and fecal samples in the chimpanzee (*Pan troglodytes troglodytes*). *Folia Primatologica*, **71**, 177–180.

Greenwood, A.D., and Pääbo, S. (1999). Nuclear insertion sequences of mitochondrial DNA predominate in hair but not in blood of elephants. *Molecular Ecology*, **8**, 133–137.

Groves, C.P. (1967). Ecology and taxonomy of the gorilla. *Nature*, **213**, 890–893.

Groves, C.P. (1970). Population systematics for the gorilla. *Journal of Zoology, London*, **161**, 287–300.

Groves, C.P. (2001). *Gorilla Taxonomy*. Washington, D.C.: Smithsonian Institution Press.

Groves, C.P. and Stott, K.W., Jr. (1979). Systematic relationships of gorillas from Kahuzi, Tshiaberimu and Kayonza. *Folia Primatologica*, **32**, 161–179.

Grubb, P. (1990). Primate geography in the Afro-tropical forest biome. In *Vertebrates in the Tropics*, eds. G. Peters and R. Hutterer, pp. 187–214. Bonn: Museum Albert Koenig.

Harcourt, A.H., Stewart, K.J., and Inahoro, I.M. (1989). Nigeria's gorillas: A survey and recommendations. *Primate Conservation*, **10**, 73–76.

Harris, E.E. and Disotell, T.R. (1998). Nuclear gene trees and the phylogenetic relationships of the mangabeys (Primates, Papionini). *Molecular Biology and Evolution*, **15**, 892–900.

Higuchi, R., von Beroldingen, C.H., Senasbaugh, G.F., and Erlich, H.A. (1988). DNA typing from single hairs. *Nature*, **332**, 543–546.

Hillis, D.M., Moritz, C., and Maple, B.K. (1996). *Molecular Systematics*. Sunderland, MA: Sinauer Associates.

Jensen-Seaman, M.J., and Kidd, K.K. (2001). Mitochondrial DNA variation and biogeography of eastern gorillas. *Molecular Evolution*, **10**, 2241–2247.

Kaessmann, H., Wiebe, V., and Pääbo, S. (2000). Extensive nuclear DNA sequence diversity among chimpanzees. *Science*, **286**, 1159–1162.

Kingdon, J. (1990). *Island Africa*. London: Collins.

Kocher, T.D., Thomas, W.K., Meyer, A., Edwards, S.V., Pääbo, S., Villablanca, F.X., and Wilson, A.C. (1989). Dynamics of mitochondrial DNA evolution in animals: Amplification and sequencing with conserved primers. *Proceedings of the National Academy of Sciences U.S.A.*, **86**, 6196–6200.

Kohn, M.H., York, E.C., Kamradt, D.A., Haught, G., Sauvajot, R.M., and Wayne, R.K. (1999). Estimating population size by genotyping feces. *Proceedings of the Royal Society of London, Series B*, **266**, 657–663.

Lanfranchi, R. and Schwartz, D. (1990). *Paysages Quaternaires de l'Afrique Centrale Atlantique*. Paris: ORSTOM.

Lee, P.C., Thornback J., and Bennett, E.L. (1988). *Threatened Primates of Africa: The IUCN Data Book*. Gland, Switzerland: IUCN.

Maley, J. (1996). The African rainforest: Main characteristics of changes in vegetation and climate change from the Upper Cretaceous to the Quaternary. *Proceedings of the Royal Society, Edinburgh*, **104B**, 31–73.

Morin, P.A., Moore J.J., and Woodruff, D.S. (1992). Identification of chimpanzee subspecies with DNA from hair and allele-specific probes. *Proceedings of the Royal Society of London, Series B,* **24,** 293–297.

Morin, P.A., Wallis, J., Moore, J.J., Chakraborthy, R., and Woodruff, D.S. (1993). Noninvasive sampling and DNA amplification for paternity exclusion, community structure and phylogeography in wild chimpanzees. *Primates,* **34,** 347–356.

Morin, P.A., Moore, J.J., Chakraborthy, R., Jin, L., Goodall, J., and Woodruff, D.S. (1994). Kin selection, social structure, gene flow, and the evolution of chimpanzees. *Science,* **265,** 1193–1201.

Moritz, C. (1995). Uses of molecular phylogenies for conservation. *Philosophical Transactions of the Royal Society of London, Series B,* **349,** 113–118.

Oates J.F., McFarland, K.L., Stumpf, R.M., Fleagle, J.G., and Disotell, T.R. (1999). New findings on the distinctive gorillas of the Nigerian–Cameroon border region. *American Journal of Physical Anthropology,* **28,** 213–214.

O'Brien, S.J. (1994). A role for molecular genetics in biological conservation. *Proceedings of the National Academy of Sciences U.S.A.,* **91,** 5748–5755.

Ruvolo, M., Disotell, T.R., Allard, M.W., Brown W.M., and Honeycutt, R.L. (1991). Resolution of the African hominoid trichotomy by use of a mitochondrial gene sequence. *Proceedings of the National Academy of Sciences U.S.A.,* **88,** 1570–1574.

Ruvolo, M., Pan, D., Zehr, S., Goldberg, T., Disotell, T.R., and von Dornum, M. (1994). Gene trees and hominoid phylogeny. *Proceedings of the National Academy of Sciences U.S.A.,* **91,** 8900–8904.

Ryder, O.A. (1986). Species conservation and systematics: The dilemma of subspecies. *Trends in Ecology and Evolution,* **1,** 9–10.

Saiki, R.K., Scharf, S., Faloona, F., Mullis, K.B., Horn, G.T., Erlich, H.A., and Arnheim, N. (1985). Enzymatic amplification of beta-globin genomic sequences and restriction site analysis for diagnosis of sickle cell anemia. *Science,* **230,** 1350–1354.

Saltonstall, K.G., Amato, G., and Powell, J. (1998). Mitochondrial DNA variability in Grauer's gorillas of Kahuzi-Biega National Park. *Journal of Heredity,* **69,** 129–135.

Santiago, M.L., Rodenburg, C.M., Kamenya, S., Bibollet-Ruche, F., Gao, F., Bailes, E., Meleth, S., Soong. S.J., Kilby, J.M., Moldoveanu, Z., Fahey, B., Muller, M.N., Ayouba, A., Nerrienet, E., McClure, H.M., Heeney, J.L., Pusey, A.E., Collins, D.A., Boesch, C., Wrangham, R.W., Goodall, J., Sharp, P.M., Shaw, G.M., and Hahn, B.H. (2002). SIV cpz in wild chimpanzees. *Science,* **295,** 465.

Sarich, V.M. (1977). Rates, sample sizes and the neutrality hypothesis for electrophoresis in evolutionary studies. *Nature,* **265,** 24–28.

Sarmiento, E.E. and Butynski, T. (1996). Present problems in gorilla taxonomy. *Gorilla Journal,* **12,** 5–7.

Schaller, G.B. (1963). *The Mountain Gorilla: Ecology and Behavior.* Chicago, IL: University of Chicago Press.

Seaman, M.L., Deinard, A.S., and Kidd, K.K.. (1999). Incongruence between mitochondrial and nuclear DNA estimated of divergence between gorilla subspecies. *American Journal of Physical Anthropology, Supplement 28,* 259.

Smith, K.L., Alberts, S.C., Bayes, M.K., Bruford, M.W., Altmann, J., & Ober, C. (2000). Cross-species amplification, non-invasive genotyping, and non-Mendelian inheritence of human STRPs in Savannah baboons. *American Journal of Primatology*, **51**, 249–227.

Stumpf, R.M., Fleagle, J.G., Jungers, W.L., Oates, J.F., and Groves, C.P. (1998). Morphological distinctiveness of Nigerian gorilla crania. *American Journal of Physical Anthropology, Supplement 26*, 213.

Swofford, D.L. (1998). *PAUP\*. Phylogenetic Analysis using Parsimony (\* and Other Methods)*, v.4. Sunderland, MA: Sinauer Associates.

Taberlet, P., Griffin, S., Goossens, B., Questiau, S., Manceau, V., Escaravage, N., Waits, L.P., and Bouvet, J. (1996). Reliable genotyping of samples with very low DNA quantities using PCR. *Nucleic Acid Research*, **24**, 3189–3194.

Tautz, T. (1993). Notes on the definition and nomenclature of tandemly repetitive DNA sequences. In *DNA Fingerprinting: State of the Science*, eds. S.D.J. Pena, R. Chakraborty, J.T. Epplen, and A.J. Jeffreys, pp. 21–28. Berlin: Birkhaüser.

Thompson, J.D., Higgins, D.G., and Gibson, T.J. (1994). CLUSTAL W: Improving the sensitivity of progressive multiple sequence alignment through sequence weighting, positions-specific gap penalties and weight matrix choice. *Nucleic Acids Research*, **22**, 4673–4680.

Tutin, C.E.G., Parnell, R.J., White, L.J.T., and Fernandez, M. (1995). Nest building by lowland gorillas in the Lopé Reserve, Gabon: Environmental influences and implications for censusing. *International Journal of Primatology*, **16**, 53–76.

Waits, S.L., Taberlet, P., Swenson, J.E., Sandegren, F., and Franzen, R. (2000). Nuclear DNA microsatellite analysis of genetic diversity and gene flow in the Scandinavian brown bear (*Ursus arctos*). *Molecular Ecology*, **9**, 421–431.

Walsh, P.S., Metzger, D.A., and Higuchi, R. (1991). Chelex 100 as a medium for simple extraction of DNA for PCR-based typing from forensic material. *Biotechniques*, **10**, 506–513.

Weber, J.L. and May, P.E. (1989). Abundant class of human DNA polymorphisms which can be typed using the polymerase chain reaction. *American Journal of Human Genetics*, **44**, 388–396.

Woodruff, D.S. (1993). Noninvasive genotyping of primates. *Primates*, **34**, 333–346.

Zischler, H., Geisert, H., von Haeseler, A., and Pääbo, S. (1995). A nuclear "fossil" of the mitochondrial D-loop and the origin of modern humans. *Nature*, **378**, 489.

# Part 3
## *Behavioral ecology*

# 11  An introductory perspective: Behavioral ecology of gorillas

CAROLINE E.G. TUTIN

Gorillas are probably the most intensely studied primate in terms of the cumulative number of years that has been spent collecting data on populations in different habitats across Equatorial Africa. The fact that they have been so well studied permits comparisons of aspects of ecology, social structure, and behavior of gorillas belonging to different subspecies and living in different habitats. Such comparisons allow hypotheses to be tested concerning the influence of a range of factors on social organization. These factors include predation pressure, the distribution of resources in time and space, contest and scramble competition for food, as well as mating strategies of males and females. In addition, integration of data from captive gorillas, as well as comparisons between gorillas and chimpanzees at sites where the two apes occur sympatrically, can add to our understanding of the selective forces that have shaped the evolution of primate social systems.

The four chapters in this section all use a comparative approach to address aspects of gorilla behavioral ecology. Different levels of comparisons are used: Groups within the same population (Goldsmith, this volume; Watts, this volume); neighboring, or more distant, populations living in different habitats (Goldsmith, this volume; Yamagiwa, this volume); the same population in different seasons (Remis, this volume; Yamagiwa, this volume); and some comparisons include data on sympatric chimpanzees (Goldsmith, this volume; Yamagiwa, this volume) or captive gorillas (Remis, this volume). These comparisons shed new light on the ecological flexibility of the genus *Gorilla* and, more specifically, on the influence of diet and mating strategies on social structure and behavior. The chapters also identify hypotheses that can be addressed with future research.

The long-term data on individual life histories, ecology, and social relationships of mountain gorillas collected at the Karisoke Research Center in Rwanda since the mid 1960s are unique. Studies of other gorilla populations began later and have not achieved the same level of behavioral detail because habituation has proven difficult, fewer groups have been studied, or data have been collected over relatively short periods. As ecological data (on diet, ranging, and food availability) can be collected by indirect means supplemented by direct

295

observations, comparisons of gorilla ecology are possible across the geographical range of the genus. The same degree of comparison is not yet possible for social behavior. While data on group size and group composition (at least in terms of the number of fully adult, silverback males per group) exist from a number of sites, methodological problems remain and no consensus has yet emerged as to the nature and extent of differences in group dynamics or social relationships between studied gorilla populations. The common theme of the chapters in this section is variability: It is clearly established that the diet of gorillas varies considerably with respect to the relative importance of fruit, but does similar variation exist in grouping patterns and social organization? The authors address this question from slightly different angles.

While mountain gorillas at Karisoke have a totally folivorous diet, gorillas living at lower altitudes in both the east (Kahuzi and Bwindi) and west (Bai Hokou) of the range eat fruit. The availability of fruit varies between sites, depending on plant species composition, and at the same site depending on season. Comparisons between sites and between neighboring populations at different altitudes all indicate that the importance of fruit in the diet of different gorilla populations is clearly a function of its availability (Tutin *et al.*, 1991; Goldsmith, this volume; Yamagiwa, this volume). Gorillas living in the botanically diverse lowland tropical forests of Gabon, Central African Republic, and Congo are the most frugivorous. They eat fruit of 80–120 different species; fruit remains are found in almost all fecal samples (90–100%) and 40–75% of feeding observations involve consumption of fruit (Table 15.1 in Remis, this volume). However, even the most frugivorous populations of gorillas also eat nonfruit foods regularly, particularly young leaves, stems, and pith of herbaceous plants. So while fruit is certainly significant in the diet of some populations, are gorillas accurately described as frugivores?

Chimpanzees are frugivores and they occur sympatrically with gorillas in all but the high-altitude montane habitats. This allows direct comparisons of how these two closely related apes exploit identical food resources within a shared habitat. Counter to the prediction of clear niche differentiation, comparative studies reveal broad overlap of foods eaten by the two ape species, particularly fruit foods of which 60–80% are shared at Kahuzi and at Lopé (Tutin and Fernandez, 1993; Yamagiwa, this volume). This raises the possibility that competition between gorillas and chimpanzees may occur. If interspecific competition for limited resources does indeed exist it could influence gorilla social structure or behavior, confounding comparisons between lowland and mountain gorillas. However, differences in foraging strategies exist in the way gorillas and chimpanzees exploit fruit (Tutin *et al.*, 1991; Yamagiwa, this volume). Aggressive interactions between the two species have never been observed, while peaceful mingling and even feeding together in the same tree have been reported from several sites (Goldsmith, this volume), suggesting that interspecific competition is not significant.

However, although the extensive dietary overlap superficially suggests a lack of niche differentiation, studies of sympatric gorillas and chimpanzees have all found that the two species respond in different ways to fruit scarcity. Gorillas shift diets and switch to a predominantly or wholly folivorous diet, while chimpanzees continue to eat fruit but show a social response by reducing the size of foraging groups. This suggests an ecologically important distinction between the two species, which reduces or eliminates competition between them when fruit is in short supply.

The distinction between frugivores and folivores is not merely a question of categorizing species into neat compartments. As a source of food, fleshy fruit differs in many ways from folivorous foods (e.g., leaves, stems, and bark) and these differences are believed to have influenced almost every aspect of their consumers, from morphology to behavior. Among plant foods, fruit stands in a class of its own, differing from other plant parts in two important respects: Palatability and patterns of availability in both time and space within the habitat.

Fruit flesh is the only part of a plant that is "designed" to be eaten. It has evolved as the succulent magnet that encourages animals to swallow the seeds that are attached to or embedded in it. The reproductive strategy of plants with fleshy fruit depends on mobile animals carrying seeds away from the parent, and fruit flesh is the reward offered for this essential dispersal service (Voysey *et al.*, 1999). Plants have evolved a variety of chemical and mechanical ways to protect every part of themselves except their fruit flesh or arils from hungry animals (whether they be caterpillars or apes). Leaves, stems, pith, bark, and the seeds themselves are all hard to access and/or difficult to digest. They are protected by secondary compounds such as tannins or by fibrous or woody layers of cells to discourage feeding by animals. Fruit flesh, on the other hand, being sugary and low in fiber is easy to digest and is, as demonstrated by taste-trial experiments, a highly preferred food of gorillas as well as chimpanzees (Remis, this volume).

Compared to chimpanzees, gorillas appear somewhat contradictory with respect to what Remis terms "digestive strategy". Gorillas have some morphological adaptations to folivory (e.g., large body size, enlarged hindgut, enlarged post-canine dentition relative to incisors, and some enhanced masticatory features) and some characters shared with frugivores (e.g., astute taste sensitivity and preference for sugary, low-fiber foods). This suggests that lowland gorillas are "hybrids", spanning the frugivore–folivore divide somewhere between the mountain gorillas of Karisoke and bonobos (see Fig. 14.1 in Goldsmith, this volume), and this intermediate position fits the field observations of dietary flexibility.

Fruit and folivorous foods also differ fundamentally in terms of their availability. Most plants that produce succulent fruit do so seasonally, meaning that the availability of fruit varies over time. In addition, the plants of tropical ecosystems show community-wide cycles of fruiting influenced by various climatic factors and resulting in seasons of fruit abundance and scarcity (van Schaik *et al.*, 1993). Not only does fruit availability vary seasonally but also large interannual differences have been recorded at all sites resulting in some bumper, or "good", fruit years and some "bad" fruit years when many crops fail (Tutin and White, 1998; Goldsmith, this volume; Remis, this volume). In addition, in tropical forest habitats where gorillas are most frugivorous, plant species diversity is high, meaning that individual fruiting trees are often rare and widely dispersed. In contrast to fruit, the availability of many folivorous foods (e.g., mature leaves, stems, pith, bark) shows little seasonal or interannual variation; while many trees produce young leaves in seasonal bursts, common herbaceous plants produce new leaves and shoots continuously.

Compared to constantly available abundant herbs, fruit is a challenging resource to exploit because it is patchy and often limited in quantity. In models of primate sociality, this crucial difference between frugivory and folivory is expressed in terms of levels of feeding competition between members of the same group (Wrangham, 1986; van Schaik, 1989; Sterck *et al.*, 1997). Competition, either contest or scramble, has been identified as the major cost of group living and overall levels of within-group competition will be higher for frugivores than for folivores. Models of primate sociality predict that as diet influences costs of group living, frugivorous populations of gorillas should have smaller groups than folivorous populations (Goldsmith, this volume; Yamagiwa, this volume). Another possible social adaptation to the increased cost of competition when feeding on fruit includes the chimpanzee-like strategy of fission–fusion through the formation of temporary subgroups in frugivorous populations of gorillas (Goldsmith, 1996, this volume; Remis, 1997).

Data available to date do not support the prediction that average group size is smaller in highly frugivorous populations of gorillas, but there is a suggestion that maximum group size is constrained by frugivory as the largest reported group sizes are in folivorous, high-altitude populations (Tutin, 1996; Goldsmith, this volume; Yamagiwa, this volume). There is also clear evidence from all study sites that gorillas travel further when fruit is abundant than when it is rare. The proposed impacts of frugivory on gorilla social structure and relationships, however, still require confirmation. Although great progress has been made, it is not yet possible to substantiate the suggestion that the occasional separation of groups into subgroups as seen at Bai Hokou is related to frugivory; it could also be explained as part of the gradual process of group fission (Remis, 1997; Goldsmith, this volume; Yamagiwa, this volume).

The analysis of the long-term data from Karisoke illustrates the complexity of explanations of group dynamics and social relationships (Watts, 1992, 1994, this volume). Decisions concerning group transfer made by both male and female gorillas are influenced by reproductive tactics; group size plays a part, but so too do the number and identity of males in the group and the available alternatives. The three decades of research at Karisoke have provided a wealth of detail on individual (and group) life histories in this population. Understanding of the factors that influence social dynamics emerges only when placed in the context of known family relationships and past histories of individual gorillas. Among Karisoke gorillas, males interact with females to minimize competitive interactions and maintain relative equality, thus buffering females from negative impacts of scramble competition. Therefore, Watts suggests that no major differences will exist in group dynamics and social relationships related to increased frugivory.

In summary, frugivory has been shown to influence some aspects of gorilla behavior such as ranging, but its relevance to social relationships remains unproven apart from the strong suggestion of frugivory constraining optimum group size to a lower level in populations that exploit fruit regularly. A return to the frugivore–folivore divide is in order to frame questions for future research. We have seen that niche separation between gorillas and chimpanzees is subtle, only becoming clear-cut when fruit is scarce. Similarly, it seems likely that competition between group members (or with other species) for fruit is likely to impose significant costs only when fruit is scarce. Models emphasize the patchy, ephemeral nature of fruit compared to other foods and "average" measures confirm that fruit availability is lower than that of folivorous foods in all gorilla habitats. The reality, however, is that when fruit is available, it is often abundant due to the plant community's response to climate. From the perspective of consumers, patterns of fruit availability are "boom" or "bust" cycles.

Several lines of evidence presented in these chapters suggest that when considering the frugivore–folivore divide, gorillas are folivores who like fruit rather than true frugivores in the ecological sense of the term. When the going gets tough and fruit supplies dwindle, gorillas switch with great ease to a diet dominated by abundant food (e.g., leaves, pith, bark, and fibrous fruit of common plant species). Morphological features, notably large body size, enlarged hindgut and cellulose-digesting ciliates (Remis, this volume), as well as larger facial size and a more robust mandible (Taylor, this volume), allow them to do this, thus avoiding high costs of group living. A comparison of reproductive profiles of wild gorillas and chimpanzees shows that gorillas may have "found" a better solution than chimpanzees as female gorillas give birth for the first time at a younger age and have shorter interbirth intervals despite being larger than chimpanzees (Tutin, 1994). Permanent association with adult male(s) not only protects female gorillas from predation and harassment by conspecifics (Watts, this volume), but also extends this constant protection to infants, thus reducing the burden of postnatal investment compared to chimpanzee mothers who have to go it alone.

The revelation of the extent of gorillas' frugivory was the most surprising result to emerge from the early years of research on gorilla populations in lower-altitude habitats as it contrasted so strongly with findings from the pioneering research at Karisoke. Placed in an ecological perspective, this dietary flexibility is not so surprising because succulent fruit is such a high-quality food (in terms of nutrients and digestibility) that it would be astonishing if gorillas ignored such "treats". Indeed, even elephants, archetypal folivores, become opportunistic frugivores in lowland tropical forests (White *et al.*, 1993).

The challenges that remain for researchers working with gorillas throughout Africa include the collection of data to test hypotheses and further our understanding of variability within the genus, but, as in the case of Karisoke, it is also of primary importance to tackle broader questions related to conservation. It is clear that the presence of researchers has made a major contribution to the survival of the Virunga gorillas. At several sites in the lowland forests of west–central Africa, and at Bwindi in Uganda, research began because habituation of gorilla groups for ecotourism had been identified as a conservation priority. Similarly, data on gorillas' ecological roles as high-quality seed dispersers in lowland tropical rainforests can be harnessed to emphasize the importance of protecting gorilla populations not only in national parks but also in timber-production forests (Tutin *et al.*, 2001).

## References

Goldsmith, M.L. (1996). Ecological influences on the ranging and grouping behavior of western lowland gorillas at Bai Hokou in the Central African Republic. PhD thesis, State University of New York, Stony Brook, NY.

Remis, M.J. (1997). Ranging and grouping patterns of a western lowland gorilla group at Bai Hokou, Central African Republic. *American Journal of Primatology*, **43**, 111–133.

Sterck, E.H.M., Watts, D.P., and van Schaik, C.P. (1997). The evolution of social relationships in female primates. *Behavioral Ecology and Sociobiology*, **41**, 291–309.

Tutin, C.E.G. (1994). Reproductive success story: Variability among chimpanzees and comparisons with gorillas. In *Chimpanzee Cultures*, eds. R.W. Wrangham, W.C. McGrew, F.B.M. de Waal, and P.G. Heltne, pp. 181–194. Cambridge, MA: Harvard University Press.

Tutin, C.E.G. (1996). Ranging and social structure of lowland gorillas in the Lopé Reserve, Gabon. In *Great Ape Societies*, eds. W.C. McGrew, T. Nishida, and L.A. Marchant, pp. 58–70. Cambridge, U.K.: Cambridge University Press.

Tutin, C.E.G. and Fernandez, M. (1993). Composition of the diet of chimpanzees and comparisons with that of sympatric lowland gorillas in the Lopé Reserve, Gabon. *American Journal of Primatology*, **30**, 195–211.

Tutin, C.E.G. and White, L.J.T. (1998). Primates, phenology and frugivory: Present, past and future patterns in the Lopé Reserve, Gabon. In *Dynamics of Tropical Communities*, eds. D.M. Newbery, H.H.T. Prins, and N. Brown, pp. 309–338. Oxford, U.K.: Blackwell Scientific Publications.

Tutin, C.E.G., Fernandez, M., Rogers, M.E., Williamson, E.A., and McGrew, W.C. (1991). Foraging profiles of sympatric lowland gorillas and chimpanzees in the Lopé Reserve, Gabon. *Philosophical Transactions of the Royal Society, Series B*, **334**, 179–186.

Tutin, C.E.G., Porteous, I.S., Wilkie, D.S., and Nasi, R. (2001). *Comment minimiser l'impact de l'exploitation forestière sur la faune dans le Bassin du Congo*. Les Dossiers de l'ADIE, Série Forêt No. 1. Libreville: Association pour le Développement de l'Information Environnementale.

van Schaik, C.P. (1989). The ecology of social relationships among female primates. In *Comparative Socioecology: The Behavioural Ecology of Humans and Other Mammals*, eds. V. Standen and R.A. Foley, pp. 195–218. Oxford, U.K.: Blackwell Scientific Publications.

van Schaik, C.P., Terborgh, J.W., and Wright, S.J. (1993). The phenology of tropical forests: Adaptive significance and consequences for primary consumers. *Annual Review of Ecology and Systematics*, **24**, 353–377.

Voysey, B.C., McDonald, K.E., Rogers, M.E., Tutin, C.E.G., and Parnell, R.J. (1999). Gorillas and seed dispersal in the Lopé Reserve, Gabon. I: Gorilla acquisition by trees. *Journal of Tropical Ecology*, **15**, 23–38.

Watts, D.P. (1992). Social relationships of immigrant and resident female mountain gorillas. I: Male–female relationships. *American Journal of Primatology*, **28**, 159–181.

Watts, D.P. (1994). Social relationships of immigrant and resident female mountain gorillas. II: Relatedness, residence, and relationships between females. *American Journal of Primatology*, **32**, 13–30.

White, L.J.T., Tutin, C.E.G., and Fernandez, M. (1993). Group composition and diet of forest elephants, *Loxodonta africana cyclotis* Matschie 1900, in the Lopé Reserve, Gabon. *African Journal of Ecology*, **31**, 181–199.

Wrangham, R.W. (1986). Ecology and social relationships in two species of chimpanzee. In *Ecological Aspects of Social Evolution*, eds. D.I. Rubenstein and R.W. Wrangham, pp. 352–378. Princeton, NJ: Princeton University Press.

# 12 Gorilla social relationships: A comparative overview

DAVID P. WATTS

## Introduction

Research on mountain gorillas (*Gorilla gorilla beringei*) in the Virunga Volcanoes region of Rwanda, Uganda, and the Democratic Republic of Congo provides one of the best case studies of nonhuman primate behavior in the wild. Data on mountain gorilla ecology, life-history tactics, and social relationships, collected mainly at the Karisoke Research Center in Rwanda's Parc National des Volcans, have contributed prominently to the development of theory in primate socioecology (e.g., van Schaik, 1989; Sterck *et al.*, 1997). For example, the association of female transfer with low levels of contest competition for food in mountain gorillas (Stewart and Harcourt, 1987; Watts, 1990*a*, 1994*a*, 1996, 1997) supports ideas on the ecology of female social relationships (van Schaik, 1989). Also, the influence of infanticide on transfer decisions (Watts, 1989, 1996) supports current ideas on male–female association and social relationships (Sterck *et al.*, 1997). Important generalizations from the Virunga population seem to apply to other gorilla populations and subspecies, but we know that feeding ecology varies considerably across habitats and subspecies, which raises intriguing questions about variation in social organization and social relationships. However, while good comparative data on feeding ecology exist for eastern lowland gorillas (*G. g. graueri*; reviewed in Yamagiwa *et al.*, 1996) and western lowland gorillas (*G. g. gorilla*; reviewed in Doran and McNeilage, 1998), we lack comparably detailed data on social behavior, demography, and life-history tactics.

In this chapter, I review results of research on the social relationships and life histories of mountain gorillas in the Karisoke study population from the perspective of models that seek to explain variation in female–female and female–male relationships in primates (van Schaik, 1989; Sterck *et al.*, 1997). I also briefly discuss male life-history tactics and co-operation between males. I then consider several issues regarding variation in gorilla socioecology, and compare mountain gorillas with other primates and nonprimates that show socioecological resemblances to gorillas.

302

## Gorilla ecology

Doran and McNeilage (1998) provided an excellent review of gorilla feeding ecology (see also Kuroda *et al.*, 1996; Yamagiwa *et al.*, 1996), and I will merely note two major contrasts and similarities among populations. First, the Virunga mountain gorillas are almost entirely herbivorous and mostly eat perennially available herbs and vines that are densely and evenly distributed in most of their habitat (Fossey and Harcourt, 1977; Watts, 1984, 1991*a*; Plumptre, 1995; McNeilage, 1996). Gorillas eat much fruit in habitats where pulpy fruit is abundant, at least seasonally, and their diets converge with those of sympatric chimpanzees (e.g., Lopé: Tutin *et al.*, 1991; Ndoki: Kuroda *et al.*, 1996; lowland areas of Kahuzi-Biega: Yamagiwa *et al.*, 1996). Second, all gorillas nevertheless eat large quantities of perennially available plant items high in structural carbohydrates, especially leaves and the pith and stems of herbaceous and some woody plants. These materials are year-round staples and in many habitats act as fallbacks when fruit is scarce (Yamagiwa *et al.*, 1996; Remis, 1997*a*; Doran and McNeilage, 1998; Goldsmith, 1999). The even distribution of such foods in space and time lowers the ecological cost of grouping for females and underlies a fundamental unity in gorilla social systems.

## The ecological model of primate female relationships

The ecological model of primate social relationships (van Schaik, 1989; cf. Sterck *et al.*, 1997) proposes that predation risk and food distribution determine the relative benefits of gregariousness to females. The intensity and relative importance of contest and scramble competition for food then determines the nature of social relationships among females. For most diurnal primates, predation risk leads to group living, which inevitably leads to feeding competition. Food in discrete, high-quality patches too small to accommodate all group members becomes the object of within-group contest competition. When gains or losses from this competition can influence female fitness significantly, females establish decided dyadic agonistic relationships (i.e., dominance relationships). In some cases – notably macaques, savanna baboons, and vervet monkeys – they also establish stable, linear dominance hierarchies, facilitated by alliances between relatives and also by agonistic support between some nonrelatives that probably represents mutualism. Dominance hierarchies are strict in some species, while high-ranking females are relatively tolerant of subordinates in others, possibly because they need co-operation from subordinates in between-group contest feeding competition. Female dispersal may have ecological costs (e.g., reduced foraging efficiency in unfamiliar or marginal habitat), to which high within-group contest competition adds social costs. Therefore, females trying to immigrate would face co-operative resistance from female group residents and would have no female allies in their attempts to counter that resistance.

When most food is evenly distributed, at high or low density, and its quality varies little, within-group contest competition has little effect on female fitness and social relationships between females depend predominantly on the potential for scramble competition. Agonistic relationships are individualistic and more egalitarian, and dominance hierarchies and even dyadic dominance relationships may be absent. Females cannot notably improve their foraging efficiency by co-operatively defending feeding sites against other group members or other groups. They forfeit little by leaving female relatives, especially if by leaving they experience less scramble competition, or by resisting female immigrants, so long as group size is within tolerable limits. High home range overlap and low variance in habitat quality can reduce the ecological costs of female dispersal; transfer is then a viable strategic option, given its low social costs.

### Mountain gorilla females and the ecological model

Data on mountain gorillas give the ecological model some of its strongest support (Sterck *et al.*, 1997). Mountain gorillas mostly eat densely and evenly distributed, perennially available food (Fossey and Harcourt, 1977; Vedder, 1984; Watts, 1984, 1991*a*; McNeilage, 1996). Contests over food are infrequent and rarely give one female access to food at another's expense or seriously interrupt feeding (Watts, 1985, 1994*a*). In many female dyads, most contests are undecided, and targets of aggression usually ignore or retaliate against aggressors. Many dyads lack decided agonistic relationships, and dominance hierarchies are weak or nonexistent (Watts, 1994*a*). No contest effects on female reproductive success occur (Watts, 1996; Sterck *et al.*, 1997).

Female natal and secondary transfers are common, but not universal (Harcourt, 1978; Stewart and Harcourt, 1987; Watts, 1990*a*, 1996). Most co-resident females are unrelated and have neutral to hostile relationships with each other: they groom little, if at all, are often intolerant of close proximity, and often interact aggressively (Harcourt, 1979*a*; Stewart and Harcourt, 1987; Watts, 1991*b*, 1994*a,b*, 1995*a,b*). However, adult female relatives often reside together, either in their natal groups or when they transfer to the same groups, and social relationships among females are then differentiated along lines of relatedness. Females associate and groom with female kin more than with non-kin.

Aggression rates are lower between kin, at least for serious aggression, and damaging fights are common between non-kin but rare between kin (Stewart and Harcourt, 1987; Watts, 1994*a,b*, 1995*a*). Kin form alliances against other females, and give each other agonistic support as often as females in some female-resident cercopithecines with strong, nepotistic dominance hierarchies (Watts, 1997). Some unrelated females sometimes groom and occasionally form coalitions, but such "friendships" are unusual and lack the long-term persistence of social bonds between kin. The positive effects of relatedness generally are greater for maternal than for paternal relatives (Watts, 1994*b*, 1996) and persist for years in most dyads (Watts, 2001). However, female alliances do not consistently affect dyadic agonistic relationships; support from kin does not necessarily enable individuals to establish dominance over any opponents, nor do unrelated females consistently support each other against third parties (Harcourt & Stewart, 1987, 1989; Watts, 2001). Benefits of social bonds between kin are insufficient to keep them together permanently, if at all. Nor are relationships between females important enough to promote direct mechanisms for peaceful conflict management and resolution. Even close relatives typically do not reconcile after aggressive interactions. Instead, females often avoid each other and/or use proximity to and interactions with males to manage conflicts with other females (Watts, 1995*a*).

Both ecological and social costs of female dispersal are low, as shown by the lack of a significant difference in age at first reproduction between females still in their natal groups and those who underwent natal transfer (Watts, 1996). Habitat quality varies relatively little (McNeilage, 1996), and home ranges are neither fixed nor exclusive (Watts, 1998, 2000*a*). Females can undergo partial locational dispersal (e.g., home range shifts: Isbell and Van Vuren, 1996) without changing groups and can change groups without completely changing foraging areas (Watts, 2000*a*). Payoffs to success in feeding contests and to co-operation among female relatives are low, and, except in large groups, immigrants meet little or no resistance from resident females (Harcourt, 1978; Watts, 1991*b*).

### *Male influence on female grouping*

The ecological model enjoys broad support with respect to relationships between females (reviewed in Sterck *et al.*, 1997; see also Koenig *et al.*, 1998), but does not explicitly incorporate the implications of reproductive conflicts of interest between males and females (Sterck *et al.*, 1997). It does not account for permanent male–female associations, for variation in male–female social relationships, for possible male influence on social relationships between females, and for the possibility that mate choice is the main influence on female transfer. The revised socioecological model (Sterck *et al.*, 1997) stresses the importance to females of male protection against other males, particularly against infanticide. The threat of infanticide favors female grouping and permanent association with protector males whose services multiple females can share if feeding competition does not preclude multi-female groups. It also means that females in multi-male groups can benefit by mating with more than one male if all mates offer protection.

Mountain gorillas fit this model well (Sterck *et al.*, 1997). Indeed, Wrangham (1979) anticipated the Sterck *et al.* (1997) argument by calling mountain gorilla groups "permanent consortships" in which multiple females shared a male's protection against harassment by other males in an ecological context where grouping did little to reduce foraging efficiency (cf. Watts, 1989, 1990*b*). Infanticide is a major threat, and females depend crucially on male protection of infants; those without mature male protectors try to defend infants against potentially infanticidal males, but persistent attackers easily overwhelm their efforts (Watts, 1989). Most infanticide cases satisfy three predictions of the sexual selection model (Hrdy, 1979), which should mean that infanticide increases male reproductive success (Watts, 1989). First, infanticide shortens interbirth intervals. Second, females who lose infants usually transfer to the infanticidal males, or stay with them when they take over groups after the deaths of previous resident males. Males do not aggressively take over groups and evict resident males; for infanticide to be advantageous, male success at killing or protecting infants must influence female residence decisions. Third, females usually subsequently mate with the infanticidal males. In addition, females in multi-male groups face much lower infanticide risk than females in one-male groups (Robbins, 1995; Watts, 2000*b*), because the death of a single male in a multi-male group does not leave females unprotected, whereas male deaths in one-male groups do (Watts, 1989).

Data on female transfer decisions and reproductive success in relation to group size and to the number of males per group also support the revised socioecological model. The earlier ecological model predicts that females transfer predominantly from larger to smaller groups, from multi-male to one-male groups (which are smaller, on average), and from groups to lone males, to minimize scramble feeding competition. However, group size has no consistent effect on transfer decisions or reproductive success (Watts, 1996; Sterck *et al.*, 1997). The socioecological model predicts that the number of males affects transfer decisions independently of group size because females seek to maximize male protection. Consistent with this prediction, females transfer from one-male to multi-male groups significantly more often than expected given the proportion of one-male and multi-male groups encountered (Watts, 2000*b*).

### Social relationships between males and females

Males are central in the social lives of mountain gorilla females. Females without adult female kin groom mostly with their own offspring and with adult males, and spend more time in close proximity to adult males than to other adult females. Even those who reside with adult female kin interact affiliatively with adult males at high rates and spend more time in close proximity to them than with any females except, perhaps, their maternal kin (Harcourt, 1979b; Watts, 1994b). Males often display at females and often direct other low-level aggression at them, but male–female aggression rarely involves serious physical contact. Males rarely wound females in within-group aggression, whereas females commonly wound each other (Harcourt, 1979b; Watts, 1994b, 1995a). Females commonly reconcile with males after receiving male aggression, and often seek reassurance from males after aggressive interactions with other females (Watts, 1995a).

Male–female relationships are differentiated in multi-male groups (Watts, 1992; Sicotte, 1994). Females typically interact affiliatively with their group's top-ranking male much more than with any other adult female, although females who are close maternal relatives sometimes engage in more affiliative interactions with each other than with the male. Some females also commonly associate and groom with subordinate males, who take much of the initiative to maintain these relationships. Many females in multi-male groups solicit matings with more than one male, and mate with subordinate males despite attempts by higher-ranking males to prevent this (Watts, 1990b; Robbins, 1999). Mating with multiple males may reduce infanticide risk. For example, when Group 4's single mature male died, an adolescent male killed the infant of a female with whom he had not mated, but behaved paternally towards the infant of another female with whom he had mated (Watts, 1989). No infanticides, either within-group or by outside males, followed the death of Group 5's dominant male in 1993; three other mature males in the group either had mated with females who had infants or were closely related to them (one male/female pair were mother and son, and another were full siblings).

### Male influences on relationships between females

Much aggression between females concerns proximity to, and social access to, adult males (Watts, 1994a). This may indicate competition for safety and/or competition for protection from aggression by other females. Males commonly intervene in conflicts between females, especially when these fights are serious. They mostly make control interventions (stop conflicts without supporting either opponent) or support targets of aggression, rather than initiators (Watts, 1991*b*, 1997). This typically ends conflicts without either female showing submission. Females often engage in prolonged sequences of aggression in the absence of male intervention, but usually cease this aggression when a male intervenes. Males intervene especially often in polyadic conflicts, which renders much coalition-based aggression ineffective; coalition partners usually cannot elicit submission from outnumbered opponents when males intervene (Watts, 1995*b*, 1997). Males thus reduce competitive asymmetries between females and make it harder for relatives to influence each other's agonistic relationships consistently. This reduces the incentive for them to stay together and may help males to retain females who otherwise would do poorly in competition with other females. It also allows males to curb resistance to immigrants effectively so long as groups are not too large, which may help persuade immigrants to stay (Watts, 1991*b*, 1997). Male control of female aggression could thus be a form of mating effort.

### Male–male relationships and male life histories

The Virunga population includes one-male and multi-male groups, solitary males, and, at times, all-male groups (Schaller, 1963; Yamagiwa, 1987; Robbins, 2001). Most multi-male groups form when one or more natal males ("followers") stay as adults, although some form when all-male groups gain females (Robbins, 1995; Watts, 2000*b*). In most cases, the males are related (e.g., fathers and sons, paternal and/or maternal brothers), although the two adult males (TS and BE) in one Karisoke research group are unrelated (Robbins, 1995, 1999, 2001; Watts, 2000*b*). Follower males have considerably higher expected lifetime reproductive success than "bachelors" who become solitary or who join all-male groups (Robbins, 1995; Watts, 2000*a*). Many bachelors fail to gain females, and followers typically start their reproductive careers with more females than successful bachelors do and have longer group tenures (Watts, 2000*b*). Demographic variation helps to explain why many males nevertheless follow the low-payoff strategy (Watts, 2000*b*). Many become bachelors because their groups' single mature male dies and females join other mature males. Maturing males near the bottom of male mating queues, those in groups with relatively few females per male, and those in groups with young dominant males, are also particularly likely to become bachelors. Males who reach maturity in large groups with many females and only single older males with short expected future tenure are more likely to succeed as followers and to have high reproductive success (Watts, 2000*b*).

Males in multi-male groups establish dominance relationships that influence access to estrous females. Dominant males often disrupt mating attempts by subordinates, except those involving their own presumed daughters (Watts, 1990*b*; Robbins, 1995, 1999, 2001). They are often intolerant of subordinates and direct aggression to them at high rates, but tolerance can vary in groups with more than two sexually mature males (Harcourt and Stewart, 1981; Watts and Pusey, 1993; Watts, 1997).

Given such within-group mating competition, the main reason that males tolerate followers seems to be that both related and unrelated males in multi-male groups cooperate against extra-group males (Harcourt, 1978; Fossey, 1984; Watts, 1989, 2000*a*; Sicotte, 1993; Robbins, 2001, personal observation). Males are antagonistic to all outside males. Encounters are not always overtly aggressive, but males usually at least exchange displays and often fight, sometimes with fatal consequences. Co-operation against outside males should lower the risks of fighting, help with mate attraction and retention, and help with defense against infanticide.

## An analysis of long-term consistency and change in male–female relationships: A case study

Karisoke data allow uniquely valuable analysis of long-term consistency and changes in social relationships. For example, Watts (2001) showed that related females in Group 5 maintained social bonds characterized by relatively high rates of affilative interaction and by alliance formation between 1984 and 1992. Indeed, stable, long-term social bonds had already been evident in some of these same dyads before 1984 (Stewart and Harcourt, 1987). Robbins (2001) argued that data on proximity and rates of affiliation between males and females in Group 5 prior to group fission predicted reasonably well which females joined which males after the fission occurred. Here, I examine the history of male–female relationships in this group from a longer-term perspective. The data largely corroborate Robbins's analysis, but they also show that variation in the quality of male–female relationships did not correctly predict all residence decisions and that relationship quality can change over time.

The data are from 20 months of observation in 1986 and 1987 (Watts, 1992) and from two study periods of 2.5 months each in 1991 and 1992 (Watts, 1995*a*,*b*). The 1991 and 1992 periods overlapped Robbins's study period, but our estimates of rates of social interaction are independent because we derived these from independent focal samples. Group 5 had from one to four fully mature silverbacks, from zero to three adolescent males, from 10 to 12 adult females, and from one to three adolescent females during these study periods (Table 12.1). Dominant male ZZ died in 1993. After his death, the group divided; six females joined adult male SH in his one-male group, while six stayed with adult males PB and CA (Robbins, 2001). PB was initially the dominant male in

Table 12.1. *Composition of Group 5 during the study periods included in the analysis of stability in social relationships. Composition in 1993, after the death of the group's dominant adult male (ZZ), is also given*

| Period | Silverbacks[a] | Adolescent males[a] | Adult females | Adolescent females |
|--------|----------------|---------------------|---------------|--------------------|
| 1986–7 | 1 (ZZ) | 3 (PB, SH, CA) | 11 | 1 |
| 1991 | 4 (ZZ, PB, SH, CA) | 0 | 10 | 2 |
| 1992 | 4 (ZZ, PB, SH, CA) | 0 | 10 | 3 |
| 1993 | 3 (PB, SH, CA) | 2 | 12 | 2 |

[a] Initials in columns for males (e.g., ZZ) are standard Karisoke abbreviations for named individuals.

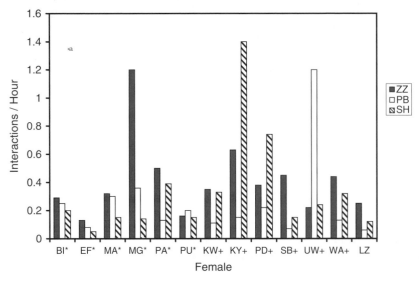

Fig. 12.1. Rates of male-female affiliative interaction in Group 5 at Karisoke from 1991 to 1992. Data on adult males ZZ, PB, and SH included. Females are ordered alphabetically and according to whether they joined PB's Group (*) in 1993, joined SH's Group (+), or had died (LZ) by 1993. Females BI, MA, and UW were adolescents; other females were adults. Most females who joined PB had higher rates of interaction with him than with SH prior to the split, and most who joined SH had higher rates of interaction with him, but the overall differences were not significant (Wilcoxon matched pairs test, $T^+ = 15$, ns, for females who stayed with Pablo; $T^+ = 16$, ns, for females who stayed with SH; $n = 6$ in both cases).

his group, but CA has since reversed rank with him (L. Williamson, personal communication).

Data on rates of affiliative interaction and on proximity between males and females during the three years prior to the fission predicted which dominant male females joined reasonably well, but not perfectly. Robbins (2001) found that

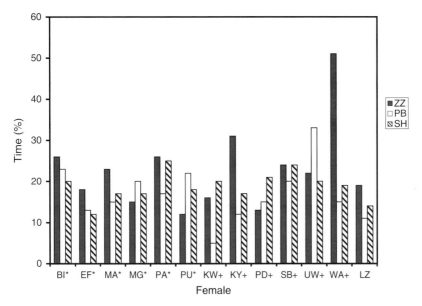

Fig. 12.2. Percentage of time that Group 5 females spent within 5 m of males ZZ, PB, and SH from 1991 to 1992. Females ordered as in Fig. 12.1. Females BI, MA, and UW were adolescents; other females were adults. Most females who joined PB spent more time near him than near SH prior to the split, and most who joined SH spent more time near him, but the overall differences were not significant (Wilcoxon matched pairs test, $T^+ = 15$, $n = 6$, ns, in both cases).

the females who joined PB's group had significantly higher rates of affiliative interaction with him than with SH before Group 5 split. In contrast, females who joined SH's group interacted affiliatively with SH and with PB at rates that did not differ significantly. My data give similar results: five of six females who joined PB had higher rates of affiliative interaction with him in 1991 and 1992, while five of six who joined SH had higher rates of affiliative interaction with him (Fig. 12.1). The overall difference was not significant, however, because two females (PA and UW in Fig. 12.1) interacted at high rates with the male they did not subsequently join. As Robbins (2001) noted, one female exception, PA, was SH's presumed full sister; she had a relatively high rate of affiliative interaction with him but joined PB's group. The other, UW, had immigrated in early 1991; she was an adult in 1993, but still an adolescent in 1991–2. Robbins (2001) found that four of six females who joined PB had spent more time within 5 m of him than within 5 m of SH prior to the split; the reverse was true for five of six females who joined SH. My data give similar results: five of six females who joined PB had spent more time within 5 m of him than within 5 m of SH in 1991–2; female PA was again the only exception (Fig. 12.2). Five of six females

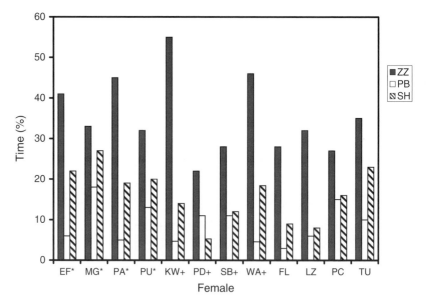

Fig. 12.3. Percentage of time that females in Group 5 spent within 5 m of adult male ZZ and adolescent males PB and SH from 1986 to 1987. Females ordered alphabetically and according to whether they joined PB's Group (*) in 1993, joined SH's Group (+), or had either emigrated (PC, TU) or died (FL, LZ) by 1993. Female MG was an adolescent; other females were adults. Females were marginally more likely to be within close proximity of SH than PB during 1986–7 (Wilcoxon matched pairs test, $T^+ = 58, 0.10 > p > 0.05$). Those who joined SH had not necessarily spent more time near him than near PB in the earlier study period (Wilcoxon matched pairs test, $T^+ = 8, n = 4$, ns). Those who joined PB marginally more time near him than near SH in 1986–7 (Wilcoxon matched pairs test, $T^+ = 10, n = 4$, $0.10 > p > 0.05$). Those who joined SH had not spent significantly more time near him during the earlier period (Wilcoxon matched pairs test, $T^+ = 8$, ns).

who joined SH had spent more time within 5 m of him than within 5 m of PB in 1991–2; female UW was again the only exception (Fig. 12.2). However, because PA had the highest proximity score with SH of all females and UW had the highest score with PB, differences were not statistically significant (Fig. 12.2).

All Group 5 females spent more time in close proximity to ZZ than to PB in 1986–7 (Watts, 1992) (Fig. 12.3). In fact, all females except PD also spent more time in close proximity to SH than to PB (Fig. 12.3), although this might have reflected ZZ's tolerance for SH and his intolerance of PB, rather than independent attraction of females to SH. Most females also spent more time within 5 m of ZZ than within 5 m of PB in 1991 and 1992. However, two of 10 adults (PU and MG) and one of three adolescents (UW) spent more time within 5 m of PB (Fig. 12.2). PU and MG, who were maternal sisters, eventually joined PB's group, but UW joined SH's group.

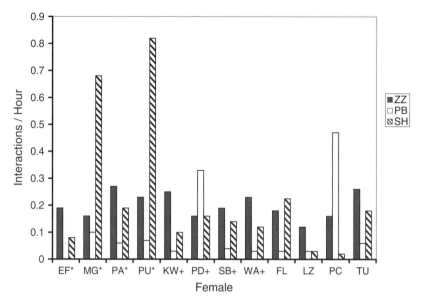

Fig. 12.4. Rates of affiliative interaction between females and adult males ZZ, PB, and SH in Group 5 from 1986 to 1987. Females ordered as in Fig. 12.3. Female MG was an adolescent; other females were adults. Rates were marginally higher with SH than with PB (Wilcoxon matched pairs test, $T^+ = 58, 0.10 > p > 0.05$). Those who joined PB interacted at marginally lower rates with him than with SH in 1986–7 (Wilcoxon matched pairs test, $T^+ = 10, n = 4, 0.10 > p > 0.05$). Rates at which those who joined SH had interacted with him did not differ significantly from their rates of interaction with PB (Wilcoxon matched pairs test, $T^+ = 6, n = 4$, ns).

Most females also had more affiliative interactions with ZZ than with PB in 1986–7 (Watts, 1992) (Fig. 12.4). However, two adult females (PC and PD) who had recently immigrated were striking exceptions (Watts, 1992). PC and PD had the second and fourth highest proximity scores with PB, respectively (Fig. 12.3). They groomed with PB more than they did with any other female and more than any female groomed with ZZ. Both received more grooming from PB than they gave to him (Watts, 1992). Despite PB's considerable grooming effort, however, PC emigrated in 1989, and PD joined SH's group in 1993. PD groomed infrequently with PB by 1991, and actually groomed more, and otherwise engaged in more affiliative behavior, with SH than with PB.

Eight of 11 adult females and the single adolescent female in Group 5 had higher rates of affiliative interaction with SH than with PB in 1986–7, although the overall difference in interaction rates was only marginally significant (Fig. 12.4). The contrast again might have reflected the fact that it was easier for SH than for PB to stay near ZZ during rest periods. Also, play accounted for most interactions with SH, but females did not play with either him or PB in 1991–2. Eight of these 12 females were still in the group when it fissioned; seven of these had higher rates of affiliative interaction with SH than with PB in 1986–7. Of these seven, only three (KW, SB, and WA in Fig. 12.4) joined SH's group in 1993. These three were immigrants who had arrived in 1984 or 1985. Overall, females who joined SH did not have significantly higher rates of affiliative interaction with him than with PB in 1991–2 (Fig. 12.1). The four who joined PB had marginally higher rates of affiliative interaction with SH. They included natal females MG, PA, and PU, plus EF, who was an adult when first seen in Group 5 in 1966; her natal group was unknown.

Male CA was the youngest and lowest-ranking of the four adult males at the time of ZZ's death. He was a young adolescent in 1986–7, and, like SH, spent more time within 5 m of almost all females than PB did and interacted affiliatively with almost all at higher rates. Only females PD and PC had higher proximity scores and interaction rates with PB (Fig. 12.4). All four females who later joined SH spent more time in close proximity to him and engaged in more affiliative interaction with him than with CA. In 1991–2, CA spent less time in close proximity and interacted less often with almost all females than either PB or SH did. The exceptional females were all maternally related to him and included his mother, PU, his grandmother, EF, and EF's daughters MG and MA.

These data offer some support for the idea that females are more likely to stay with follower males with whom they have relatively strong social bonds, or "good" relationships, than with those with whom their bonds are not so strong. Although the Group 5 fission was a unique event, relationship quality may have broad importance because females often face choices between staying with maturing or fully adult follower males or emigrating. However, relationship quality is not the sole influence on such decisions, nor is it always a good predictor in individual cases. For example, neither of PB's top grooming partners in 1986–7 later joined his group; one had already emigrated, and the second groomed more with SH by 1991–2 and joined his group. Also, as Robbins (2001) noted, four of six females who joined PB's group belonged to EF's matriline. CA, initially the subordinate silverback in this group, also belonged to this matriline. These females were allies against unrelated females in the group (Watts, 1997) and might have benefited by staying together. PB's group had two fully adult males at the time of the fission and SH's group had only one; this might also have influenced residence choices independently of relationship quality. Also, the degree of relatedness between females and males in their natal groups presumably influences group residence decisions (Harcourt, 1978). In this case, PA had a stronger social bond to her maternal, and presumed full, brother SH than to PB, but stayed with PB. PU stayed in her son's group, but this group also had another fully mature male with whom she had previously copulated.

## Discussion

### Comparisons with other gorilla subspecies

*Social structure, social relationships, and life histories*
Multi-female groups with stable long-term membership, multi-male groups, and solitary males occur in all known gorilla populations (Doran and McNeilage, 1998). Encounters between groups usually involve antagonism between males (e.g., Lopé: Tutin, 1996; but see below), and Tutin (1996) describes prolonged pursuits of groups by solitary males similar to those observed at Karisoke (e.g., Watts, 1994c). Female transfer occurs in western lowland gorillas at Lopé (Tutin, 1996) and eastern lowland gorillas at Kahuzi-Biega (Yamagiwa, 1983) and is probably a species-typical characteristic. However, we have few comparative data on social relationships. Female transfer leads to the expectation that grooming and agonistic support are uncommon in most female dyads; that many female dyads lack dominance relationships; and that social relationships between females are differentiated along kinship lines when kin reside together. Male–female grooming, female competition for social access to males, and male interventions in contests between females are probably relatively common. Data from Lopé on western lowland gorillas (Tutin, 1996) tentatively support these predictions.

Ecological costs of grouping are probably low for all gorillas. Even where fruit provides much of the gorillas' diet, females can avoid contests in trees too small to accommodate all group members by eating other foods on the ground instead (Kuroda *et al.*, 1996; Doran and McNeilage, 1998). They also can reduce competition by forming subgroups and adjusting subgroup size to food patch size (Remis, 1997b; Doran and McNeilage, 1998; Goldsmith, 1999). Nevertheless, some females could face disadvantages if they consistently avoid others at good feeding sites, as can happen in cercopithecines with strong female dominance hierarchies (e.g., vervets: Whitten, 1983; baboons: Barton and Whiten, 1993). Goldsmith (1999) found no evidence for scramble effects in her analysis of the relationship between day journey length and group size for western lowland gorillas at Bai Hokou, and contest effects on foraging efficiency also seem unlikely, although whether any occur is unknown.

### Possible differences from mountain gorillas
Several possible differences between mountain gorillas and other subspecies deserve comment. First, researchers at several sites have reported temporary division of western lowland gorilla groups into subgroups that may stay apart for several days (Olejniczak, 1996; Remis, 1997b; Doran and McNeilage, 1998; Goldsmith, 1999). This occurs mostly during peaks in fruit abundance

and presumably reduces scramble and contest competition for fruit. This led Doran and McNeilage (1998) to argue that western lowland gorillas have a social system more like that of chimpanzees than of mountain gorillas in some respects. However, gorilla subgroups apparently always contain adult males (Doran and McNeilage, 1998), a fundamental difference from the chimpanzee system. Feeding competition constrains female grouping and association with males much more in chimpanzees than in gorillas (Wrangham, 1979, 2000). Females are often in parties that lack adult males, and anestrous females are often alone (e.g., Gombe: Goodall, 1986; Mahale: Nishida, 1968; Kibale: Wrangham *et al.*, 1992; Pepper *et al.*, 1999). Fission–fusion foraging by the one-male units of a hamadryas baboon (*Papio hamadryas*) clan (Kummer, 1968) may be a better analogy with gorilla subgrouping. Some other primates also form temporary subgroups to use scattered fruit patches and thereby reduce feeding competition (e.g., longtailed macaques: van Noordwijk and van Schaik, 1987; grey-cheeked mangabey: Waser, 1977). As in gorillas, females and immatures from these other primate taxa are not out of contact with males for extended periods. Being in a peripheral longtailed macaque subgroup nevertheless increases exposure to predators (van Noordwijk and van Schaik, 1987). Subgrouping may also impose safety costs on female gorillas by decreasing group-size-related dilution and vigilance effects and by depriving them of protection by multiple males, but staying with at least one male should minimize such costs (cf. Doran and McNeilage, 1998). Members of western lowland gorilla groups also sometimes spread out widely while foraging, which could expose females to risks that they trade off against greater foraging efficiency (Doran and McNeilage, 1998). However, mountain gorilla group spread can also be very wide and females can be out of sight of males during foraging (Watts, 1996, personal observation).

Subgrouping raises the issue of infanticide by males, which is only definitely known from the Virungas. However, most primates, including gorillas, have slow life histories and low female reproductive rates, and males can probably increase their reproductive success by killing unrelated infants when the costs are sufficiently low and when this improves their chances of mating with the infants' mothers (van Schaik, 1996). Even in the Virungas, infanticide is infrequent except after the death of protector males (Watts, 1989). That no one has yet seen infanticide elsewhere could be an artifact of much less intense and shorter-term sampling and of the absence of habituation (see Mitani and Watts, 1999, for a case in point, concerning hunting by chimpanzees). As Doran and McNeilage (1998) noted, male western lowland gorillas behave as if extragroup males are threats, and infanticide risk could explain why females always associate with males, even when in temporary subgroups. Yamagiwa's (1999) report of transfer by female eastern lowland gorillas with dependent infants who survived the transfer is more difficult to reconcile with the socioecological model. However, it is unclear whether infanticide would have shortened post-partum amenorrhea or whether males were related to the infants in these cases.

Tutin (1996) saw between-group contests over access to fruit crops at Lopé. Between-group contests over food are unimportant for the Virunga gorillas, except perhaps encounters in bamboo forest when new bamboo shoots are appearing. Females face energy costs if their groups move unusually far after intergroup encounters. Pursuits by solitary males also impose energy costs (Watts, 1994c). However, most contests between groups reflect male efforts to gain or retain mates and to protect offspring, not competition over food (Harcourt, 1978; Sicotte, 1993; Watts, 1994c; Doran and McNeilage, 1998). Several researchers have seen many peaceful encounters between two or more western lowland gorilla groups simultaneously feeding in large swamp clearings ("bais") (Olejnizcak, 1994, 1996; Tutin, 1996; Magliocca and Querouil, 1997). Bais are large, perennial food patches that many groups use and within which individual food sources are abundant and evenly distributed. The absence of competition for access to them is unsurprising, especially given that each group spends only a small amount of its time in a given bai (Olejnizcak, 1996). Still, such high tolerance between males is surprising if infanticide is a threat. Not all encounters between mountain gorilla groups are overtly hostile, either (Harcourt, 1978; Fossey, 1983; Sicotte, 1993, personal observation), and immature mountain gorillas from different groups sometimes play together during encounters, as happens at Mbeli Bai (Olejniczak, 1996). The high potential costs of male combat may facilitate tolerance in bais, given the high encounter rates there.

### Gorillas and other mammals

#### Other primates
The widespread influence of male sexual coercion, especially infanticide, on primate social evolution (Smuts and Smuts, 1993; van Schaik, 1996) means that some similarity to gorilla male–female relationships occurs in many primate taxa, including some in which female transfer is rare or absent (Sterck *et al.*, 1997). Male protection against other males is combined with male limitation of female aggression in some of these cases. For example, female baboons groom and mate preferentially with their male "friends" in exchange for protection against coercion, including infanticide, by other males, and sometimes also against the effects of contest competition among females (Smuts, 1985; Palombit *et al.*, 1997; Castles and Whiten, 1998). However, similarities to

mountain gorillas, especially in female–female social relationships, are strongest in other primates with common female transfer, including some colobine monkeys, mountain baboons, and hamadryas baboons.

Female transfer occurs in many colobines, although we have detailed data on feeding ecology, feeding competition, and social relationships only for Thomas's langurs (*Presbytis thomasi*). Female Thomas's langurs have egalitarian agonistic relationships. They are highly folivorous and their food is more evenly distributed than that of sympatric longtailed macaques (Sterck and Steenbeek, 1997). Within-group contest competition influences female reproductive success in the macaques, but not the langurs (Sterck and Steenbeek, 1997; Sterck *et al.*, 1997; van Noordwijk and van Schaik, 2000). Also, female Thomas's langurs can minimize scramble effects by transferring (Sterck, 1997; Sterck and Steenbeek, 1997; Sterck *et al.*, 1997). Like female mountain gorillas, however, they sometimes join larger groups. Males are infanticidal, and infanticide risk influences transfer decisions: most transfers are by nulliparae or by parous females without dependent infants, including some whose infants outside males have just killed (Steenbeek, 1997, 2000; Sterck, 1997). Females who lose infants to infanticide tend to join the killers (Sterck, 1997), and females apparently test the protective abilities of new resident males by approaching extra-group males during encounters (Steenbeek, 2000). Also like in mountain gorillas (Watts, 1989), females with infants in a group whose male has died try to avoid outside males, who are infanticide threats, but those without infants join such males (Steenbeek, 1997).

Male policing of female aggression may be less common in colobines than in mountain gorillas, perhaps because most show less body-size dimorphism. However, males sometimes intervene in female contests in Thomas's langurs (Sterck, 1997), and male control interventions in female contests were common in two captive groups of spectacled langurs (*Trachypithecus obscurus*) (Arnold and Barton, 2001).

Chacma baboons (*Papio ursinus*) in high-altitude populations mostly eat sparsely, but evenly, distributed food. Contest feeding competition is low and female agonistic relationships are egalitarian. Grooming is uncommon between females and relatively common between males and females, and female transfer is common (Anderson, 1982; Byrne *et al.*, 1990). Groups are small and most have single males, although some contain subordinate male followers who help to guard females and to protect infants against outside males (Byrne *et al.*, 1987). Female hamadryas baboons also face low contest competition for food and do not always have decided dominance relationships. Females groom other females infrequently, but commonly groom males and sometimes compete for social access to them (Kummer, 1968; Abegglen, 1984). Males intervene in many female conflicts; like in mountain gorillas, this reduces competitive asymmetries between females (Kummer, 1968; Colmenares and Lazaro-Perea, 1994; Watts *et al.*, 2000). Males also protect their females against harassment from members of other one-male units and against infanticide (Watts *et al.*, 2000). Female hamadryas transfer from their natal one-male units as adults, but mostly to one-male units in the same band, not between bands (Sigg *et al.*, 1982), so locational dispersal is less common than in mountain gorillas (Isbell and Van Vuren, 1996).

*Group-living equids*

Partly because of ecological convergence, social relationships and life-history tactics in group-living equids notably resemble those in mountain gorillas, although important contrasts exist (Watts, 1994*a*, 1996, 1997; Sterck *et al.*, 1997; Linklater *et al.*, 1999). Equid bands typically have a single stallion and several mares. Like the social groups of most primates, but unlike those of most other group-living ungulates, male–female association is permanent and band composition shows long-term stability (Linklater *et al.*, 1999). Bands use evenly distributed food resources in widely overlapping home ranges, and contest competition over food is low. Female coalitions are rare, and females in the same band may not have dominance relationships or form dominance hierarchies. Even when they do, rank-related variation in reproductive success mediated through variation in foraging efficiency usually does not occur (Lloyd and Rasa, 1989; Rutberg and Greenberg, 1990; Duncan, 1992; Monard and Duncan, 1996), although artificially elevated contest feeding competition can produce rank effects (Schilder and Boer, 1987). Female natal transfer is common and does not impose reproductive costs (Berger, 1986; Rutberg, 1990; Monard and Duncan, 1996). However, aggression to immigrant mares by residents and male harassment of females may cumulatively decrease female reproductive success after multiple transfers (Rutberg and Greenberg, 1990; Linklater *et al.*, 1999). We do not know whether lifetime reproductive success varies with the number of transfers by mountain gorilla females. Some nulliparous female plains and mountain zebra form bonds, and eventually breeding groups, with stallions in bachelor groups they join temporarily, although most transfer to established breeding groups (Klingel, 1967; Lloyd and Rasa, 1989). Emigrating female mountain gorillas usually join solitary males or breeding groups, although one juvenile female in the Karisoke sample joined an all-male group; she later transferred to a bisexual group while still nulliparous.

Stallions intervene in some contests between females (Klingel, 1967; Pennzhorn, 1984; Berger, 1986), although how much this affects female agonistic relationships is unknown. Infanticide by males seems to be a pervasive threat in horse societies (Linklater *et al.*, 1999), and males sometimes induce abortions during band takeovers (Berger, 1986). Harassment by extra-band males can indirectly lower reproductive success by reducing female foraging efficiency (Berger, 1986; Rubenstein, 1986; Linklater *et al.*, 1999). Male protection against infanticide and harassment influences female residence decisions (Stevens, 1990; Linklater *et al.*, 1999). Linklater *et al.* (1999; cf. Rubenstein, 1986) proposed that long-term bonds ("consortships") between mares and stallions who protect them against harassment and infanticide underlie band stability. This resembles Wrangham's (1979) description of a gorilla group as a permanent consortship between a male and multiple females and the description of the mountain gorilla social system in terms of the socioecological model. However, Linklater *et al.* (1999) found that females had lower reproductive success in multi-stallion bands than in one-stallion bands in the Kaimanawa feral horse population. They attributed this to much higher within-band aggression from males in multi-stallion bands and argued that this is common in feral horses. Few comparative data are available, but female reproductive success was higher in multi-stallion bands in the Camargue feral horse population (Feh, 1999). Within-group male aggression probably does not outweigh the advantage of better protection against infanticide in multi-male mountain gorilla groups.

Male followers in equids sometimes help dominant stallions to protect females against harassment by outside males and to keep outside males from mating with band females (Miller, 1981; Berger, 1986; Stevens, 1990; Asa, 1999; Feh, 1999). Multi-stallion bands can be more stable than one-stallion bands (Stevens, 1990). Feh (1999) saw pairs of unrelated Camargue horse stallions, low-ranking in their herd's male dominance hierarchy, form alliances that allowed them to hold females successfully. Like followers and dominant males in mountain gorilla groups, stallions in multi-male bands thus may cooperate at the between-group level despite within-band mating competition. Although followers and some allies fare poorly in mating competition, they obtain higher short-term reproductive payoffs than bachelor males gain by trying to steal copulations and may have higher lifetime reproductive success (Feh, 1990, 1999; Asa, 1999). However, Linklater *et al.* (1999) argued that the negative effects of within-band male harassment on females indicate that multi-stallion bands are byproducts of intense competition among multiple males simultaneously trying to establish permanent consortships with dispersing females, rather than outcomes of male–male co-operation. The merger of a mountain gorilla group that had lost its male with an all-male group led to intense competition among the males and the eventual eviction of most of them. However, two unrelated males with whom most of the females stayed formed a stable multi-male group and co-operated against outside males (Watts, 1989, 2000b; Robbins, 2001). Mountain gorilla males who tolerate followers can benefit directly because followers help with mate retention and protection of infants, unlike Kaimanawa horses but like those in the Camargue herd (Feh, 1999). In contrast to stallions, most also benefit indirectly because followers are close relatives (Watts, 2000*b*).

## Summary and conclusions

The revised socioecological model of primate social relationships, despite its lacunae (Koenig and Borries, 2001), applies well to Virunga mountain gorilla social groups in that populations form when females transfer to solitary males or, rarely, when multiple males join each other in all-male groups. Female transfer is associated with relatively even food distribution and little or no effect of contest feeding competition on female reproductive success. Female social relationships are strongly differentiated along lines of relatedness when female kin reside together. Those between kin are relatively friendly, and kin form alliances, whereas relationships between non-kin are mostly indifferent to hostile. Social relationships between females and adult males are centrally important. Males influence female agonistic relationships, which are egalitarian, by intervening in conflicts among females. The need for protection of females and infants by males, notably against the threat of infanticide by other males, may be the main cause of grouping. Some differentiation of male–female relationships occurs in groups with multiple sexually mature males. The quality of male–female relationships may influence females' decisions whether to transfer or to breed in natal groups and their residence decisions in the event of group fissions. Male life-history tactics vary, in association with variation in the number of males, their relative ages, and the number of females per group.

Social relationships and female life-history tactics in mountain gorillas resemble those in other primates and some other mammals with similar regimes of feeding competition and generally similar modes of mating competition among males (e.g., Thomas's langurs, group-living equids). Notable ecological variation exists across, and within, gorilla subspecies and populations. Reliance on perennial plant foods high in structural carbohydrates as fallbacks superimposes some uniformity on this variation and may also lead to fundamental similarity in social relationships and life-history tactics. Some data from wild populations of eastern and western lowland gorilla populations support this view, but other data raise important questions about the degree of social variation. Notable examples concern the dynamics of subgroup formation in populations more frugivorous than that in the Virungas, the role of infanticide, and the possible effects of competition for access to fruit on relationships among females. Only long-term research on better-habituated groups can answer such questions. We can only hope that conservation efforts are successful enough, and the political and economic futures of countries where gorillas live sufficiently positive, for this to be possible.

## References

Abegglen, J.J. (1984). *On Socialization in Hamadryas Baboons*. Lewisburg, PA: Bucknell University Press.

Anderson, C.M. (1982). Baboons below the Tropic of Capricorn. *Journal of Human Evolution*, **11**, 205–217.

Arnold, K. and Barton, R.A. (2001). Postconflict behavior of spectacled leaf monkeys (*Trachypithecus obscurus*). II: Contact with third parties. *International Journal of Primatology*, **22**, 267–286.

Asa, C.S. (1999). Male reproductive success in free-ranging feral horses. *Behavioral Ecology and Sociobiology*, **46**, 89–93.

Barton, R.A. and Whiten, A. (1993). Feeding competition among female olive baboons (*Papio anubis*). *Animal Behaviour*, **46**, 777–789.

Berger, J. (1986). *Wild Horses of the Great Basin*. Chicago, IL: University of Chicago Press.

Byrne, R.B., Whiten, A., and Henzi, S.P. (1987). One-male groups and intergroup interactions of mountain baboons (*Papio ursinus*). *International Journal of Primatology*, **8**, 615–633.

Byrne, R.B., Whiten, A., and Henzi, S.P. (1990). Social relationships of mountain baboons: Leadership and affiliation in a non-female bonded monkey. *American Journal of Primatology*, **18**, 191–207.

Castles, D. and Whiten, A. (1998). Post-conflict behavior of wild olive baboons. II: Stress and self-directed behavior. *Ethology*, **104**, 148–160.

Colmenares, F. and Lazaro-Perea, C. (1994). Greeting and grooming during social conflicts in baboons: Strategic uses and social functions. In *Current Primatology*, vol. 2, *Social Development, Learning, and Behavior*, eds. J.J. Roeder, B. Thierry, J.R. Anderson, and N. Herrenschmidt, pp. 165–174. Strasbourg, France: Université Louis Pasteur.

Doran, D.M. and McNeilage, A. (1998). Gorilla ecology and behavior. *Evolutionary Anthropology*, **6**, 120–131.

Duncan, P. (1992). *Horses and Grasses*. New York: Springer-Verlag.

Feh, C. (1990). Long-term paternity data in relation to different aspects of rank for Camargue stallions, *Equus caballus*. *Animal Behaviour*, **40**, 995–996.

Feh, C. (1999). Alliances and male reproductive success in Camargue stallions. *Animal Behaviour*, **57**, 705–713.

Fossey, D. (1983). *Gorillas in the Mist*. Boston, MA: Houghton Mifflin.

Fossey, D. (1984). Infanticide in mountain gorillas (*Gorilla gorilla beringei*) with comparative notes on chimpanzees. In *Infanticide: Comparative and Evolutionary Perspectives*, eds. G. Hausfater and S.B. Hrdy, pp. 217–236. Chicago, IL: Aldine Press.

Fossey, D. and Harcourt, A.H. (1977). Feeding ecology of free-ranging mountain gorillas (*Gorilla gorilla beringei*). In *Primate Feeding Ecology*, ed. T.H. Clutton-Brock, pp. 415–447. New York: Academic Press.

Goldsmith, M.L. (1999). Ecological constraints on the foraging effort of western gorillas (*Gorilla gorilla gorilla*) at Bai Hokou, Central African Republic. *International Journal of Primatology*, **20**, 1–24.

Goodall, J. (1986). *The Chimpanzees of Gombe*. Cambridge, MA: Harvard University Press.

Harcourt, A.H. (1978). Strategies of emigration and transfer by female primates, with special reference to mountain gorillas. *Zeitschrift für Tierpsychologie*, **48**, 401–420.

Harcourt, A.H. (1979*a*). Social relationships among adult female mountain gorillas. *Animal Behaviour*, **27**, 251–264.

Harcourt, A.H. (1979*b*). Social relationships between adult male and female mountain gorillas. *Animal Behaviour*, **27**, 325–342.

Harcourt, A.H. and Stewart, K.J. (1981). Gorilla male relationships: Can differences during immaturity lead to contrasting reproductive tactics during adulthood? *Animal Behaviour*, **29**, 206–210.

Harcourt, A.H. and Stewart, K.J. (1987). The influence of help in contests on dominance rank in primates: Hints from gorillas. *Animal Behaviour*, **35**, 182–190.

Harcourt, A.H. and Stewart, K.J. (1989). Functions of alliances in contests within wild gorilla groups. *Behaviour*, **109**, 176–190.

Hrdy, S.B. (1979). Infanticide among animals: A review, classification, and examination of the implications for the reproductive strategies of females. *Ethology and Sociobiology*, **1**, 13–40.

Isbell, L.A. and Van Vuren, D. (1996). Differential costs of locational and social dispersal and their consequences for female group-living primates. *Behaviour*, **133**, 1–36.

Klingel, J.H. (1967). Sociale Organisation und verhaltung freilebender Steppenzebras (*Equus quagga*). *Zeitschrift für Tierpsychologie*, **24**, 580–624.

Koenig, A. and Borries, C. (2001). Socioecology of hanuman langurs: The story of their success. *Evolutionary Anthropology*, **10**, 122–137.

Koenig, A., Beise, J., Chalise, M., and Ganzhorn, J. (1998). When females should contest for food: Testing hypotheses about resource density, distribution, size, and quality with hanuman langurs. *Behavioral Ecology and Sociobiology*, **42**, 225–237.

Kummer, H. (1968). *Social Organization of Hamadryas Baboons*. Chicago, IL: University of Chicago Press.

Kuroda, S., Nishihara, T., Suzuki, S., and Oko, R.A. (1996). Sympatric gorillas and chimpanzees in the Ndoki Forest, Congo. In *Great Ape Societies*, eds. W.C. McGrew, L.A. Marchant, and T. Nishida, pp. 71–81. Cambridge, U.K.: Cambridge University Press.

Linklater, W.L., Cameron, E.Z., Minot, E.O., and Stafford, K.J. (1999). Stallion harassment and the mating system of horses. *Animal Behaviour*, **58**, 295–306.

Lloyd, P.H. and Rasa, O.A.E. (1989). Status, reproductive success, and fitness in Cape Mountain zebras (*Equus zebra zebra*). *Behavioral Ecology and Sociobiology*, **25**, 411–420.

Magliocca, F. and Querouil, S. (1997). Preliminary report on the use of the Maya-Maya North Saline (Odzala National Park, Congo) by lowland gorillas. *Gorilla Conservation News*, **11**, 5.

McNeilage, A.R. (1996). Mountain gorillas in the Virunga Volcanos: Feeding ecology and carrying capacity. PhD thesis, University of Bristol, U.K.

Miller, R.K. (1981). Male aggression, dominance, and breeding behavior in feral horses. *Zeitschrift für Tierpsychologie*, **57**, 340–351.

Mitani, J.C. and Watts, D.P. (1999). Demographic influences on the hunting behavior of chimpanzees. *American Journal of Physical Anthropology*, **109**, 439–454.

Monard, A. and Duncan, P. (1996). Consequences of natal dispersal in female horses. *Animal Behaviour*, **52**, 565–579.

Nishida, T. (1968). The social group of wild chimpanzees in the Mahale Mountains. *Primates*, **9**, 167–224.

Olejniczak, C. (1994). Report on a pilot study of western lowland gorillas at Mbeli Bai, Nouabale-Ndoki Reserve, Northern Congo. *Gorilla Conservation News*, **8**, 9–11.

Olejniczak, C. (1996). Update on the Mbeli Bai gorilla study, Nouabala-Ndoki National Park, northern Congo. *Gorilla Conservation News*, **10**, 5–8.

Palombit, R.A., Seyfarth, R.M., and Cheney, D.L. (1997). The adaptive value of 'friendships' to female baboons: Experimental and observational evidence. *Animal Behaviour*, **54**, 599–614.

Pennzhorn, B.L. (1984). A long-term study of social organization and behavior of cape mountain zebra (*Equus zebra zebra*). *Zeitschrift für Tierpsychologie*, **64**, 97–146.

Pepper, J.W., Mitani, J.C., and Watts, D.P. (1999). General gregariousness and specific partner preference among wild chimpanzees. *International Journal of Primatology*, **20**, 613–632.

Plumptre, A. (1995). The chemical composition of montane plants and its influence on the diets of large mammalian herbivores in the Parc National des Volcans, Rwanda. *Journal of Zoology, London*, **235**, 323–337.

Remis, M.J. (1997*a*). Western lowland gorillas (*Gorilla gorilla gorilla*) as seasonal frugivores: Use of variable resources. *American Journal of Primatology*, **43**, 87–109.

Remis, M.J. (1997*b*). Ranging and grouping patterns of a western lowland group at Bai Hokou, Central African Republic. *American Journal of Primatology*, **43**, 111–133.

Robbins, M.M. (1995). A demographic analysis of male life history and social structure of mountain gorillas. *Behaviour*, **132**, 21–47.

Robbins, M.M. (1999). Male mating patterns in wild multimale mountain gorilla groups. *Animal Behaviour*, **57**, 1013–1020.

Robbins, M.M. (2001). Variation in the social system of mountain gorillas: The male perspective. In *The Mountain Gorillas of Karisoke*, eds. M.M. Robbins, P. Sicotte, and K.J. Stewart, pp. 29–58. Cambridge, U.K.: Cambridge University Press.

Rubenstein, D.I. (1986). Ecology and sociality in zebras and horses. In *Ecological Aspects of Social Evolution*, eds. D.I. Rubenstein and R.W. Wrangham, pp. 469–87. Princeton, NJ: Princeton University Press.

Rutberg, A.T. (1990). Inter-group transfer in Assateague pony mares. *Animal Behaviour*, **40**, 945–952.

Rutberg, A.T. and Greenberg, S.A. (1990). Dominance, aggression frequencies, and modes of aggressive competition in feral pony mares. *Animal Behaviour*, **40**, 322–331.

Schaller, G.S. (1963). *The Mountain Gorilla: Ecology and Behavior*. Chicago, IL: University of Chicago Press.

Schilder, M.B.H. and Boer, P.L. (1987). Ethological investigations on a herd of plains zebra in a safari park: Time budgets, reproduction, and food competition. *Applied Animal Behavioral Science*, **18**, 45–56.

Sicotte, P. (1993). Intergroup encounters and female transfer in mountain gorillas: Influence of group composition on male behavior. *American Journal of Primatology*, **30**, 21–36.

Sicotte, P. (1994). Effects of male competition on male–female relationships in bi-male groups of mountain gorillas. *Ethology*, **97**, 47–64.

Sigg, H., Stolba, A., Abegglen, J.J., and Dasser, V. (1982). Life history of hamadryas baboons: Physical development, infant mortality, reproductive parameters, and family relationships. *Primates*, **23**, 473–487.

Smuts, B.B. (1985). *Sex and Friendship in Baboons*. Chicago, IL: Aldine Press.

Smuts, B.B. and Smuts, R. (1993). Male aggression and sexual coercion of females in nonhuman primates and other mammals: Evidence and theoretical implications. *Advances in the Study of Behavior*, **22**, 1–63.

Steenbeek, R. (1997). What a maleless group can tell us about the constraints on female transfer in Thomas's langurs. *Folia Primatologica*, **67**, 169–181.

Steenbeek, R. (2000). Infanticide by males and female choice in wild Thomas's langurs. In *Infanticide by Males and Its Implications*, eds. C.P. van Schaik and C.H. Janson, pp. 153–177. Cambridge, U.K.: Cambridge University Press.

Sterck, E.H.M. (1997). Determinants of female dispersal in Thomas's langurs. *American Journal of Primatology*, **42**, 179–198.

Sterck, E.H.M. and Steenbeek, R. (1997). Female dominance relationships and food competition in the sympatric Thomas's langur and longtailed macaque. *Behaviour*, **134**, 749–774.

Sterck, E.H.M., Watts, D.P., and van Schaik, C.P. (1997). The evolution of social relationships in female primates. *Behavioral Ecology and Sociobiology*, **41**, 291–309.

Stevens, E.F. (1990). Instability of harems of feral horses in relation to season and presence of subordinate stallions. *Behaviour*, **112**, 149–161.

Stewart, K.J. and Harcourt, A.H. (1987). Gorillas: Variation in female relationships. In *Primate Societies*, eds. B.B. Smuts, D.L. Cheney, R.M. Seyfarth, R.W. Wrangham, and T.T. Struhsaker, pp. 155–164. Chicago, IL: University of Chicago Press.

Tutin, C.E.G. (1996). Ranging and social structure of lowland gorillas in the Lopé Reserve, Gabon. In *Great Ape Societies*, eds. W.C. McGrew, T. Nishida, and L.A. Marchant, pp. 58–70. Cambridge, U.K.: Cambridge University Press.

Tutin, C.E.G., Fernandez, M., Rogers, M.E., Williamson, E.A., and McGrew, W.C. (1991). Foraging profiles of sympatric lowland gorillas and chimpanzees in the Lopé Reserve, Gabon. *Philosophical Transactions of the Royal Society, Series B*, **334**, 179–186.

van Noordwijk, M. and van Schaik, C.P. (1987). Competition among female longtailed macaques, *Macaca fascicularis*. *Animal Behaviour*, **35**, 577–589.

van Noordwijk, M. and van Schaik, C.P. (2000). The effect of dominance rank and group size on female lifetime reproductive success in wild longtailed macaques, *Macaca fascicularis*. *Primates*, **40**, 105–130.

van Schaik, C.P. (1989). The ecology of social relationships amongst female primates. In *Comparative Socioecology*, eds. V. Standen and R. Foley, pp. 195–218. Oxford, U.K.: Blackwell Scientific Publications.

van Schaik, C.P. (1996). Social evolution in primates: The role of ecological factors and male behavior. *Proceedings of the British Academy*, **88**, 9–31.

Vedder, A.L. (1984). Movement patterns of a group of free ranging mountain gorillas (*Gorilla gorilla beringei*) and their relation to food availability. *American Journal of Primatology*, **7**, 73–88.

Waser, P. (1977). Feeding, ranging, and group size in the mangabey, *Cercocebus albigena*. In *Primate Ecology: Studies of Feeding and Ranging Behavior in Lemurs, Monkeys, and Apes*, ed. T. Clutton-Brock, pp. 183–222. New York: Academic Press.

Watts, D.P. (1984). Composition and variability of mountain gorilla diets in the central Virungas. *American Journal of Primatology*, **7**, 325–356.

Watts, D.P. (1985). Relations between group size and composition and feeding competition in mountain gorilla groups. *Animal Behaviour*, **33**, 72–85.

Watts, D.P. (1989). Infanticide in mountain gorillas: New cases and a reconsideration of the evidence. *Ethology*, **81**, 1–18.

Watts, D.P. (1990a). Ecology of gorillas and its relationship to female transfer in mountain gorillas. *International Journal of Primatology*, **11**, 21–45.

Watts, D.P. (1990b). Mountain gorilla life histories, reproductive competition, and some sociosexual behavior and some implications for captive husbandry. *Zoo Biology*, **9**, 185–200.

Watts, D.P. (1991a). Strategies of habitat use by mountain gorillas. *Folia Primatologica*, **56**, 1–16.

Watts, D.P. (1991b). Harassment of immigrant female mountain gorillas by resident females. *Ethology*, **89**, 135–153.

Watts, D.P. (1992). Social relationships of immigrant and resident female mountain gorillas. I: Male–female relationships. *American Journal of Primatology*, **28**, 159–181.

Watts, D.P. (1994a). Agonistic relationships of female mountain gorillas. *Behavioral Ecology and Sociobiology*, **34**, 347–358.

Watts, D.P. (1994b). Social relationships of immigrant and resident female mountain gorillas. II: Relatedness, residence, and relationships between females. *American Journal of Primatology*, **32**, 13–30.

Watts, D.P. (1994c). The influence of male mating tactics on habitat use in mountain gorillas (*Gorilla gorilla beringei*). *Primates*, **35**, 35–47.

Watts, D.P. (1995a). Post-conflict social events in wild mountain gorillas. 1: Social interactions between opponents. *Ethology*, **100**, 158–174.

Watts, D.P. (1995b). Post-conflict social events in wild mountain gorillas. 2: Redirection and consolation. *Ethology*, **100**, 175–191.

Watts, D.P. (1996). Comparative socioecology of gorillas. In *Great Ape Societies*, eds. W.C. McGrew, T. Nishida, and L.A. Marchant, pp. 16–28. Cambridge, U.K.: Cambridge University Press.

Watts, D.P. (1997). Agonistic interventions in wild mountain gorilla groups. *Behaviour*, **134**, 23–57.

Watts, D.P. (1998). Long-term habitat use by mountain gorillas (*Gorilla gorilla beringei*). I: Consistency, variation, and home range size and stability. *International Journal of Primatology*, **19**, 651–680.

Watts, D.P. (2000*a*). Mountain gorilla habitat use strategies and group movements. In *On the Move: How and Why Animals Travel in Groups*, eds. S. Boinski and P. Garber, pp. 351–374. Cambridge, U.K.: Cambridge University Press.

Watts, D.P. (2000*b*). Causes and consequences of variation in the number of males in mountain gorilla groups. In *Primate Males: Causes and Consequences of Variation in Group Composition*, ed. P. Kappeler, pp. 169–179. Cambridge, U.K.: Cambridge University Press.

Watts, D.P. (2001). Social relationships of female mountain gorillas. In *Mountain Gorillas: Three Decades of Research at Karisoke*, eds. M. Robbins, P. Sicotte, and K.J. Stewart, pp. 215–240. Cambridge, U.K.: Cambridge University Press.

Watts, D.P. and Pusey, A.E. (1993). Behavior of juvenile and adolescent great apes. In *Juvenile Primates: Life History, Development, and Behavior*, eds. M.E. Pereira and L.A. Fairbanks, pp. 148–167. New York: Oxford University Press.

Watts, D.P., Colmenares, F., and Arnold, K. (2000). Redirection, consolation, and male policing: How targets of aggression interact with bystanders. In *Natural Conflict Resolution*, eds. F.B.M. de Waal and F. Aureli, pp. 281–301. Berkeley, CA: University of California Press.

Whitten, P.L. (1983). Diet and dominance among female vervet monkeys (*Cercopithecus aethiops*). *American Journal of Primatology*, **5**, 139–159.

Wrangham, R.W. (1979). On the evolution of ape social systems. *Social Science Information*, **1**, 334–368.

Wrangham, R.W. (2000). Why are male chimpanzees more gregarious than mothers? A scramble competition hypothesis. In *Primate Males: Causes and Consequences of Variation in Group Composition*, ed. P. Kappeler, pp. 248–258. Cambridge, U.K.: Cambridge University Press.

Wrangham, R.W., Clark, A.P., and Isibirye-Basuta, G. (1992). Female social relationships and social organization of the Kibale Forest chimpanzees. In *Topics in Primatology*, Vol. 1, *Human Origins*, eds. T. Nishida, W.C. McGrew, P. Marler, M. Pickford, and F.B.M. de Waal, pp. 81–98. Tokyo: University of Tokyo Press.

Yamagiwa, J. (1983). Diachronic changes in two eastern lowland gorilla groups (*Gorilla gorilla graueri*) in the Mt. Kahuzi region, Zaïre. *Primates*, **24**, 174–183.

Yamagiwa, J. (1987). Male life history and the social structure of wild mountain gorillas (*Gorilla gorilla beringei*). In *Evolution and Coadaptation in Biotic Communities*, eds. S. Kawano, J.H. Connell, and T. Hidaka, pp. 31–51. Tokyo: Tokyo University Press.

Yamagiwa, J. (1999). Socioecological factors influencing population structure of gorillas and chimpanzees. *Primates*, **40**, 87–104.

Yamagiwa, J., Maruhashi, T., Yumoto, T., and Mwanza, N. (1996). Dietary and ranging overlap in sympatric gorillas and chimpanzees in Kahuzi-Biega National Park, Zaïre. In *Great Ape Societies*, eds. W.C. McGrew, T. Nishida, and L.A. Marchant, pp. 82–100. Cambridge, U.K.: Cambridge University Press.

# 13 Within-group feeding competition and socioecological factors influencing social organization of gorillas in the Kahuzi-Biega National Park, Democratic Republic of Congo

JUICHI YAMAGIWA, KANYUNYI BASABOSE, KISWELE KALEME, AND TAKAKAZU YUMOTO

## Introduction

Competition over food and predation pressure plays an important role in shaping the social system of group-living primates (Wrangham, 1980, 1987; van Schaik, 1983, 1989; van Schaik and van Hooff, 1983; Terborgh and Janson, 1986; Dunbar, 1988). When a high risk of predation forces primates to live in cohesive groups, within-group scramble competition may stimulate females of folivorous primates to develop individualistic and egalitarian ranking systems (van Schaik, 1989; Barton *et al.*, 1996; Sterck *et al.*, 1997). When the risk of predation is low, females no longer form cohesive groups and tend to disperse to forage alone. Both food distribution and male mating strategy may influence female grouping patterns.

Gorillas have a folivorous diet and an individualistic ranking system (Fossey and Harcourt, 1977; Stewart and Harcourt, 1987). By contrast, chimpanzees have a fission–fusion social system based on individual foraging, and their party size varies with fruit abundance and availability (Goodall, 1968; Nishida, 1970; Wrangham, 1980). Male chimpanzees tend to associate with each other and to show territorial behavior against other groups of males (Goodall *et al.*, 1979; Nishida *et al.*, 1985). The various mating patterns (promiscuous, possessive or consort) adopted by males may influence the size and composition of temporary parties (Tutin, 1979; Goodall, 1986). These observations seem to support previous arguments that food distribution and male mating strategy effect social relationships among females (Wrangham, 1987; Dunbar, 1988; van Schaik, 1989).

However, recent findings on gorilla ecology reveal considerable variation in diet among gorillas of differing habitats, in spite of consistent similarities in

328

social structure across habitats. Western lowland gorillas exhibit a frugivorous diet as do chimpanzees, but the former usually lives in groups of similar size to those of folivorous mountain gorillas (Tutin and Fernandez, 1993; Nishihara, 1995; Goldsmith, 1996; Tutin, 1996). Although they occasionally form subgroups, the cohesiveness of gorilla groups is much higher than that of chimpanzees in spite of the former's frugivorous diet (Remis, 1994; Goldsmith, 1999). Eastern lowland gorillas seasonally show frugivorous characteristics but tend to form cohesive groups throughout the year (Yamagiwa *et al.*, 1994, 1996*a*). Both western and eastern lowland gorillas prominently increase their day journey length during the period of fruit abundance (Yamagiwa and Mwanza, 1994; Goldsmith, 1999). Thus, a frugivorous diet may not affect group size but does appear to influence day range in gorillas. How do gorillas cope with the costs of within-group competition in a large cohesive group while visiting fruit patches?

Recent observations of chimpanzee groupings suggest that predation affects their social organization. Chimpanzees have presumably been killed (based on hair in feces of predators) by lions and leopards in Mahale and Taï (Boesch, 1991; Tsukahara, 1993). Chimpanzees tend to form large parties where lions are found in their range in dry and open habitats (Itani and Suzuki, 1967; Tutin *et al.*, 1983). Local variation has been noted in the degree of cohesiveness of groups and the dispersal patterns of sexes, and the presence or absence of neighboring groups is the primary factor influencing male–male association and their dispersal patterns (Sugiyama, 1999). Nevertheless, chimpanzees show a higher degree of fission–fusion in grouping than gorillas in any habitat. Why do these two ape species differ in their grouping patterns despite similarities in diet and in female dispersal patterns? How does within-group competition force these apes to develop different social organizations with similar ecological features?

Sympatry may be one of the important factors explaining these differences. In equatorial Africa, gorillas and chimpanzees are widely distributed sympatrically and overlap extensively in both diet and ranging patterns (Jones and Sabater Pi, 1971; Tutin and Fernandez, 1992, 1993; Remis, 1994; Kuroda *et al.*, 1996; Tutin, 1996; Yamagiwa *et al.*, 1996*a,b*). Due to regional and seasonal similarities in diet, gorillas and chimpanzees occasionally encounter each other at fruiting trees but they do not respond aggressively and tend to avoid contacts with each other (Suzuki and Nishihara, 1992; Yamagiwa *et al.*, 1996*a*). However, when fruit is scarce, gorillas tend to increase the folivorous portion of their diet prominently, while chimpanzees persistently continue to seek fruits (Tutin and Fernandez, 1993; Yamagiwa *et al.*, 1996*a*; Remis, 1997*a*, this volume). Terrestrial herbaceous vegetation (THV) and fibrous fruits are frequently eaten by western and eastern lowland gorillas, while fig fruit assumes the role of a fallback food for chimpanzees during lean fruiting seasons (Kuroda *et al.*, 1996; Wrangham *et al.*, 1996; Tutin *et al.*, 1997). Tool-use also enables chimpanzees to feed on high-quality foods (nuts and pith of oil-palm) protected by hard shells during times of fruit scarcity (Yamakoshi, 1998). These differences may result in the marked reduction in day journey length for gorillas and in the reductions in party size and range expansion for chimpanzees (Goldsmith, 1999; Yamagiwa, 1999).

Thus, interspecies competition may have promoted niche separation between gorillas and chimpanzees and stimulated different foraging strategies. The observed variation in foraging strategies and dietary divergence is supported to some extent morphologically. Despite similarities in skull morphology among the African apes (Shea, 1983), gorillas and chimpanzees differ in some craniodental proportions (Kay, 1975; Shea, 1983) and mandibular morphology (Taylor, 2002, this volume). These differences may be linked to dietary variation and, in particular, to a trend towards extreme folivory in the mountain gorilla.

Since 1987, we have conducted ecological studies on sympatric populations of gorillas and chimpanzees in lowland and montane tropical forests of Kahuzi-Biega National Park, Democratic Republic of Congo. Gorillas live at similar group densities in both lowland and montane forests, while chimpanzees live at lower densities in montane forest (where the amount and diversity of fruits is low). A group of gorillas and a unit-group of chimpanzees have been semi-habituated and followed daily by us or field assistants, and their dietary composition and ranging behavior have been continuously recorded. Here we compare diet, group size, and day journey length of gorillas between two different habitats to assess the influence of within-group competition on their foraging strategies. Socioecological factors influencing intraspecific variation in social organization of gorillas are discussed, including sympatry with chimpanzees and infanticide.

## Study site and methods

The Kahuzi-Biega National Park is located to the west of Lake Kivu and covers an area of 6000 km² at an altitude of 600–3308 m (Fig. 13.1). The Park consists of the Kahuzi highland (600 km²) and Itebero lowland regions (5400 km²), which are interconnected by a corridor of forest. Forty-four species of larger mammals (including ten primate species) are found in the highland region and 56 species (including 14 primate species) exist in the lowland region (Mankoto *et al.*, 1994). Gorillas inhabiting the Itebero and Kahuzi regions are classified as the same subspecies (*Gorilla gorilla graueri*) because the two populations have had genetic exchange until recently through the corridor. Both populations are also sympatric with chimpanzees (*Pan troglodytes schweinfurthii*).

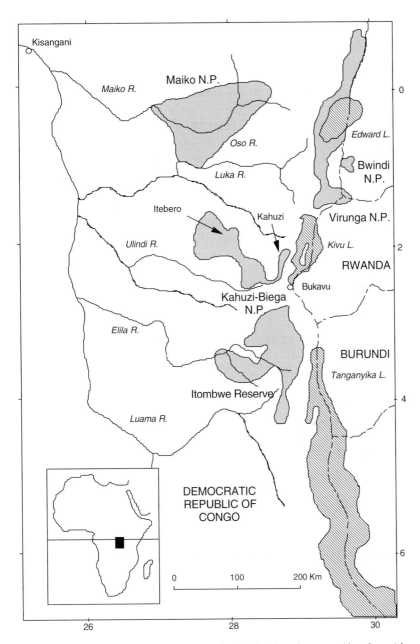

Fig. 13.1. Map showing the location of national parks and a reserve (dotted areas) in the distribution of eastern lowland gorillas. The slashed areas show lakes. Our study sites are located at Kahuzi (highland) and Itebero (lowland) in the Kahuzi-Biega National Park.

## Ecology of the Kahuzi and Itebero sites

The highland Kahuzi region is made up of bamboo (*Arundinaria alpina*) forest (37%), primary montane forest (28%) (in western and northern parts), secondary montane forest (20%) (in the eastern part), *Cyperus latifolius* swamp (7%), and other vegetation (8%) as described by Casimir (1975) and Goodall (1977). Dominant species of trees include *Podocarpus usambarensis* (Podocarpaceae), *Symphonia globulifera* (Guttiferae), and *Carapa grandiflora* (Meliaceae) in the primary forest; *Hagenia abyssinica* (Rosaceae), *Myrianthus holstii* (Moraceae), and *Vernonia* spp. (Compositae) in the secondary forest; *Hypericum revolutum* (Guttiferae) and *Rapanea melanophloeos* (Myrsinaceae) in the *Cyperus* swamps; and *Symphonia globulifera* (Guttiferae) and *Syzygium parvifolium* (Myrtaceae) in and around the swamp areas. Herbs, vines, and ferns (*Urera hypselodendron, Basella alba, Lactuca* sp., *Pteridium aquilinum*, etc.) constitute the dense terrestrial vegetation of the secondary forest (Yumoto *et al.*, 1994).

The lowland Itebero region is situated in the extended part of the park at an altitude of 600–1300 m. It is covered with tropical forest, which includes primary forest, secondary forest, abandoned fields, and ancient secondary forest. The primary forests are characterized by Caesalpiniaceae: *Michelsonia microphylla, Gilbertiodendron dewevrei, Julbernardia seretii, Dialium polyanthum,* and *Cynometra alexandri*; Mimosaceae: *Piptadeniastrum africanum*; and Myristicaceae: *Staudtia gabonensis* and *Pycnanthus angolensis*. The Zingiberaceae species include *Aframomum* spp. and *Costus afer*, with the Marantaceae species forming the generally scarce terrestrial vegetation. The secondary forests and abandoned fields cultivated prior to 1985 have been invaded by *Musanga cecropioides* (Moraceae) and *Macaranga spinosa* (Euphorbiaceae), and herbaceous plants are dense. The ancient secondary forests are the result of deforestation by a mining company in the colonial era and subsequent successional regeneration. *Ficus sur, Uapaca guineensis,* and *Celtis brieyi* are commonly found in this forest. *Halopegia azurea* is one of the dominant herbaceous plants. *Uapaca corbisierii* is occasionally found in swamps (Yumoto *et al.*, 1994).

In 1989 and 1994, we made vegetation surveys using a line-transect method (Yamagiwa *et al.*, 1993*a*, 1996*b*) to estimate the diversity and density of tree species in the Itebero lowlands and Kahuzi highlands, respectively. We also measured the diameter at breast height (DBH) of all trees above 10 cm along the transects. In the Itebero region, we counted 6922 trees consisting of about 150 species in a belt transect 10 × 8000 m (Table 13.1). In the Kahuzi region, we counted 2033 trees consisting of about 50 species in a belt transect 20 × 5000 m. Tree density and species diversity in the Itebero lowland region were about three times higher than those in the Kahuzi highland region. More species of tree fruits are also eaten by apes in the Itebero region (Yamagiwa *et al.*, 1994; Yumoto *et al.*, 1994). The ten most abundant species made up more than half

Table 13.1. *Comparison of vegetation between lowland and highland in Kahuzi-Biega National Park*

|  | Lowland (Itebero) | Highland (Kahuzi) |
|---|---|---|
| Altitude (above sea level) | 650 m | 2200 m |
| Transect (length × width) | 8000 × 10 m | 5000 × 20 m |
| Number of trees (over 10 cm DBH) | 6922 | 2033 |
| Number of tree species | *c.* 150 | 50 |
|    Number of gorilla fruit species | 46 | 16 |
|    Number of chimpanzee fruit species | 35 | 19 |
|    Number of both ape fruit species | 30 | 12 |
| Percentage of top ten species in the total basal area | 61.2 | 59.9 |
| Number of ape fruit species in top ten | 4 | 3 |
| Percentage of ape fruit species in the top ten basal area | 22.4 | 34.0 |

of the total basal area, and a few fruit species eaten by apes are in the top ten species in both regions. In the Kahuzi region, however, fruit species eaten by apes occupied a larger basal area proportion of the top ten species. In the highland habitat, tree fruits are less diversified and more sparsely distributed than in the lowland habitat, but a few particular fruit species eaten by apes are relatively densely distributed.

Plant specimens were identified by one of us (T. Yumoto) as matching the identified specimen types in the herbaria of the Jardin Botanique National de Belgique, Brussels, the Musée National d'Histoire Naturelle, Paris, and Kew Gardens, London. Among 329 plant species fully identified (220 in Itebero, 118 in Kahuzi), only nine species were collected in both regions (Yumoto *et al.*, 1994). More than 90% of the plant foods eaten by gorillas are only available in one of the two regions (Yamagiwa *et al.*, 1994).

Since August 1994, we have monitored fruiting, flowering, and leafing of each tree (above 10 cm DBH) within the transect in the Kahuzi region. To compare the abundance of fruit monthly, we calculated a monthly fruit index ($F_m$) as

$$F_m = \sum_{k=1}^{s} P_{km} B_k,$$

where $P_{km}$ denotes the proportion of the number of trees in fruit for species $k$ in month $m$, and $B_k$ denotes the total basal area per ha for species $k$. The fruit index was calculated monthly and its seasonal fluctuation was compared between primary and secondary forests.

Meteorological data are available from the Meteorological Station at Centre de Recherches en Sciences Naturelles (1600 m above sea level), which is located beside the highland sector of the Park. The mean annual rainfall from 1982 to 1997 was 1607 mm (range: 1419–1856 mm) with a distinct dry season in June,

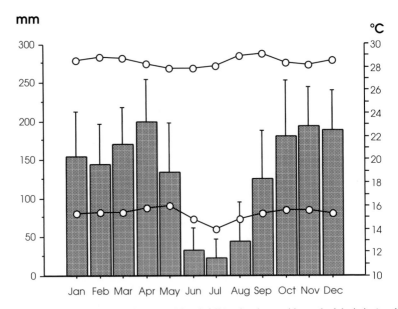

Fig. 13.2. Cumulative monthly rainfall (each column with standard deviation) and temperature (line graphs showing maximum and minimum) in the highland sector of Kahuzi-Biega National Park (1982–97).

July, and August, during which the mean rainfall was below 50 mm (Fig. 13.2). The mean monthly temperature was 19.7 °C and had been constant for 16 years.

### Ape habituation and density estimates

From counting the number of nests in fresh nest sites (up to two days old), we estimated 0.27–0.32 gorillas/km$^2$ and 0.27–0.33 chimpanzees/km$^2$ in 308 km$^2$ in the Itebero region, and 0.43–0.47 gorillas/km$^2$ and 0.13 chimpanzees/km$^2$ in 600 km$^2$ in the Kahuzi region (Yamagiwa *et al.*, 1989, 1992*a*, 1993*b*). Another survey was made in the lowland sector using the belt transect method (Hall *et al.*, 1998), which estimated 0.81–1.78 gorillas/km$^2$ and 0.26–0.58 chimpanzees/km$^2$, higher than our estimates.

From 1987 to 1991, we tried to habituate gorillas and chimpanzees in the Itebero region, but were unsuccessful. In 1991, we followed four groups of gorillas and a single unit-group (Ka group) of chimpanzees in about a 30 km$^2$ area of the Kahuzi region. These groups had extensive overlapping ranging areas (Yamagiwa *et al.*, 1996*a,b*). Since 1994, a group of gorillas (Ga group) and a unit-group of chimpanzees (Ka group) have been semi-habituated and occasionally tolerate the presence of human observers when we stay at a distance of 20–50 m. We have continuously collected data on ranging and diet of the Ga and the Ka groups by direct observations, evaluation of feeding remains along fresh trails, and fecal analysis.

### Ape habitat use and diet

Gorillas of the Ga group and chimpanzees of the Ka group have been followed daily and their travel routes recorded on a 1:25 000 scale map with 250 × 250 m grids. A square was considered to have been visited by an animal if the animal entered by at least 150 m. The location of their night nests was also recorded on the map along with type of vegetation, topography, and altitude. The day journey length for the Ga group was measured by following fresh (up to two days old) trails with a pedometer between consecutive nest sites.

The composition of diet for gorillas and chimpanzees was estimated from direct observations, evaluation of feeding remains along fresh trails, and fecal analysis. Fresh (up to one day old) feces were collected mainly in nest sites, washed in 1-mm mesh sieves, dried in sunlight, and stored in plastic bags. The contents of each sample were examined macroscopically and listed as seeds, fruit skins, fiber, leaves, fragments of insects, and other matter. Large seeds were counted and small seeds were rated on a four-point scale of abundance (abundant, common, few, rare). Fruit seeds and skin were identified at species level macroscopically. Fruit parts (seeds, skin, fiber) can be readily distinguished from non-fruit remnants and their volume percentage for each fecal sample was recorded at a 5% level of abundance scale. The mean proportion of fruit remains was calculated from the abundance scale of fruit parts in each fecal sample collected in a month.

### Results

#### Phenology of fruits and diet of gorillas

From August 1994 to July 1996, species of trees bearing ripe fruits were recorded monthly within the transect including 2033 individual trees from 50 species. Among them, the number of fruit food species eaten by gorillas fluctuated between eight and 13 monthly (Fig. 13.3). We collected a total of 5317 fresh fecal samples of gorillas in this period (monthly mean 221.5, range 68–361). The total number of fruit species included in fecal samples per month fluctuated between one and ten. Using Pearson's product moment correlation, no significant correlation was found between the number of fruit food species per month in the transect and that in the fecal samples ($r^2 = 0.01$, $p > 0.05$). The mean number of fruit species per fecal sample also fluctuated monthly and tended to increase in the dry season. The mean number of fruit species

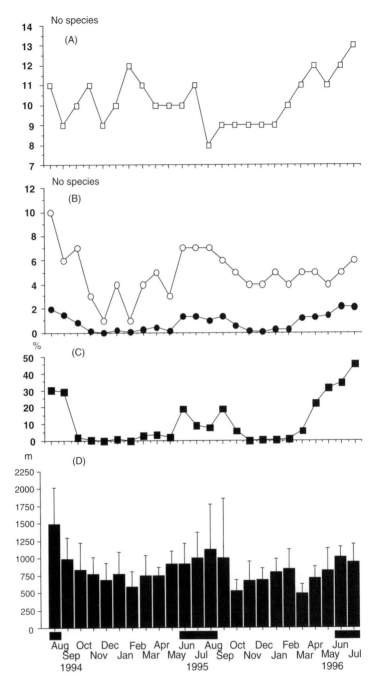

Fig. 13.3. (A) Monthly changes in total number of fruit food species in the transect; (B) total number of fruit species (O) and mean number of fruit species (●) in fecal samples of gorillas; (C) mean % of fruit remains in fecal samples of gorillas; and (D) mean day journey length in the Kahuzi-Biega National Park. Black horizontal bar indicates the dry season.

was significantly positively correlated with both the total number of fruit food species per month in the transect ($r^2 = 0.175$, $p < 0.05$) and the total number of fruit species in the fecal samples ($r^2 = 0.56$, $p < 0.001$). These observations indicate that gorillas did not usually eat more fruit species according to an increase in the number of fruit food species in their habitat, but rather tended to consume more kinds of fruit daily when the diversity of fruit was high.

A fruit index was calculated from 14 fruit food species consumed by gorillas. We also calculated a common fruit index from six tree species (*Myrianthus holstii*, *Bridelia bridelifolia*, *Ficus thonningii*, *Ficus oreadryadum*, *Syzygium parvifolium*, *Allophyllus* sp.) that were defined as species whose seeds were found in more than 1% of the total fecal samples. These consist of species eaten by gorillas in a small amount during the whole year (*Myrianthus holstii*, *Ficus* spp.) and in a large amount during a short period (*Bridelia bridelifolia*, *Syzygium parvifolium*, *Allophyllus* sp.). The fruit index was always higher in the primary forest than in the secondary forest and tended to be high from June to August (dry season) and low from January to May (rainy season) (Fig. 13.4). A marked decrease was observed in March and April 1995 in the primary forest. The common fruit index fluctuated in relation to the fruit index in the secondary forest, while it constantly kept to a lower level than the fruit index in the primary forest. In two months (August 1994 and July 1996) during the dry season, the common fruit index was higher in the secondary forest than in the primary forest. The abundance of common fruits was not greatly different between the two types of vegetation.

The degree of fruit consumption by gorillas in each month was estimated from the mean proportion of fruit remains in the fecal samples. It tended to be high during or around the dry season but kept to a very low level between November and February (rainy season) (Fig. 13.3C). However, the fruit index indicates that fruit abundance was not the lowest during these months (Fig. 13.4). In addition, the monthly change in the mean proportion of fruit remains in the fecal samples is not significantly correlated with the fruit index or the common fruit index in either the secondary forest or primary forest, respectively ($r^2 = 0.056, 0.066, 0.021, 0.012$; $p > 0.05$). These data suggest gorillas did not change the degree of fruit consumption simply in response to the fluctuation in abundance of fruit in their habitat

### *Group size and fruit abundance*

In the Itebero region, a gorilla population census was conducted from July to October 1987, and ten groups and six solitary males were confirmed from the analysis of fresh nest sites (Yamagiwa *et al.*, 1989). The mean nest-group size (the mean number of weaned individuals per group) was 4.8 ($n = 10$,

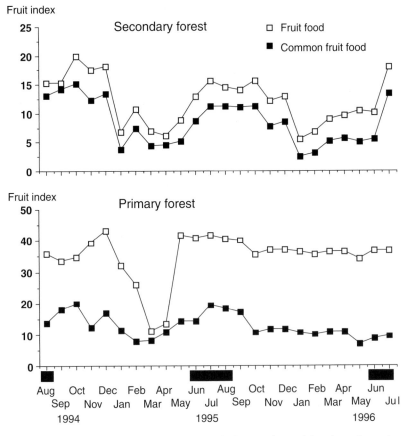

Fig. 13.4. Monthly changes in fruit index in the secondary and the primary forests in the Kahuzi-Biega National Park. □, value calculated from all fruit foods of gorillas; ■, value calculated from six common fruit foods. Black horizontal bar indicates the dry season.

range 2–17). In the Kahuzi region, a census was made from September to November 1990, and 25 groups and nine solitary males were found (Yamagiwa *et al.*, 1993*b*). The mean nest-group size was 5.7 ($n = 25$, range 2–21). No significant difference ($p > 0.05$) was found in mean group size between the two regions.

We tried to follow the selected groups continuously to habituate them in the Itebero region. The age and sex composition of gorilla groups was estimated from fresh nest sites. The size and estimated composition of these nest-groups did not change during at least four consecutive nights. From September 1989 to February 1990 and from April to July 1991, we followed four groups sporadically in the Itebero region. These periods cover both the major fruiting season

(September–February) and the minor fruiting season (April–July), which were defined by fecal analysis (Yamagiwa *et al.*, 1994). We did not obtain any evidence of subgrouping from nest counts in either season.

In the Kahuzi region, age and sex composition of groups was also stable for several consecutive days during the census in 1990. However, the composition sometimes fluctuated. From 1994 to 1996, we followed the Ga group on a daily basis and collected data on 285 night nest sites. The Ga group consisted of a silverback, a maturing silverback, eight to ten adults, five to six independent immatures and three to five dependent immatures. They usually built 11 to 15 nests, as the majority of immatures did not build nests. In September 1994, we counted 20 and 24 nests on two consecutive days, when the Ga group built nests with another group. In November 1995, the Ga group encountered a neighboring group and 24 nests were counted in one night. From June to August 1996, the Ga group frequently dispersed into more than two groups and the maximum seven to nine nests were counted at a nest site on 16 out of 28 days. This subgrouping behavior appeared to be linked to a maturing silverback and a few females, who often slept apart from the main group's nest sites, and can best be regarded as a process of group fission derived from the mating strategy of the maturing male. However, it should be noted that subgrouping occurred during the fruiting season (the dry season: June–August) in Kahuzi. By contrast, subgrouping rarely occurred during the non-fruiting season (the rainy season: September–May), and two groups occasionally slept together in the same nest sites during the bamboo season (September–December), when bamboo shoots were available. Clumped fruit distribution may stimulate group fission when a group includes two or more males.

The differences in the stability of group composition between the Itebero and the Kahuzi regions can be attributed to the number of silverback males within groups. In the Itebero region, all ten groups included only one silverback male. The one-male group composition may not allow subgrouping. When subgrouping occurred in Kahuzi, each subgroup always included a silverback male. The increasing competition between two silverback males over mating partners may be the primary cause of subgrouping. Feeding competition among females for fruit also limits feeding group size and may promote subgrouping during the fruiting season in the group with two or more males.

### Day range

From August 1994 to July 1996, we measured the complete day journey length of the study group (Ga group) consisting of 18–23 gorillas between consecutive night nest sites for 225 days (monthly mean 9.4 days, range 5–19). The mean length of day journey was 850.8 m (range 239–3570 m), and tended to increase during the dry season (Fig. 13.3D). No significant correlation was found between the mean length of day journey and the number of different tree species fruiting ($r^2 = 0.001$, $p > 0.05$). However, mean day journey length did show a significant positive correlation with the total number of fruit species per month in the fecal samples ($r^2 = 0.495$, $p < 0.0001$), the mean number of fruit species in fecal samples per month ($r^2 = 0.357$, $p < 0.01$) and the mean proportion of fruit remains in fecal samples per month ($r^2 = 0.286$, $p < 0.01$). The diversity of fruit in their habitat may not directly affect their day range. However, gorillas tended to increase day journey length with an increase in abundance and diversity of fruit available for consumption. These observations suggest that gorillas did not eat fruits opportunistically but purposefully sought fruiting trees when they relied on fruit. Actually, the Ga group used a wider range than usual to shift their nesting site daily to different primary forest patches where more fruits were available during the dry (fruiting) season in 1994 (Yamagiwa *et al.*, 1996*b*).

In the Itebero region, we only measured two complete day journey lengths (2155 m and 2735 m) for different groups, and eight complete day journey lengths (mean 1531 m, range 142–3439 m) for three solitary males (Yamagiwa *et al.*, 1992*b*; Yamagiwa and Mwanza, 1994). These figures are greater than the mean length measured in montane forests of Kahuzi (this study) and Virungas ($\bar{x} = 472$ m for seven groups: Schaller, 1963; 378 m for a solitary male: Yamagiwa and Mwanza, 1994) and are close to those calculated in lowland tropical forests of Lopé ($\bar{x} = 1238$ m for a single group: Tutin *et al.*, 1992) and Bai Hokou ($\bar{x} = 2590$ m for several groups: Goldsmith, 1996). Solitary males tended to increase day journey length during the fruiting season by increasing the mean distance between consecutive feeding sites in the Itebero region (Yamagiwa and Mwanza, 1994). A frugivorous diet may stimulate gorillas to search for fruits in a wide area and may consequently enlarge their day range.

### Intergroup relationships

In both lowland (Itebero) and highland (Kahuzi) habitats, ranging areas of neighboring gorilla groups overlapped extensively. Furthermore, groups frequently entered the auditory range of other groups but tended to avoid direct contact with each other. We have rarely seen gorillas in the Itebero region, but occasionally have heard chest-beating from different directions at night. We usually found two or three nest sites of different groups the next morning in such cases. These nest sites were 100–500 m apart from each other, and their fresh trails came from different directions, indicating that two or more groups had encountered each other at least within auditory ranges on the previous day. Evidence of physical contact between groups was not found along their fresh trails. However, feeding remains and feces suggested that more than one group used the same fruiting trees on the same day on several occasions.

Range overlap is also extensive among gorilla groups in the Kahuzi region. Four groups of gorillas including the Ga group usually ranged within an area of about 30 km$^2$ of the study area (Yamagiwa *et al.*, 1996*a*). Another two to four groups were occasionally seen within the study area from August 1994 to July 1996. When other groups were in the vicinity of the Ga group, silverbacks of both groups exchanged chest-beating displays, as observed for mountain gorillas in the Virungas (Fossey, 1983). During five encounters that we observed directly, silverbacks made explosive displays, and in two cases they came into aggressive contact with each other. Bloody hairs scattered in the encounter sites indicated severe fights between these males. By contrast, no females were found wounded during such encounters.

Among 27 possible encounters between the study groups and the other groups from August 1994 to July 1996, 16 of these encounters occurred during the dry seasons (June–August). The high diversity of fruit foods during the dry season may stimulate gorillas to range widely and to gather around particular fruiting trees, and thus increase the opportunity for encounters.

### Relationships with sympatric chimpanzees

Although overlap in diet and ranging between gorillas and chimpanzees was extensive, the relationships between them were more peaceful than between groups of the same species in both the Itebero and the Kahuzi regions. In both regions, fruit species showed the highest ratio of overlap between gorillas and chimpanzees; more than 80% of fruit species eaten by gorillas were also eaten by chimpanzees (Yamagiwa *et al.*, 1992*b*, 1996*a*).

In the Itebero region, gorillas and chimpanzees ranged in the same area, but used various types of vegetation differently for feeding and resting. Their fresh trails and nest sites indicated that both gorillas and chimpanzees visited primary forest most frequently, probably because the diversity and abundance of fruits were highest in the primary forest. Chimpanzees prominently increased visits in the primary forest during the rainy season, while gorillas continuously used the secondary forests to feed on vegetative foods, such as Zingiberaceae and Marantaceae (e.g., THV) (Mwanza *et al.*, 1992; Yamagiwa *et al.*, 1992*b*). Gorillas also used swamps where such THV is densely distributed in both the rainy and dry seasons, while no trails or nests of chimpanzees were found in swamps.

In the Kahuzi region, a unit-group (Ka group) consisting of 22 to 24 chimpanzees was always found in the study area. We recorded their fresh trails for 342 days from August 1994 to July 1996 (monthly mean 14.3 days, range 8–27). The total number of grids in which chimpanzees entered by more than 150 m was 144 (9 km$^2$). The greatest part (73%) of their home range overlapped with that of the gorilla study group (Ga group), which was estimated to be 22.9 km$^2$ during the same period.

In the Kahuzi region, chimpanzees occasionally visited the *Cyperus* swamps, but they passed through the swamps without eating foods, and no chimpanzee nests were found in them. Chimpanzees frequently visited primary forest patches, which were scattered fragmentarily in the secondary forest. Auditory and visual encounters between gorillas and chimpanzees were occasionally observed in the primary forest. Most such encounters occurred around the fruiting trees in which gorillas or chimpanzees were feeding on fruits. No agonistic interactions were observed during these encounters, and each species seemed to avoid contact with the other (Yamagiwa *et al.*, 1996*a*).

The home range of the chimpanzee unit-group (Ka group) did not overlap with those of neighboring unit-groups of chimpanzees. The nearest unit-group usually ranged more than 10 km away from the Ka group's home range. There was a blank area of chimpanzees between the home ranges of the two neighboring groups during the study period. This suggests that chimpanzees continuously range in small areas and avoid entering the neighboring group's home range. The strong hostility between unit-groups may possibly prevent them from extending their range.

By contrast, chimpanzees did not seem actively to avoid encounters with gorillas. Different foraging strategies may prevent agonistic encounters. During fruiting season, chimpanzees persistently used particular fruiting trees for days. Gorillas increased day range and visited several small patches of primary vegetation daily to avoid reuse of previous ranging areas (Yamagiwa *et al.*, 1996*b*). During fruit scarcity, chimpanzees tended to disperse to seek fruits in a wide area. Gorillas tended to stay in a small area to feed mainly on vegetative foods. Such differences may reduce contest competition and support their coexistence sympatrically.

## Discussion

### *Group size and frugivory in gorillas*

Earlier socioecological theories predict that predation risk and food competition limit primate group size (van Schaik *et al.*, 1983; Stacey, 1986; Terborgh and Janson, 1986; Dunbar, 1988; Janson and van Schaik, 1988). Living in larger groups reduces predation risk, while increasing group size raises within-group competition and results in reduction of individual foraging efficiency. However, in the presence of predators, some folivorous primates live in small groups even though the costs of feeding competition are low or absent (Isbell, 1991; Janson and Goldsmith, 1995). The upper limit to group size in these folivorous species is not decided by ecological factors, but rather by social factors such as male harassment or infanticide (Treves and Chapman, 1996; Crockett and Janson, 2000; Steenbeek and van Schaik, 2001).

Dietary variation in gorillas does not seem to affect gorilla group size. There are striking differences in diet between gorillas of highland and lowland habitats. Mountain gorillas are regarded as terrestrial folivores (Schaller, 1963; Fossey and Harcourt, 1977; Watts, 1984), while western lowland gorillas consume various kinds of fruits and regularly feed on insects (Sabater Pi, 1977; Tutin and Fernandez, 1992, 1993; Kuroda *et al.*, 1996; Remis, 1997*a*). The diversity of fruits consumed by gorillas occasionally exceeds that of sympatric chimpanzees (Tutin and Fernandez, 1993). These findings cannot be attributed to inherent differences between subspecies. Eastern lowland gorillas inhabit both highland and lowland tropical forests and show distinct dietary differences related to altitude and seasonal changes (Yamagiwa *et al.*, 1994, 1996*a*). Based on these observations, socioecological theory predicts that gorilla group size in lowland habitats will be smaller than in highland habitats. In fact, Harcourt *et al.* (1981) compared group size and composition of gorillas between East Africa (highlands: Virunga, Kahuzi, Bwindi, and Tshiaberium) and West Africa (lowlands: Alen, Abuminzok) and found that group size in West Africa was significantly smaller than in East Africa. Harcourt *et al.* (1981) suggested that the difference in group size may be related to dietary variation.

However, recent studies of western lowland gorillas reported similar group sizes to those of gorillas in the highland habitats of East Africa (Remis, 1994; Goldsmith, 1996; Tutin, 1996). Our data also show similarities in group size in comparisons between highland and lowland gorillas of the same subspecies. Olejniczak (1996) calculated the average group size to be 5.5 weaned gorillas from direct observation of 13 groups in an open swamp at Mbeli Bai, Congo. This is close to the average nest-group size (number of weaned individuals) of eastern lowland gorillas in the Itebero (4.8) and the Kahuzi (5.7) regions obtained in our study. Tutin (1996) calculated the median group size to be 10 from eight groups (range 4–16) during a 6.5-year survey at Lopé, Gabon. Goldsmith (1999) found six groups ranging from three to 18 gorillas during her 17-month survey at Bai Hokou, Central African Republic. These results are not consistent with the previous socioecological theory. Janson and Goldsmith (1995) also found a weak relationship between group size and feeding competition for folivorous primate species and suggested this might be explained by the energetic constraints of a leafy diet or by limits to group size imposed by infanticide as a habitual male reproductive strategy.

Both ecological and social factors may influence group size in gorillas. Watts (1990) found that spatial displacement rates for individuals associated with feeding were positively correlated with group size in Virunga mountain gorillas. Such foraging costs fall particularly on females, since the number of females per group is negatively correlated with birth rate (Watts, 1991). However, immigration and emigration rates were found to be independent of group size, and some females transferred from small to very large groups. Such female dispersal patterns may not be promoted by ecological factors but by social factors, such as infanticide (Fossey, 1984; Stewart and Harcourt, 1987; Watts, 1989). Female group size was positively correlated with the number of males per group (Robbins, 1995). Female mountain gorillas may transfer between groups to seek male protection against infanticide, as observed in Thomas's langurs (Watts, 1989, 1996, 2000*a*; Steenbeek, 1996; Sterck, 1997).

Foraging costs, however, may limit maximum group size and this, along with the number of adult males within a group, may be the main difference between highland and lowland populations (Goldsmith, 1996; Doran and McNeilage, 1998, 2001). Maximum group size in highland habitats ($n = 30$ in the Virungas: Schaller, 1963; $n = 37$ in Kahuzi: Murnyak, 1981; $n = 42$ in Kahuzi: Yamagiwa, 1983; $n > 40$ in the Virungas: L. Williamson, personal communication) was much larger than that in the lowland habitats ($n = 12$ in Equatorial Guinea: Jones and Sabater Pi, 1971; $n = 16$ at Lopé: Tutin, 1996; $n = 17$ in Itebero: this study; $n = 18$ at Bai Hokou: Goldsmith, 1999). In an exception, a large group consisting of 32 gorillas was found in the lowland habitat at Lossi, Congo, where high densities of both herbs and gorillas are reported (Bermejo, 1997).

Multi-male groups are found more frequently in mountain gorillas than in western and eastern lowland gorillas. In mountain gorillas of the Virungas and Bwindi, about 40% of the groups have been reported to contain two or more adult males (Robbins, 1995, 2001; McNeilage *et al.*, 1998). This contrasts with the eastern lowland gorillas living in the montane forest of Kahuzi. In 1978, only two of 14 Kahuzi groups (14%) were found to include two silverbacks (Murnyak, 1981), and in 1990, only two of 25 Kahuzi groups (8%) were observed to contain two silverbacks (Yamagiwa *et al.*, 1993*b*). We also found that in the lowland forest of Itebero, all ten groups observed contained only one adult male each. In western lowland gorillas, multi-male groups constituted only 8% ($n = 13$) at Mbeli Bai (Olejniczak, 1996) and 0% ($n = 31$) at Maya Maya (Magliocca *et al.*, 1999) in the groups observed. These findings suggest that the difference in the proportion of multi-male groups in the population may not be linked to differences in habitat, but rather may reflect unique group formation specific to mountain gorillas.

Formation of multi-male groups may possibly reflect female choice at transfer. If females seek stronger protection by males against infanticide, females are expected to transfer to multi-male groups where a coalition among kin-related males (mostly between a father and his maturing sons) assures them more effective vigilance and protection during intergroup encounters (Sicotte, 1993).

Males that remain in their natal group after maturity appear to start reproduction at an earlier age than males that disperse to form new groups (Robbins, 1995; Watts, 2000*a*). These benefits for both females and males may promote large multi-male groups in the Virunga mountain gorillas.

However, this is not the case for eastern and western lowland gorillas. Large groups do not always contain more than one male in Kahuzi, where infanticide has not been reported (Murnyak, 1981; Yamagiwa *et al.*, 1993*b*). The absence of infanticide in Kahuzi may stimulate group fission by maturing males emigrating with young females and may prevent gorillas from forming large multi-male groups (Yamagiwa and Kahekwa, 2001). Our study suggests that group fission may occur during the dry season, when gorillas exhibit frugivorous characteristics. A folivorous diet may facilitate the formation of large gorilla groups, but a frugivorous diet may promote group fission and may hinder the formation of multi-male groups in the absence of infanticide in Kahuzi.

Western lowland gorillas, when highly frugivorous, form temporary subgroups and sleep separately from one another (Goldsmith, 1996; Remis, 1997*b*). No infanticide has been reported in their habitats. Females occasionally keep several hundred meters away from a male at different sites (Doran and McNeilage, 2001). Two groups with a total of 40 individuals reportedly nested together, forming a supergroup near a preferred fruit tree (*Dialium* sp.) at Lossi, Congo (Bermejo, 1997). Frugivory in gorillas may set an upper limit to group size, and the clumped distribution of fruits may increase group fluidity by increasing within-group scramble competition or by attracting several groups to a large fruiting tree forming a supergroup in the lowland habitats. These socioecological factors (larger group spread and less hostility in intergroup encounters) might reduce the likelihood of infanticide (Doran and McNeilage, 2001).

### Within-group competition and foraging strategy of gorillas

Dietary differences in gorillas between habitats do not strongly influence group size but may influence their ranging behavior. Daily travel distance is frequently used as a useful estimate of foraging efforts by primates in their natural habitats (Terborgh, 1983; Chapman, 1990; Janson and Goldsmith, 1995). Recent studies of western and eastern lowland gorillas show a longer day journey length in lowland gorillas as compared to gorillas of highland habitats (Yamagiwa *et al.*, 1992*b*; Yamagiwa and Mwanza, 1994; Tutin, 1996; Remis, 1997*b*; Goldsmith, 1999). This difference is possibly caused by the availability of fruit foods. Gorillas tend to travel further when fruit is abundant than when it is scarce (Yamagiwa and Mwanza, 1994; Tutin, 1996; Goldsmith, 1999; Doran and McNeilage, 2001).

Day journey length is possibly influenced by fruit density and its spatial patterns (Goldsmith, 1999), but it is difficult to identify direct relationships between these factors. Our study found no significant correlation between monthly mean day journey length and the number of food species fruiting each month in the montane forest of Kahuzi (Fig. 13.3). The monthly number of fruit food species in the gorillas' habitat shows no correlation with the monthly total number of fruit species in their fecal samples. These observations suggest that gorillas may not respond to increasing fruit diversity in their habitat simply by increasing consumption. Our study also showed that gorillas possibly changed the degree of fruit consumption in response to the fluctuations in abundance of fruits in secondary forest, but not in primary forest. These gorillas tended to increase their day journey length while increasing the abundance and diversity of fruit they consumed. This suggests that gorillas are selective fruit-feeders, consistent with the findings of Remis (this volume), and may not respond to the abundance of fruits simply by extending or contracting daily travel. At Bai Hokou, day journey length of frugivorous western lowland gorillas was not correlated with fruit patch size in their habitat (Goldsmith, 1999). Within-group scramble competition may not be a primary cause of long day journey length.

Intergroup relationships among gorillas may reflect weak feeding competition and strong competition between males over mates. Home range overlap between neighboring gorilla groups has been observed in all habitats of gorillas (Schaller, 1963; Tutin *et al.*, 1992; Tutin, 1996; Olejniczak, 1996, 1997; Yamagiwa *et al.*, 1996*a*; Watts, 1998*a*; Magliocca *et al.*, 1999). Gorilla foraging strategy is possibly characterized by weak between-group contest competition over foods and limited site fidelity irrespective of dietary differences. This may reduce ecological costs of female transfer, as predicted for mountain gorillas and other primates in which female transfer occurs (Isbell and Van Vuren, 1996; Watts, 1996; Sterck *et al.*, 1997). However, the lack of territoriality does not mean peaceful intergroup relationships. Our study showed the occasional occurrence of agonistic encounters between groups in both highland and lowland habitats. Such encounters tended to occur around particular fruiting trees during the fruiting season in the Kahuzi region. The limited distribution of fruiting trees possibly increased the opportunity for intergroup encounters, which were usually caused by exchanges of adult male explosive displays and appeared to be strongly motivated by mating strategies. Mountain gorillas occasionally shift their range after such agonistic encounters with another group (Fossey, 1974; Watts, 2000*b*). A shift in range has also been observed in western and eastern lowland gorillas (Remis, 1994; Tutin, 1996; Yamagiwa *et al.*, 1996*b*). Although peaceful intermingling of different groups is expected to be more common in western lowland as compared to mountain gorillas (Magliocca and Querouil,

1997; Doran and McNeilage, 2001), agonistic intergroup encounters around fruit trees are also common in western lowland gorillas (Tutin, 1996). Finally, ecological factors such as clumped distribution of fruit and sparse distribution of THV may increase day journey length. Enlarged day range may provide more opportunity for interunit encounters, and social factors related to male mating strategies may cause agonistic interactions between groups and further extend their day journey length in the lowland habitats.

### Sympatry with chimpanzees

Chimpanzees form social groups based on female transfer as do gorillas, but they respond in different ways to seasonal changes in diversity and abundance of fruit. Although day journey length of a group is difficult to estimate because of their fission–fusion grouping patterns, individual day journey length for both sexes in chimpanzees may be longer than those of gorilla groups in lowland habitats (Wrangham, 1975). This difference is attributed to the greater degree of frugivory in chimpanzees as compared to gorillas (Goldsmith, 1999; Yamagiwa, 1999). Party size in chimpanzees decreased during fruit-poor seasons in Kibale, Uganda (Wrangham *et al.*, 1992, 1996; Chapman *et al.*, 1995). In Kahuzi, gorillas increased day journey length to visit several patches of primary forest where fruit diversity and abundance were high during the fruiting season. By contrast, chimpanzees stayed in a small area to visit particular fruiting trees continuously for days during the same period (Yamagiwa *et al.*, 1996*b*). Frugivorous primates tend to repeatedly use individual fruiting trees to harvest newly ripened fruits (MacKinnon, 1974; Terborgh, 1983; Nunes, 1995). Chimpanzees apparently adopt this frugivorous foraging strategy.

Differences in foraging strategies may support the sympatry of gorillas and chimpanzees. Gorillas visit fruiting trees in cohesive groups for a short time and leave them even when ripe fruits are still available. They revisit previous ranging areas after a long interval, probably because they avoid trampled areas (Watts, 1998*b*). Chimpanzees change party size and degree of gregariousness according to fruit abundance and repeatedly use particular fruiting trees until all of the ripe fruit disappears (Chapman and Wrangham, 1993; Yamagiwa *et al.*, 1996*a*). Although the clumped distribution of fruits may increase the opportunity for gorillas and chimpanzees to encounter each other in the same fruiting trees during the fruiting season, they need not face each other for a prolonged period, perhaps due to the differences in their foraging behaviors. Some observations of their encounters at fruiting trees reported calm co-feeding in Ndoki, Congo (Suzuki and Nishihara, 1992) and mutual avoidance without aggressive responses in Kahuzi (Yamagiwa *et al.*, 1996*a*). These observations suggest that relationships between gorillas and chimpanzees are more peaceful than intergroup relationships within both species.

The high degree of dietary overlap suggests ecological competition between gorillas and chimpanzees may influence their foraging strategies. When fruit is scarce, the diets of the two ape species show the greatest divergence in lowland (Tutin and Fernandez, 1993; Kuroda *et al.*, 1996) and highland habitats (Yamagiwa *et al.*, 1996*b*). Gorillas increase the proportion of vegetative foods in their diets while decreasing day and monthly range, whereas chimpanzees persistently seek fruits, expand their range and decrease party size. These different responses to the phenology of fruit foods may have developed in the past dry and cold weather during the glacial epoch. As the area of tropical forest decreased along with a decline in fruit availability, gorillas and chimpanzees might have been forced to develop foraging strategies to reduce within-group feeding competition under sympatric conditions in the refuge forests. Despite morphological and ecological similarities, gorillas and chimpanzees differ in some aspects of masticatory form that may be linked to increasing folivory, particularly in the Virunga mountain gorilla (Taylor, 2002, this volume). The presence of multiple species of ape may have promoted niche divergence between them. The more severe seasonal fluctuation in fruit production during the glacial epoch may have resulted in differences in foraging strategies between chimpanzees and gorillas with some specialization of the masticatory complex possibly related to extreme folivory in the mountain gorilla, while both species still maintained similar diets in the tropical forest.

Different linkage of socioecological factors between gorillas and chimpanzees may reflect their density and distribution in equatorial Africa (Yamagiwa, 1999). The population density and home range sizes of gorillas are relatively constant across various forest habitats, while the density of chimpanzees is highly flexible according to habitat type (lowland forest, medium-altitude forest, high montane forest, woodland and savanna). Chimpanzees can survive in dry savannas at very low densities with an extremely large home range (Kano, 1972; Baldwin *et al.*, 1982). Chimpanzees may alter home range size according to food availability, and population densities may not affect female dispersal and reproductive success because female chimpanzees can choose to join various parties within a unit-group and transfer into neighboring groups at any time of their cycling (Nishida, 1979; Goodall, 1986). By contrast, female gorillas usually associate with one or more adult males to form a cohesive group and transfer into other groups or to solitary males only at interunit encounters (Harcourt, 1978; Yamagiwa, 1987). Low population density or short day range of gorillas will decrease rates of interunit encounters, thereby decreasing the chance of female transfer, and consequently preventing female dispersal. On the other hand, high population density or long day range will increase interunit encounters, stimulate female transfers, and may increase rates of formation of new groups, which usually interact aggressively with other groups. Escalated intermale competition over mates occasionally resulted in fierce fights between silverbacks or infanticide, as observed in the Virunga gorilla population (Fossey, 1983; Watts, 1984).

## Conclusions

We examined how ecological and social factors shape the social organization of gorillas and result in intraspecific variation. Sympatry with chimpanzees may have promoted a folivorous foraging strategy in gorillas. When high-quality foods such as fruit decline, gorillas minimize the expenditure of energy at expense of food quality (Shoener, 1971; Clutton-Brock, 1977). In this strategy, within-group or between-group competition over food has a weak effect on group size. Instead, day journey length may change according to degree of frugivory or folivory which gorillas demonstrate seasonally or locally. The density of gorillas and day journey length may greatly influence the rate of intergroup encounters, thereby impacting on the decision of females to transfer. Since female transfer occurs only at encounters with other groups or with solitary males, mating strategies of both males and females also affect the rate of encounters and consequently their foraging patterns. As a first response to temporal decline of food quality, gorillas reduce day range and decrease the rate of encounters. However, if both males and females seek contact with other groups, they gradually increase in density within their habitats. When groups become too crowded within such populations, they disperse to reduce contact. The flexibility in diet and the lack of territoriality enable gorillas to regulate such modifications, and this may result in the similar densities of gorillas across habitats.

Infanticide also has a significant influence on female movements in primate species where female dispersal is common, including the red colobus monkey (Struhsaker and Leland, 1985), Thomas's langur (Sterck, 1997), and the mountain gorilla (Watts, 1989). Females have been observed to disperse soon after their infants are killed by extra-group males during intergroup encounters. After the death of the resident male, females with infants avoided strange males that were potentially infanticidal (Watts, 1989; Steenbeek, 1996). Thus, avoidance of infanticide may prevent females with infants from transferring into other groups or groups with solitary males, and may stimulate females to seek stronger protection from males. This female strategy may not set the upper limit to group size but may promote multi-male group formation in which kin-related males form a coalition against extra-group males in Virungas. The absence of infanticide may facilitate group fission by maturing males with young females, as observed in Kahuzi (Yamagiwa and Kahekwa, 2001), and consequently contribute to the high proportion of one-male groups within populations of eastern and western lowland gorillas. Frugivory in gorillas of the lowland habitats may set the upper limits to their group size, and prevent males from remaining in their natal group after maturity due to the strong competition among males over mates within a small group. Frugivory also provides female gorillas with a wide range of choice in the absence of infanticide and may result in greater group fluidity as observed in western lowland gorillas (Doran and McNeilage, 2001).

Predation may also set the lower limits to group size and prevent female gorillas from traveling alone. After a leading silverback died, female and immature gorillas prominently increased arboreal nests in Kahuzi, probably because of their vulnerability to large predators (Yamagiwa, 2001). Leopards are known to attack gorillas in both lowland and montane forests (Schaller, 1963; Tutin *et al.*, 1992; Fay *et al.*, 1995), and a fragment of gorilla bone was found in the feces of leopards at Ndoki (T. Nishihara, personal communication). The absence of infanticide may increase the distance between female gorillas and their protector male to allow frequent subgrouping, but predation risk may prevent them from dispersing widely without males in lowland habitats.

Folivorous foraging strategies of gorillas have developed under sympatric conditions with frugivorous chimpanzees. However, gorillas occasionally exhibit frugivory, and distribution and availability of fruit may set the upper limit of their group size. Predation risk may increase female's gregariousness around a protector male while the probability of infanticide may enhance the reliance of females on protection from males. This may facilitate female transfer into a group including two or more males. Intraspecific variation in group size and composition may reflect the combination of differences in ecological and social factors in each habitat.

### References

Baldwin, P.J., McGrew, W.C., and Tutin, C.E.G. (1982). Wide-ranging chimpanzees at Mt. Assirik, Senegal. *International Journal of Primatology*, **3**, 367–385.

Bermejo, M. (1997). Study of western lowland gorillas in the Lossi Forest of North Congo and a pilot gorilla tourism plan. *Gorilla Conservation News*, **11**, 6–7.

Barton, R.A., Byrne, R.W., and Whiten, A. (1996). Ecology, feeding competition and social structure in baboons. *Behavioral Ecology and Sociobiology*, **38**, 321–329.

Boesch, C. (1991). The effects of leopard predation on grouping patterns in forest chimpanzees. *Behaviour*, **117**, 220–242.

Casimir, M.J. (1975). Feeding ecology and nutrition of an eastern gorilla group in the Mt. Kahuzi region (République du Zaïre). *Folia Primatologica*, **24**, 1–36.

Chapman, C.A. (1990). Ecological constraints on group size in three species of neotropical primates. *Folia Primatologica*, **55**, 1–9.

Chapman, C.A. and Wrangham, R.W. (1993). Range use of the forest chimpanzees of Kibale: Implications for the understanding of chimpanzee social organization. *American Journal of Primatology*, **31**, 263–273.

Chapman, C.A., Wrangham, R.W., and Chapman, L.J. (1995). Ecological constraints on group size: An analysis of spider monkey and chimpanzee subgroups. *Behavioral Ecology and Sociobiology*, **36**, 59–70.

Clutton-Brock, T.H. (1977). Some aspects of intraspecific variation in feeding and ranging behaviour in primates. In *Primate Ecology*, ed. T.H. Clutton-Brock, pp. 539–556. London: Academic Press.

Crockett, C.M. and Janson, C.H. (2000). Infanticide in red howlers: Female group size, male membership, and a possible link to folivory. In *Infanticide by Males*, eds. C.P. van Schaik and C.H. Janson, pp. 75–98. Cambridge, U.K.: Cambridge University Press.

Doran, D. and McNeilage, A. (1998). Gorilla ecology and behavior. *Evolutionary Anthropology*, **6**, 120–131.

Doran, D. and McNeilage, A. (2001). Subspecific variation in gorilla behavior: The influence of ecological and social factors. In *Mountain Gorillas: Three Decades of Research at Karisoke*, eds. M.M. Robbins, P. Sicotte, and K.J. Stewart, pp. 123–149. Cambridge, U.K.: Cambridge University Press.

Dunbar, R.I.M. (1988). *Primate Social Systems*. London: Croom Helm.

Fay, M.J., Carroll, R.W., Paterhans, C.K., and Harris, D. (1995). Leopard attack on and consumption of gorillas in the Central African Republic. *Journal of Human Evolution*, **29**, 93–99.

Fossey, D. (1974). Observations on the home range of one group of mountain gorillas (*Gorilla gorilla beringei*). *Animal Behaviour*, **22**, 568–581.

Fossey, D. (1983). *Gorillas in the Mist*. Boston, MA: Houghton Mifflin.

Fossey, D. (1984). Infanticide in mountain gorillas (*Gorilla gorilla beringei*) with comparative notes on chimpanzees. In *Infanticide: Comparative and Evolutionary Perspectives*, eds. G. Hausfater and S. Hrdy, pp. 217–236. Hawthorne, NY: Aldine Press.

Fossey, D. and Harcourt, A.H. (1977). Feeding ecology of free-ranging mountain gorillas (*Gorilla gorilla beringei*). In *Primate Ecology*, ed. T.H. Clutton-Brock, pp. 415–447. New York: Academic Press.

Goldsmith, M.L. (1996). Ecological influences on the ranging and grouping behavior of western lowland gorillas at Bai Hokou in the Central African Republic. PhD thesis, State University of New York, Stony Brook, NY.

Goldsmith, M.L. (1999). Ecological constraints on the foraging effort of western gorillas (*Gorilla gorilla gorilla*) at Bai Hokou, Central African Republic. *International Journal of Primatology*, **20**, 1–23.

Goodall, A.G. (1977). Feeding and ranging behavior of a mountain gorilla group (*Gorilla gorilla beringei*) in the Tshibinda-Kahuzi region (Zaïre). In *Primate Ecology*, ed. T.H. Clutton-Brock, pp. 450–479. New York: Academic Press.

Goodall, J. (1968). The behavior of free-living chimpanzees in the Gombe Stream Reserve. *Animal Behaviour Monographs*, **1**, 161–331.

Goodall, J. (1986). *Chimpanzees of Gombe: Patterns of Behavior*. Cambridge, MA: Harvard University Press.

Goodall, J., Bandora, A., Bergmann, E., Busse, C., Matama, H., Mpongo, E., Pierce, A., and Riss, D. (1979). Intercommunity interactions in the chimpanzee population of the Gombe National Park. In *The Great Apes*, eds. D.A. Hamburg and E.R. McCown, pp. 13–53. Menlo Park, CA: Benjamin/Cummings.

Hall, J.S., White, L.J., Inogwabini, B.I., Omari, I., Morland, H.S., Williamson, E.A., Saltonstall, K., Walsh, P., Sikubwabo, C., Bonny, D., Kiswele, K.P., Vedder, A., and Freeman, K. (1998). Survey of Grauer's gorillas (*Gorilla gorilla graueri*) and eastern chimpanzees (*Pan troglodytes schweinfurthii*) in the Kahuzi-Biega National Park lowland sector and adjacent forest in eastern Democratic Republic of Congo. *International Journal of Primatology*, **19**, 207–235.

Harcourt, A.H. (1978). Strategies of emigration and transfer by primates, with particular reference to gorillas. *Zeitschrift für Tierpsychologie*, **48**, 401–420.

Harcourt, A.H., Fossey, D., and Sabater Pi, J. (1981). Demography of *Gorilla gorilla*. *Journal of Zoology, London*, **195**, 215–233.

Isbell, L.A. (1991). Contest and scramble competition: Patterns of female aggression and ranging behaviour among primates. *Behavioral Ecology and Sociobiology*, **2**, 143–155.

Isbell, L.A. and Van Vuren, D. (1996). Differential costs of locational and social dispersal and their consequences for female group-living primates. *Behaviour*, **133**, 1–36.

Itani, J. and Suzuki, A. (1967). The social unit of chimpanzees. *Primates*, **8**, 355–381.

Janson, C.H. and Goldsmith, M.L. (1995). Predicting group size in primates: Foraging costs and predation risks. *Behavioral Ecology*, **6**, 326–336.

Janson, C.H. and van Schaik, C.P. (1988). Recognizing the many faces of primate food competition: Methods. *Behaviour*, **105**, 165–186.

Jones, C. and Sabater Pi, J. (1971). Comparative ecology of *Gorilla gorilla* (Savage and Wyman) and *Pan troglodytes* (Blumenbach) in Rio Muni, West Africa. *Bibliotheca Primatologica*, **13**, 1–96.

Kano, T. (1972). Distribution and adaptation of the chimpanzee on the eastern shore of Lake Tanganyika. *Kyoto University African Studies*, **7**, 37–129.

Kay, R.F. (1975). The functional adaptation of primate molar teeth. *American Journal of Physical Anthropology*, **43**, 195–216.

Kuroda, S., Nishihara, T., Suzuki, S., and Oko, R.A. (1996). Sympatric chimpanzees and gorillas in the Ndoki Forest, Congo. In *Great Ape Societies*, eds. W.C. McGrew, L.F. Marchant, and T. Nishida, pp. 71–81. Cambridge, U.K.: Cambridge University Press.

MacKinnon, J.R. (1974). The ecology and behavior of wild orangutans (*Pongo pygmaeus*). *Animal Behaviour*, **22**, 3–74.

Magliocca, F. and Querouil, S. (1997). Preliminary report on the use of the Maya Maya North Saline (Odzala National Park, Congo) by lowland gorillas. *Gorilla Conservation News*, **11**, 5.

Magliocca, F., Querouil, S., and Gautier-Hion, A. (1999). Population structure and group composition of western lowland gorillas in north-western Republic of Congo. *American Journal of Primatology*, **48**, 1–14.

Mankoto, M.O., Yamagiwa, J., Steinhauer, B.B., Mwanza, N., Maruhashi, T., and Yumoto, T. (1994). Conservation of eastern lowland gorillas in the Kahuzi-Biega National Park, Zaïre. In *Current Primatology*, Vol. 1, *Ecology and Evolution*, eds. B. Thierry, J.R. Anderson, J.J. Roeder, and N. Herrenschmidt, pp. 113–122. Strasbourg, France: Université Louis Pasteur.

McNeilage, A., Plumptre, A.J., Vedder, A., and Brock-Doyle, A. (1998). *Bwindi Impenetrable National Park, Uganda Gorilla and Large Mammal Census, 1997*. Wildlife Conservation Society Working Paper no. 14. Washington, D.C.: Wildlife Conservation Society.

Murnyak, D.F. (1981). Censusing the gorillas in Kahuzi-Biega National Park. *Biological Conservation*, **21**, 163–176.

Mwanza, N., Yamagiwa, J., Yumoto, T., and Maruhashi, T. (1992). Distribution and range utilization of eastern lowland gorillas. In *Topics in Primatology*, Vol. 2, *Behavior, Ecology and Conservation*, eds. N. Itoigawa, Y. Sugiyama, G.P. Sackett, and R.K.R. Thompson, pp. 283–300. Tokyo: University of Tokyo Press.

Nishida, T. (1970). Social behavior and relationship among wild chimpanzees of the Mahale mountains. *Primates*, **11**, 47–87.

Nishida, T. (1979). The social structure of chimpanzees of the Mahale Mountains. In *The Great Apes*, eds. D.A. Hamburg and E.R. McCown, pp. 73–121. Menlo Park, CA: Benjamin/Cummings.

Nishida, T., Hiraiwa-Hasegawa, M., Hasegawa, T., and Takahata, Y. (1985). Group extinction and female transfer in wild chimpanzees in the Mahale Mountains. *Zeitschrift für Tierpsychologie*, **67**, 284–301.

Nishihara, T. (1995). Feeding ecology of western lowland gorillas in the Nouabale-Ndoki National Park, Congo. *Primates*, **36**, 151–168.

Nunes, A. (1995). Foraging and ranging behavior in white-bellied spider monkeys. *Folia Primatologica*, **65**, 85–99.

Olejniczak, C. (1996). Update on the Mbeli Bai gorilla study, Nouabale-Ndoki National Park, northern Congo. *Gorilla Conservation News*, **10**, 5–8.

Olejniczak, C. (1997). Update on the Mbeli Bai gorilla study, Nouabale-Ndoki National Park, northern Congo. *Gorilla Conservation News*, **11**, 7–10.

Remis, M.J. (1994). Feeding ecology and positional behavior of western lowland gorillas (*Gorilla gorilla gorilla*) in the Central African Republic. PhD thesis, Yale University, New Haven, CT.

Remis, M.J. (1997*a*). Western lowland gorillas (*Gorilla gorilla gorilla*) as seasonal frugivores: Use of variable resources. *American Journal of Primatology*, **43**, 87–109.

Remis, M.J. (1997*b*). Ranging and grouping patterns of a western lowland gorilla group at Bai Hokou, Central African Republic. *American Journal of Primatology*, **43**, 111–133.

Robbins, M.M. (1995). A demographic analysis of male life history and social structure of mountain gorillas. *Behaviour*, **132**, 21–47.

Robbins, M.M. (2001). Variation in the social system of mountain gorillas: The male perspective. In *Mountain Gorillas: Three Decades of Research at Karisoke*, eds. M.M. Robbins, P. Sicotte, and K. Stewart, pp. 29–58. Cambridge, U.K.: Cambridge University Press.

Sabater Pi, J. (1977). Contribution to the study of alimentation of lowland gorillas in the natural state, in Rio Muni, Republic of Equatorial Guinea (West Africa). *Primates*, **18**, 183–204.

Schaller, G.B. (1963). *The Mountain Gorilla: Ecology and Behavior*. Chicago, IL: University of Chicago Press.

Schoener, T.W. (1971). Theory of feeding strategies. *Annual Review of Ecology and Systematics*, **2**, 369–404.

Shea, B.T. (1983). Size and diet in the evolution of African ape craniodental form. *Folia Primatologica*, **40**, 32–68.

Sicotte, P. (1993). Inter-group encounters and female transfer in mountain gorillas: Influence of group composition on male behavior. *American Journal of Primatology*, **30**, 21–36.

Stacey, P.B. (1986). Group size and foraging efficiency in yellow baboons. *Behavioral Ecology and Sociobiology*, **18**, 175–187.

Steenbeek, R. (1996). What a maleless group can tell us about the constraints on female transfer in Thomas's langurs (*Presbytis thomasi*). *Folia Primatologica*, **67**, 169–181.

Steenbeek, R. and van Schaik, C.P. (2001). Competition and group size in Thomas's langurs (*Presbytis thomasi*): The folivore paradox revisited. *Behavioral Ecology and Sociobiology*, **49**, 100–110.

Sterck, E.L. (1997). Determinants of female dispersal in Thomas's langurs. *American Journal of Primatology*, **42**, 179–198.

Sterck, E.H.M., Watts, D.P., and van Schaik, C.P. (1997). The evolution of female social relationships in nonhuman primates. *Behavioral Ecology and Sociobiology*, **41**, 291–309.

Stewart, K.J. and Harcourt, A.H. (1987). Variation in female relationships. In *Primate Societies*, ed. B.B. Smuts, D.L. Cheney, R.M. Seyfarth, R.W. Wrangham, and T.T. Struhsaker, pp. 155–164, Chicago, IL: University of Chicago Press.

Struhsaker, T.T. and Leland, L. (1985). Infanticide in a patrilineal society of red colobus monkeys. *Zeitschrift für Tierpsychologie*, **69**, 89–132.

Sugiyama, Y. (1999). Socioecological factors of male chimpanzee migration at Bossou, Guinea. *Primates*, **40**, 61–68.

Suzuki, S. and Nishihara, T. (1992). Feeding strategies of sympatric gorillas and chimpanzees in the Ndoki-Nouabale forest, with special reference to co-feeding behavior by both species. In *Abstracts of the 16th Congress of the International Primatological Society*, p. 86. Strasbourg, France: Société Francophone de Primatologie.

Taylor, A.B. (2002). Masticatory form and function in the African apes. *American Journal of Physical Anthropology*, **117**, 133–156.

Terborgh, J. (1983). *Five New World Primates: A Study of Comparative Ecology*. Princeton, NJ: Princeton University Press.

Terborgh, J. and Janson, C. (1986). The socioecology of primate groups. *Annual Review of Ecology and Systematics*, **17**, 111–135.

Treves, A. and Chapman, C.A. (1996). Conspecific threat, predation avoidance, and resource defence: Implications for grouping in langurs. *Behavioral Ecology and Sociobiology*, **39**, 43–53.

Tsukahara, T. (1993). Lions eat chimpanzees: The first evidence of predation by lions on wild chimpanzees. *American Journal of Primatology*, **29**, 1–11.

Tutin, C.E.G. (1979). Mating patterns and reproductive strategies in a community of wild chimpanzees (*Pan troglodytes schweinfurthii*). *Behavioral Ecology and Sociobiology*, **6**, 29–38.

Tutin, C.E.G. (1996). Ranging and social structure of lowland gorillas in the Lopé Reserve, Gabon. In *Great Ape Societies*, eds. W.C. McGrew, L.F. Marchant, and T. Nishida, pp. 58–70. Cambridge, U.K.: Cambridge University Press.

Tutin, C.E.G. and Fernandez, M. (1992). Insect-eating by sympatric lowland gorillas (*Gorilla g. gorilla*) and chimpanzees (*Pan t. troglodytes*) in the Lopé Reserve, Gabon. *American Journal of Primatology*, **28**, 29–40.

Tutin, C.E.G. and Fernandez, M. (1993). Composition of the diet of chimpanzees and comparisons with that of sympatric lowland gorillas in the Lopé Reserve, Gabon. *American Journal of Primatology*, **30**, 195–211.

Tutin, C.E.G., McGrew, W.C., and Baldwin, P.J. (1983). Social organization of savanna-dwelling chimpanzees, *Pan troglodytes verus*, at Mt. Assirik, Senegal. *Primates*, **24**, 154–173.

Tutin, C.E.G., Fernandez, M., Rogers, M.E., and Williamson, E.A. (1992). A preliminary analysis of the social structure of lowland gorillas in the Lopé Reserve, Gabon. In *Topics in Primatology*, Vol. 2, *Behavior, Ecology and Conservation*, eds. N. Itoigawa, Y. Sugiyama, G.P. Sackett, and R.K.R. Thompson, pp. 245–254. Tokyo: University of Tokyo Press.

Tutin, C.E.G., Ham, R.M., White, L.J.T., and Harrison, M.J.S. (1997). The primate community of the Lopé Reserve, Gabon: Diets, responses to fruit scarcity, and effects on biomass. *American Journal of Primatology*, **42**, 1–24.

van Schaik, C.P. (1983). Why are diurnal primates living in groups? *Behaviour*, **87**, 120–144.

van Schaik, C.P. (1989). The ecology of social relationships among female primates. In *Comparative Socioecology: The Behavioural Ecology of Humans and Other Mammals*, eds. V. Standen and R.A. Foley, pp. 195–218, Oxford, U.K.: Blackwell Scientific Publications.

van Schaik, C.P. and van Hooff, J.A.R.A.M. (1983). On the ultimate causes of primate social systems. *Behaviour*, **85**, 91–117.

van Schaik, C.P., van Noordwijk, M.A., de Boer, R.J., and den Tonkelaar, T. (1983). The effect of group size on time budgets and social behaviour in wild longtailed macaques (*Macaca fascicularis*). *Behavioral Ecology and Sociobiology*, **13**, 173–181.

Watts, D.P. (1984). Composition and variability of mountain gorilla diets in the Central Virungas. *American Journal of Primatology*, **7**, 323–356.

Watts, D.P. (1989). Infanticide in mountain gorillas: New cases and a reconsideration of the evidence. *Ethology*, **81**, 1–18.

Watts, D.P. (1990). Ecology of gorillas and its relationship of female transfer in mountain gorillas. *International Journal of Primatology*, **11**, 21–45.

Watts, D.P. (1991). Habitat use strategies of mountain gorillas. *Folia Primatologica*, **56**, 1–16.

Watts, D.P. (1996). Comparative socioecology of gorillas. In *Great Ape Societies*, eds. W.C. McGrew, L.F. Marchant, and T. Nishida, pp. 16–28. Cambridge, U.K.: Cambridge University Press.

Watts, D.P. (1998*a*). Long-term habitat use by mountain gorillas (*Gorilla gorilla beringei*). II: Reuse of foraging areas in relation to resource abundance, quality, and depletion. *International journal of Primatology*, **19**, 681–702.

Watts, D.P. (1998*b*). Long-term habitat use by mountain gorillas (*Gorilla gorilla beringei*). I: Consistency, variation, and home range size and stability. *International Journal of Primatology*, **19**, 651–680.

Watts, D.P. (2000*a*). Mountain gorilla habitat use strategies and group movements. In *Group Movements: Patterns, Processes, and Cognitive Implications in Primates and Other Animals*, eds. S. Boinski and P. Kappeler, pp. 351–374. Cambridge, U.K.: Cambridge University Press.

Watts, D.P. (2000*b*). Causes and consequences of variation in male mountain gorilla life histories and group membership. In *Primate Males*, ed. P.M. Kappeler, pp. 169–179. Cambridge, U.K.: Cambridge University Press.

Wrangham, R.W. (1975). The behavioral ecology of chimpanzees in the Gombe National Park, Tanzania. PhD thesis, University of Cambridge, Cambridge, U.K.

Wrangham, R.W. (1980). An ecological model of female-bonded primate groups. *Behaviour*, **75**, 262–300.

Wrangham, R.W. (1987). Evolution and social structure. In *Primate Societies*, eds. B.B. Smuts, D.L. Cheney, R.M. Seyfarth, R.W. Wrangham, and T.T. Struhsaker, pp. 282–296. Chicago, IL: University of Chicago Press.

Wrangham, R.W., Clark, A.P., and Isabirye-Basuta, G. (1992). Female social relationships and social organization of Kibale Forest chimpanzees. In *Topics in Primatology*, Vol. 1, *Human Origins*, eds. T. Nishida, W.C. McGrew, P. Marler, M. Pickford, and F.B.M. de Waal, pp. 81–98. Tokyo: University of Tokyo Press.

Wrangham, R.W., Chapman, C.A., Clark-Arcadi, A.P., and Isabyrye-Basuta, G. (1996). Social ecology of Kanyawara chimpanzees: Implications for understanding the costs of great ape groups. In *Great Ape Societies*, eds. W.C. McGrew, L.F. Marchant, and T. Nishida, pp. 45–57. Cambridge, U.K.: Cambridge University Press.

Yamagiwa, J. (1983). Diachronic changes in two eastern lowland gorilla groups (*Gorilla gorilla graueri*) in the Mt. Kahuzi region, Zaïre. *Primates*, **24**, 174–183.

Yamagiwa, J. (1987). Male life history and social structure of wild mountain gorillas (*Gorilla gorilla beringei*). In *Evolution and Coadaptation in Biotic Communities*, eds. S. Kawano, J.H. Connell, and T. Hidaka, pp. 31–51. Tokyo: University of Tokyo Press.

Yamagiwa, J. (1999). Socioecological factors influencing population structure of gorillas and chimpanzees. *Primates*, **40**, 87–104.

Yamagiwa, J. (2001). Factors influencing the formation of ground nests by eastern lowland gorillas in Kahuzi-Biega National Park: Some evolutionary implications of nesting behavior. *Journal of Human Evolution*, **40**, 99–109.

Yamagiwa, J. and Kahekwa, J. (2001). Dispersal patterns, group structure and reproductive parameters of eastern lowland gorillas at Kahuzi in the absence of infanticide. In *Mountain Gorillas: Three Decades of Research at Karisoke*, eds. M.M. Robbins, P. Sicotte, and K. Stewart, pp. 89–122. Cambridge, U.K.: Cambridge University Press.

Yamagiwa, J. and Mwanza, N. (1994). Day-journey length and daily diet of solitary male gorillas in lowland and highland habitats. *International Journal of Primatolology*, **15**, 207–224.

Yamagiwa, J., Mwanza, N., Yumoto, T., and Maruhashi, T. (1989). A preliminary survey on sympatric populations of *Gorilla g. graueri* and *Pan t. schweinfurthii* in eastern Zaïre. In *Grant-in-Aid for Overseas Scientific Research Report: Interspecies Relationships of Primates in the Tropical and Montane Forest*, Vol. 1, pp. 23–40. Inuyama, Japan: Primate Research Institute of Kyoto University.

Yamagiwa, J., Mwanza, N., Spangenberg, A., Maruhashi, T., Yumoto, T., Fischer, A., Steinhauer, B.B., and Refisch, J. (1992*a*). Population density and ranging pattern of chimpanzees in Kahuzi-Biega National Park, Zaïre: A comparison with a sympatric population of gorillas. *African Study Monographs*, **13**, 217–230.

Yamagiwa, J., Mwanza, N., Yumoto, T., and Maruhashi, T. (1992*b*). Travel distances and food habits of eastern lowland gorillas: A comparative analysis. In *Topics in Primatology*, Vol. 2, *Behavior, Ecology and Conservation*, eds. N. Itoigawa, Y. Sugiyama, G.P. Sackett, and R.K.R. Thompson, pp. 267–281. Tokyo: University of Tokyo Press.

Yamagiwa, J., Yumoto, T., Maruhashi, T., and Mwanza, N. (1993*a*). Field methodology for analyzing diets of eastern lowland gorillas in Kahuzi-Biega National Park, Zaïre. *Tropics*, **2**, 209–218.

Yamagiwa, J., Mwanza, N., Spangenberg, A., Maruhashi, T., Yumoto, T., Fischer, A., and Steinhauer, B.B. (1993*b*). A census of the eastern lowland gorilla (*Gorilla gorilla graueri*) in Kahuzi-Biega National Park with reference to mountain gorilla (*Gorilla gorilla beringei*) in the Virunga region, Zaïre. *Biological Conservation*, **64**, 83–89.

Yamagiwa, J., Mwanza, N., Yumoto, Y., and Maruhashi, T. (1994). Seasonal change in the composition of the diet of eastern lowland gorillas. *Primates*, **35**, 1–14.

Yamagiwa, J., Maruhashi, T., Yumoto, T., and Mwanza, N. (1996*a*). Dietary and ranging overlap in sympatric gorillas and chimpanzees in Kahuzi-Biega National Park, Zaïre. In *Great Ape Societies*, eds. W.C. McGrew, L.F. Marchant, and T. Nishida, pp. 82–98. Cambridge, U.K.: Cambridge University Press.

Yamagiwa, J., Kaleme, K., Milynganyo, M., and Basabose, K. (1996*b*). Food density and ranging patterns of gorillas and chimpanzees in the Kahuzi-Biega National Park, Zaïre. *Tropics*, **6**, 65–77.

Yamakoshi, G. (1998). Dietary responses to fruit scarcity of wild chimpanzees at Bossou, Guinea: Possible implications for ecological importance of tool use. *American Journal of Physical Anthropology*, **106**, 283–295.

Yumoto, T., Yamagiwa, J., Mwanza, N., and Maruhashi, T. (1994). List of plant species identified in Kahuzi-Biega National Park, Zaïre. *Tropics*, **3**, 295–308.

# 14 Comparative behavioral ecology of a lowland and highland gorilla population: Where do Bwindi gorillas fit?

MICHELE L. GOLDSMITH

## Introduction

Primatologists have long sought explanations for the diversity of social systems found among and within primate species. Apes have received much attention given their close relationship to *Homo*, and the fact that within this group there are as many social systems as there are genera. Differences among genera, species, subspecies, and in some cases populations within a subspecies are presumed to be due to adaptations to varied environments.

An important way in which the environment varies is in the spatiotemporal availability of different food sources. Folivorous items, such as mature leaves and terrestrial herbaceous vegetation (THV), are often difficult to digest, low in energy and high in protein. In addition, they are generally available year-round, abundant and evenly distributed in the environment. On the other hand, fruits (and in most cases, young leaves) are often easily digested, high in energy, perhaps low in protein, only seasonally available, and are rarely evenly distributed in the environment (see Remis, this volume for nutritional information).

The inconsistent and unpredictable availability of fruit presents a number of obstacles for primates. For one, it requires them to travel farther each day to fulfill their nutritional requirements than folivores (Milton and May, 1976; Clutton-Brock and Harvey, 1977). Furthermore, the distance traveled is intrinsically related to the size of the social group (Wrangham *et al.*, 1993; Janson and Goldsmith, 1995). As more members are added to the group, competition within the group increases and individuals need to travel farther, which acts to constrain group size (Janson and Goldsmith, 1995). This may potentially explain why all frugivorous apes either live in small social groups (e.g., monogamous gibbons and solitary orangutans) or display grouping patterns that allow group size to fluctuate (e.g., fission–fusion system in chimpanzees and bonobos). These systems may have evolved to reduce energetic costs that gregarious, large-bodied primates face when feeding on fruit, suggesting that foraging effort may be a major selective force on group size and social structure in primarily frugivorous primates. Although some folivores experience within-group feeding competition, these costs are often small and individuals usually respond by increasing group spread rather than the total distance traveled (Watts, 1991). Therefore, foraging effort may be less of a selective pressure on folivorous primates.

### Highland versus lowland gorillas

Altitude strongly influences the availability and distribution of food. In general, temperature decreases and wind speed tends to increase in a linear fashion with altitude (Lieberman *et al.*, 1996). Although rainfall tends to decrease (though not linearly), excessive moisture occurs from fog cover at higher altitudes (Grubb and Whitmore, 1966). These general trends result in altitudinal differences in plant structure and availability, with canopy height and species diversity being greatest at lower elevations, and forests that are shorter in stature and poorer in both species and family richness at higher elevations (Richards, 1952; Lieberman *et al.*, 1996). Therefore, plants that provide fruit such as dicot trees, palms, and lianas occur at greater densities and diversities in lowland forests, while fruit-poor vegetation, such as tree ferns and hemi-epiphytes occur at greater densities in montane forests.

As their common names imply, altitude is the major distinction between mountain and lowland gorillas. Virunga mountain gorillas live at the upper limit of their altitudinal range between 2700 and 3700 m above sea level (Watts, 1998). In this region, although rainfall does fluctuate seasonally, most gorilla foods (e.g., THV) are perennially available (Fossey and Harcourt, 1977) except for seasonal use of bamboo (Vedder, 1984). Fruit is rare and the folivorous diet of the Virunga gorillas results in short daily path lengths and relatively large, stable groups with low levels of within-group feeding competition (Watts, 1991).

In contrast to over 35 years of study of the Virunga mountain gorillas, very little is known concerning the other mountain gorilla population living 25 km to the north in Bwindi-Impenetrable National Park, Uganda. Although these gorillas have been part of an active tourism program since 1993, studies on their ecology and behavior are limited (Butynski, 1984; Achoka, 1994; Berry, 1998), and long-term systematic studies have just recently been initiated (Goldsmith *et al.*, 1998; Goldsmith, 2000; Robbins, 2000). Although their status as mountain gorillas has been questioned based on morphological and ecological evidence (Sarmiento *et al.*, 1996), these findings run counter to molecular studies of mitochondrial DNA that indicate they are no different from the Virunga population (Garner and Ryder, 1996; Jensen-Seaman *et al.*, this volume). Perhaps the most influential difference between the two mountain gorilla populations is that Bwindi lies at a lower altitude between 1400 and 2300 m above sea level (Butynski, 1984). At this elevation, there is a greater diversity and density of fruit available than in the Virungas. However, it is not yet known how frugivorous Bwindi gorillas are or how the availability of fruit may influence their foraging effort and grouping patterns.

Although western lowland gorilla populations are distributed over a much larger area than either mountain gorilla population, they all generally live between 200 and 800 m above sea level. In Nigeria, they actually reach heights similar to those found in Bwindi, sometimes surpassing 1500 m (Oates *et al.*, this volume). Lowland gorillas use both secondary and primary forest and rely heavily on fruit (Tutin and Fernandez, 1985; Rogers *et al.*, 1990; Williamson *et al.*, 1990; Yamagiwa *et al.*, 1994; Nishihara, 1995; Goldsmith, 1996; Remis, 1997*a*). Their fruit consumption, however, varies seasonally and the importance of herbaceous and woody vegetation (e.g., bark and stems) in their diet should not be underestimated (Tutin and Fernandez, 1985; Goldsmith, 1996). Their frugivorous behavior leads to both longer daily path lengths (Tutin, 1996; Remis, 1997*b*; Goldsmith, 1999; Doran and McNeilage, 1999) and more arboreal behavior (Remis, 1999) than has been observed in Virunga gorillas. Some researchers suggest their grouping patterns and social behavior are similar to that found in the Virungas (i.e., groups are relatively cohesive and stable in size) (Tutin *et al.*, 1992; Tutin, 1996). Others propose that lowland gorilla social units are constrained by within-group feeding competition and are therefore less cohesive (i.e., they may form temporary subgroups on occasion) (Mitani, 1992; Remis, 1994; Kuroda *et al.*, 1996; Goldsmith, 1996).

### *Predicting comparative behavioral ecology of lowland and mountain gorillas: The bonobo/chimpanzee analogy*

How do different degrees of frugivory versus folivory influence gorilla foraging effort and grouping patterns? Do dietary differences influence the degree of within-group feeding competition, and if so, what effect does that have on group cohesion? The comparative behavioral ecology of bonobos (*Pan paniscus*) and chimpanzees (*Pan troglodytes*) provides a good model for comparisons among gorilla populations. Although chimpanzees and bonobos both display a fission–fusion system, chimpanzee females often forage alone and rarely affiliate with each other and males interact more with other males than with females (except during estrus) (Nishida, 1979; Wrangham and Smuts, 1980; Goodall, 1986; Wrangham *et al.*, 1992). On the other hand, bonobo parties are somewhat larger, more stable and often mixed, and females affiliate with each other and with males (Kuroda, 1979; Kano, 1982; Badrian and Badrian, 1984; White, 1988).

The primary explanation as to why these species differ is resource based and in general suggests that within-group feeding competition in bonobos is somehow relaxed compared to chimpanzees, thereby allowing for larger, more stable party sizes (Wrangham, 1986). Many resource-based hypotheses have been advanced to explain the difference in gregariousness, or group cohesion, between the two *Pan* species. For example, it has been suggested that bonobos have better access to fruit because of greater species diversity (Kano and Mulavwa, 1984), larger patch size (White and Wrangham, 1988), and/or wider seasonal availability (Malenky, 1990). However, as reviewed in Wrangham *et al.* (1996), these hypotheses have been weakened by demonstrations that differences in patch size do not really exist when differences in habitat are properly controlled (Chapman *et al.*, 1994). In addition, it has been argued that all sites along the equator should experience the same seasonality (van Schaik *et al.*, 1993).

Hypotheses on the consumption and availability of THV and how this resource acts to reduce competition have had more support. The fact that bonobos rely more heavily on THV than do chimpanzees (Badrian and Malenky, 1984) led Wrangham (1986) to suggest that it was the abundant and even distribution of THV that relaxed competition, allowing for more stable and cohesive bonobo groups. Although THV is ubiquitously distributed, bonobos are quite selective about which herbs they eat (Malenky and Stiles, 1991); consequently, the actual density and distribution of THV may not be as important as its nutritional quality (Wrangham *et al*, 1996).

While I do not expect differences among gorilla subspecies to fit neatly into a *Pan* model, it is useful to apply this model as a working analogy. For example, gorillas, like *Pan*, vary in their degree of frugivory versus folivory (i.e., dietary ratio). Figure 14.1 demonstrates this ratio and models the general effect this ratio has on group cohesion in African great apes. Chimpanzees are obligate frugivores, whereas bonobos fall back on THV when fruits are scarce. As a result, chimpanzees experience greater within-group competition, which results in smaller, less cohesive groups. Western lowland gorillas, like bonobos, eat fruit and THV, whereas Virunga mountain gorillas eat almost exclusively THV. As a result, western lowland gorilla groups should experience greater within-group competition and less cohesive groups than Virunga gorillas. Where, then, do Bwindi gorillas fit? Are they more similar in diet and behavior to their Virunga cousins or to lowland populations? The answer is probably related to their dietary ratio of frugivory and folivory, which is most likely influenced by altitude.

To compare differential effects of frugivory and folivory on gorilla behavioral ecology, I studied a population of western lowland gorillas in Bai Hokou, Central African Republic and a population of mountain gorillas in

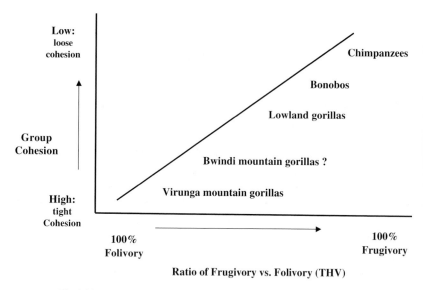

Fig. 14.1. A model of bonobo, chimpanzee, and gorilla group cohesion as it relates to the ratio of frugivory and folivory: Where do Bwindi gorillas fit?

Bwindi-Impenetrable National Park, Uganda. Comparisons are made both between subspecies and among groups within the same population (within Bwindi). More attention has been given to these latter comparisons as we continue to note differences within populations and even among individuals within social groups (perhaps explaining the emerging field of "cultural primatology"). To compare the behavioral ecology of these two subspecies, I examined relationships among the availability and distribution of plant foods in their environment, and dietary behavior, foraging effort, group size, and grouping pattern.

## Methods

### Study sites and subjects

The Bwindi Forest in Southwestern Uganda became Bwindi-Impenetrable National Park (331 km$^2$) in 1991 (latitude ranges from 0° 53′ S to 1° 8′ S and longitude from 29° 35′ E to 29° 50′ E). Temperature averages 13 °C with a range from 7 to 20 °C, and rainfall is estimated at 1400 mm per year with two dry (December–January and June–August) and two rainy (February–May and September–November) seasons (Butynski, 1984). An estimated 292 gorillas live in Bwindi-Impenetrable National Park (McNeilage *et al.*, 1998). Data from Bwindi are preliminary and were collected during periodic visits from 1996 to 1999 on five habituated groups including: the Mubare, Katendegere, and Habinyanja tourist groups in Buhoma (1450 m), the Nkuringo tourist group in Kashasha (1650 m), and the Kyaguliro research group in Mubwindi Swamp (2200 m) (Fig. 14.2A). Data collection on the Katendegere and Habinyanja are limited so results concentrate on the remaining three groups.

Data were collected daily on western lowland gorillas from July 1993 through February 1995 at the Bai Hokou Research Center (50 km$^2$) in the Dzanga Sector (495 km$^2$) of the Dzanga-Ndoki National Park (gazetted in 1990) in southeastern Central African Republic (2° 50′ N, 16° 28′ E) (Fig. 14.2B). The terrain at Bai Hokou is relatively flat at about 300 m and is seasonally fruit-rich. It is primarily composed of secondary (46%) and mixed-open (26%) forest, with an annual rainfall for 1994 of 1952 mm with less than 120 mm of rain falling during each of the two dry seasons (Goldsmith, 1996). Gorillas within this site were not habituated to the presence of humans and occur at rather high densities (0.89 to 1.45 individuals/km$^2$: Carroll, 1986). Subjects included at least three lone males and five groups, all of which had some part of their home range within the study site.

### Data collection

Many comparisons drawn between gorilla species or subspecies have been conducted in a *post-hoc* fashion, where an investigator set out to study one population, and then compared his or her findings to those obtained by another investigator. Often, these investigators have employed different methodologies, which may introduce uncontrolled variables into the comparisons that can confound findings. In this study, I investigate the behavioral ecology of two different populations of gorillas, one at Bai Hokou, the other at Bwindi, and employ the same methods of analysis in preplanned comparisons, in an effort to control for some of these variables.

Regardless of differences in habituation level, methods used to collect data on diet, ranging and grouping behavior were similar between sites (i.e., they were primarily determined indirectly from dung, feeding sites, and nesting areas). Supplementary visual observations of feeding and grouping behavior, arboreal behavior, and social structure were collected opportunistically. Methods are briefly summarized below (for a more complete description see Goldsmith, 1996, 1999). For all comparisons, Student's $t$-test and analysis of variance (ANOVA) were used to compare groups when assumptions for normality and homogeneity of variance were met, and nonparametric tests employed when assumptions were not met.

To determine the diversity and density of THV within the study areas, 1-m$^2$ quadrats were randomly displaced 1–2 m to either the left or right side

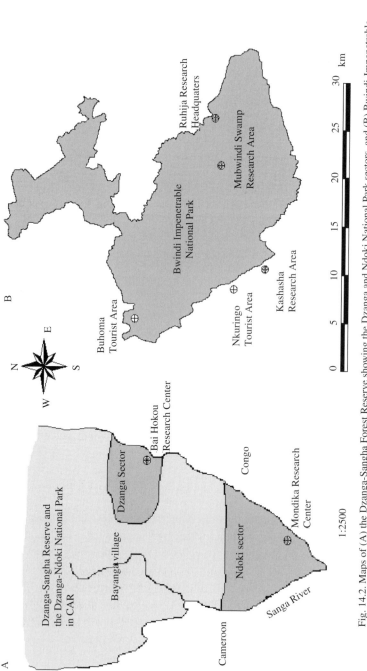

Fig. 14.2. Maps of (A) the Dzanga and Ndoki National Park sectors, and (B) Bwindi-Impenetrable National Park, which is situated within the southeastern edge of Uganda along the eastern border of the Democratic Republic of Congo (DRC).

of a 5-km (Bai Hokou) or a 3-km (Bwindi) line transect at 5-m intervals (cf. Rogers and Williamson, 1987). Within each quadrat, species were identified and stem counts taken. Diet was determined by feeding trail remains (foods left along fresh gorilla feeding trails were identified and noted) and fecal samples (collected at nest sites, washed and dried, and fruit remnants identified and weighed). Daily path length was measured as the complete distance the gorillas traveled between sleeping nest sites using a hand counter and compass. Sleeping nest site areas were used to estimate group size, composition, and cohesion (nest area measures follow Williamson, 1988).

## Results of between-subspecies comparisons

### *Vegetation diversity/density and diet*

Species of THV were significantly more diverse ($t_{[1,1118]} = 718$, $p < 0.001$) and more densely distributed ($t_{[1,1118]} = 2030$, $p < 0.001$) at the higher elevation in Bwindi (3.4 species and 9.9 stems/m$^2$ plot) than in Bai Hokou (0.4 species and 1.1 stems/m$^2$ plot). In Bai Hokou, all THV within herb plots represented gorilla foods; this was not the case, however, in Bwindi. When considering only those herbaceous plants known to date to be gorilla foods, THV density was estimated to be to 3.1 stems/m$^2$ plot in Bwindi (however, this figure could increase as we learn more about the dietary behavior of this population).

Table 14.1 compares THV density with altitude for a number of African ape populations. In general, western lowland gorillas, chimpanzees, and bonobos average only 1–2 stems per plot, while Virunga mountain gorillas at the highest elevation have an average of 9 stems per plot. Although the relationship is not exactly linear, there does appear to be a trend towards a greater density of herbaceous vegetation as altitude increases. Known exceptions are Kibale Forest in Uganda, which is 60% moist evergreen forest and is transitional between lowland and montane forests (Struhsaker, 1975), and Lopé where THV might be influenced by past logging in the area.

Fruit represented on average 38% of dried dung weight for western lowland gorillas and fruit remains were found in 96% of all fecal samples. Bwindi gorillas also ate fruit, but it represented only 10% of the dried dung weight and was found in only 36% of all fecal samples (Fig. 14.3A). Both subspecies ate significantly more fruit during the rainy season months (Fig. 14.3B). In addition to eating more fruit, western lowland gorillas also ate a greater diversity of fruit with an average of 3.4 species in each dung sample. By contrast, Bwindi gorillas had, on average, only one fruit species in each dung sample (Fig. 14.3C). A greater diversity of fruit species was found in the diet of both subspecies during

Table 14.1. *The relationship between THV density and altitude for a number of African ape sites*

| Site | Species | Altitude (m above sea level, estimated) | THV density (1 m$^2$) |
|---|---|---|---|
| Bai Hokou, CAR[a] | *Gorilla g. gorilla* | 300 | 1.1 |
| Ndoki, Congo[b] | *Gorilla g. gorilla* | 350 | 2.25 |
| Lopé, Gabon[c] | *Gorilla g. gorilla* | 200–500 | 7.66 |
| Lomako, DRC[b] | *Pan paniscus* | 400 | 2.02 |
| Kibale, Uganda[b] | *Pan t. schweinfurthii* | 1500 | 0.89 |
| Bwindi, Uganda[d] | *Gorilla g. beringei* | 2000 | 9.9 |
| | *Pan t. schweinfurthii* | | (3.1 – gorilla foods) |
| Karisoke, Rwanda[e] | *Gorilla g. beringei* | 3300 | 8.81 |

[a] Bai Hokou: Goldsmith (1996) – plots include only gorilla foods.
[b] Ndoki, Lomako, and Kibale: Malenky *et al.* (1994).
[c] Lopé: Rogers and Williamson (1987) – plots include all herbs in the Zingiberaceae and Marantaceae families.
[d] Data from this study; 9.9 is all THV, but when only known gorilla foods are considered the density per plot decreases to 3.1 stems.
[e] Karisoke: Watts (1984) weighted mean across eight vegetation zones ranging in densities from 0.8 to 18.5 stems/m$^2$. All food within plots represents gorilla foods.

the rainy season months (Fig. 14.3D). These findings support the prediction that diet is related to altitude with higher levels of frugivory at lower altitudes. However, it also reveals that, unlike the Virunga mountain gorillas, Bwindi gorillas do incorporate fruit into their diet, especially during the rainy season; this dietary difference may influence other aspects of their behavior.

Very few foods overlapped between the two subspecies. Two plant species, however, that were important sources for both populations were the fruit and pith of the herb *Aframomum* (Zingiberaceae) and fruit from the tree *Myrianthus* (Moraceae). These two foods are also popular among other ape populations, especially herbs from the Zingiberaceae family (bonobos: Malenky and Stiles, 1991; chimpanzees: Wrangham *et al.*, 1991).

### *Foraging effort*

Western lowland gorillas traveled significantly ($p < 0.05$) farther each day (mean 2.59 km) than did Bwindi mountain gorillas (mean 0.78 km) (Fig. 14.4A). This is in comparison to an average of 0.57 km per day for Virunga gorillas (Group 4: Watts, 1991). Although western lowland gorillas traveled farther

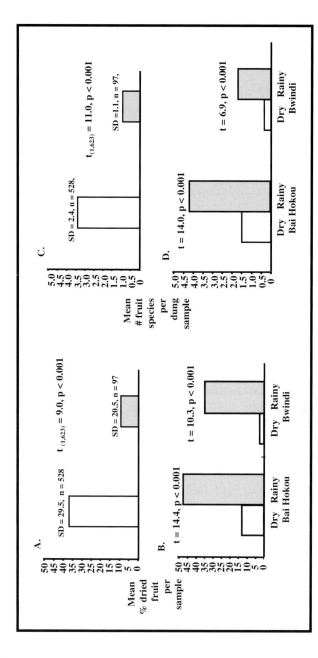

Fig. 14.3. (A) The mean abundance of fruit and (B) its seasonal influence, and (C) mean fruit species diversity and (D) its seasonal influence in the diet of Bai Hokou and Bwindi gorillas.

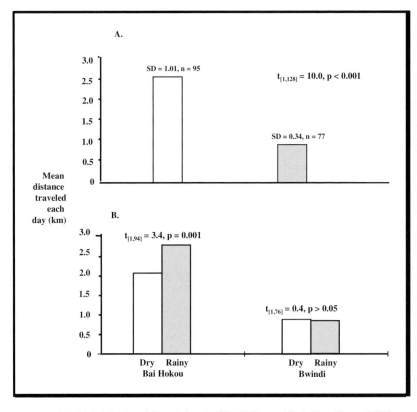

Fig. 14.4. (A) Mean daily path length of Bai Hokou and Bwindi gorillas and (B) its seasonal influence.

in the rainy (mean 2.81 km, $n = 66$) versus the dry season (mean 2.08 km, $n = 29$), no statistical difference was found for Bwindi gorillas (Fig. 14.4B). Like western lowland gorillas, Bwindi gorillas spent a large amount of time foraging for foods arboreally. Males and females of all ages were found to forage in trees, with silverback males of both subspecies being observed at heights as great as 20 m. When foraging in the trees, gorillas of both subspecies were rarely seen eating fruit; instead, they were most often observed eating the leaves, bark, and pith of lianas. In Bwindi, arboreal gorillas showed a great liking for succulent epiphytes. Both subspecies also nested in trees; however, nests in Bwindi were often in small trees and shrubs (average nest height 1.5 m off the ground), while in Bai Hokou, lowland gorillas nested quite high, on average 13 m above the ground (Remis, 1993). In the Virungas, juveniles and to a lesser extent adult females climb trees (Schaller, 1963) but not as often or as high as Bwindi or Bai Hokou gorillas.

## Group size and composition

Five groups lived within the Bai Hokou study area ranging in size (based on mean nest counts) from three to 19 individuals, with a mean group size based on all recorded nest sites of 9.4 ($n = 232$, SD $= 4.1$). This estimated mean group size is high compared to other counts in this area (cf. Carroll, 1988; Remis, 1993, 1997*b*). In Bwindi, the five study groups ranged in size from three to 25 with a mean group size of 9.9 individuals ($n = 139$, SD $= 3.0$). This estimate is similar to the group size estimate of 9.8 (SD $= 6.2$) based on a 1997 census of an estimated 28 groups within Bwindi with a maximum group size of 23 (McNeilage *et al.*, 1998). Mean group size estimates for both Bai Hokou and Bwindi gorillas are similar to those for Virunga gorillas, which range from two to 34 with a mean of 9.2 ($n = 32$) (Sholley, 1991).

Results of this study showed that both Bai Hokou and Bwindi gorillas had a high number of multi-male groups (51.8% and 40%, respectively). McNeilage and others (1998) also found a high percentage of multi-male groups in Bwindi (46%). These estimates are higher than those found in the Virungas (28%) during the latest census by Sholley (1991) but similar to those found by Robbins (1995). The number of silverback males within each group was related to group size for both Bai Hokou and Bwindi gorillas. In Bai Hokou, the number of silverback males was a good predictor of overall group size ($r^2 = 0.47$, $F_{[1,110]} = 95.8$, $p < 0.001$). Groups with one silverback male averaged 5.9 individuals ($n = 23$, SD $= 2.8$), groups with two silverbacks averaged 12.3 individuals ($n = 81$, SD $= 3.2$), and groups with three silverbacks averaged 16.1 individuals ($n = 8$, SD $= 2.2$) ($F_{[2,109]} = 50.4$, $p < 0.001$).

To determine whether a predictable relationship also existed within Bwindi, I performed a regression between group size and the number of silverback males using the census data from McNeilage *et al.* (1998) (excluding the estimated number of infants, which are not reflected in nest counts). As in Bai Hokou, the number of silverback males in Bwindi significantly predicted group size ($r^2 = 0.27$, $F_{[1,27]} = 9.6$, $p = 0.005$). Groups with one silverback male averaged 5.7 individuals ($n = 22$), groups with two males averaged 9.8 individuals ($n = 8$), and groups with three males averaged 15.4 individuals ($n = 5$) ($F_{[2,25]} = 3.7$, $p = 0.04$) (findings between the number of silverbacks and group size hold regardless of whether or not they are included in the total group count).

## Group cohesion

The first measure of group cohesion was determined by examining the nearest neighbor of nesting individuals. Bai Hokou lowland gorillas had a significantly smaller mean area around each nest ($34.5$ m$^2$) than did Bwindi mountain gorillas ($45.5$ m$^2$; $t_{[1,304]} = 5.0$, $p = 0.03$). Differences in the cohesion of sleeping individuals between sites may result from distinctions in habitat and predation risk. In Bai Hokou, gorillas preferred to nest in small light-gaps and other rare forest types (Remis, 1993). In addition, there is risk of predation by both leopards (Fay *et al.*, 1995) and humans in Bai Hokou which may warrant tighter cohesion while sleeping. In contrast, the looser cohesion of sleeping individuals in Bwindi may be due to the fact that they often nest in open areas along mountainsides, in a protected area with no natural predators and low poaching pressure.

Although there were no seasonal differences in group cohesion in Bwindi (dry season mean $48.3$ m$^2$, rainy season mean $56.4$ m$^2$; $t_{[1,121]} = 0.83$, $p = 0.36$), Bai Hokou gorillas had a smaller mean area around each nest in the dry ($24$ m$^2$) versus the rainy ($39.6$ m$^2$) season ($t_{[1,178]} = 5.8$, $p = 0.02$). Seasonal differences in nest site cohesion in Bai Hokou gorillas may be due to the fact that they were less cohesive when foraging in the rainy season (see below) and also nested in trees more often (Williamson, 1988; Remis, 1993), both of which may increase the spread of the nest site.

To determine another measure of group cohesion, the group spread of traveling individuals, I employed several unconventional methods. For example, I estimated group spread from traveling individuals by measuring the distance between prints when crossing major paths. For Bai Hokou gorillas, group spread was not related to group size ($r^2 = 0.08$, $F_{[1,19]} = 0.6$, $p = 0.20$) and averaged $333.0$ m ($n = 21$, range 95-842 m, SD $= 202$) with traveling individuals being more cohesive in the dry (mean $226.7$, $n = 7$, range 95-645, SD $= 188.5$) versus the rainy season (mean $386.1$ m, $n = 14$, range 105-842 m, SD $= 193.9$ m; Wilcoxon sign test, $\chi^2 = 4.06$, $p = 0.04$). The mean distance between any two traveling individuals (group spread/group size) was $32.0$ m ($n = 21$, SD $= 8.8$, range 7.3-69.6) with a smaller mean distance in the dry (mean 18.8, range $=$ 7.3-43.0 m) versus the rainy season (mean 38.6, range 11.7-69.6 m, $\chi^2 = 6.4$, $p = 0.01$). These data were more difficult to obtain for Bwindi gorillas, since they rarely crossed over well-cut paths. On the few occasions it was noted, group spread averaged only 60 m ($n = 3$, group size $= 14$, range 38-79 m) with a mean of 4.3 m between each individual.

I also compared the number of gorillas observed when we contacted the group after following their trail out from their morning nests. In Bai Hokou, when groups were contacted only 24.7% of nesting individuals were observed ($t_{[1,30]} = 8.4$, $p < 0.001$, mean difference 7.2) suggesting that group members were spread out over a rather large area. Although the mean estimated group size from nest sites was 9.4, the mean number of individuals observed at any one time was only 2.1 (SD $=1.6$) and this did not differ with season.

It was interesting to note the composition of the Bai Hokou groups when contacted. Females, either alone or with other females were contacted 32% of the time. During these observations, males were not found to be in the near vicinity. This is surprising since females are rarely found foraging on their own in other gorilla populations, especially in the Virungas. All other observations of Bai Hokou gorillas were of females associated with males (one female/one male 16%, multiple females/one male 19%), with the rarest observation being a group male with no females (13%) (all other observations were of lone males, 19%). In Bwindi, all study groups were habituated and when observed, most individuals were noted as being within 100–200 m of each other. The most frequently observed individuals were infants and juveniles who were almost always found playing on or around the alpha silverback male.

Finally, in Bai Hokou, I estimated group spread by timing how long it took for a silverback male to charge after a member of the group was contacted. After contacting a lone female, I timed the interval between her communication with the male (via tree-slap, chest-beat or both) and when the male charged. Time interval ($n = 7$) ranged from 50 s through 3 min 30 s, with a mean of 2.2 min. This lag in time may suggest that males were often quite far away from the females in the group. In Bwindi, silverback males very seldom charged and were rarely out of visual contact with the rest of the group.

## The formation of temporary subgroups

Along with increasing group spread, another efficient way to reduce competition is to form smaller temporary subgroups (defined as when morning and evening nest counts differ by three or more individuals). Evidence of subgrouping occurred in both subspecies. In Bai Hokou, there were significant group size differences with regard to season. During the dry season, when gorillas were primarily folivorous, mean estimated group size was significantly larger (10.1 individuals, $n = 125$, SD $= 4.0$) than when they were primarily frugivorous (8.7 individuals, $n = 107$, SD $= 4.2$, $t_{[1,230]} = 2.3$, $p = 0.02$). This finding suggests perhaps that larger groups were splitting into smaller subgroups more frequently in the rainy season, when feeding on fruits, than in the dry season. In contrast, in Bwindi there was no difference between dry (10.2 individuals, $n = 69$, SD $= 3.8$) and rainy (9.7 individuals, $n = 70$, SD $= 2.2$) season group size ($t_{[1,137]} = 0.98$, $p > 0.05$). However, two of the three study groups in Bwindi were one-male groups and, as noted below, multiple males are necessary if smaller subgroups are to form.

In Bai Hokou, groups were found to form subgroups on 29 occasions (31%, $n = 93$). Although groups were relatively unstable throughout the year, subgroups formed significantly more often during the rainy (38%) than the dry season (17%) (likelihood-ratio $\chi^2$ test, $\chi^2 = 4.08$, df $= 1$, $p = 0.04$). When the minimum number of nights a given subgroup slept apart could be estimated (as determined by consecutive day follows when subgroups formed, $n = 17$), subgroups slept separately for one night on 13 occasions (76%) and two nights on four occasions (24%).

During the first two years of study within Bwindi there was no evidence of subgrouping. In December 1998, however, two newly habituated groups became available for study and both had multiple males. The Nkuringo group has been observed to split and nest more than 40 m from each other on 15 occasions (based on preliminary data from 2001 not included in this study). When more than 40 m apart, the mean distance from the two nesting silverbacks was 165 m ($n = 15$, range 40–370 m). This grouping behavior, however, may be the result of the Nkuringo group experiencing a true group fissioning, with the younger male getting ready to leave. The second group in Bwindi, Habinyanja, was habituated for tourism in July 1998, but actual group size and composition are still unknown. Nest counts of this group varied from 15 to 25 ($n = 7$) with anywhere from two to three silverbacks and four to six infants. On at least two occasions, this group formed two smaller groups that slept separately from each other. However, further work with this group is needed to determine the reasons for this behavior.

I have suggested elsewhere that it is the number of silverback males that allows for this fluid behavior in gorillas (Goldsmith, 1996). If a group has just one silverback male, subgrouping can not occur. Therefore, a prerequisite for this behavior is a minimum of two mature males within the group. As noted above, the number of males within a group is positively correlated with group size in both subspecies. The fact that larger, rather than smaller, groups fission is some indication that subgrouping results from increased competition within groups.

## Results of within-population comparisons

To further demonstrate flexibility in gorilla behavioral ecology, I present comparisons among the three study populations in Bwindi. The Mubare group was studied at the Buhoma tourist site, which represents the lowest-elevation site in the study area at 1450 m. At low elevations, gorilla and chimpanzee densities are relatively high and fruits are seasonally readily available. Mubwindi, the research site where the Kyaguilo group is habituated, lies at the highest elevation at about 2100 m. Chimpanzee and fruit densities are much reduced compared to lower elevations. Preliminary data are available from a third site, Kashasha, where the Nkuringo group is now habituated for tourism, though tourism has not yet begun. This site lies at about 1650 m (see Fig. 14.2A).

The diversity and density of terrestrial vegetation differed among sites within Bwindi, but were not necessarily related to altitude (Fig. 14.5A). Although

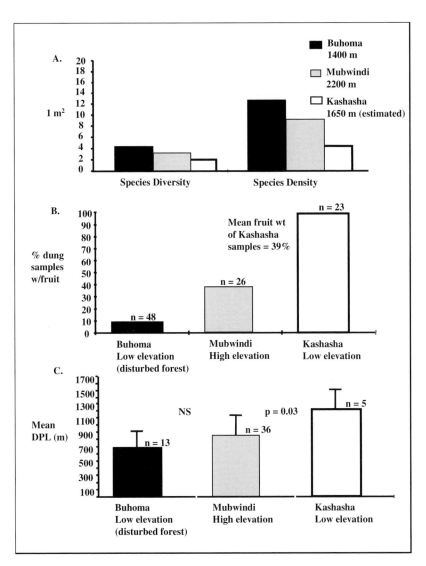

Fig. 14.5. (A) The availability of terrestrial vegetation, (B) the amount of fruit within the diet, and (C) the daily path length of three mountain gorilla groups within Bwindi.

predicted to have the lowest density of vegetation, Buhoma actually had the highest density with 12.2 stems per plot and a diversity of 3.5 species per plot. However, the likely reason for this is that Buhoma gorillas spent most of their time outside the park in artificially dense secondary growth (e.g., in January 1997 they spent only three of 29 days within the forest), and line transects for vegetation analyses were placed within these core areas. Mubwindi, the highest elevation, shows the same diversity with 3.3 species per plot, but a lower density of 8.5 stems per plot. In Kashasha, subjective observation suggests species diversity and density were relatively low (no data available from transects to date). However, during my last field season (data not included in this analysis), the Nkuringo group in Kashasha spent more than 95% of their time outside the park boundary in and around agricultural fields (Goldsmith, 2000). Because gorillas in the low elevation of Buhoma and Kashasha are spending more time outside of the park, they may experience a higher density of THV due to artificial cutting and increases in secondary growth than those groups at higher altitudes limited to within the park boundary. Further study is needed comparing the density of THV inside versus outside the park boundary (Goldsmith, 2000).

Bwindi gorillas ate fruits from 10 of the 46 plant species they consumed. Only 17% of dung samples from Buhoma gorillas, which spent little time within the fruit-rich forest, had fruit remnants, and when fruit was present, it represented only 1.5% of the total dried dung weight. At the high elevation in Mubwindi where gorillas were confined within the forest boundary, 35% of samples had fruit remains and fruit represented 2.8% of the dried weight. In the low-elevation site of Kashasha, when the study group spent all of its time inside the park boundary and THV was at low densities, 92% of all samples contained fruit and dried fruit remains represented 39% of the dung weight (Fig. 14.5B).

Important foods for Bwindi gorillas that were not eaten by western lowland gorillas (Goldsmith, 1996) include: *Chrysophyllum* fruits (Sapotaceae), *Brillantasia* pith (Acanthaceae), *Basella* leaves (Basellaceae), *Rubis* leaves and fruit (Rosaceae), *Triumfetta* bark and leaves (Tiliaceae), *Mimulopsis* bark and leaves (Acanthaceae), *Urera* bark and leaves (Urticaceae), and *Cyathea* pith (Cyathaceae).

Daily path length among sites appears to follow trends in frugivory (Fig. 14.5C). Buhoma gorillas, who were the least frugivorous, traveled the shortest distance with a mean of 697 m ($n = 13$, SD $= 238$), Mubwindi gorillas traveled 867 m per day ($n = 40$, SD $= 291$), while the most frugivorous group in Kashasha traveled 1013 m per day ($n = 5$, SD $= 504$). These differences, however, were not statistically significant ($F_{[2,56]} = 2.6$, $p = 0.10$). In addition, one note of caution is needed. When data were being collected on the Nkuringo group in Kashasha, they were undergoing habituation so it is difficult to determine whether their longer path lengths were due to that experience or to their greater frugivory.

Group size differed among the three populations. The high elevation Mubwindi gorillas had a significantly smaller nest count (mean 8.7 nests/site), than either Buhoma or Kashasha (mean 10.5 and 11.6 nests/site, respectively; $F_{[2,104]} = 14.66$, $p = 0.001$; *post-hoc* Tukey $p < 0.05$). The mean area around each nest was also significantly different among sites. The Buhoma tourist group was the most cohesive (mean 16.7 m$^2$ around each nest) with the Kashasha and Mubwindi groups being much less cohesive in their nesting sites (74.9 m$^2$ and 66 m$^2$, respectively) ($F_{[2,103]} = 3.3$, $p = 0.04$). In a *post-hoc* Tukey test, the only significant difference occurs between the Mubwindi and Buhoma groups ($p = 0.04$). Perhaps individuals in the Buhoma group slept closer to one another because they nested outside of the park boundary, where they might be vulnerable to harassment or predation by humans (Goldsmith, 2000).

**Discussion**

The fact that frugivory influences primate behavior differently from folivory has become critical to our understanding of the diversity being discovered among gorilla populations. Each subspecies demonstrates a different ratio of frugivory to folivory, a ratio that appears to be related to altitude. Following differences in altitude, Bwindi mountain gorillas were found to be more frugivorous than Virunga mountain gorillas living at a higher elevations, but less frugivorous than were lowland gorillas. However, their level of frugivory did not seem to radically affect their foraging effort and grouping patterns, suggesting that they fall closer to their Virunga relatives along the continuum in Fig. 14.1 (contrary to suggestions by Sarmiento *et al.*, 1996). These data, however, are somewhat preliminary (compared to information from Bai Hokou and the Virungas) and longer-term observations on this population are needed.

Although fruit availability was related to altitude, the relationship of THV to altitude was less clear. Watts (1984) demonstrates nicely how the density of THV in the Virungas is related to vegetation zone or habitat. For example, the density of THV in Lobelia and Afro-Alpine zones was less than 1 stem per plot, whereas in the Nettle zone it increased to 18.5 stems per plot. Goldsmith (1996) and White *et al.* (1995) demonstrate similar findings in the habitats of western lowland gorillas in Bai Hokou and Lopé, respectively. White *et al.* (1995) also suggest that biomass, rather than overall density, might be most meaningful because different herbs represent various proportions of food to gorillas.

Overall, THV was more densely distributed in the Virunga region where a greater reliance on this more evenly distributed and perennially available food may allow for larger and more stable groups to form in the Virungas (Watts, 1984). Wrangham (1979) had suggested that because of their frugivorous diet, western lowland gorillas should live in smaller groups than mountain gorillas. Early estimates comparing Equatorial Guinea lowland gorillas did suggest that groups were smaller than those of Virunga mountain gorillas (Harcourt *et al.*, 1981; but see criticisms raised by Tutin *et al.*, 1992). However, more recent data from the Congo (Olejniczak, 1996) and Bai Hokou (Goldsmith, 1996) suggest groups of 7.2 and 9.4 individuals, respectively (excluding lone males), very similar to that of Virunga and Bwindi mountain gorillas. More likely, it is maximum group size, rather than mean group size, that is restricted in lowland gorillas, as compared to mountain gorillas (see also Yamagiwa *et al.*, this volume). For example, a group of 34 individuals was known to form in the Virungas (Doran and McNeilage, 1999), whereas in Bai Hokou, the largest group is only about 20 individuals. In the more frugivorous Bwindi region gorilla maximum group size is 25 (or 23 per the 1997 census: McNeilage *et al.*, 1998). However, recently in Lossi, Congo, Bermejo (1997) has recorded western lowland groups as large as 20 and 30, but suggests that these larger groups are a result of low levels of frugivory in this area.

In addition to ecological factors, it is important to note that habituation may cause groups to increase in size. A study of the entire Virunga population demonstrated that fecundity was significantly higher in those groups monitored by humans on a regular basis (McNeilage, 1996). It is possible that added protection from predation and poaching, as well as veterinary intervention, has offset the balance of the "natural" ecological and social factors influencing group size (especially in the Virungas).

Both herbivory and habituation, which might help to promote large groups, may be counterbalanced by social factors that act to control optimal size. For example, infanticide may constrain the size of folivorous Virunga gorilla groups (Watts, 1989; Crockett and Janson, 2000). Although infanticide was found in 34 of 37 primate species with female dispersal (van Schaik, 2000), this behavior has only been observed in gorillas living in the Virungas. While observations are limited, there have been no documented or suspected cases of infanticide in either lowland gorillas or in Bwindi mountain gorillas. In fact, peaceful co-mingling among western lowland gorillas is often seen in open clearings (Oljecnick, 1994, 1996) and evidence of super-grouping (two groups coming together and sleeping in the same site) also exists (Goldsmith, 1996; Bermejo, 1997).

Where frugivory is high and maximum group size is constrained, groups respond by increasing their foraging effort. In addition, they increase their group spread when traveling and may be pressured to form smaller, temporary subgroups. More than five years of study at Bai Hokou has documented that when feeding on fruits, gorillas travel farther, have a larger group spread, and form temporary subgroups more often (Goldsmith, 1996; Remis, 1997*b*). In contrast, findings from Lopé suggest groups are similar in degree of cohesion to those in the Virungas (Tutin, 1996). Table 14.1 shows an unusually high density of THV in Lopé as compared to other western lowland gorilla sites, which might reduce within-group feeding competition further in this population and allow for larger, more stable groups.

The number of silverback males in a group influences flexibility in group size and cohesion. All nests recorded in Bai Hokou and Bwindi have at least one silverback male present, and this is generally true for all gorilla populations. This suggests that males play an important role in protection, especially at night when interactions with leopards are more common. The number of males in both Bai Hokou and Bwindi was related to group size and the presence of multiple males was a prerequisite to subgroup formation.

### Bwindi gorillas

This study provides some of our first insight into the behavioral ecology of the poorly known Bwindi population. These gorillas were found to eat fruit throughout the year, but more so during the rainy season when more fruit were available. In this study they tended to concentrate on one or two species of fruit at a time, but more data are needed on fruit availability and diet choice to determine levels of selectivity. Although Bwindi gorillas do consume fruit, their staple foods are primarily herbaceous stems, leaves, and bark. They also consumed large amounts of dead wood and often foraged on ants. Arboreally, they mostly ate lianas and appeared to prefer feeding on a variety of epiphyte species. When given the chance, they would also strip entire eucalyptus trees of their bark. Unlike the primarily terrestrial Virunga population, Bwindi gorillas of all ages spent a great deal of time in the trees foraging, resting, and playing.

Although the distance Bwindi groups traveled each day was related to their level of frugivory, their overall daily path length (780 m) was not drastically different from that of the Virunga gorillas (570 m for Group 4) (Watts, 1991). Results on the relationship between foraging effort, group cohesion, and fru-givory are preliminary and more data collected during the rainy season and corresponding to phenological cycles are needed. Multi-male groups tended to be less cohesive than one-male groups. It is unclear if this behavior is a result of increased within-group feeding competition or whether the Nkuringo Group in Kashasha is in the process of fissioning (as suggested for some western lowland gorilla groups by Remis, 1994). Besides degree of frugivory, Bwindi gorillas also differ from those in the Virungas by living sympatrically with chimpanzees and elephants.

Long-term sympatric studies in Lopé, Gabon, show that most gorilla plant foods are harvested arboreally (69%), and that dietary overlap between gorillas and chimpanzees is high (60–80%). In addition, the diet of Lopé gorillas is much closer to that of chimpanzees than it is to Virunga gorillas (Rogers *et al.*, 1990; Tutin and Fernandez, 1993). However, gorillas were more likely to feed on THV than chimpanzees and to concentrate on this resource when fruit was scarce. Chimpanzee and gorilla diets diverge most at times when fruit is not abundant, although it is mainly gorillas that shift foraging strategies, while chimpanzees continue to forage for sparsely available ripe fruits. In 1997, the Bwindi Impenetrable Great Ape Project (BIGAPE) began a similar study on the comparative behavioral ecology of sympatric gorillas and chimpanzees (e.g., Goldsmith and Stanford, 1997). Gorillas and chimpanzees do share some fruit sources and gorillas do eat more herbaceous vegetation than chimpanzees (Goldsmith *et al.*, 1998), but more data are still needed. Direct competition between the two species has not been observed; in fact, groups have been seen co-feeding in trees (personal observation). Similar findings of chimpanzees and gorillas feeding in trees come from Bai Hokou (Remis, 1994; Goldsmith, 1996) and Ndoki Forest, Congo (Kuroda *et al.*, 1996). Due to peaceful co-mingling and the fact that some study groups spend large amounts of time outside of the park (not sympatric with chimpanzees), it is likely that scramble competition between the two species is not a major selective force on foraging effort and grouping patterns for Bwindi's gorillas.

Comparisons among the three populations within Bwindi were not highly enlightening. Although attempts were made to compare altitudinal differences in the availability and distribution of THV and fruit, no clear differences emerged. Findings were probably confounded by the fact that altitudinal differences among Bwindi sites are not a real phenomenon, since mountains within each site provide multiple elevations in which gorillas can forage. In addition, a more confounding effect on diet and behavior among sites was the fact that some groups spent most of their time outside the forest in and around agricultural areas, artificially influencing their dietary choices. More data are needed on what factors attract gorillas to these areas and what effect spending time outside the park has on their foraging and grouping patterns (Goldsmith, 2000).

### *Lowland/highland gorillas and the bonobo and chimpanzee analogy: Where do Bwindi gorillas fit?*

In Fig. 14.1 the obligate frugivorous diet of chimpanzees and the primarily herbaceous diet of the Virunga mountain gorillas place them at either extreme of the model; all other African apes fall somewhere along this continuum. In practicality, the dietary ratio of frugivory to folivory provides an image of the environment by determining what kind of food patches are available. If you are an obligate frugivore, there are few patches; if you consume THV, there are many patches. Size and distance between all possible food patches influences foraging and grouping patterns. This is similar to Wrangham's (2000) "feed as you go" theory. Bonobos, and Bai Hokou and Bwindi gorillas all feed as they go. Bonobos feed on THV in their continued search for fleshy fruits.

This reduces their foraging effort (there are more patches and shorter distances between them), thereby minimizing competition and allowing for larger groups as compared to chimpanzees. Bai Hokou lowland gorillas, with their large body size, are able to accommodate a more fibrous diet. So when they "feed as they go" they not only nibble on THV, they also include large amounts of woody vegetation (e.g., bark, pith, and leaves of lianas) as well as termites. This adds more patches and shortens the distance among them. As a result, Bai Hokou groups are larger and more stable than those of bonobos, but similar to bonobos in that grouping patterns can fluctuate in relation to resource availability. Given the high density of THV in Lopé the "feed as you go" theory suggests even larger and closer patches, again possibly explaining more cohesive groups similar to those in the Virungas.

The Bwindi gorillas, living at lower elevations where fruit is available, fall in between lowland and Virunga mountain gorillas in their dietary ratio and group cohesion. Fruit is less plentiful and diverse in Bwindi than in lowland areas, and as a result less fruit is found in their diet. Although data are preliminary, fruit does not seem to influence their foraging effort. Costs to grouping are probably low and the greater ratio of THV to fruit probably places them closer to the Virunga gorillas along the continuum rather than the lowland gorilla. Although there is some evidence of reduced cohesion and the formation of subgroups, more data are needed to reveal the costs of Bwindi gorilla foraging.

**Acknowledgments**

For permission and logistical help while in the Central African Republic, I thank the Ministères des Eaux et Forêts, the Ministères de Recherches Scientifiques, and the World Wildlife Fund, US, and in Uganda I thank the National Council of Science and Technology, Ugandan Wildlife Authority, and the Institute of Tropical Forest Conservation. For financial support in the Central African Republic I acknowledge the J. William Fulbright Foreign Scholarship Board, the Wildlife Conservation Society, the Wenner–Gren Anthropological Research Foundation, and the L.S.B. Leakey Foundation. For financial support in Uganda I acknowledge grants received from the National Geographic Society, the L.S.B Leakey Foundation, Primate Conservation Inc., and the Claire Goodman Fund of Dartmouth College. I gratefully acknowledge the many local field assistants (too many to list) who made my research in both countries possible and my colleagues Melissa Remis, Etienne Ndoulongbe, Danielle Wilson, Cathy Clark, Craig Stanford, Alexa Hanke, and John Bosco Nkurunungi. I greatly appreciate the review of this manuscript by two anonymous reviewers, Allen Rutberg, and Andrea Taylor. A special thank you to Andrea, my co-editor on this volume, for her support throughout the process.

## References

Achoka, I. (1994). Home range, group size, and group composition of mountain gorillas in the Bwindi-Impenetrable National Park, southwest Uganda. MSc thesis, Makerere University, Kampala, Uganda.

Badrian, A. and Badrian, N. (1984). Social organization of *Pan paniscus* in the Lomako Forest, Zaïre. In *The Pygmy Chimpanzee: Evolutionary Biology and Behavior*, ed. R.L. Susman, pp. 325–346. New York: Plenum Press.

Badrian, N. and Malenky, R. (1984). Feeding ecology of *Pan paniscus* in the Lomako Forest, Zaïre. In *The Pygmy Chimpanzee: Evolutionary Biology and Behavior*, ed. R.L. Susman, pp. 275–299. New York: Plenum Press.

Bermejo, M. (1997). Study of western lowland gorillas in the Lossi Forest of North Congo and a pilot gorilla tourism plan. *Gorilla Conservation News*, **11**, 6–7.

Berry, J.P. (1998). The chemical ecology of mountain gorillas (*Gorilla gorilla beringei*): With special reference to antimicrobial constituents in the diet. PhD thesis, Cornell University, Ithaca, NY.

Butynski, T.M. (1984). *Ecological Survey of the Impenetrable (Bwindi) Forest, Uganda, and Recommendations for its Conservation and Management.* Unpublished report to the Ministers of Tourism and Wildlife, Kampala.

Carroll, R.W. (1986). Status of the lowland gorilla and other wildlife in the Dzanga-Sangha region of the Southwestern Central African Republic. *Primate Conservation* **7**, 38–41.

Carroll, R.W. (1988). Relative density, range extension, and conservation potential of the lowland gorilla (*Gorilla g. gorilla*) in the Dzanga-Sangha region of the southwestern Central African Republic. *Mammalia*, **52**, 309–323.

Chapman, C.A., White, F.J., and Wrangham, R.W. (1994). Party size in chimpanzees and bonobos: A reevaluation of theory based on two similarly forested sites. In *Chimpanzee Cultures*, eds. W.C. McGrew, R.W. Wrangham, F.B.M. de Waal, and P.G. Heltne, pp. 41–58. Cambridge, MA: Harvard University Press.

Clutton-Brock, T.H. and Harvey, P.H. (1977). Primate ecology and social organization. *Journal of Zoology, London*, **183**, 1–39.

Crockett, C.M. and Janson, C.H. (2000). Infanticide in red howlers: Female group size, male membership, and a possible link to folivory. In *Infanticide By Males and Its Implications*, eds. C.P. van Schaik and C.H. Janson, pp. 75–98. Cambridge, U.K.: Cambridge University Press.

Doran, D.M. and McNeilage, A. (1999). Gorilla ecology and behavior. *Evolutionary Anthropology*, **6**, 120–131.

Fay, J.M., Carroll, R., Kerbis-Peterhans, J.C., and Harris, D. (1995). Leopard attack on and consumption of gorillas in the Central African Republic. *Journal of Human Evolution*, **29**, 93–99.

Fossey, D. and Harcourt, A.H. (1977). Feeding ecology of free-ranging mountain gorillas (*Gorilla gorilla beringei*). In *Primate Ecology: Studies of Feeding and Ranging Behaviour in Lemurs, Monkeys and Apes*, ed. T.H. Clutton-Brock, pp. 415–447. London: Academic Press.

Garner, K.J. and Ryder, O.A. (1996). Mitochondrial DNA diversity in gorillas. *Molecular Phylogenetics and Evolution*, **6**, 39–48.

Goldsmith, M.L. (1996). Ecological influences on the ranging and grouping behavior of western lowland gorillas at Bai Hokou in the Central African Republic. PhD thesis, State University of New York, Stony Brook, NY.

Goldsmith, M.L. (1999). Ecological constraints on the foraging effort of western gorillas (*Gorilla gorilla gorilla*) at Bai Hokou, Central African Republic. *International Journal of Primatology*, **20**, 1–24.

Goldsmith, M.L. (2000). Effects of ecotourism on behavioral ecology of Bwindi gorillas, Uganda: Preliminary results. *American Journal of Physcial Anthropology, Supplement* **30**, 161.

Goldsmith, M.L. and Stanford, C.B. (1997). Bwindi Great Ape Project, Uganda. *Gorilla Conservation News*, **11**, 17–18.

Goldsmith, M.L., Hanke, A., Nkurunungi, J.B., and Stanford, C.B. (1998). Comparative behavioral ecology of Bwindi gorillas and chimpanzees, Uganda: Preliminary results. *American Journal of Physical Anthropology, Supplement* **26**, 101.

Goodall, J. (1986). *The Chimpanzees of Gombe: Patterns of Behavior*. Cambridge, MA: Harvard University Press.

Grubb, P.J. and Whitmore, T.C. (1966). A comparison of montane and lowland rain forest in Equador. II: Climate and its effects on the distribution and physiognomy of the forests. *Journal of Ecology*, **54**, 303–333.

Harcourt, A.H., Fossey, D., and Sabater Pi, J. (1981). Demography of *Gorilla gorilla*. *Journal of Zoology*, London, **195**, 215–233.

Janson, C.H. and Goldsmith, M.L. (1995). Predicting group size in primates: Foraging costs and predation risks. *Behavioral Ecology*, **6**, 326–336.

Kano, T. (1982). The social group of pygmy chimpanzees (*Pan paniscus*) of Wamba. *Primates*, **23**, 171–188.

Kano, T. and Mulavwa, M. (1984). Feeding ecology of the pygmy chimpanzee (*Pan paniscus*) of Wamba. In *The Pygmy Chimpanzee: Evolutionary Biology and Behavior*, ed. R.L. Susman, pp. 233–273. New York: Plenum Press.

Kuroda, S. (1979). Grouping of the pygmy chimpanzee. *Primates*, **20**, 161–183.

Kuroda, S., Nishihara, T., Suzuki, S., and Oko, R.A. (1996). Sympatric chimpanzees and gorillas in the Ndoki Forest, Congo. In *Great Ape Societies*, eds. W.C. McGrew, L.T. Marchant, and T. Nishida, pp. 71–81. Cambridge, U.K.: Cambridge University Press.

Lieberman, D., Lieberman, M., Peralta, R., and Harshorn, G.S. (1996). Tropical forest structure and composition on a large-scale altitudinal gradient in Costa Rica. *Journal of Ecology*, **84**, 137–152.

Malenky, R.K. (1990). Ecological factors affecting food choice and social organization in *Pan paniscus*. PhD thesis, State University of New York, Stony Brook, NY.

Malenky, R.K. and Stiles, E.W. (1991). Distribution of terrestrial herbaceous vegetation and its consumption by *Pan paniscus* in the Lomako Forest, Zaïre. *American Journal of Primatology*, **23**, 153–169.

Malenky, R.K., Kuroda, S., Vineber, E.O., and Wrangham, R.W. (1994). The significance of terrestrial hevbaceous foods for bonobos, chimpanzees and gorillas. In *Chimpanzee Cultures*, eds. R.W. Wrangham, W.C. McGrew, F.B.M. de Waal, and P.G. Heltne, pp. 59–75. Cambridge, MA: Harvard University Press.

McNeilage, A. (1996). Ecotourism and mountain gorillas in the Virunga Volcanoes. In *The Exploitation of Mammal Populations*, eds. V.J. Taylor and N. Dunstone, pp. 334–344. London: Chapman & Hall.

McNeilage, A., Plumptre, A., Brock-Doyle, A., and Vedder, A. (1998). Bwindi-Impenetrable National Park, Uganda: Gorilla and large mammal census, 1997. New York: Wildlife Conservation Society.

Milton, K. and May, M.L. (1976). Body weight, diet and home range area in primates. *Nature*, **259**, 459–462.

Mitani, M. (1992). Preliminary results of the studies of wild western lowland gorillas and other sympatric diurnal primates in the Ndoki Forest, Northern Congo. In *Topics in Primatology*, Vol. 2, *Ecology and Conservation*, eds. N. Itoigawa, Y. Sugiyama, G.P. Sackett, and R.K.R. Thompson, pp. 215–224. Tokyo: University of Tokyo Press.

Nishida, T. (1979). The social structure of chimpanzees of the Mahale Mountains. In *The Great Apes*, eds. D.A. Hamburg and E.R. McCown, pp. 73–122. Menlo Park, CA: Benjamin/Cummings.

Nishihara, T. (1995). Feeding ecology of western lowland gorillas in the Nouabalé-Ndoki National Park, Congo. *Primates*, **36**, 151–168.

Olejniczak, C. (1994). Report on a pilot study of western lowland gorllas at Mbeli Bai, Nouablé-Ndoki Reserve, northern Congo. *Gorilla Conservation News*, **8**, 9–11.

Olejniczak, C. (1996). Update on the Mbeli Bai gorilla study, Nouablé-Ndoki National Park, northern Congo. *Gorilla Conservation News*, **10**, 5–8.

Remis, M.J. (1993). Nesting behavior of lowland gorillas in the Dzanga-Sangha Reserve, Central African Republic: Implications for population estimates and understandings of group dynamics. *Tropics*, **2**, 245–255.

Remis, M.J. (1994). Feeding ecology and positional behavior of western lowland gorillas (*Gorilla gorilla gorilla*) in the Central African Republic. PhD thesis, Yale University, New Haven, CT.

Remis, M.J. (1997a). Western lowland gorillas (*Gorilla gorilla gorilla*) as seasonal frugivores: Use of variable resources. *American Journal of Primatology*, **43**, 87–109.

Remis, M.J. (1997b). Ranging and grouping patterns of a western lowland gorilla group at Bai Hoköu, Central African Republic. *American Journal of Primatology*, **43**, 111–133.

Remis, M.J. (1999). Tree structure and sex differences in arboreality among western lowland gorillas (*Gorilla gorilla gorilla*) at Bai Hokou, Central African Republic. *Primates*, **40**, 383–96.

Richards, J.W. (1952). *The Tropical Rain Forest: An Ecological Study*. London: Cambridge University Press.

Robbins, M.M. (1995). A demographic analysis of male life history and social structure of mountain gorillas. *Behaviour*, **132**, 21–47.

Robbins, M.M. (2000). Behavioral ecology of the (other) mountain gorillas of Bwindi Impenetrable National Park, Uganda, Africa. In *Proceedings of conference "The Apes: Challenges for the 21ˢᵗ Century"*, p. 56. Chicago, IL: Brookfield Zoological Association.

Rogers, M.E. and Williamson, E.A. (1987). Density of herbaceous plants eaten by gorillas in Gabon: Some preliminary data. *Biotropica*, **19**, 278–281.

Rogers, M.E., Maisels, F., Williamson, E.A., Fernandez, M., and Tutin, C.E.G. (1990). Gorilla diet in the Lopé Reserve, Gabon: A nutritional analysis. *Oecologia*, **84**, 326–339.

Sarmiento, E.E., Butynski, T.M., and Kalina, J. (1996). Gorillas of Bwindi-Impenetrable Forest and the Virunga Volcanoes: Taxonomic implications of morphological and ecological differences. *American Journal of Primatology*, **40**, 1–21.

Schaller, G.B. (1963). *The Mountain Gorilla: Ecology and Behavior*. Chicago, IL: University of Chicago Press.

Sholley, C.R. (1991). Conserving gorillas in the midst of guerrillas. In *American Association of Zoological Parks and Aquariums, Annual Conference Proceedings*, pp. 30–37.

Struhsaker, T.T. (1975). *The Red Colobus Monkey*. Chicago, IL: University of Chicago Press.

Tutin, C.E.G. (1996). Ranging and social structure of lowland gorillas in the Lopé Reserve, Gabon. In *Great Ape Societies*, eds. W.C. McGrew, L.F. Marchant, and T. Nishida, pp. 58–70. Cambridge, U.K.: Cambridge University Press.

Tutin, C.E.G. and Fernandez, M. (1985). Foods consumed by sympatric populations of *Gorilla g. gorilla* and *Pan t. troglodytes* in Gabon: Some preliminary data. *International Journal of Primatology*, **6**, 27–43.

Tutin, C.E.G. and Fernandez, M. (1993). Composition of the diet of chimpanzees and comparisons with that of sympatric lowland gorillas in the Lopé Reserve, Gabon. *American Journal of Primatology*, **30**, 195–211.

Tutin, C.E.G., Fernandez, M., Rogers, M.E., and Williamson, E.A. (1992). A preliminary analysis of the social structure of lowland gorillas in the Lopé Reserve, Gabon. In *Topics in Primatology*, Vol. 2, *Ecology and Conservation*, eds. N. Itoigawa, Y. Sugiyama, G.P. Sackett, and R.K.R. Thompson, pp. 245–253. Tokyo: University of Tokyo Press.

van Schaik, C.P. (2000). Infanticide by male primates: The sexual selection hypothesis. In *Infanticide by Males and Its Implications*, eds. C.P. van Schaik and C.H. Janson, pp. 27–59. Cambridge, U.K.: Cambridge University Press.

van Schaik, C.P., Terborgh, J.W., and Wright, S.J. (1993). The phenology of tropical forests: Adaptive significance and consequences for primary consumers. *Annual Review of Ecology and Systematics*, **24**, 353–377.

Vedder, A.L. (1984). Movement patterns of a group of free-ranging mountain gorillas (*Gorilla gorilla beringei*) and their relation to food availability. *American Journal of Primatology*, **7**, 73–88.

Watts, D.P. (1984). Composition and variability of mountain gorilla diets in the central Virungas. *American Journal of Primatology*, **7**, 323–356.

Watts, D.P. (1989). Infanticide in mountain gorillas: New cases and a reconstruction of the evidence. *Ethology*, **81**, 1–18.

Watts, D.P. (1991). Strategies of habitat use by mountain gorillas. *Folia Primatologica*, **56**, 1–16.

Watts, D.P. (1998). Seasonality in the ecology and life histories of mountain gorillas (*Gorilla gorilla gorilla*). *International Journal of Primatology*, **19**, 929–948.

White, F.J. (1988). Party composition and dynamics in *Pan paniscus*. *International Journal of Primatology*, **9**, 179–193.

White, F.J. and Wrangham, R.W. (1988). Feeding competition and patch size in the chimpanzee species *Pan paniscus* and *Pan troglodytes*. *Behaviour*, **105**, 148–164.

White, L.J.T., Rogers, M.E., Tutin, C.E.G., Williamson, E.A., and Fernandez, M. (1995). Herbaceous vegetation in different forest types in the Lopé Reserve, Gabon: Implications for keystone food availability. *African Journal of Ecology*, **33**, 124–141.

Williamson, E.A. (1988). Behavioural ecology of western lowland gorillas in gabon. PhD thesis, University of Stirling, Stirling, U.K.

Williamson, E.A., Tutin, C.E.G., Rogers, M.E., and Fernandez, M. (1990). Composition of the diet of lowland gorillas at Lopé in Gabon. *American Journal of Primatology*, **21**, 265–277.

Wrangham, R.W. (1979). On the evolution of ape social systems. *Social Science Information*, **18**, 335–368.

Wrangham, R.W. (1986). Ecology and social relationships in two species of chimpanzee. In *Ecological Aspects of Social Evolution*, eds. D.I. Rubenstein and R.W. Wrangham, pp. 352–378. Princeton, NJ: Princeton University Press.

Wrangham, R.W. (2000). Why are male chimpanzees more gregarious than mothers? A scramble competition hypothesis. In Primate Males: *Causes and Consequences of Variation in Group Composition*, ed. P.M. Kappeler, pp. 248–258. Cambridge, U.K.: Cambridge University Press.

Wrangham, R.W. and Smuts, B.B. (1980). Sex differences in the behavioral ecology of chimpanzees in the Gombe National Park, Tanzania. *Journal of Reproduction and Fertility, Supplement* **28**, 13–31.

Wrangham, R.W., Conklin, N.L., Chapman, C.A., and Hunt, K.D. (1991). The significance of fibrous foods for the Kibale Forest chimpanzees. *Philosophical Transactions of the Royal Society of London, Series B*, **334**, 171–178.

Wrangham, R.W., Clark, A., and Isabirye-Basuta, G. (1992). Female social relationships and social organization of Kibale forest chimpanzees. In *Topics in Primatology*, Vol. 1, *Human Origins*, eds. T. Nishida, W.C. McGrew, P. Marler, M. Pickford, and F.B.M. de Waal, pp. 81–98. Tokyo: University of Tokyo Press.

Wrangham, R.W., Gittleman, J.L., and Chapman, C.A. (1993). Constraints on group size in primates and carnivores: Population density and day-range as assays of exploitation competition. *Behavioral Ecology and Sociobiology*, **32**, 199–209.

Wrangham, R.W., Chapman, C.A., Clark-Arcadi, A.P., and Isabirye-Basuta, G. (1996). Social ecology of Kanyawara chimpanzees: Implications for understanding the costs of great ape groups. In *Great Ape Societies*, eds. W.C. McGrew, L.F. Marchant, and T. Nishida, pp. 45–57. Cambridge, U.K.: Cambridge University Press.

Yamagiwa, J., Mwanza, N., Yumoto, T., and Maruhashi, T. (1994). Seasonal change in the composition of the diet of eastern lowland gorillas. *Primates*, **35**, 1–14.

# 15 Are gorillas vacuum cleaners of the forest floor? The roles of body size, habitat, and food preferences on dietary flexibility and nutrition

MELISSA J. REMIS

## Introduction

Gorilla diet varies seasonally, geographically, and with altitude among eastern (*Gorilla gorilla beringei* and *Gorilla gorilla graueri*) and western (*Gorilla gorilla gorilla*) populations. Recent research on gorillas has highlighted the importance of fruit consumption and dietary flexibility at most sites (Rogers *et al.*, 1990; Tutin *et al.*, 1991, 1997; Remis, 1994, 1997*a,b*; Nishihara, 1995; Goldsmith, 1996, 1999 this volume; Doran and McNeilage, 1998; McFarland, 2000). Traditional characterizations of gorillas as folivores (Schaller, 1963; Fossey and Harcourt, 1977) have now shifted to include seasonal frugivory at all but the highest elevation sites (Yamagiwa *et al.*, 1996; Remis, 1997*a*; Robbins, 2000). Nevertheless, gorillas have been argued to be opportunistic frugivores relative to the smaller and more persistently frugivorous chimpanzee (Nishihara, 1995; Kuroda *et al.*, 1996), and some might characterize them as vacuum cleaners of the forest floor. I examine the consequences of large body size and habitat on dietary flexibility among gorillas as well as interpopulation variation in frugivory and nutrient intake. I further integrate research in captivity on taste sensitivity and food preferences with field data to explore some of the physiological and behavioral bases for dietary flexibility of gorillas in their native habitats.

Gorillas are most often described as herbivore–folivores, primarily as a consequence of their large size and spacious colons and cecums that contain a high number of cellulose-digesting ciliates (Collet *et al.*, 1984). Gorillas have also been presumed to be able to retain low-quality foods in the gut to maximize absorption of nutrients (Milton, 1984; Demment and van Soest, 1985; Remis, 2000). This view has been largely shaped by study of the Karisoke mountain gorillas whose diet is herbivorous and characterized by little diversity or seasonal variation (Watts, 1996, 1998), and whose teeth and jaws are adapted for shearing leaves and minimizing fatigue failure from repetitive chewing, respectively (Groves, 1986; Uchida, 1998; Taylor, 2002, this volume). Study of other gorilla populations who consume larger amounts of fruit, and whose teeth more closely resemble those of frugivorous chimpanzees (Shea, 1983; Uchida, 1998) mandates re-evaluation of this model.

Variation in dietary selectivity and food preferences among primates likely relates to habitat variability and niche separation among closely related species (Ganzhorn, 1989; Wrangham *et al.*, 1998). Species differences are presumed to result from differences in digestive strategies for metabolizing the nutrients and secondary compounds found in plants (Freeland and Janzen, 1974; Glander, 1981; Chivers and Langer, 1994; Milton, 1998). Folivorous primates often consume foods that are high in digestion inhibitors, such as lignin and secondary plant compounds (Oates *et al.*, 1977; Mole and Waterman, 1987; Freeland, 1991; Waterman and Kool, 1994) but also high in protein (Hladik, 1978; Waterman *et al.*, 1983; Kay and Davies, 1994).

Gorillas have an enlarged hindgut associated with colic–cecal fermentation (Bauchop, 1978; Parra, 1978; Chivers and Hladik, 1980; Milton and Demment, 1988). They may thus be well equipped to digest fiber and tolerate tannin-rich foods that are diluted among voluminous gut contents (Rogers *et al.*, 1990; Cork and Foley, 1992). Despite their large gut size and ability to consume a tough, low-quality diet (Strait, 1997), gorillas are selective eaters (Casimir, 1975). Karisoke mountain gorillas are primarily herbivorous and choose vegetation that is relatively high in protein content and low in acid detergent fiber and condensed tannins relative to plant parts consumed by gorillas in lowland forests with a more varied diet (Waterman *et al.*, 1983; Rogers *et al.*, 1990; Plumptre, 1995; Popovich *et al.*, 1997). Low-altitude populations of gorillas are more frugivorous, but they have proved difficult to study. We now have a rapidly growing appreciation of the variability of gorilla diet between and within sites and subspecies. Nevertheless, our efforts to understand the effects of body size on the ecological adaptation of gorillas relative to chimpanzees and their evolution are complicated by the roles of habitat and altitude in shaping intraspecific variation in diet. Such efforts have been hampered further by an inadequate understanding of the physiological foundations of ape dietary adaptations.

A comparison of the chemical content of foods eaten during seasons of fleshy fruit scarcity by gorillas at Bai Hokou, Central African Republic, with nutrient profiles from other gorilla sites (Waterman *et al.*, 1983; Calvert, 1985; Rogers *et al.*, 1990), highlights population variability in gorilla diet. This information, along with preliminary results of experimental studies of food preferences and taste perception among captive gorillas, provides further insights into the physiological correlates of dietary selectivity and flexibility among gorillas. It permits exploration of the location of gorillas along a continuum of hominoid adaptations to frugivory.

## Methods

### Field research at Bai Hokou, Central African Republic

Ecological monitoring and studies of gorilla foraging ecology have been conducted at Bai Hokou, Central African Republic (2° 57′ N; 16° 22′ E) since 1984 (Remis, 1994, 1997*a,b*, 1999; Goldsmith, 1996; Carroll, 1997). Overall, rainfall averages 1365 mm annually with a single three-month dry season generally from December through March (Remis, 1997*a*).

Observational data on gorilla feeding behavior were collected opportunistically on semi-habituated focal animals at Bai Hokou in the periods September–November 1988, August 1990–October 1992, June–December 1995, and June–August 1998. Dietary information collected by Goldsmith (1996) from September 1993 to January 1995 is also included in this study. Data are presented for each wet and dry season using a feeding-visit method whereby each individual seen foraging is only scored once per observational period of feeding behavior (for additional details see Goldsmith, 1996; Carroll, 1997; Remis, 1997*a*). Nutritional information combines the analysis of gorilla foods collected on Bai Hokou gorilla feeding trails and during observations in March 1989 (late dry season), with those from the same study groups during the fruiting or wet season, June–August 1998.

Long-term records on the monthly phenological patterns of 973 marked trees (>10 cm diameter at breast height) of 152 species on 19 km of cut north–south transects at Bai Hokou allow variation in resource availability during the study periods to be assessed (Remis, 1994, 1997*a*; Goldsmith, 1996; Carroll, 1997). These trees are not restricted to gorilla food species, but provide a broader window into forest-wide resource availability. Phenology data were collected monthly on the presence or absence of ripe or unripe fruit, flowers, young or mature leaves and are available for the following months: June 1988–May 1989, March 1990–October 1992, September 1993–January 1995, June–December 1995, June 1997, and July 1998. Monthly percentages are averaged for each nine-month wet and three-month dry season. Foods partially consumed and discarded by gorillas (or foods similar to those consumed by the gorillas) were collected and dried or fixed in ethanol. Both sets of samples were analyzed by Dr. Ellen Dierenfeld (Wildlife Conservation Society, New York), for nutritional and fiber content (van Soest *et al.*, 1991; Association of Official Analytical Chemists, 1996) and Dr. Christopher Mowry (Berry College, Georgia) for condensed, hydrolyzable and total tannin content. Plant parts were compared using Mann–Whitney $U$ tests (see Remis *et al.*, 2001 for details).

## *Captive research at the San Francisco Zoological Gardens*

Ongoing experimental captive taste sensitivity and food preference trials on the six gorillas and four chimpanzees housed at the San Francisco Zoological Gardens further our understanding of dietary strategies and food choices of animals in the wild. Trials were conducted on the gorillas while they were maintained on their typical zoo diet of primate chow, domestic produce and browse (total dietary fiber approximately 23%: E. Dierenfeld, personal communication). Separate food preference and taste sensitivity trials were conducted as paired-choice experiments (Benz *et al.*, 1992; Simmen, 1994). During the trials, individual gorillas and chimpanzees were presented with (1) two domestic vegetable or fruit items varying in nutritional content, or (2) two drinking bottles ("Lixit" Corporation 1-liter bottles with sipper tubes) that contained either water or various concentrations of fructose (3–300 mM), or tannin–fructose mixtures (0.25–6 mM tannic acid ($C_{76}H_{52}O_{46}$, mw: 1701.23 Fluka) in 100 mM fructose or 300 mM fructose solutions). For example 1 l of 100 mM fructose solution contained 20 g fructose and water to make 1 l (Remis, 2002; Remis and Kerr, 2002). Each combination of foods was presented at least 10 times to each individual ($n = 3100$ pairs); each solution concentration and water pair was presented at least 10 times to group members ($n = 270$ trials). Preference was determined by order of selection, consumption, and mean differences in amounts consumed (Remis, 2002; Remis and Kerr, 2002). In the paired-food preference tests, food items were ranked according to the frequency of first selection, using standard methods that produce individual and species-specific food preference scores for each food item (Nunnally, 1978). Individual and species preferences were tested for correlations with each other and with specific nutrients. Principle components analysis was used to cluster the variation in nutrient composition of the foods and factor scores were used in subsequent correlations between preference and nutrient profiles (Remis, 2002).

In the bottle experiments (Simmen, 1994; Laska *et al.*, 1999), conservative estimates of taste detection abilities were made using the smallest mean concentration at which consumption of sugary solutions differed significantly from water (Bonferroni correction, paired $t$-test, $p < 0.004$). Tannins were judged to have an inhibitory effect on consumption when amounts of the tannin–fructose solution consumed differed (paired $t$-tests, $p < 0.004$) from similar concentration of fructose alone (see Remis and Kerr, 2002).

## Results

### Food availability and gorilla diet at Bai Hokou

Previously published analyses of the phenological records at Bai Hokou report two seasons of higher and lower availability of ripe fruit correlated with seasonal wet and dry patterns of rainfall (Goldsmith, 1996; Carroll, 1997; Remis, 1997a). The present longitudinal combined data set reveals a similar pattern with more ripe fruit available in wet months (especially July and August) than dry months when sample sizes are sufficient to reach significance based on Student's $t$-test (e.g., wet season months in 1991: $t = 2.4$, df $= 9$, $p = 0.039$; and in 1994: $t = 3.6$, df $= 8$; $p = 0.007$). During most dry-season months, ripe fleshy fruits are scarce, and although dehiscent or fibrous fruits are present, they are not a main source of the gorillas' diet (Remis, 1997a; Tutin *et al.*, 1997). Even more marked than seasonal distinctions in fruit availability is the yearly variation in fruit production at Bai Hokou, especially during wet-season months (availability in wet-season months by year based on analysis of variance: $F = 6.45$, $p = 0.000$), resulting in "good" and "poor" fruit years. For example, during sample collection in 1989, fewer species on phenology trails set fruit than in subsequent "good" fruit wet-season months or years (Remis *et al.*, 2001). Sample collection in 1998 also occurred during a "poor" fruit wet season. While in "good" fruit years at Bai Hokou 35–72% marked species have set fruit (Remis, 1997a, 1999), in August 1998 following an El Niño event, only 7% of species contained fruit, with only 4% having ripe fruit ($n = 720$) (Remis *et al.*, 2001). Figure 15.1A depicts the seasonal mean for the percentage of tree species on phenology trails that produced ripe fruit (measured monthly) in the wet and dry seasons when researchers were present during the years 1988 through 1998.

Gorillas at Bai Hokou consume at least 230 plant parts from over 129 plant species, including more than 89 species of fruits (Goldsmith, 1996; Carroll, 1997; Remis, 1997a). Western lowland gorilla diet varies in response to fruit availability. The amount of fruit eaten by gorillas (determined via gorilla fecal remains and feeding sites) was positively correlated with rainfall and the availability of ripe fruit trees on phenology trails (Goldsmith, 1996; Remis, 1997a). When fruit abundance is high, gorillas consume fruit in 59% of feeding visit observations (42–75%, $n = 176$ feeding visits, 3208 feeding minutes, during seven wet seasons) and consumption of fiber declines. During seasonal periods of low fleshy fruit abundance, observation of consumption of fleshy fruits is reduced (0–10% of feeding visit observations, during three dry seasons), although consumption of fibrous fruits that are more available in the dry season continues, especially fallen fruits of *Duboscia macrocarpa*. Observational data

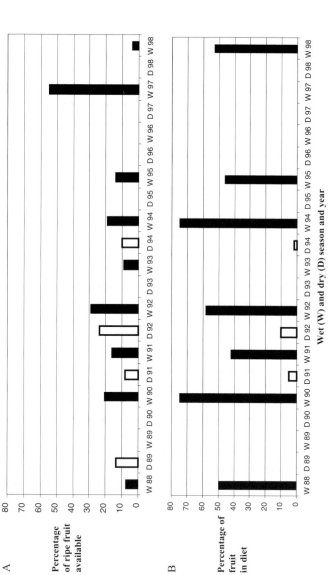

Fig. 15.1. Seasonal variation in ripe fruit availability and consumption at Bai Hokou. (A) Mean monthly proportion of ripe fruit species available on phenology trails (973 trees, 152 species) during wet (black bars) and dry (white bars) seasons. Data were only collected in June 1997 and July 1998 for those years. (Data from Goldsmith, 1996; Carroll, 1997; Remis, 1997*a* and unpublished data.) (B) Proportion of fruit eaten by gorillas during observations of feeding bouts in the wet (black bars) and dry (white bars) seasons. Data represent 176 feeding visits and 3208 minutes of observations. (Observational data for October–November 1988, August 1990–October 1992, June–December 1995, and July–August 1998 collected by Remis; data for October 1993–January 1995 from Goldsmith, 1996.)

presented in Fig. 15.1B depict the proportion of fruit consumed by gorillas during observations of feeding visits by Remis or Goldsmith (1996), in the wet or dry seasons during 1988–98. These data show that the proportion of fruit contained in the gorilla diet is correlated with season in this combined sample ($r = 0.800$, df $= 9$, $p = 0.005$). These observations are biased in favor of arboreal animals during rainy seasons but results are supported by fecal analysis during individual study periods that show stronger correlations between fruit availability (as measured by the forest-wide sample of trees on the phenology trails described here) and gorilla diet (Goldsmith, 1996; Carroll, 1997; Remis, 1997*a*). Future analyses will further explore the relationship between gorilla food species availability and diet.

### Nutritional analyses of foliage and fruits: Consequences of dietary flexibility

A total of 68 plant samples was collected and phytochemically analyzed for nutrient, including fiber, minerals, and secondary compound content. Items analyzed came from at least 20 different plant families and 35 species. Plant parts analyzed included ripe and unripe fruits, seeds, leaves, stems, and vines consumed by gorillas; uneaten seeds, rinds, etc. were discarded. Because plants were collected during poor fruit seasons and years, many fleshy fruits commonly consumed by gorillas were not analyzed. These data are more representative of the nutritional profile of gorillas during seasons of fruit scarcity than overall gorilla diet at the Bai Hokou site.

Figure 15.2 demonstrates that the foliage (herbs and leaves of woody species, including vines) analyzed in this sample was higher in crude protein and ash content than fruits ($p < 0.001$ for both crude protein and ash). There were no significant differences in the crude protein or ash content of herbaceous stems (stems) and woody leaves. On the other hand, the fruits analyzed in this sample contained variable amounts of soluble sugars (1.4–48.9%) and more total tannins ($p = 0.0046$), or condensed tannins ($p = 0.0039$) than the combined foliage sample, with similar amounts of neutral and acid detergent fiber (neutral, $p = 0.104$, acid, $p = 0.364$). Although fruits are commonly referred to as high-quality foods, the fruits in this study show much more variation in antifeedant (fiber and tannin) content between species than either herbs or leaves. Moderate tannin content did not deter gorillas from eating fruit that also contained a sugar reward. Fruit and foliage may contain similar amounts of fiber, especially in this sample where fruit availability was poor. Nevertheless, they appear to serve as complementary food sources for gorillas, each providing different essential energy and nutrients.

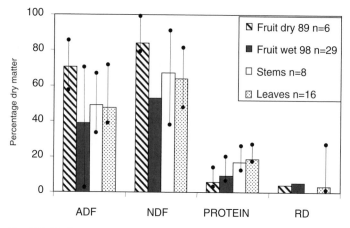

Fig. 15.2. Nutrient composition of plant parts consumed by gorillas at Bai Hokou during seasons of scarcity (dry season of 1989 and poor wet season of 1998). ADF = acid detergent fiber, NDF = neutral detergent fiber, and PROTEIN = condensed protein are expressed as percentage of dry matter. RD = total tannins are expressed as percentage of dry weight quebracho tannin equivalents.

## *Comparison of nutritional composition of gorilla diet across sites*

Table 15.1 demonstrates that across sites, gorillas consume large amounts of fruit and herbaceous stems, but at Karisoke and other high elevation sites, the vegetation is montane-adapted; fruit is scarce (Watts, 1998; Remis, 1998), and gorilla diet is largely herbaceous. Table 15.2 compares nutritional content across gorilla sites. Herbaceous stems eaten across mountain and lowland sites are similar in nutrient content and low in condensed tannins, but those eaten at Bai Hokou have more crude protein than those consumed elsewhere. Mountain and lowland gorillas at lower altitudes have more diverse diets than those at high-altitude Karisoke (Williamson *et al.*, 1990; Yamagiwa *et al.*, 1994, 1996; Remis, 1997*a*; Goldsmith, 1999, this volume; Robbins, 2000), and include a wide variety of fruits and leaves from woody species. As might be expected from the relationship of plant chemistry to soils, foods consumed by gorillas at all the lowland sites examined here have more condensed tannins than foods eaten at Karisoke (Waterman *et al.*, 1983; McNeilage, 1995; Plumptre, 1995). Compared to other western lowland sites that report yearly averages, fruits eaten during seasons of scarcity at Bai Hokou are high in crude protein and total and condensed tannins, though leaves appear lower in tannins than those analyzed from Lopé, Gabon or Campo, Cameroon (Calvert, 1985; Rogers *et al.*, 1990). High fiber content (acid detergent fiber) distinguishes the leaf and fruit portion of the diet at both Bai Hokou and Campo from Lopé. Overall, foods

Table 15.1. *Comparison of the relative importance of fruit in gorilla diets across sites and species*

| Species | Site | Study | Altitude (m) | Duration of study (months) | Number of species eaten | Number of fruit species eaten | Percentage fruit in feeding observations | Percentage fecal samples containing fruit |
|---|---|---|---|---|---|---|---|---|
| *Gorilla gorilla beringei* | Karisoke | Watts, 1984 | 2800–3600 | 17 | 38 | 4 | .3 | na |
|  | Virungas | McNeilage, 1995 | 2000–4000 | 28 | 35–44 | 1–13 | <2 | na |
|  | Bwindi | Goldsmith, this volume | 1400–2300 | 11 | 46 | 10 | na | 36 |
|  |  | Robbins, 2000 | 2100–2300 | 10 | na | 9 | 11 (4–25%) | na |
|  |  | Berry, 1998 | 1400–2300 | 13 | 81 | 12 | na | na |
| *Gorilla gorilla graueri* | Itebero | Yamagiwa et al., 1994 | 600–1300 | 11 | 121 | 48 | na | 89 |
|  | Kahuzi | Yamagiwa et al., 1996 | 2050–2350 | 12+ | 126 | 20 | na | 96.5 |
|  |  | Goodall, 1977 |  | 7 | 78 | 3 | na | na |
|  |  | Casimir, 1975 |  | 30 | 57 | 2 | na | na |
| *Gorilla gorilla gorilla* | Rio Muni | Sabater Pi, 1977 | 300–800 | 10 | 92 | 54 | na | na |
|  | Campo | Calvert, 1985 | 200 | 15 | 51 | 21 | na | 50 |
|  | Lopé | Tutin et al., 1997 | 200–500 | 120 | 220 | 100 | 47 | 98 |
|  | Bai Hokou | Remis, this study | 300 | 37 | 129 | 89 | 42–75 | 99 |
|  |  | Goldsmith, 1996 | 300 | 17 | 150 | 97 | 0; 75[a] | 96 |
|  | Ndoki | Nishihara, 1995 | 300–600 | 12 | 152 | 115 | na | na |
|  | Mondika | Doran and McNeilage, 1999 | 300–600 | 35 | 100 | 70 | 50 | 99.8 |
|  | Afi | Oates, this volume | 350–1000 | 32 | 125+ | 83 | na | 90.2 |

na, not available.

[a] 0% = the dry season; 75% = the rainy season.

Table 15.2. *Nutrient content of gorilla foods across sites*

| Site | Subspecies | Condensed tannins | Neutral detergent fiber | Acid detergent fiber | Crude protein | Lignins | Number of samples | Study |
|---|---|---|---|---|---|---|---|---|
| *Leaves* | | | | | | | | |
| Bai Hokou, CAR | *G. g. gorilla* | 4.5 | 64 | 47.6 | 18.9 | 25.2 | 16 | Remis, this study |
| Lopé, Gabon | *G. g. gorilla* | 14.6 | na | 28.9 | 18.4 | na | 16 | Rogers *et al.*, 1990 |
| Campo, Cameroon | *G. g. gorilla* | 7.3 | 46.1 | 42.6 | 16.6 | 19.4 | 8 | Calvert, 1985 |
| Karisoke, Rwanda | *G. g. beringei* | 1.1 | na | 35.5 | 15.5 | na | 21 | Waterman *et al.*, 1983 |
| Bwindi, Uganda | *G. g. beringei* | 2 | 20 | 15 | 6 | na | 9 | Berry, 1998 |
| *Fruit* | | | | | | | | |
| Bai Hokou, CAR | *G. g. gorilla* | 12.3 | 58.4 | 44.4 | 8.9 | 23.9 | 29 | Remis, this study |
| Lopé, Gabon | *G. g. gorilla* | 8.8 | na | 23.7 | 5.2 | na | 46 | Rogers *et al.*, 1990 |
| Campo, Cameroon | *G. g. gorilla* | 2 | 64.6 | 44.8 | 6.3 | 26.9 | 8 | Calvert, 1985 |
| Karisoke, Rwanda | *G. g. beringei* | na | na | na | na | na | na | Waterman *et al.*, 1983 |
| Bwindi, Uganda | *G. g. beringei* | 5 | 14 | 14 | 2 | na | 3 | Berry, 1998 |
| *Stems (Pith only)* | | | | | | | | |
| Bai Hokou, CAR | *G. g. gorilla* | 0.6 | 67.4 | 49.2 | 16.9 | 26 | 7 | Remis, this study |
| Lopé, Gabon | *G. g. gorilla* | 1.7 | na | 48.6 | 5.1 | na | 6 | Rogers *et al.*, 1990 |
| Campo, Cameroon | *G. g. gorilla* | 0.5 | 55.9 | 44.4 | 6.7 | 11.3 | 11 | Calvert, 1985 |
| Karisoke, Rwanda | *G. g. beringei* | 0.72 | na | 49.3 | 6.2 | na | 12 | Waterman *et al.*, 1983 |
| Bwindi, Uganda | *G. g. beringei* | 1 | 33 | 30 | 2 | na | 5 | Berry, 1998 |

na, not available.

consumed by the gorillas during seasons of fleshy fruit scarcity at Bai Hokou have more crude protein and fiber than foods consumed, on average, at other sites.

### Captive research on food preferences and taste sensitivity

In paired-choice food preference trials, captive individual gorillas and chimpanzees were markedly consistent in their preferences, and mean species-specific food preference scores were significantly correlated based on Pearson's product moment correlation analysis ($r = 0.844$, $p = 0.035$). Individual gorillas and chimpanzees consistently preferred domestic fruits to vegetable foods, and high-sugar to low-sugar fruits. Overall, gorilla food choices were correlated with sugar content ($r = 0.785$, $p = 0.012$), and sugar to fiber ratios ($r = 0.668$, $p = 0.049$). Additional analyses showed their food preferences were inversely correlated with principle components analyses factor 1 scores, which accounted for 41% of the variation in nutrients, and which clustered foods high in fiber and protein ($r = -0.676$, $p = 0.008$) (Remis, 2002). Tannins were not common in the domestic foods tested from the typical gorilla diets. Tannins were present at moderate levels in both preferred novel foods (mangoes) and less preferred novel foods (tamarinds). Thus, tannins did not appear to greatly influence ape preferences (Remis, 2002). Additional research will further explore the role of fats and tannins in food choices.

In paired-choice taste detection experiments, gorillas consistently preferred fructose solutions to water. The same individuals appeared to be able to detect even dilute concentrations of fructose, preferentially consuming very weak fructose (6 mM fructose) solutions to water (mean individual consumption based on paired $t$-test: $t = 2.289$, df $= 8$, $p = 0.048$). Nevertheless, given the small sample size and interindividual variation, conservative statistical procedures were used. By these methods, the taste threshold, defined as the lowest level at which there were significant differences between the consumption of fructose and water, was provisionally determined to be 50 mM fructose (paired $t$-test with Bonferroni correction: $t = 6.607$, $p < 0.001$).

Tannins were broadly tolerated at the concentrations administered, and consumption was higher when presented in the higher–fructose solutions. The taste inhibition threshold (when individuals consumed less tannin–fructose solution than fructose solution) was reached at 4 mM tannic acid (paired $t$-test with Bonferroni correction, 4 mmol tannic acid in 100 mM fructose solution versus 100 mM fructose; $t = -3.729$, df $= 9$, $p = 0.005$). Gorillas readily consumed sweet tannin mixtures at much higher tannin concentrations than smaller primates studied to date (Remis and Kerr, 2002).

## Discussion

### *Gorillas as seasonal frugivores*

Integration of field data on food availability, diet, and nutrition with captive studies on food preferences and taste allows us to further explore the nature of gorilla dietary adaptation and the possible physiological correlates of their dietary selectivity and flexibility in the wild. The field data indicate that among primates, gorillas consume more fruit than might be predicted from their body and gut size, suggesting general adherence to an ape pattern of frugivory. Nevertheless, as predicted, they also appear to consume more fibrous foods than do their closest relatives, the smaller-bodied and more persistently frugivorous chimpanzees; the role of lipids in food selection remains unknown. Unlike chimpanzees, gorillas engage in dietary switching whenever and wherever fruit is scarce, hence the suggestion that their diet reflects broad resource availability more than selectivity (Nishihara, 1995; Kuroda *et al.*, 1996; Yamagiwa *et al.*, 1996; Remis, 1997*a*; Wrangham *et al.*, 1998; M.J. Remis *et al.*, unpublished data). The experimental data suggest a relationship between variability in gorilla diet and response to resource availability and their broad taste sensitivity that results in a preference for sugary foods yet tolerance for foods containing tannins or fiber.

Herbaceous stems, high in protein and low in tannins, are important staples or fallback foods for most gorillas, and account for the overwhelming bulk of their diet at the highest altitudes (Watts, 1984). At most low- and medium-altitude field sites, gorilla diet shifts along a seasonal and interannual gradient with high variability in proportions of fleshy fruit in the diet (Rogers *et al.*, 1990; Remis, 1997*a*) (Table 15.1). The Bai Hokou sample shows considerable interannual variability in availability of ripe fruit, similar to several other data sets from Africa that have recently been published (Newbery *et al.*, 1998; Tutin and White, 1998; Chapman *et al.*, 1999; but see van Schaik *et al.*, 1993). At Bai Hokou, the between-year variability sometimes overrides the seasonal variability in fruit availability. More research needs to be done on phenological patterns in these tropical forest sites as well as on the nutritional consequences of seasonal diets high in ripe fleshy fruits. Nevertheless, many fruits and leafy foods may contain similar proportions of fiber, especially during seasons of scarcity such as the ones described here. Thus, within- or between-site differences in nutrient intake may be lower than seasonal, annual, or geographic variability in diet might suggest. For example, within the Bai Hokou site, dry-season fruits and leaves were nutritionally similar, while in a between-site comparison, herbs were nutritionally similar at Karisoke and Bai Hokou (Table 15.2).

This study confirms earlier findings that western and eastern lowland gorillas in primary and old secondary forests (including low- and medium-altitude sites) have a more diverse diet than high-altitude populations (Waterman *et al.*, 1983; Calvert, 1985; Rogers *et al.*, 1990; Yamagiwa *et al.*, 1994; Watts, 1998; Goldsmith, 1999, this volume; McFarland, 2000; Robbins, 2000). Further, western and eastern lowland gorillas consume both more fruit and more secondary compounds than high-altitude populations, even when ripe fruit is scarce (Rogers *et al.*, 1990; Popovich *et al.*, 1997). The Bai Hokou findings highlight the importance of high between-year as well as seasonal variability in fruit availability in forests in the Central African Republic and its influences on western gorilla frugivory. The dietary data described here provide a window into a nutritional profile of gorillas during limited periods of scarcity, when we expect adaptations related to body size to be most important. During periods of fleshy fruit scarcity, the Bai Hokou gorillas consumed more leafy foods and fruits with higher average levels of fiber and secondary compounds than averages of foods analyzed from other gorilla populations to date.

The gorillas' large body size and gut morphology lend them flexibility to cope with scarcity of preferred fruits by consuming and processing large amounts of fiber relative to smaller species (e.g., Wrangham *et al.*, 1998). Research on the mean gut retention times of the captive gorillas studied here showed them to retain foods longer in their gastrointestinal tracts (50 hours: Remis, 2000; see also Caton, 1999) than a similar study of zoo chimpanzees (31 hours: Lambert, 1997). Large surface area of the colon and cecum along with longer gut retention times should allow animals to maximize absorption of nutrients (Chivers and Hladik, 1980), regardless of any possible differences in gut flora between captive and wild individuals. Despite the digestive consequences of large body and gut size, gorillas are not restricted to a high-fiber diet. Whenever ripe fruit is abundant, the foliage component of their diet is reduced.

While the large body size and digestive physiology of gorillas apparently allows them to tolerate a high-fiber diet, gorillas at Bai Hokou appeared to avoid those leaves high in condensed tannins (Remis *et al.*, 2001). Carbon-based tannins are more prominent in gorilla fruits than in leafy foods at Bai Hokou (see also Nijboer *et al.*, 1997). The gorilla foods analyzed here contain higher amounts of fiber and tannins than the foods consumed by other gorillas at Karisoke, Lopé, or Gabon or the primates at Kibale, Uganda (Wrangham *et al.*, 1998). Site-specific differences in plant biochemistry have been found for regions across the tropics and habitat differences shape the diets of the primates in different locations, complicating our ability to identify taxon specific adaptations.

## *Are gorillas vacuum cleaners of the forest floor?*

Both phylogeny and body size shape physiological and nutritional requirements (Martin *et al.*, 1985; van Soest, 1994), and taste preferences and abilities (Hladik and Simmen, 1996; Simmen and Hladik, 1998; Laska *et al.*, 1999). Despite the ecological implications of the evolution of body-size differences among the African apes (Clutton-Brock, 1977; Shea, 1983), both captive and wild chimpanzees and gorillas prefer low-fiber, high-sugar foods or those containing high protein or sugar to tannin ratios (Conklin-Brittain *et al.*, 1998; Reynolds *et al.*, 1998; Remis *et al.*, 2001; Remis, 2002). Gorillas pursue sugary fleshy fruits and avoid leaves high in digestion inhibitors. Seasonal, or habitat-specific, frugivory influences gorilla arboreality, ranging patterns, and behavior (Remis, 1997*a*,*b*, 1999; Doran and McNeilage, 1999). Thus, gorillas are selective in their frugivory and not aptly described as "vacuum cleaners".

Experimental ongoing captive research on gorillas and chimpanzees suggests that less persistent frugivory and higher dietary flexibility among wild gorillas may well be influenced by a body-size-related high sensitivity to sugars, coupled with a high threshold for tannins (M.J. Remis, unpublished data). Gorillas appear to detect very dilute concentrations of fructose, and future research may show taste perception to be even more astute than in the smaller primates studied to date (Hladik and Simmen, 1996; Laska *et al.*, 1999; Simmen *et al.*, 1999; Remis and Kerr, 2002). The tannic acid used in the captive trials described here may not be directly comparable to the more complex array of tannins found in wild foods. Nevertheless, the high taste-inhibition threshold for tannic acid found in this study may be related to body size and diet, as consumption of large amounts of fiber would dilute the relative concentration of tannins in the gut (Rogers *et al.*, 1990; Cork and Foley, 1992). In addition, as has been suggested for other primates, gorillas may have protein-rich saliva that helps to dilute tannins (Milton, 1998). A high tannin-taste threshold that increases with sugar content likely relates to wild gorillas' tolerance for appreciable levels of condensed and total tannins in native fruits but not leaves (Simmen and Hladik, 1998; Remis and Kerr, 2002).

Across sites and subspecies, gorillas share a generalized hominoid morphology and physiology and a penchant for fleshy fruit with the smaller-bodied apes. Dietary differences among western and eastern gorilla populations cannot be easily divided along taxonomic groups as they appear primarily related to altitude and reflect resource availability. Gorillas are selective frugivores whenever possible, preferring sugary foods, although the roles of fats in food selection by apes should be further explored. Nevertheless, gorilla body size and related adaptations, especially large guts and taste perception, may lead them to view a broad range of sugary, fibrous and tannin-rich foods as palatable. These adaptations lend them the flexibility to consume alternate foods as staples or fallbacks whenever fleshy fruit is scarce. Consumption of fibrous foods likely facilitates gorilla group cohesion (Wrangham, 1979; Remis, 1997*a*) as well as coexistence with sympatric chimpanzees (Yamagiwa *et al.*, this volume). Moreover, it is likely that dietary flexibility permitted gorilla ancestors to expand their range into high-altitude areas (Groves, 1986; Yamagiwa *et al.*, 1996) where both fruit and chimpanzees are scarce.

## Acknowledgments

This research was conducted under the guidance of the Purdue Animal Care and Use Committee (97-099), and was supported by a National Science Foundation Grant (BCS-9815841) and funds from Purdue University. I would like to thank the government of the Central African Republic for permission to conduct research at Bai Hokou, Dzanga-Sangha Dense Forest Reserve and the Reserve staff for logistical assistance over the years. Many of the plant samples were collected by Jean Bosco Kpanou and research assistants at the Bai Hokou site. The laboratory analyses were conducted and partially supported by Ellen Dierenfeld at the Wildlife Nutrition Laboratory, Wildlife Conservation Society, New York and Christopher Mowry at Berry College, Georgia. I thank Carroll, Dierenfeld and Mowry for productive collaboration and the editors of this volume for inviting my contribution. The San Francisco Zoological Gardens granted permission for the captive research. Mary Kerr and other zoo staff facilitated and collaborated on this effort, greatly assisting data collection efforts.

## References

Association of Official Analytical Chemists (1996). *Official Methods of Analysis of AOAC International*, 16th edn. Gaithersburg, MD: AOAC International.

Bauchop, T. (1978). Digestion of leaves in vertebrate arboreal folivores. In *The Ecology of Arboreal Folivores*, ed. G.G. Montgomery, pp. 193–204. Washington, D.C.: Smithsonian Institution Press.

Benz J.J., Leger D.W., and French, J.A. (1992). Relation between food preference and food-elicited vocalizations in golden lion tamarins (*Leontopithecus rosalia*). *Journal of Comparative Psychology*, **106**, 142–149.

Berry, J.P. (1998). The chemical ecology of mountain gorillas (*Gorilla gorilla beringei*): With special reference to antimicrobial constituents in the diet. PhD thesis, Cornell University, Ithaca, NY.

Calvert, J.J. (1985). Food selection by western gorillas (*Gorilla gorilla gorilla*) in relation to food chemistry. *Oecologia*, **65**, 236–246.

Carroll, R.W. (1997). Feeding ecology of lowland gorillas (*Gorilla gorilla gorilla*) in the Dzanga-Sangha Dense Forest Reserve of the Central African Republic. PhD thesis, Yale University, New Haven, CT.

Casimir, M.J. (1975). Feeding ecology and nutrition of an eastern gorilla group in the Mt. Kahuzi region (Republic of Zaïre). *Folia Primatologica*, **24**, 81–136.

Caton, J.M. (1999). A preliminary report on the digestive strategy of the western lowland gorilla. *Australasian Primatology*, **13**(4), 2–7.

Chapman, C.A., Wrangham, R.W., Chapman, L.J., Kennard, D.K., and Zanne, A.E. (1999). Fruit and flower phenology at two sites in Kibale National Park, Uganda. *Journal of Tropical Ecology*, **15**, 189–211.

Chivers, D.J. and Hladik, C.M. (1980). Morphology of the gastrointestinal tract in primates: Comparisons with other mammals in relation to diet. *Journal of Morphology*, **166**, 337–386.

Chivers, D.J. and Langer, P. (eds) (1994). *The Digestive System in Mammals: Food Form and Function*. Cambridge, U.K.: Cambridge University Press.

Clutton-Brock, T.H. (1977). Species differences in feeding and ranging behavior in primates. In *Primate Ecology*, ed. T.H. Clutton-Brock, pp. 557–588. London: Academic Press.

Collet, J., Bourreau, E., Cooper, R.W., Tutin, C.E.G., and Fernandez, M. (1984). Experimental demonstration of cellulose digestion by *Troglodytella gorillae*, an intestinal ciliate of lowland gorillas. *International Journal of Primatology*, **5**, 328.

Conklin-Brittain, N.L., Wrangham, R.W., and Hunt, K.D. (1998). Dietary response of chimpanzees and cercopithecines to seasonal variation in fruit abundance. II: Macronutrients. *International Journal of Primatology*, **19**, 971–999.

Cork, S.J. and Foley, W.J. (1992). Digestive and metabolic strategies of arboreal mammalian folivores in relation to chemical defenses in temperate and tropical forests. In *Plant Defenses against Mammalian Herbivory*, eds. R.T. Palo and C.T. Robbins, pp. Boca Raton, FL: CRC Press.

Demment, M.W. and van Soest, P.J. (1985). A nutritional explanation of body-size patterns of ruminant and non-ruminant herbivores. *American Naturalist*, **125**, 641–672.

Doran, D. and McNeilage, A. (1998). Gorilla ecology and behavior. *Evolutionary Anthropology*, **6**, 120–131.

Doran, D. and McNeilage, A. (1999). Diet of western lowland gorillas in south-west Central African Republic: Implications for subspecific variation in gorilla grouping and ranging patterns. *American Journal of Physical Anthropology, Supplement* **27**, 121.

Fossey, D. and Harcourt, A.H. (1977). Feeding ecology of free-ranging mountain gorillas. In *Primate Ecology*, ed. T.H. Clutton-Brock, pp. 415–449. London: Academic Press.

Freeland, W.J. (1991). Plant secondary metabolites: Biochemical coevolution with herbivores. In *Plant Defenses against Mammalian Herbivory*, eds. R.T. Palo and C.T. Robbins, pp. 61–81. Boca Raton, FL: CRC Press.

Freeland, W.J. and Janzen, D.H. (1974). Strategies in herbivory by mammals: The role of plant secondary compounds. *American Naturalist*, **108**, 269–289.

Ganzhorn, J.U. (1989). Primate species separation in relation to secondary plant chemicals. *Human Evolution*, **4**, 125–132.

Glander, K.E. (1981). The impact of plant secondary compounds on primate feeding behavior. *Yearbook of Physical Anthropology*, **25**, 1–18.

Goldsmith, M.L. (1996). Ecological influences on the ranging and grouping behavior of western lowland gorillas at Bai Hokou, Central African Republic. PhD thesis, State University of New York, Stony Brook, NY.

Goldsmith, M.L. (1999). Gorilla socioecology. In *The Nonhuman Primates*, eds. P. Dolhinow and A. Fuentes, pp. 58–63. Mountain View, CA: Mayfield Publishing.

Goodall, A.G. (1977). Feeding and ranging behavior of a mountain gorilla group (*Gorilla gorilla beringei*) in the Tshiabinda–Kahuzi region (Zaïre). In *Primate Ecology*, ed. T.H. Clutton-Brock, pp. 449–479. London: Academic Press.

Groves, C.P. (1986). Systematics of the great apes. In *Comparative Primate Biology*, Vol. 1, *Systematics, Evolution and Anatomy*, eds. D.R. Swindler and J. Erwin, pp. 187–217. New York: A.R. Liss.

Hladik, C.M. (1978). Adaptive strategies of primates in relation to leaf-eating. In *The Ecology of Arboreal Folivores*, ed. G.G. Montgomery, pp. 373–395. Washington, D.C.: Smithsonian Institution Press.

Hladik, C.M. and Simmen, B. (1996). Taste perception and feeding behavior in nonhuman primates and human populations. *Evolutionary Anthropology*, **5**, 58–71.

Kay, R.F. and Davies, A.G. (1994). Digestive physiology. In *Colobine Monkeys: Their Ecology, Behaviour and Evolution*, eds. A.G. Davies and J.F. Oates, pp. 229–250. Cambridge, U.K.: Cambridge University Press.

Kuroda, S., Nishihara, T., Suzuki, S., and Oko, R.A. (1996). Sympatric chimpanzees and gorillas in the Ndoki Forest, Congo. In *Great Ape Societies*, eds. W.C. McGrew, L.F. Marchant, and T. Nishida, pp. 71–81. Cambridge, U.K.: Cambridge University Press.

Lambert, J.E. (1997). Digestive Strategies, Fruit Processing, and Seed Dispersal in the Chimpanzees (*Pan troglodytes*) and Redtail Monkeys (*Cercopithecus ascanius*) of Kibale National park, Uganda. PhD thesis, University of Illinois, Urbana–Champaign, IL.

Laska, M., Schuell, E., and Scheuber, H.P. (1999). Taste preference thresholds for food-associated sugars in baboons (*Papio hamadryas anubis*). *International Journal of Primatology*, **20**, 25–34.

Martin, R.D., Chivers, D.J., Maclarnon, A.M., and Hladik, C.M. (1985). Gastrointestinal allometry in primates and other mammals. In *Size and Scaling in Nonhuman Primates*, ed. W.L. Jungers, pp. 61–89. New York: Plenum Press.

McFarland, K. (2000). Natural history of the Afi Mountain (Nigeria) gorilla population. In *Proceedings of "The Apes: Challenges for the 21$^{st}$ Century"*, 10–13 May 2000, p. 49. Chicago, IL: Brookfield Zoological Society.

McNeilage, A.J. (1995). Mountain gorillas in the Virunga Volcanoes: Ecology and carrying capacity. PhD thesis, University of Bristol, Bristol, U.K.

Milton, K. (1984). The role of food-processing factors in primate food choice. In *Adaptations for Foraging in Nonhuman Primates*, eds. P. S. Rodman and J.G.H. Cant, pp. 249–279. New York: Columbia University Press.

Milton, K. (1998). Physiological ecology of howlers (*Alouatta*): Energetic and digestive considerations and comparison with the Colobinae. *International Journal of Primatology*, **19**, 513–548.

Milton, K. and Demment, M.W. (1988). Digestion and passage kinetics of chimpanzees fed high and low fiber diets and comparison with human data. *Journal of Nutrition*, **118**, 1082–1088.

Mole, S. and Waterman, P.G. (1987). Tannins as antifeedants to mammalian herbivores: Still an open question? In *Allelochemicals: Role in Agriculture and Forestry*, ed. G.R. Waller, pp. 572–587. Washington, D.C.: American Chemical Society Press.

Newbery, D.M., Songwe, N.C., and Chuyong, G.B. (1998). Phenology and dynamics of an African rainforest at Korup, Cameroon. In *Dynamics of Tropical Communities*, eds. D.M. Newbery, H.H.T. Prins, and N.D. Brown, pp. 267–308. Oxford, U.K.: Blackwell Scientific Publications.

Nijboer, J., Dierenfeld, E.S., Yeager, C.P., Bennett, E.L., Bleish, W., and Mitchell, A.H. (1997). Chemical composition of SouthEast Asian colobine foods. In *Proceedings of the AZA Nutrition Advisory Group Conference*, Fort Worth, Texas. Washington, D.C.: American Zoological Association.

Nishihara, T. (1995). Feeding ecology of western lowland gorillas in the Nouabale-Ndoki National Park, northern Congo. *Primates*, **36**, 151–168.

Nunnally J.C. (1978). *Psychometric Theory*, 2nd edn. New York: McGraw-Hill.

Oates, J.F., Swain, T., and Zantovska, J. (1977). Secondary compounds and food selection by colobus monkeys. *Biochemical Systematics and Ecology*, **5**, 317–321.

Parra, R. (1978). Comparison of foregut and hindgut fermentation in herbivores. In *The Ecology of Arboreal Folivores*, ed. G.G. Montgomery, pp. 205–229. Washington, D.C.: Smithsonian Institution Press.

Plumptre, A.J. (1995). The chemical composition of montane plants and its influence on the diet of large mammalian herbivores in the Parc National des Volcans, Rwanda. *Journal of Zoology, London*, **235**, 323–337.

Popovich, D.G., Jenkins, D.J.A., Kendall, C.W.C., Dierenfeld, E.S., Carroll, R.W., Tariq, N., and Vidgen, E. (1997). The western lowland gorilla diet has implications for the health of humans and other hominoids. *Journal of Nutrition*, **127**, 2000–2006.

Remis, M.J. (1994). Feeding ecology and positional behavior of lowland gorillas in the Central African Republic. PhD thesis, Yale University, New Haven, CT.

Remis, M.J. (1997a). Gorillas as seasonal frugivores: Use of variable resources. *American Journal of Primatology*, **43**, 87–109.

Remis, M.J. (1997b). Ranging and grouping patterns of a lowland gorilla group. *American Journal of Primatology*, **43**, 110–130.

Remis, M.J. (1998). The gorilla paradox: Effects of habitat and body size on the positional behavior of lowland and mountain gorillas. In *Primate Locomotion*, eds. E. Strasser, J.G.H. Fleagle, A. Rosenberger, and H. McHenry, pp. 95–106, New York: Plenum Press.

Remis, M.J. (1999). Tree structure and sex differences in arboreality among western lowland gorillas (*Gorilla gorilla gorilla*) at Bai Hokou, Central African Republic. *Primates*, **40**, 383–96.

Remis, M.J. (2000). Initial studies on the contributions of body size and gastrointesti-nal passage times to dietary flexibility among gorillas (*Gorilla gorilla gorilla*). *American Journal of Physical Anthropology*, **112**, 171–180.

Remis, M.J. (2002). Food preferences of captive gorillas (*Gorilla gorilla gorilla*) and chimpanzees (*Pan troglodytes*). *International Journal of Primatology*, **23**, 231–249.

Remis, M.J. and Kerr, M.E. (2002). Taste responses of gorillas to fructose and tannic acid solutions. *International Journal of Primatology*, **23**, 251–261.

Remis, M.J., Dierenfeld, E.S., Mowry, C.B., and Carroll, R.W. (2001). Nutritional as-pects of western lowland gorilla diet during seasons of scarcity at Bai Hokou, Central African Republic. *International Journal of Primatology*, **40**, 106–136.

Reynolds, V., Plumptre, A.J., and Greenham, J. (1998). Condensed tannins and sugars in the diet of chimpanzees (*Pan troglodytes schweinfurthii*) in the Budongo Forest, Uganda. *Oecologia*, **115**, 331–336.

Robbins, M.M. (2000). Behavioral ecology of the (other) mountain gorillas of Bwindi Impenetrable National Park, Uganda, Africa. In *Proceedings of "The Apes: Chal-lenges for the 21st Century"*, 10–13 May 2000, p. 56. Chicago, IL: Brookfield Zoological Association.

Rogers, M.E., Williamson, E.A., Tutin, C.E.G., and Fernandez, M. (1990). Gorilla diet in the Lopé Reserve, Gabon: A nutritional analysis. *Oecologia*, **84**, 326–339.

Sabater Pi, J. (1977). Contribution to the study of the alimentation of lowland gorillas in the natural state of Rio Muni Republic of Equatorial Guinea (W. Africa). *Primates*, **18**, 183–204.

Schaller, G.B. (1963). *The Mountain Gorilla: Ecology and Behavior*. Chicago, IL: Uni-versity of Chicago Press.

Shea, B.T. (1983). Size and diet in the evolution of African ape craniodental form. *Folia Primatologica*, **40**, 22–66.

Simmen, B. (1994). Taste discrimination and diet differentiation in New World primates. In *The Digestive System in Mammals*, eds. D.J. Chivers and P. Langer, pp. 150–165. Cambridge, U.K.: Cambridge University Press.

Simmen, B. and Hladik, C.M. (1998). Sweet and bitter taste discrimination in primates: Scaling effects across species. *Folia Primatologica*, **69**, 129–138.

Simmen, B., Josseaume, B., and Atramentowicz, M. (1999). Frugivory and taste re-sponses to fructose and tannic acid in a prosimian primate and a didelphid marsu-pial. *Journal of Chemical Ecology*, **25**, 331–346.

Strait, S. (1997). Tooth use and the physical properties of food. *Evolutionary Anthro-pology*, **5**, 199–211.

Taylor, A.B. (2002). Masticatory form and function in the African apes. *American Jour-nal of Physical Anthropology*, **117**, 133–156.

Tutin, C.E.G. and White, L.J.T. (1998). Primates, phenology and frugivory: Present, past and future patterns in the Lopé Reserve, Gabon. In *Dynamics of Tropical Communities*, eds. D.M. Newbery, H.H.T. Prins, and N.D. Brown, pp. 309–338. Oxford, U.K.: Blackwell Scientific Publications.

Tutin, C.E.G., Fernandez, M., Rogers, M.E., Williamson, E.A., and McGrew, W.C. (1991). Foraging profiles of sympatric lowland gorillas and chimpanzees in the Lopé Reserve, Gabon. *Philosophical Transactions of the Royal Society, Series B*, **334**, 179–186.

Tutin, C.E.G., Ham, R.M., White, L.J.T., and Harrison, M.J.S. (1997). The primate community of the Lopé Reserve, Gabon: Diets, responses to fruit scarcity and effects on biomass. *American Journal of Primatology*, **42**, 1–24.

Uchida, A. (1998). Variation in tooth morphology of *Gorilla gorilla*. *Journal of Human Evolution*, **34**, 55–70.

van Schaik, C.P., Terborgh, J.W., and Wright, S.J. (1993). The phenology of tropical forests: Adaptive significance and consequences for primary consumers. *Annual Review of Ecology and Systematics*, **24**, 353–377.

van Soest, P.J. (1994). *Nutritional Ecology of the Ruminant*, 2nd edn. Ithaca, NY: Cornell University Press.

van Soest, P.J., Robertson, J.B., and Lewis, B.A. (1991). Methods for dietary fiber, neutral detergent fiber and nonstarch polysaccharides in relation to animal nutrition. *Journal of Dairy Science*, **74**, 3583–3597.

Waterman, P.G. and Kool, K. (1994). Colobine food selection and plant chemistry. In *Colobine Monkeys: Their Ecology, Behaviour and Evolution*, eds. A.G. Davies and J.F. Oates, pp. 251–284. Cambridge, U.K.: Cambridge University Press.

Waterman, P.G., Choo, G.M., Vedder, A.L., and Watts, D.P. (1983). Digestibility, digestion-inhibitors, and nutrients of herbaceous foliage and green stems from an African montane flora and comparison with other tropical flora. *Oecologia*, **60**, 244–249.

Watts, D.P. (1984). Composition and variation of mountain gorilla diets in the Central Virungas. *American Journal of Primatology*, **7**, 323–356.

Watts, D.P. (1996). Comparative socio-ecology of gorillas. In *Great Ape Societies*, eds. W.C. McGrew, L.F. Marchant, and T. Nishida, pp. 16–28. Cambridge, U.K.: Cambridge University Press.

Watts, D.P. (1998). Seasonality in the ecology and life histories of mountain gorillas (*Gorilla gorilla beringei*). *International Journal of Primatology*, **19**, 929–948.

Williamson, E.A., Tutin, C.E.G., Rogers, M.E., and Fernandez, M. (1990). Composition of the diet of lowland gorillas at Lopé, Gabon. *American Journal of Primatology*, **21**, 265–277.

Wrangham R.W. (1979). On the evolution of ape social systems. *Social Science Information*, **18**, 334–368.

Wrangham, R.W., Conklin-Brittain, N.L., and Hunt, K.D. (1998). Dietary response of chimpanzees and cercopithecines to seasonal variation in fruit abundance. I: Antifeedants. *International Journal of Primatology*, **19**, 949–970.

Yamagiwa, J., Mwanza, N., Yumoto, T., and Maruhashi, T. (1994). Seasonal changes in the composition of the diet of eastern lowland gorillas. *Primates*, **35**, 1–14.

Yamagiwa, J., Maruhashi, T., Yumoto, T., and Mwanza, N. (1996). Dietary and ranging overlap in sympatric gorillas and chimpanzees in Kahuzi-Biega National Park, Zaïre. In *Great Ape Societies*, eds. W.C. McGrew, L.F. Marchant, and T. Nishida, pp. 82–98. Cambridge, U.K.: Cambridge University Press.

# Part 4
*Gorilla conservation*

# 16 An introductory perspective: Gorilla conservation

ALEXANDER H. HARCOURT

To conserve a species we need to know the threats it faces, and the biology of its potential reaction to those threats. Many other areas of knowledge are necessary too, of course. But the threats and the biology are a start, and these are the topics of Plumptre *et al.*'s and Sarmiento's comprehensive chapters. Oates et al. complete the section with the equivalent of a worked example, a valuable description for one small population of the integration of survey work, natural history, biology, and conservation management.

Plumptre *et al.* describe the four horsemen of the conservation apocalypse as they affect gorillas – habitat loss and modification, hunting, disease, and war. Currently, the majority of gorillas, about 80% of them, live outside of protected areas (Harcourt, 1996). Although creation and maintenance of protected areas is one of the main means of conserving gorilla populations, Plumptre et al. also provide an enlightening account of how conservationists might help conservation outside protected areas by, for example, working in collaboration with logging companies. In other words, it is as often the biologists themselves, as much or more than their biological findings, that are important to conservation. The biologists are the ones in the wilderness more than is anyone else, and thus they are usually the first to see the problem, to warn of the problem, to substantiate the problem, and to stimulate action (Harcourt, 2000; Chapman and Peres, 2001).

Depressing as the state of affairs in gorilla conservation is, my reading of the status of gorillas is a little more sanguine than many others'. The *Red List* (IUCN, 2000) classifies all gorilla species and subspecies as "Endangered" (it lists more than one species). However, I classify the gorilla and the western gorilla subspecies *G. g. gorilla* as "Vulnerable" by the same IUCN *Red List* criteria (Harcourt, 1996). Among some scientists, the *Red List* is falling into disrepute because despite its claims for quantified scientific objectivity, not only does subjectivity so clearly enter the analysis, but far more problematically, IUCN will not release the data and analysis behind their classifications (e.g., Mrosovsky, 1997).

While objectively the gorilla might be only "Vulnerable", and thus warrant less attention than many other primate species, an argument can be made for paying special attention to our closest animal relatives, and hence ignoring the quantified IUCN criteria. The US Endangered Species Act effectively does just this in listing the gorilla as one of only three taxa whose populations number more than 100 000. (The other two are the chimpanzee and elephant.) But if IUCN and those of us who study gorillas are going to do the same, we must be explicit about it, letting it be known that we are giving the apes special consideration, and making our analysis available for scrutiny.

For decades, good conservation managers have asked to be supplied with good biology, and students particularly will find useful Sarmiento's basic compendium of the sort of biology that might be applicable. He gives exhaustive information on geographic distribution, with a very useful gazetteer of gorilla sites and associated data, such as climate, habitat, and gorilla densities. He also provides basic information on censusing, taxonomy, and ecology. I say "students particularly", because much of Sarmiento's chapter is written for the student.

Students might, however, need a little extra information about what exactly Sarmiento is talking about in his section on alpha taxonomy. By "alpha taxonomy", Sarmiento means the description (and naming) of types of organisms as separate species or subspecies, what might be termed classification. Alpha taxonomy is distinguished from beta taxonomy, which is the analysis of the evolutionary relationships among groups of organisms (Groves, 2001). Sarmiento argues that alpha taxonomy should use morphological traits and coding nuclear DNA because traits are likely to be used in species recognition and speciation, and the DNA is likely to code for them. Of course, only some of the traits are so used, and in the absence of precise knowledge of exactly which traits are used by the animals themselves in species recognition, and which traits are fixed as opposed to varying with habitat, taxonomists use any trait that will allow difference and similarity to be measured. Groves (2001) uses, for example, shaggy scalp hair as a distinctive feature of mountain gorillas (the *beringei* subspecies) in the absence of any indication that gorillas pay attention to the pelage in mating decisions, or that shagginess is anything but an annual response to the environment.

Furthermore, ever since Darwin, classification has had to depend on the relationships the taxonomist sees among organisms. Indeed, Darwin argues that classification should depend on those relationships. In Darwin's words, "As descent has universally been used in classing together the individuals of the same species ... and as it has been used in classing varieties ..." (Darwin, 1859: Ch. 13). And the same goes for modern primate taxonomy itself (Groves, 2001: Ch. 1). Our decision about whether to class, say, the Bwindi population

of gorillas with the Virunga population (*beringei*), or with the eastern lowland populations (*graueri*), or into its own subspecies, usefully depends on the
historical relationships that we detect among the populations, which are very
usefully determined by neutral traits. If differences between the Bwindi population and the other two populations are minor, and genetic analysis indicates that
the Bwindi population is closer to the Virunga population than to other eastern
lowland populations, it might be sensible to class the Bwindi population in one
subspecies with the Virunga population. That having been said, we need to heed
strongly Sarmiento's warnings about, on the one hand, how little we know about
the partitioning of mtDNA or other noncoding DNA among populations, and
on the other hand, the loose linkage among traits, whether phenotypic, genetic,
under selection, or neutral, and hence the inadvisability, indeed invalidity, of
using any one trait for classification.

Understanding behavior is, of course, vital to conservation management
(e.g., Harcourt, 1986; Sutherland, 1996, 1998; Caro, 1998). Sarmiento appears
to argue that the most vital behavior to understand is that concerned with energy
balance, and makes the interesting suggestion that behavior is usefully analyzed
by morphology. However, without explicitly saying so, he also incorporates behavior under genetics, because genetics can sometimes be used to elucidate
behavior in a more efficient way than can behavioral observation. Genetics
can relatively quickly provide information on, for instance, mating system and
dispersal. Among other uses of the information so gained, Sarmiento suggests
testing population viability analyses (PVAs). Such testing is particularly important because the models often demand an unrealistic level and time-span
of detail, especially for as long-lived a species as a primate (Harrison, 1994;
Harcourt, 1995; Ludwig, 1999). PVAs, it turns out, are usually more usefully
employed as sensitivity analyses than as predictive tools (for instance, pinpoint-
ing stages of the life cycle that most strongly affect survival of the population)
(Crouse *et al.*, 1987; Wennergren *et al.*, 1995).

Both Plumptre *et al.* and Sarmiento spend some time on disease, a topic that
is becoming increasingly important in conservation (Dobson and May, 1986;
McCullum and Dobson, 1995; MacPhee and Marx, 1997; Daszak *et al.*, 2000).
Nevertheless, I wonder if humans, especially tourists, being a cause of disease
in wild gorillas (e.g., Butynski and Kalina, 1998) is less well substantiated than
is sometimes implied. Sarmiento writes, "Gombe stands as a grim reminder
as to how projects without conservation research and long-term goals despite
their ability to generate money and world-wide publicity can be disastrous to
animals (Goodall, 1986; Wallis and Lee, 1999)."

Surely, most readers will strongly disagree with this opinion. If it were
not for Jane Goodall and the research center that she helped establish, with the

long-term goal of understanding the biology of chimpanzees and furthering their conservation, there could well be no Gombe Stream National Park at the site, and arguably, there might therefore be no Gombe chimpanzees. Furthermore, I could find no statement in Goodall (1986) indicting humans as a cause of any of the epidemics among the Gombe chimpanzees. And Wallis and Lee (1999) specifically state that neither the polio outbreaks nor the respiratory epidemic that the Gombe chimpanzees suffered could be traced to humans. While they stated that the scabies outbreak could be linked to humans, they gave only a "personal communication" with an e-mail address as substantiation. But, even if a disease were demonstrably caused by humans, the benefits of research and ecotourism clearly have to be weighed against such a cost. Thus, as for the Gombe chimpanzees, so for the Virunga gorillas of Rwanda, Uganda, and Zaïre, it is very arguable that were it not for research and tourism, no gorillas would be there to be infected by anybody's diseases (Harcourt, 1986). However, one has only to look at the human tragedy in tropical countries now to see the role that disease plays, and there can be no doubt that Plumptre et al. and Sarmiento are correct in arguing that an extremely close watch needs to be kept on disease in the increasingly confined populations of wild gorillas.

In my analysis of the conservation status of gorillas, the Cross River population of Nigeria and Cameroon described by Oates et al. was one of two that I classified as "Critically Endangered", as does IUCN now (Harcourt, 1996; IUCN, 2000). I have suggested elsewhere that much conservation biology being done now might be currently irrelevant, because conservation is mostly done by, in effect, simply putting fences around conservation areas (Harcourt, 2000). At the same time, I also suggested that in the future, when the threats are so intense that nothing survives except with very active management, then the biology that we are doing now will become vital (Harcourt, 2000). If any gorilla population has already reached the stage that nothing survives except with very active management, and therefore application of biology to conservation management is vital, the Cross River gorilla population is surely it.

The Cross River gorillas exist in a tiny area, surrounded by a dense human population. Not only that, but as Oates et al. describe, the gorillas survive not as one population, but as about nine largely or completely separated subpopulations, none of which numbers more than 30 animals. This is surely not a population that can be saved by putting a fence around it. To prevent inbreeding, or to help recovery from a subpopulation crash, are we going to have to start moving individuals between subpopulations? What are we going to do if an epidemic breaks out in one subpopulation? Are we going to try to equalize numbers of breeding males and females in each subpopulation? What are we going to do if dry-season food supply runs out for one of the subpopulations?

Thankfully, the gorilla is one of the few tropical species for which we might know enough biology to answer some of these questions.

Nevertheless, here we have our third closest relative, a mammal weighing around 100 kg. And only just now, as Oates *et al.* tell us, have we discovered a hitherto unknown (to science) population between the Cross River population that they describe in their chapter, and what we used to think was the next nearest gorilla population, 250 km away. Oates *et al.*, like Sarmiento, go into some detail on gorilla taxonomy. Their discussions reflect some of the taxonomic debate that exists in primatology, as primatology goes through a phase of splitting (compare Groves (1993) with Groves (2001)). If any gorilla population is a candidate for new subspecific status, it is surely the Cross River population that Oates *et al.* describe in such cohesive detail.

As we start this century, only about 20% of gorillas live inside protected areas (Harcourt, 1996). In about 100 years time, if things continue as now, all gorillas will survive only inside protected areas, and indeed in some countries they already do so (Harcourt, 1996). Oates *et al.* talk of tropical "paper parks", as do many other conservationists, meaning that the protected areas exist only on paper. However, Bruner *et al.* (2001) nicely showed that in 93 parks from 22 tropical countries representing all four tropical areas of the world (Africa, Americas, Asia, Madagascar), these so-called paper parks had retained or re-covered natural vegetation, prevented land clearing, fires, and grazing, and maintained game populations and commercial timber species.

Even if Africa's and other tropical regions' parks are more than paper parks and are protecting wilderness, Africa suffers massively from political instability, as Plumptre *et al.* describe, and years of conservation management can be destroyed in an instant (Oates, 1999). The Interahamwe rebels hiding in the Virunga forests are now apparently eating gorillas, the first time ever that gorillas have been eaten in this region as far as I know. But poaching, ephemeral official protection of wilderness, and the oft-criticized nepotism and corruption of officials in tropical countries are, of course, very far from being exclusive to tropical countries.

Plumptre *et al.* and Oates *et al.* end their chapters on optimistic notes. And surely one must be optimistic about conservation in a continent in which some nations spend on their national parks as great a proportion of their governmental budget as do many western nations (Wilkie *et al.*, 2001). Even the continued existence of gorillas in the Virunga region of Zaïre, Uganda, and Rwanda is a beacon of hope, let alone the possibility described by Plumptre *et al.* that the gorilla population has increased in this war-torn region. A great many Africans are fighting (some giving their lives indeed) for their wilderness under conditions of hardship unimaginable in the comfortable west. The three chapters in this section indicate how we might continue to contribute to that effort.

## References

Bruner, A.G., Gullison, R.E., Rice, R.E., and da Foncesca, G.A.B. (2001). Effectiveness of parks in protecting tropical biodiversity. *Science*, **291**, 125–128.

Butynski, T.M. and Kalina, J. (1998). Gorilla tourism: A critical look. In *Conservation of Biological Resources*, eds. E.J. Milner-Gulland and R. Mace, pp. 280–300. Oxford, U.K.: Blackwell Scientific Publications.

Caro, T.M. (ed.) (1998). *Behavioral Ecology and Conservation Biology*. New York: Oxford University Press.

Chapman, C.A. and Peres, C.A. (2001). Primate conservation in the new millenium: The role of scientists. *Evolutionary Anthropology*, **10**, 16–33.

Crouse, D.H., Crowder, L.B., and Caswell, H. (1987). A stage-based population model for loggerhead sea turtles and implications for conservation. *Ecology*, **68**, 1412–1423.

Darwin, C. (1859). *The Origin of Species*. London: John Murray.

Daszak, P., Cunningham, A.A., and Hyatt, A.D. (2000). Emerging infectious diseases of wildlife: Threats to biodiversity and human health. *Science*, **287**, 443–449.

Dobson, A.P. and May, R.M. (1986) Disease and conservation. In *Conservation Biology: The Science of Scarcity and Diversity*, ed. M.E. Soulé, pp. 345–365. Sunderland, MA: Sinauer Associates.

Goodall, J. (1986). *The Chimpanzees of Gombe*. Cambridge, MA: Harvard University Press.

Groves, C.P. (1993). Order Primates. In *Mammal Species of the World: A Taxonomic and Geographic Reference*, eds. D.E. Wilson and D.M. Reeder, pp. 243–277. Washington, D.C.: Smithsonian Institution Press.

Groves, C.P. (2001). *Primate Taxonomy*. Washington, D.C.: Smithsonian Institute Press.

Harcourt, A.H. (1986). Gorilla conservation: Anatomy of a campaign. In *Primates: The Road to Self-Sustaining Populations*, ed. K. Benirschke, pp. 31–46. New York: Springer-Verlag.

Harcourt, A.H. (1995). Population viability estimates: Theory and practice for a wild gorilla population. *Conservation Biology*, **9**, 134–142.

Harcourt, A.H. (1996). Is the gorilla a threatened species? How should we judge? *Biological Conservation*, **75**, 165–176.

Harcourt, A.H. (2000). Conservation in practice. *Evolutionary Anthropology*, **15**, 258–265.

Harrison, S. (1994). Metapopulations and conservation. In *Large-Scale Ecology and Conservation Biology*, eds. P.J. Edwards, R.M. May, and N.R. Webb, pp. 111–128. Oxford, U.K.: Blackwell Scientific Publication.

IUCN (2000). *2000 IUCN Red List of Threatened Species*. Gland, Switzerland: International Union for Conservation of Nature and Natural Resources. http://www.redlist.org/.

Ludwig, D. (1999). Is it meaningful to estimate a probability of extinction? *Ecology*, **80**, 298–310.

MacPhee, R.D.E. and Marx, P.A. (1997). The 40 000–year plague: Humans, hyper-disease, and first-contact extinctions. In *Natural Change and Human Impact in*

*Madagascar*, eds. S.M. Goodman and B.D. Patterson, pp. 169–217. Washington, D.C.: Smithsonian Institution Press.

McCullum, H. and Dobson, A.P. (1995). Detecting disease and parasite threats to endangered species and ecosystems. *Trends in Ecology and Evolution*, **10**, 190–194.

Mrosovsky, N. (1997). IUCN's credibility critically endangered. *Nature*, **389**, 436.

Oates, J.F. (1999). *Myth and Reality in the Rain Forest*. Berkeley, CA: University of California Press.

Sutherland, W.J. (1996). *From Individual Behaviour to Population Ecology*. Oxford, U.K.: Oxford University Press.

Sutherland, W.J. (1998). The importance of behavioural studies in conservation biology. *Animal Behaviour*, **56**, 801–809.

Wallis, J. and Lee, D.R. (1999). Primate conservation: The prevention of disease transmission. *International Journal of Primatology*, **20**, 803–826.

Wennergren, U., Ruckelshaus, M., and Kareiva, P. (1995). The promise and limitations of spatial models in conservation biology. *Oikos*, **74**, 349–356.

Wilkie, D.S., Carpenter, J.F., and Zhang, Q. (2001). The under-financing of protected areas in the Congo Basin: So many parks and so little willingness-to-pay. *Biodiversity and Conservation*, **10**, 691–709.

# 17 The current status of gorillas and threats to their existence at the beginning of a new millennium

ANDREW J. PLUMPTRE, ALASTAIR McNEILAGE, JEFFERSON S. HALL, AND ELIZABETH A. WILLIAMSON

## The status of gorilla populations

Gorilla numbers across Africa have declined dramatically over the last century. Some of the earliest conservation efforts on the continent were aimed at protecting mountain gorillas, but destruction of habitat, hunting, and human disturbance have all contributed to the reduction of gorilla populations. All gorilla subspecies are classified as "Endangered" according to the International Union for Conservation of Nature (IUCN) criteria (Oates, 1996; IUCN, 2000). However, in recent years, innovative new conservation ideas have been applied, which have shown some success in slowing the decline, and have even allowed some populations to start to recover. In this chapter, we review the current threats to gorilla populations, and outline some of the strategies that may hold the key to their survival in the twenty-first century.

Gorillas are found in the forests of central Africa, from Nigeria in the west to Rwanda and Uganda in the east, with most animals occurring in the Congo Basin (Cameroon to eastern Democratic Republic of Congo) (DRC) (Fig. 17.1). Until recently there were thought to be three subspecies of gorilla, the mountain gorilla (*Gorilla gorilla beringei*), the Grauer's gorilla (*G. gorilla graueri*) and the western lowland gorilla (*G. gorilla gorilla*). Recent taxonomic research is now suggesting that there are more subspecies (Sarmiento *et al.*, 1996; Oates *et al.*, this volume) and possibly that western and eastern populations should be given separate specific status, *G. gorilla* and *G. beringei*, on the basis of genetic and morphological differences (Groves, 2001, this volume)

Where gorillas occur they are generally found at densities ranging between 0.1 and 2.5 individuals/km$^2$, occasionally as high as 10.6 locally. They tend to occur at their highest densities in secondary forest (Carroll, 1988; Fay, 1989, 1997; Hall *et al.*, 1998*a*). Extrapolating from where gorillas have been surveyed to their known range of occurrence allows us to estimate the sizes of the various populations (Table 17.1). If we use the taxonomic classification that

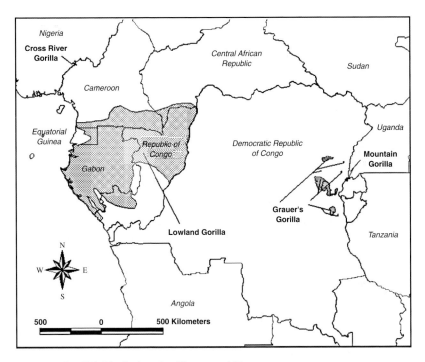

Fig. 17.1. Distribution of gorillas across Africa.

existed from the 1980s until the mid 1990s, then population sizes are as follows: *G. g. gorilla* (population of about 110 000), *G. g. graueri* (population of about 17 000), and *G. g. beringei* (population of around 650–700). If the gorillas in Bwindi-Impenetrable National Park in southwest Uganda are different from the mountain gorillas in the Virungas (Sarmiento *et al.*, 1996) then there would only be around 360–395 mountain gorillas (J. Kalpers *et al.*, unpublished data). Similarly, if the gorillas found on the Nigeria–Cameroon border around the Cross River are accepted as a new subspecies, *G. g. diehli*, then these would become the rarest subspecies of gorilla with only about 150–200 individuals (Sarmiento and Oates, 2000; Oates *et al.*, this volume).

Whether we accept these further divisions of gorilla populations or not, all gorilla populations are threatened throughout their range, primarily by human activities. As human populations increase in Africa the available habitat for gorillas is decreasing and gorillas are coming into contact with humans much more than they used to. Each of the subspecies is threatened by different factors that result from increasing human populations, and all gorilla populations are considered under IUCN criteria as "Endangered" (Harcourt, 1996).

Table 17.1. *The densities of gorillas at various sites in Africa and approximations of population size for each country*

| Country | Region | Density (individuals/km) | Population size[a] | Reference |
|---|---|---|---|---|
| *Cross River gorilla* | | | | |
| Nigeria/Cameroon | Cross River area | 0.8 | 150–200 | Harcourt *et al.*, 1989; Oates *et al.*, this volume |
| *Lowland gorilla* | | | | |
| Cameroon | Whole country | | 12 500 | Harcourt, 1996 |
| | Boumba Bek | 1.6 | | Stromayer and Ekobo, 1991 |
| | Dja Reserve | 1.7 | | Williamson and Usongo, 1995 |
| | Lac Lobéké | 2.5 | | Fay, 1997 |
| | Mongokele | 1.2 | | Fay, 1997 |
| Equatorial Guinea | | 0.2 | 3 000 | Jones and Sabater Pi, 1971 |
| Gabon | | 0.5 | 43 000 | Tutin and Fernandez, 1984 |
| | Lopé Reserve | 0.4 | | White, 1992 |
| Democratic Republic of Congo | | | 44 000 | Harcourt, 1996; Fay, 1997 |
| | Odzala | 10.0 | | Fay, 1997 |
| | Mboukou | 1.2 | | Fay, 1997 |
| | Motaba | 0.1 | | Fay, 1997 |
| | Ndoki | 0.2 | | Fay, 1997 |
| | Mbomo | 0.6 | | Fay, 1997 |
| | Garabinzam | 0.5 | | Fay, 1997 |
| | Northern Congo | 0.4 | | Fay, 1997 |
| | Southern Congo | 0.05 | | Fay, 1997 |

| | | | | |
|---|---|---|---|---|
| Central African Republic | Sangha region | 0.8–1.5 | 9000 | Carroll, 1988 |
| | Secondary forest | 10.6 | | Carroll, 1988; Fay, 1997 |
| Angola | NW Angola | very low to zero | 0 | Fay, 1997 |
| Democratic Republic of Congo | NW Democratic Republic of Congo | very low to zero | 0 | Fay, 1997 |
| *Grauer's gorilla* | | | | |
| Democratic Republic of Congo | Kahuzi-Biega National Park | 1.26 | 14 900 | Hall et al., 1998b |
| | Maiko National Park | 0.25 | 859 | Hart and Hall, 1996; Hall et al., 1998b |
| | Itombwe | | 1155 | Hall et al., 1998b |
| | Mt. Tshiaberimu | | 16 | Butynski and Sarmiento, 1995; Hall et al., 1998b |
| | Masisi | | 28 | Mwanza et al., 1988; Hall et al., 1998b |
| *Mountain gorilla* | | | | |
| Rwanda, Democratic Republic of Congo, Uganda | Virungas | 0.9 | 360–395 | J. Kalpers et al., unpublished data |
| Uganda | Bwindi | 0.9 | 300 | McNeilage et al., 1998 |

[a]The populations have been given for the three subspecies (Groves, 1967) because there are fewest of these.

The major threats to gorillas can be categorized as: (1) habitat loss, modification, or fragmentation; (2) hunting or poaching; (3) disease transmission from humans; and (4) war or political unrest. We first review these threats and then outline conservation initiatives underway to address them.

## Threats to gorilla populations

### *Habitat loss or modification*

Habitat loss due to agriculture or modification, through logging, fuel wood and nontimber forest product collection, and grazing domestic animals, is undoubtedly one of the greatest long-term threats to gorillas. Gorillas are dependent on forest habitat that, once converted to agriculture, rarely reverts back to natural vegetation unless it can be left for hundreds of years. Even where forest is not destroyed, it may be effectively lost as gorilla habitat if disturbance is severe enough.

The bulk of the western populations lives outside protected areas. The western gorillas are consequently most at threat from loss or modification of habitat. In the Congo Basin, average deforestation rates are about 0.5% per year (World Resources Institute, 2000*a*). Where human density is high, forest clearance is also high (Barnes, 1990; Harcourt, 1996), and therefore it is not surprising that gorilla populations are healthiest in the most remote areas of the Congo Basin such as northern Congo and Gabon (Table 17.1). The western populations are also most affected by selective logging. The impacts of logging vary depending on whether hunting accompanies logging operations. Logging concessions now cover 76% of the forest in Cameroon (World Resources Institute, 2000*b*) and 56% of the forest in Gabon (World Resources Institute, 2000*a*). Similar percentages occur for the Democratic Republic of Congo (J. Brunner, personal communication) and the Central African Republic (50–75%). When civil war in the DRC ends it is very probable that logging companies will quickly move into this country as well. If there is no hunting of gorillas within the concessions, gorillas probably can coexist with logging because they tend to favor areas of secondary vegetation where herbaceous food plants can be found (Harcourt, 1981; White, 1992, 1994; Fay, 1997). No study has directly compared gorilla populations before and after logging in a specific site, but significant gorilla populations have been found in previously logged forest comparable with densities in unlogged forest (White, 1994). However, where hunting does occur, large-bodied and slow-reproducing species such as gorillas are usually eliminated completely (Lahm, 1994; Wilkie and Carpenter, 1999*a*), and hunting does go on virtually everywhere that logging is carried out.

In eastern DRC, George Schaller surveyed gorilla populations in the early 1960s (Schaller, 1963), which have since been lost following forest clearance (Hall *et al.*, 1998*a*). With the current instability in DRC, we have no concrete figures, but the rate of habitat loss for Grauer's gorillas is probably the highest of any subspecies.

In 1959, Schaller carried out the first extensive research on any of the gorilla subspecies in the wild. At that time, he considered habitat loss to be the greatest danger to the survival of the mountain gorillas (Schaller, 1963). Following this prediction, about 54% of the Rwandan part of the Virungas was taken for an agricultural project between 1967 and 1970 (Weber, 1987), and the mountain gorilla population dropped from an estimated 400–500 to 252 animals (Harcourt and Fossey, 1981; Weber and Vedder, 1983). However, in recent years, habitat loss has been close to zero, although war and instabilities in the Virungas has caused some disturbance to habitat.

### Hunting or poaching

The killing of gorillas can be direct or indirect depending on whether they are specifically targeted or are unintentionally trapped by snares. The major reasons for intentional killing of gorillas are: (1) meat, (2) capture of animals for collections, and (3) body parts for trophies. The extent of hunting of gorillas for meat varies across their range according to local traditions and taboos and the effectiveness of protection measures. Infant gorillas from all subspecies have been captured for sale or attempted sale to zoological or private collections. During the process of capture, adult males and females are often killed whilst trying to protect their infants. For every infant that reaches a zoo, many other animals die. For example, in 1968–69, 18 gorillas were killed in attempts to capture gorilla infants (Fossey, 1983). Many of the infants sold to collections from the western gorilla populations may be caught when adults are hunted for meat and hence the sale of infants may at times be a byproduct of the bushmeat trade.

Gorillas were hunted directly for trophies (skulls, hands, feet, and skins) until quite recently. Oscar von Beringe shot the first mountain gorilla to be discovered and Carl Akeley shot several for the American Museum of Natural History in the early 1920s. Similarly, sports hunters during the 1800s and early 1900s shot western gorillas for skins (Du Chaillu, 1861). Occasionally, gorillas are killed because they have been raiding crops in fields (Hall *et al.*, 1998*b*) but this is less common than for the other reasons already delineated.

Snares, although not set specifically for gorillas, are known to kill them throughout their range (Hall *et al.*, 1998*a*; Noss, 1998, 2000) and can be a source of high mortality even where protection is supposedly present. Wire and rope snares are set to trap antelope, and animals that get caught struggle to get free, creating deep cuts to their hands or feet. If these cuts become infected the animal can easily die.

Among threats to western lowland gorillas, the bushmeat trade in Central Africa is one of the most serious, and has led IUCN to upgrade the status of western gorillas to endangered. An estimated 1.2 million tonnes of bushmeat are consumed each year in the Congo basin (Wilkie and Carpenter, 1999*a*), although gorillas form less than 1% of this trade. In northern Congo, around Motaba River, gorilla and chimpanzee nest densities increase greatly after about 20 km from villages because of hunting pressures (Kano and Asato, 1994). Around 5% of the gorilla population has been killed annually. With this rate of offtake, gorilla populations would be expected to decline to 50% in 10–15 years because of the gorillas' slow reproductive rates (Kano and Asato, 1994); a female gorilla only produces offspring once every three to five years (Watts, 1990). This area has one of the lowest human population densities in Africa apart from the deserts (Cincotta and Engelman, 2000), and yet even here hunting of gorillas is not sustainable. Many studies around the world of traditional hunting of wild animals now show that hunting is rarely sustainable unless the human population density is very low (Robinson and Bennett, 2000).

The logging industry can greatly exacerbate the problem of gorilla-hunting. Timber harvesting opens up access to the forest through the construction of roads and hunting almost always increases as timber companies start to extract the wood. Employees earn enough money on the timber concessions to be able to purchase and eat meat more regularly than they can if unemployed, and hence a market for bushmeat is created on the concessions (Wilkie and Carpenter, 1999*a*). Logging trucks also transport bushmeat to markets in the cities down roads that have been constructed for the timber industry.

Grauer's gorillas in the east of the Congo Basin do not appear to be eaten as much as they are in the west because of taboos among certain people against eating apes. More recently, however, the taboos in the east have been changing and gorillas are now being eaten more commonly (Hall *et al.*, 1998*b*). Recently as well, many Grauer's gorillas in the montane sector of Kahuzi-Biega National Park have been killed for meat as a result of civil strife in DRC (G. Debonnet and J. Mankoto, personal communication). Between 1998 and 2000 the population fell from around 250 to 130 individuals (O. Ilambu, personal communication).

Hunting of mountain gorillas for meat has also occurred, but only rarely. It was thought that many mountain gorillas were killed for food in Zaïre (now DRC) during fighting for independence in the 1960s (Weber and Vedder, 1983). Apart from the mortality linked to armed conflict described below, direct killings of mountain gorillas have been rare in the last 20 years because of good protection and reasonable support and funding for national parks. However, indirect killing of mountain gorillas caught in snares set for antelope still occurs.

Finally, the Democratic Republic of Congo is well-known for its mineral wealth and with the onset of civil war many protected areas have been invaded by local villagers in search of gold, diamonds and more recently columbo–tantalite (Coltan). Coltan is an ore that is used to make semiconductors and is important for electronic components of cell phones and computers. Coltan is found in scattered locations but where it occurs it can be found in fairly large quantities. During 1998–2000 coltan was selling at about $80/kg and people were able to mine several kilograms each day creating a huge influx of people into the forests looking for this mineral in a region where monthly salaries on average are less than $30. In the Nyungwe Forest in Rwanda, for example, there were three camps of over 3000 miners and only 90 guards to try to evict them. Kahuzi-Biega National Park, which contained about 86% of Grauer's gorillas in 1996, has been the most severely affected with large mining camps having been established in the lowland sector where coltan can be found. Miners have been hunting large mammals, including gorillas, to feed themselves while they are there, and to date, it has been impossible to reach this part of the park to evaluate the impacts of the mines on the gorilla population. The Wildlife Conservation Society (WCS) was involved with other nongovernmental organizations (NGOs) in publicizing the plight of the Kahuzi-Biega gorilla population following a survey that WCS undertook with the Institut Congolais pour la Conservation de la Nature and the Dian Fossey Gorilla Fund International in July 2000. Pressure has subsequently been put on electronics companies to purchase "gorilla-friendly" coltan. This has occurred to some extent and most companies are buying from Australia now. The price of coltan has fallen to about $18/kg and coltan is also being found outside forests in Rwanda, which has relieved some of the pressure on the protected areas in this region, and park staff can now control the miners in Nyungwe Forest. However, if the price should increase again in the future, there will be increased pressure on the protected areas again. Mining is still continuing in Kahuzi-Biega park because the insecurity in the region prevents park guards from accessing the lowland sector of the park and they are unable therefore to prevent it.

### Disease transmission

Gorillas are susceptible to human diseases and many in zoos are vaccinated against common human ailments. With small populations of gorillas, any infectious disease could devastate the population. Diseases transmitted to immunologically naïve populations have resulted in massive mortality in other species – up to entire populations (Thorne and Williams, 1988; Macdonald, 1996), and primates are especially vulnerable due to their slow reproductive rates (Young, 1994). Disease and subsequent deaths of habituated mountain gorillas have been caused by respiratory outbreaks and scabies (Hastings *et al.*, 1991; Kalema *et al.*, 1998) and several other pathogens have been identified in the population (Mudakikwa *et al.*, 2001). However, little data are available on the impacts of disease, particularly outside the Virunga population. In the closely related chimpanzee (*Pan troglodytes*), polio and respiratory diseases have occurred (Goodall, 1986). Recent evidence from Minkebe in northern Gabon indicates that gorillas have died in large numbers and are now very rare when in the 1970s they had been common. What killed them is not known but it is thought to be due to an unknown disease (J. M. Fay, personal communication).

Gorilla ecotourism, which started in Rwanda and Uganda in the 1960s, has increased the potential threat of disease transmission. While most of the international tourists visiting Rwanda are fairly fit, having been inoculated against certain diseases, they may be carrying new viruses for the region such as influenza. It is these illnesses to which the gorillas have never been exposed which are potentially the most dangerous (McNeilage, 1996; Butynski and Kalina, 1998). If poorly controlled, tourism can also lead to increased stress in the animals, which can increase susceptibility to disease (Hudson, 1992; McNeilage, 1996). While this was recognized as a risk at the start of the tourism program in Rwanda, the loss of habitat was considered a far greater threat to the gorillas at the time and the potential benefits of tourism were thought to outweigh the possible negative impacts on the gorillas. The tourism program was implemented with rules in place to regulate tourist visiting times and the number of tourists per group. Strict rules also exist on the behavior of park staff and tourists to try to eliminate the risk of passing on a disease to which the animals will not have been previously exposed (Homsy, 1999). For example, a minimum distance of 5 m is supposed to be maintained between people and gorillas, eating and drinking are not allowed when with the gorillas, and all human fecal material in the park should be buried. However, some of these rules are not always adhered to (Butynski and Kalina, 1998).

Homsy (1999) points out that many diseases that are dangerous to gorillas can survive in feces in the soil for up to six months or more. These include hepatitis A, poliovirus, tapeworm, and the TB bacillus. Increasing human density as a result of roads being built into the forest and people settling along them will increase the potential for human–gorilla transmission of disease. Gorillas that live near human populations have probably been exposed to human parasites for decades. People visit the forest to collect firewood, building poles, and nontimber forest products, to harvest trees for timber and to hunt, and might defecate and urinate while in the forest. Infections transmitted via feces are far more likely to occur than are respiratory diseases, and the risks from such fecal transmission may be at least as serious as those from close proximity of people to gorillas for limited periods of time (Homsy, 1999).

During the war and civil unrest that has occurred during the 1990s in the Congo Basin, people moved into the more remote forests as either refugees or rebel groups that were being chased. They were poorly nourished, living in harsh and unsanitary conditions, and many died in the forests. How these movements might have affected the health of gorilla populations is unknown, but clearly this could greatly increase the threat of disease outbreaks devastating such small populations of gorillas.

### War and political unrest

Civil wars are not a new threat to the conservation of protected areas, but the participants are now much better armed than they were in the past. The ever-shrinking forests that result from expanding human populations are ideal hiding places or retreats for armed opposition groups. Many national parks in Central Africa are associated with the presence of rebels, and conflicts extend over a much larger arena than they used to. These parks include Kahuzi-Biega National Park, Virunga National Park, and Maiko National Park, where most of the populations of eastern gorillas occur. There are hundreds of thousands of refugees throughout the range of *G. g. graueri* and in areas of *G. g. gorilla* range in the Democratic Republic of Congo. Protected areas such as the Virungas, which straddle international borders, are particularly at risk as people can move back and forth between countries easily while hiding within the forest. Refugees who fled Rwanda for Zaïre were settled near the Virunga National Park in one of the world's largest refugee camps and ended up deforesting 113 km$^2$ of the park while collecting firewood (Henquin and Blondel, 1996). Similarly, a refugee camp sited near the Kahuzi-Biega National Park increased the price of fuel wood by two to three times, and although the camp and the refugees are now gone, there is still a big fuel-wood problem for the people who live in this region (B.-I. Inogwabini, personal communication). However, in this case the park warden managed to keep the refugees and the fuel-wood collection outside the park.

Little is known about how the recent wars in central Africa have affected gorilla populations. It is known that mountain gorillas around the volcano Mikeno in DRC dropped in number between 1960 and 1971–73 (Weber and Vedder, 1983) and it is thought this may have been due to increased hunting after the civil war in the 1960s. More recently the civil war in Rwanda and DRC has made it impossible to assess how the mountain gorillas have fared. It has been possible to monitor the habituated groups of gorillas, however, and most of these have survived remarkably well. It is known that since 1995 at least 18 gorillas have been killed in the Virunga National Park (DRC) as a result of poaching with firearms or getting caught in crossfire. Remarkably, almost all habituated individuals were accounted for in Rwanda in 1999 after 14 months without any monitoring during which many rebels and refugees lived in the park. Recent evidence from known numbers of habituated animals and ranger counts of nest groups indicates that the population has risen to a minimum count of about 360 animals (Kalpers *et al.*, unpublished data).

In Kahuzi-Biega National Park there has been heavy hunting of gorillas for meat in the mountain sector as a result of war and displacement. The gorilla population in that area has been reduced from 250 to 130 individuals. Nobody from the Congolese National Parks has been able to visit the lowland sector where the bulk of the gorilla population was found in the mid 1990s (Hall *et al.*, 1998*b*) and we have no idea how they are faring. We only know that there are many people living in the forest mining for gold and diamonds, and probably living off the wild animals in the forest, including gorillas given that taboos have been changing.

## Conserving gorilla populations

Traditional protected areas are still one of the main conservation strategies to address the threats to gorillas, particularly hunting and habitat loss. However, more recently, efforts to protect gorillas outside traditionally protected areas by working with logging companies and other interested groups have met with some success in reducing hunting pressure. Recent experience in areas of conflict has shown that all is not necessarily lost when war breaks out. In this section we describe how different conservation strategies, both new and traditional, are contributing to the future conservation of gorillas.

### Protected areas

The first national park in Africa, the Virunga Park, was created in 1925 to protect the mountain gorilla. This park still exists today but in a highly degraded state as a result of human population growth and civil wars in this part of Africa. While the gorillas and the montane habitat are thought to be relatively intact, the lowland savanna parts of the park have been severely hunted and most large mammal populations have been greatly reduced. Many of the parks in eastern DRC have only been "paper parks" in that they have often had very little resources flowing to them to help the park staff maintain them. While paper parks may not be effective, well-funded, managed parks, they have historically fared better than areas that had no protected status, and people have often respected their boundaries (Hart and Hall, 1996; Oates, 1999). Most of the resources that have reached many parks have come from international NGOs. This is a criticism of most of the protected areas in the Congo Basin, whose governments have not put much revenue into supporting them. Analyses of the possibility of generating revenue from tourist receipts (Wilkie and Carpenter, 1999*b*) show that it is highly unlikely that the costs of managing the protected areas in the Congo Basin can be generated in this way. Mountain gorillas, living in a small island of habitat that is easily accessed by tourists and can be intensively protected by a relatively small number of rangers, may be able to survive through the support of tourism dollars. However, it is unlikely that tourism will be able to support the conservation of eastern and western lowland gorillas in the long term, as they live over a much wider range, many in areas difficult to access.

If gorillas are to be conserved into the next millennium, national parks are going to be essential tools for maintaining viable populations. Only parks of sufficiently large size will be able to maintain ecologically functional populations of gorillas, and given the typical densities at which gorillas are found (Table 17.1) this will generally mean at least several hundred square kilometers. The international community is going to have to support conservation efforts for this species. Through treaties, such as the 1993 Rio Convention on Biological Diversity, the developed world has committed to support conservation in the southern hemisphere countries, which hold most of the world's biodiversity. Recently one important precedent was set in this regard. In recognition of the plight of the five World Heritage Sites in DRC, the UN Foundation promised $3 million to support these protected areas (Salonga, Kahuzi-Biega, Okapi, Virunga, and Garamba) through an emergency relief fund. This precedent will hopefully lead to further support of other areas considered to be of global importance in the future, particularly during periods of conflict. There is also a move to try to establish a trust fund to support the protected areas in the Congo Basin known as the Central Africa Management of Biodiversity project (CAMBIO). The World Bank is being approached to provide approximately $90 million to establish this fund. Whether this crucial initiative will succeed or fail will depend on the popular support it receives. Another small endowment trust fund has already been set up to support conservation efforts in and around Bwindi and Mgahinga, two forests in southwest Uganda where mountain gorillas are found. The Mgahinga and Bwindi-Impenetrable Forest Conservation Trust was set up in 1995 with initial funds from the Global Environment Facility, and further support from the United States Agency for International Development and the Netherlands government, and provides a sustainable source of funding for park management, research, and community conservation projects. Eventually it is hoped that this trust will reach $10 million or more, at which point it should generate around 7% or $700 000 per year to fund its programs.

### Working with logging companies

Protected areas alone may not be enough to conserve healthy populations of gorillas. If gorillas are also going to survive outside protected areas then something must be done to curb the hunting. In all countries of the Congo Basin the hunting of gorillas is illegal but this does not stop the trade in gorilla meat. There are little or no resources to monitor and police this illegal hunting. The best places to target actions to conserve gorillas outside protected areas are in large areas of land controlled by one or a few people, where such individuals have control over what can happen on their land. Working with logging companies to conserve gorillas outside protected areas would ensure that the populations are not isolated and this may be vital if changes in the forest extent occur as a result of global warming. Logging companies are an obvious choice in this respect because they usually have access to large concessions that can be held for many years. Some of the key concessions neighbor important protected areas. Often there are few roads that enter or exit the concession so that these points can be monitored and the bushmeat trade controlled to some extent. Traditionally, conservationists have been at odds with timber extraction companies, but realistically we are unlikely to be able to raise sufficient funds to offset the opportunity costs of logging in the Congo Basin. If conservationists and timber companies can learn to collaborate then we are more likely to succeed at saving these animals over large areas of forest.

The Wildlife Conservation Society has been piloting a project that is a collaborative effort between the Ministry of Forest Economy in the Congo and the Congolaise Industrielle des Bois (CIB), a logging company. The project operates around the periphery of the Nouabalé-Ndoki National Park in northeast Congo and has been working with the company to establish mutually agreed upon guidelines for hunting, which include a ban on the hunting of apes, elephants and other endangered/vulnerable species and a ban on the export of meat from the concession. Commercial bushmeat hunting has reduced as a result through enforcement by the local community, the logging company, and the WCS project. The incidence of gorilla and chimpanzee hunting has dropped by an estimated 90% on this concession in the last two years (Elkan, 1999) and recently an employee of CIB lost his job for killing a gorilla. In addition, the company gave up a large part of its concession in an area where there are apes with little fear of humans, presumably because they have never come in contact with them. The World Bank has become interested in promoting this pilot project in other concessions in Central Africa and has been bringing together the chief executive officers of many of the companies operating in this region to encourage them to follow CIB's lead.

### Conservation in areas of conflict

War and political insecurity usually lead to the reduction or complete elimination of bilateral support. Much of the money for international conservation comes from bilateral grants and these funds are usually frozen at the smallest sign of insecurity. This action often has led to the complete cessation of projects that were being supported and the consequent loss of project staff. Experience from WCS's projects in Central Africa has shown that maintaining a presence during war at projects they have supported prior to insecurity has enabled staff to continue some actions on the ground (Plumptre *et al.*, in press). More importantly, they have also been able to negotiate with any changed leadership to ensure that protected areas continue to be recognized ( Fimbel and Fimbel, 1997; Hart and Hart, 1997; Plumptre and Williamson, 2001). Unfortunately, donors tend to withdraw funding at the first sign of instability. Bilateral agencies should consider channeling funds through international NGOs during periods of insecurity if they cannot work directly with governments for political reasons. An analysis of protected areas in Rwanda showed that during the last decade of the twentieth century, only in those protected areas where an NGO presence was maintained did the protected area survive (Plumptre *et al.*, in press).

### Conclusions

The future of the gorilla in the long term is still not secure despite the fact that the first park to protect this species was created over 75 years ago. It is inevitable that the global population of *Gorilla* will be reduced as the human population continues to increase, along with the consequent threats to gorillas, particularly in central Africa where many gorillas still exist outside protected areas. Even those within protected areas are not secure as many of the parks and reserves with gorillas lack adequate financial and institutional support to combat the recent increase in the hunting of animals for bushmeat. If the gorilla is to survive another century, individuals and institutions within the richer countries of the world will have to be prepared to provide financial support for their protection for many years to come.

However, not everything is bleak. Apes are highly charismatic species and the tourism programs in the eastern part of the gorilla's range show that people are willing to pay large sums of money to be able to visit them in the wild. Through television and books, the image of gorillas has been transformed from aggressive brutes in the nineteenth century to the "gentle giants" we know them to be at the beginning of the twenty-first century. Raising funds for gorillas is generally much easier than for many other species and the recent creation of a Great Ape Fund in the U.S.A. by Congress has ensured a certain level of support to all apes for the foreseeable future. More difficult will be to try and ensure the long-term survival of all the genetic and ecological variation within this genus, particularly the smallest populations of the most endangered subspecies. There have been recent proposals to raise the international status of apes, given their genetic and behavioral similarities to humans, possibly creating "World Heritage Species" to complement World Heritage Sites (International Primatological Society, personal communication). This would certainly benefit these charismatic animals and could raise their profile considerably. Gorillas can act as very useful "umbrella species" where they occur, so that highlighting the need for their conservation can lead to the protection of many other rainforest species. Raising their profile and ensuring their long-term protection would therefore conserve much of the other biodiversity found in the forests where they live.

## References

Barnes, R.F.W. (1990). Deforestation trends in tropical Africa. *African Journal of Ecology*, **28**, 161–173

Butynski, T.M. and Kalina, J. (1998). Gorilla tourism: A critical look. In *Conservation of Biological Resources*, eds. E.J. Milner-Gulland & R. Mace, pp. 280–300. Oxford, U.K.: Blackwell Scientific Publications.

Butynski, T.M. and Sarmiento, E. (1995). Gorilla census on Mt Tshiaberimu: Preliminary report. *Gorilla Journal*, **1**, 11.

Carroll, R. (1988). Relative density, range extension, and conservation potential of the lowland gorilla (*Gorilla gorilla gorilla*) in the Dzanga-Sangha region of southwest Central African Republic. *Mammalia*, **52**, 309–323.

Cincotta, R.P. and Engelman, R. (2000). *Nature's Place: Human Population and the Future of Biological Diversity*. Washington, D.C.: Population Action International.

Du Chaillu, P.B. (1861). *Explorations and Adventures in Equatorial Africa*. New York: Harper.

Elkan, P. (1999). Wildlife management in areas surrounding Nouabalé-Ndoki National Park. *Gnusletter*, **18**(2), 15–16.

Fay, J.M. (1989). Partial completion of a census of the lowland gorilla (*Gorilla g. gorilla* (Savage and Wyman)) in southwestern Central African Republic. *Mammalia*, **53**, 203–215.

Fay, J.M. (1997). The ecology, social organisation, populations, habitat and history of the western lowland gorilla (*Gorilla g. gorilla* Savage and Wyman, 1847). PhD thesis, Washington University, St. Louis, MO.

Fimbel, C. and Fimbel, R. (1997). Conservation and civil strife: Two perspectives from Central Africa: Rwanda: The role of local participation. *Conservation Biology*, **11**, 309–310.

Fossey, D. (1983). *Gorillas in the Mist*. Boston, MA: Houghton Mifflin.

Goodall, J. (1986). *The Chimpanzees of Gombe: Patterns of Behavior*. Cambridge, MA: Harvard University Press.

Groves, C.P. (1967). Ecology and taxonomy of the gorilla. *Nature*, **213**, 890–893.

Groves, C.P. (2001). *Primate Taxonomy*. Washington, D.C.: Smithsonian Institution Press.

Hall, J.S., White, L.J.T., Inogwabini, B.-I., Omari, I., Morland, H.S., Williamson, E.A., Saltonstall, K., Walsh, P., Sikubwabo, C., Bonny, D., Kiswele, K.P., Vedder, A., and Freeman, K. (1998*a*). Survey of Grauer's gorillas (*Gorilla gorilla graueri*) and eastern chimpanzees (*Pan troglodytes schweinfurthii*) in the Kahuzi-Biega National Park lowland sector and adjacent forest in eastern Democratic Republic of Congo. *International Journal of Primatology*, **19**, 207–235.

Hall, J.S., Saltonstall, K., Inogwabini, B-I., and Omari, I. (1998*b*) Distribution, abundance and conservation status of Grauer's gorilla. *Oryx*, **32**, 122–130.

Harcourt, A.H. (1981). Can Uganda's gorillas survive? A survey of the Bwindi Forest Reserve. *Biological Conservation*, **19**, 262–282.

Harcourt, A.H. (1996). Is the gorilla a threatened species? How should we judge? *Biological Conservation*, **75**, 165–176.

Harcourt, A.H. and Fossey, D. (1981). The Virunga gorillas: Decline of an island population. *African Journal of Ecology*, **19**, 83–97.

Harcourt, A.H., Stewart, K.J., and Inahoro, I.M. (1989). Nigeria's gorillas: A survey and recommendations. *Primate Conservation*, **10**, 73–76.

Hart, J.A. and Hall, J.S. (1996). Status of eastern Zaïre's forest parks and reserves. *Conservation Biology*, **10**, 316–324.

Hart, T. & Hart, J. (1997). Conservation and civil strife: Two perspectives from Central Africa: Zaïre: New models for an emerging state. *Conservation Biology*, **11**, 308–309.

Hastings, B.E., Kenny, D., Lowenstine, L.J., and Foster, J.W. (1991). Mountain gorillas and measles: Ontogeny of a wildlife vaccination program. *American Association of Zoological Veterinarians Annual Proceedings*, **1991**, 198–205.

Henquin, B. and Blondel, N. (1996). Etude par télédetection sur l'évolution recente de la couverture boisée du Parc National des Virunga. Unpublished report, Laboratoire d'Hydrologie et de Télédetection, Gembloux, Belgium.

Homsy, J. (1999). *Ape Tourism and Human Diseases: How Close Should We Get?* Report to the International Gorilla Conservation Program, Kigali.

Hudson, H.R. (1992). The relationship between stress and disease in orphan gorillas and its significance for gorilla tourism. *Gorilla Conservation News*, **6**, 8–10.

IUCN (2000). *Red List of Threatened Species*. Gland, Switzerland: IUCN. http://www.redlist.org/

Jones, C. and Sabater Pi, J. (1971). Comparative ecology of *Gorilla gorilla* (Savage and Wyman) and *Pan troglodytes* (Blumenbach) in Rio Muni, West Africa. *Biblioteca Primatologica*, **13**, 1–96.

Kalema, G., Kock, R.A., and Macfie, E. (1998). An outbreak of sarcoptic mange in free-ranging mountain gorillas (*Gorilla gorilla beringei*) in Bwindi Impenetrable

National Park, Southwestern Uganda. In *Joint Proceedings of the AAZV and AAWV Annual Meeting*, Omaha, Nebraska, p. 438 (abstract).

Kano, T. and Asato, R. (1994). Hunting pressure on chimpanzees and gorillas in the Motaba River area, Northeastern Congo. *African Study Monographs*, **15**, 143–162.

Lahm, S.A. (1994). *Hunting and Wildlife in Northeastern Gabon: Why Conservation Should Extend beyond Protected Areas*. Report to Institut de Recherche en Ecologie Tropicale, Makoukou, Gabon.

Macdonald, D.W. (1996). Dangerous liaisons and disease. *Nature*, **379**, 400–401.

McNeilage, A. (1996). Ecotourism and mountain gorillas in the Virungas. In *The Exploitation of Mammal Populations*, eds. V.J. Taylor and N. Dunstone, pp. 334–344. London: Chapman & Hall.

McNeilage, A., Plumptre, A.J., Brock-Doyle, A., and Vedder, A. (1998). *Bwindi-Impenetrable National Park, Uganda Gorilla and Large Mammal Census, 1997*. WCS Working Paper no 14. Washington, D.C.: Wildlife Conservation Society.

Mudakikwa, A.B., Cranfield, M.R., Sleeman, J.M., and Eilenberger, U. (2001) Clinical medicine, preventative health care and research on mountain gorillas in the Virunga Volcanoes region. In *Mountain Gorillas: Three Decades of Research at Karisoke*, eds. M.M. Robbins, P. Sicotte and K.J. Stewart, pp. 341–360. Cambridge, U.K.: Cambridge University Press.

Mwanza, N., Maruhashi, T., Yumoto, T., and Yamagiwa, J. (1988). Conservation of eastern lowland gorillas in the Masisi region, Zaïre. *Primate Conservation*, **9**, 111–114.

Noss, A.J. (1998). Cables, snares and bushmeat markets in a central African forest. *Environmental Conservation*, **25**, 228–233.

Noss, A.J. (2000). Cables, snares and nets in the Central African Republic. In *Hunting for Sustainability in Tropical Forests*, eds. J.G. Robinson and E.L. Bennett, pp. 282–304. New York: Columbia University Press.

Oates, J.F. (1996). *African Primates: Status Survey and Conservation Action Plan*. Gland, Switzerland: International Union for Conservation of Nature and Natural Resources.

Oates, J.F. (1999). *Myth and Reality in The Rainforest: How Conservation Strategies Are Failing in West Africa*. Berkeley, CA: University of California Press.

Plumptre, A.J. and Williamson, E.A. (2001). Conservation-oriented research in the Virunga region. In *Mountain Gorillas: Three Decades of Research at Karisoke*, eds. M.M. Robbins, P. Sicotte, and K.J. Stewart, pp. 361–389. Cambridge, U.K.: Cambridge University Press.

Plumptre, A.J., Masozera, M., and Vedder, A. (2001). *The Impact of Civil War on the Conservation of Protected Areas in Rwanda*. Washington, D.C.: Biodiversity Support Program.

Robinson, J.G. and Bennett, E.L. (eds.) (2000). *Hunting for Sustainability in Tropical Forests*. New York: Columbia University Press.

Sarmiento, E.E. and Oates, J.F. (2000). The Cross River gorillas: A distinct subspecies, *Gorilla gorilla diehli* Matschie 1904. *American Museum Novitates*, **3304**, 1–55.

Sarmiento, E.E., Butynski, T.M., and Kalina, J. (1996). Gorillas of Bwindi-Impenetrable Forest and the Virunga Volcanoes: Taxonomic implications of morphological and ecological differences. *American Journal of Primatology*, **40**, 1–21.

Schaller, G.B. (1963). *The Mountain Gorilla: Ecology and Behavior*. Chicago, IL: University of Chicago Press.

Stromayer, K. and Ekobo, A. (1991). *Biological Surveys of Southeastern Cameroon*. Report to European Community, Brussels.

Thorne, E.T. and Williams, E.S. (1988). Disease and endangered species: The black-footed ferret as a recent example. *Conservation Biology*, **2**, 66–74.

Tutin, C.E.G. and Fernandez, M. (1984). Nationwide census of gorilla (*Gorilla g. gorilla*) and chimpanzee (*Pan t. troglodytes*) populations in Gabon. *American Journal of Primatology*, **6**, 313–336.

Watts, D.P. (1990). Mountain gorilla life histories, reproductive competition and sociosexual behavior and some implications for captive husbandry. *Zoo Biology*, **9**, 185–200.

Weber, A.W. (1987). Ruhengeri and its resources: An environmental profile of the Ruhengeri Prefecture. Kigali: ETMA/USAID.

Weber, A.W. and Vedder, A.L. (1983). Population dynamics of the Virunga gorillas 1959–1978. *Biological Conservation*, **26**, 341–366.

White, L.J.T. (1992). Vegetation history and logging disturbance: Effects on rainforest mammals in the Lopé Reserve, Gabon (with special emphasis on elephants and apes). PhD thesis, University of Edinburgh, Edinburgh, U.K.

White, L.J.T. (1994). Biomass of rainforest mammals in the Lopé Reserve, Gabon. *Journal of Animal Ecology*, **63**, 499–512.

Wilkie, D.S. and Carpenter, J. (1999*a*). The impact of bushmeat hunting on forest fauna and local economies in the Congo Basin: A review of the literature. *Biodiversity and Conservation*, **8**, 927–955.

Wilkie, D.S. and Carpenter, J. (1999*b*). Can nature tourism help finance protected areas in the Congo Basin? *Oryx*, **33**, 332–338.

Williamson, L. and Usongo, L. (1995). *Survey of Elephants, Gorillas and Chimpanzees, Reserve de Faune de Dja, Cameroun*. Report to ECOFAC Composante Cameroun and Ministère de l'Environment, Cameroun.

World Resources Institute (2000*a*). *A First Look at Logging in Gabon*. Washington, D.C.: World Resources Institute.

World Resources Institute (2000*b*). *An Overview of Logging in Cameroon*. Washington, D.C.: World Resources Institute.

Young, T.P. (1994). Natural die-offs of large mammals: Implications for conservation. *Conservation Biology*, **8**, 410–418.

# 18  *Distribution, taxonomy, genetics, ecology, and causal links of gorilla survival: The need to develop practical knowledge for gorilla conservation*

ESTEBAN E. SARMIENTO

## Introduction

While most would agree that gorilla conservation is a worthwhile endeavor, there is less agreement as to what gorilla conservation actually entails. Ultimately, the most successful conservation measures should provide animals with independence from escalating human intervention, enable them to live out their lives by their own means, and promote self-sustaining populations. Raising animals in captive situations, or in exotic habitats in which they do not naturally occur, engenders human dependency and entrusts their survival to the whims of human economic and sociopolitical concerns. Such rearing cannot be justified as conservation unless it is a prelude to reintroduction of otherwise extinct animals into their past natural habitats, and leads to a de-escalation of human dependency.

Conservation, therefore, entails more than just the animal's protection, but also protection of future generations and of natural habitats that will support them. Short-term or "band aid" management, which over time escalates human intervention creating captive situations in what were once natural environments, must be avoided. Although veterinary care may seem especially appropriate when infirmities of free-ranging animals result from human causes, a zealousness to tackle results instead of causes can foster an unnatural dependency that may prove fatal unless human support continues. Fences and 24-hour armed security may protect animals from poachers and habitats from human trespass and exploitation; however, they are ineffective at stopping inbreeding depression, plagues, disease, water contamination, animal overuse and a multitude of other causes of habitat deterioration.

Unfortunately, aside from such obvious measures as impeding poaching, providing care to infirmed animals, and preventing destruction of forests they

432

live in, we presently know little about conserving gorillas and their habitats (Harcourt, 1995). This is not surprising given gorillas have a relatively long life cycle (Harcourt *et al.*, 1983; Harcourt, 1994), a widely varied diet (Schaller, 1963; Sabater Pi, 1977; Calvert, 1985; Rogers *et al.*, 1990), and both their discovery by the western world (Savage and Wyman, 1847; Matschie, 1904) and our interest in their conservation (Akeley, 1929, 1931) are relatively recent phenomena. It is the object of this chapter, therefore, to provide a framework for developing some of the practical knowledge necessary to conserve gorillas. The fact that half of the recognized subspecies of gorillas are critically endangered (Hilton-Taylor, 2000) and live in small isolated and/or fragmented forests (Sarmiento *et al.*, 1996; Sarmiento and Oates, 2000) underscores the urgency to do so.

What follows is a review of gorilla distribution and censusing techniques, along with aspects of taxonomy, genetics, and ecology bearing on gorilla conservation with an emphasis on those areas in which practical knowledge for conserving gorillas is deficient.

## Distribution

In large part, distribution reflects the limits imposed by the environment on an animal. As such, it relates to all aspects of the animal's biology. It is especially significant to conservation, since size of area the animal is distributed in bears directly on its conservation status. Distribution also helps to identify the types (specificity) of habitats suitable to the animal's survival. Together with estimates on animal density, identification of available habitat(s) imposes an upper limit on potential population size, and in this manner also relates to an animal's conservation status.

Gorillas are distributed in forests of tropical Africa between 6° 30′ N and 5° 30′ S latitude (Groves, 1970; Sarmiento and Oates, 2000). Longer dry seasons and less rain at latitudes farther from the equator prevent the growth of continuous forest and thus the spread of gorillas past these latitudes. The absence of gorillas south of the Congo River, and the arching course of this river and its major tributaries (i.e., Oubangui, Uele, and Lualaba Rivers), effectively separate gorillas into western and eastern groups that are nearly 900 km apart (Coolidge, 1929; Schaller, 1963; Groves, 1970; Sarmiento and Butynski, 1996; Sarmiento and Oates, 2000; Butynski, 2001) (Figs. 18.1 and 18.2).

More than a century ago LeMarinel collected three skulls in the vicinity of Bondo (3° 47′ N, 23° 47′ E), in forests along the northern bank of the Uele River approximately halfway between the present ranges of eastern and western gorillas (Schouteden, 1947). These skulls resemble those of western gorillas

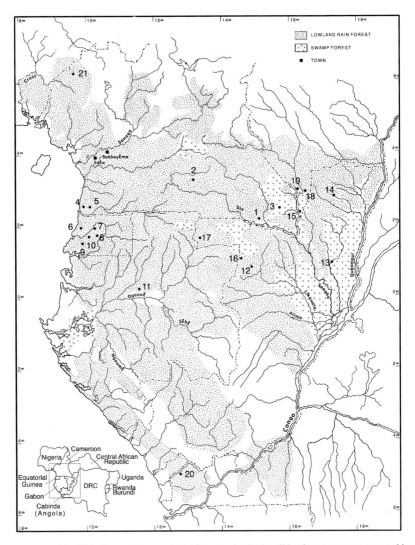

Fig. 18.1. The western equatorial forest and point localities for areas or towns noted in text and in Tables 18.1 and 18.2. (1) Boumba Bek, (2) Dja Reserve, (3) Lac Lobéké, (4) Mongokele, (5) Campo, (6) Bata, (7) Niefang, (8) Mt. Alen, (9) Mt. Okuro Biko, (10) Abuminzok-Aninzok, (11) Lopé, (12) Odzala, (13) Mbokou, (14) Motaba, (15) Ndoki, (16) Mbomo, (17) Garabinzam, (18) Bai Hoku, (19) Bayanga, (20) Mayombe Forest, (21) Cross River watershed.

in every respect (Coolidge, 1929; personal observation) and suggest that until recently gorillas had a continuous distribution across tropical Africa in forests north of the Congo and Oubangui Rivers. Unfortunately, the presence of this population has never been verified despite intermittent reports of its existence (K. Amman, personal communication).

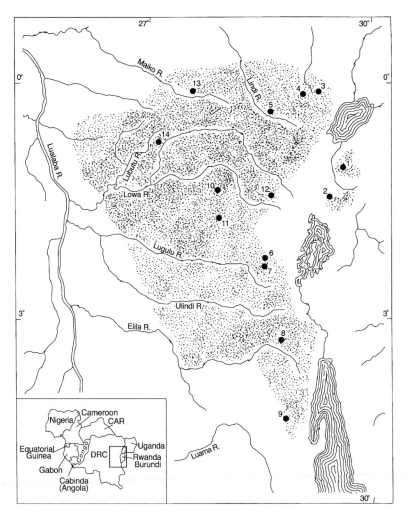

Fig. 18.2. Map of eastern gorilla distribution including localities in text and in Tables 18.1 and 18.2. (1) Bwindi-Impenetrable forest, (2) Virunga protected area, (3) Mt. Tshiaberimu, (4) Lubero, (5) Mohanga, (6) Mt. Kahuzi, (7) Tshibinda, (8) Mwenga, (9) Itombwe, (10) Walikale, (11) Utu, (12) Masisi, (13) Angumu, (14) Maiko south.

## *Western gorillas* (Gorilla gorilla)

The majority of western gorillas inhabit swamp and low-lying rain forests that comprise the western equatorial forest (White, 1983) (Fig. 18.1). This forest covers a geographic area of approximately 709 000 km$^2$ at elevations from sea level to about 1000 m (Butynski, 2001) (Table 18.1). It is bounded to the east, south, and west by the Oubangui and lower Congo Rivers, and the Atlantic

Table 18.1. *Geographic coordinates, elevation, annual rainfall, and yearly mean, minimum, maximum, and average temperatures for localities within or proximal to eastern and western gorilla habits*

| Area inhabited | Locality | Coordinates | Elevation[a] (m) | Annual rainfall (mm) | Temperature (°C) Minimum/maximum | Average | References |
|---|---|---|---|---|---|---|---|
| *Eastern gorilla* | | | | | | | |
| Bwindi-Impenetrable | | | 1500–2400 | | | | Griffiths, 1972; Sarmiento et al., 1996 |
| | Ruhiza | 1° 03′ S, 29° 46′ E | 2300 | 1440 | 13.8/18.7 | — | |
| | Buhoma | 0° 55′ S, 29° 36′ E | 1500 | 2490 | — | — | |
| | Kabale[b] | 1° 15′ S, 29° 58′ E | 1871 | 1265 | 10.0/23.3 | — | |
| Mt. Tshiaberimu | | | 2670–3000 | | | | Elevations personal observation; temperature after Schaller, 1963 |
| | Base camp | 0° 08′ S, 29° 25′ E | 2621 | — | 9.0/12.0 | — | |
| Lubero-Mbohe | | | 1400–2000 | | | | Elevations personal observation; rainfall based on 28 years CIDAT-CITLO |
| | Lubero[b] | 0° 10′ S, 29° 13′ E | 1960 | 1560 | — | — | |
| Virunga | | | 2400–4050 | | | | Watt, 1956; Bourlière and Versucheren, 1960; Schaller, 1963; Vedder, 1989 |
| | Karisoke | 1° 29′ S, 29° 29′ E | 3100 | 1980 | 3.9/14.5 | — | |
| | Kabara | 1° 29′ S, 29° 26′ E | 3019 | 1500 | 4.0/15.0 | — | |
| | Mgahinga | 1° 23′ S, 29° 40′ E | 2957 | — | 4.5/17.0 | — | |
| | Kissoro[b] | 1° 18′ S, 29° 41′ E | 1890 | 1678 | — | — | |
| | Rumangabo[b] | 1° 20′ S, 29° 22′ E | 1718 | 1760 | 13.3/25.8 | — | |

| Location | Coordinates | Elevation (m) | Rainfall (mm) | Temperature (min/max) | | Source |
|---|---|---|---|---|---|---|
| **Kahuzi-Biega, upper sector** | | 2100–2400 | | | | Casimir (1975) refers to Tshibinda weather station as Bukulumiza and lists it at a higher elevation; Tshibinda rainfall and temperature based on data over 40 and 18 years respectively CIDAT-CITLO |
| Bukulumiza hill | 2° 19′ S, 28° 45′ E | 2378 | 1790 | 10.4/17.9 | — | |
| Tshibinda | 2° 19′ S, 28° 45′ E | 2055 | 1860 | 11.1/21.3 | — | |
| **Kahuzi-Biega, lower sector** | | 600–1300 | — | 20.0/30.0 | — | Schaller, 1963; Yamagiwa et al., 1992, 1993; Hall et al., 1998 |
| Utu | 1° 37′ S, 27° 55′ E | 732 | — | — | 21.8 | |
| Irangi[b] | 1° 40′ S, 28° 20′ E | 793 | 2320 | — | — | Based on CIDAT-CITLO data 1957–59 |
| Walikale | 1° 24′ S, 28° 03′ E | 810 | 2420 | — | — | |
| Masisi | 1° 25′ S, 28° 49′ E | 1500–2000 | — | | | Mwanza et al., 1988; Yamagiwa et al., 1992 |
| **Itombwe/Sibatwe forest** | | 1200–2600 | | | | Based on CIDAT-CITLO data 1936–59; Butynski et al., 1996; Omari et al., 1999 |
| Mwenga[b] | 3° 02′ S, 28° 36′ E | 1385 | 2780 | — | — | |
| Fizi[b] | 4° 19′ S, 28° 57′ E | 1340 | 1204 | 16.1/26.6 | — | |
| Baraka[b] | 4° 06′ S, 29° 05′ E | 1570 | 1092 | — | — | |
| **Maiko** | | 700–900 | | | | Based on CIDAT-CITLO data 1952–53; Hart and Hall, 1996 |
| Angumu | 0° 09′ S, 27° 44′ E | 900 | 2563 | — | — | |
| ***Western gorilla*** | | | | | | |
| Cross River watershed | | 150–1600 | | | | Sarmiento and Oates, 2000 |
| Obudu plateau | 6° 25′ N, 9° 20′ E | 1585 | 4280 | 18.8/26.0 | — | |
| Afi mountains | 6° 20′ N, 9° 03′ E | 700 | 3346 | — | — | |

Table 18.1 (cont.)

| Area inhabited | Locality | Coordinates | Elevation[a] (m) | Annual rainfall (mm) | Temperature (°C) Minimum/maximum | Temperature (°C) Average | References |
|---|---|---|---|---|---|---|---|
| | Mamfe[b] | 5° 31′ N, 9° 37′ E | 122 | 3424 | — | — | Letouzey, 1968; Calvert, 1985; Griffiths, 1972; Usongo, 1998a,b |
| | Obudu town[b] | 6° 37′ N, 9° 08′ E | 396 | 1813 | — | — | |
| Southern Cameroon | | | | | | | |
| | Boumba Bek | 2° 02′ N, 15° 12′ E | — | — | — | — | |
| | Molondo[b] | 2° 03′ N, 15° 13′ E | — | 1492 | — | — | |
| | Dja Reserve | 3° 10′ N, 13° 10′ E | 500–800 | — | — | — | |
| | Lomi[b] | 3° 09′ N, 13° 37′ E | 624 | 1624 | 19.0/28.0 | — | |
| | Lac Lobéké | 2° 21′ N, 15° 46′ E | — | 1400 | — | — | |
| | Mongokele | 2° 20′ N, 9° 48′ E | — | — | — | — | |
| | Campo | 2° 23′ N, 10° 03′ E | 200 | 1524[c] | — | 25.7 | |
| | Kribi[b] | 2° 57′ N, 9° 54′ E | 624 | 3017 | 23.0/29.0 | — | |
| Equatorial Guinea | | | 200–800 | | | | Jones and Sabater Pi, 1971; Griffiths, 1972; Sabater Pi, unpublished data |
| | Bata | 1° 50′ N, 9° 47′ E | — | 1850 | — | 25.3 | |
| | Niefang | 1° 50′ N, 10° 15′ E | — | 3529 | — | 25.7 | |
| | Mt. Alen | 1° 40′ N, 10° 18′ E | <800 | — | 18.0/23.7 | — | |
| | Mt. Okoro Biko | 1° 28′ N, 9° 52′ E | <750 | — | — | 25.4 | |
| | Abuminzok-Aninzok | 1° 36′ N, 10° 04′ E | 200–700 | — | — | 25 | |
| | Nsork | 1° 09′ N, 11° 16′ E | — | 1987 | — | — | |
| Gabon, Lopé Reserve | | | 0–1000 | 1400–3200 | | | Griffiths, 1972; Tutin and Fernandez, 1984, 1993; Williamson et al., 1990; Voysey et al., 1999 |
| | Lopé | 0° 10′ S, 11° 35′ E | 100–500 | 1507 | 21.0/31.0 | — | |
| | Makokou | 0° 34′ N, 12° 47′ E | — | 1755 | — | — | |
| | Bitam | 2° 05′ N, 11° 29′ E | 599 | 1747 | 20.0/29.0 | — | |

| | | | | | | | |
|---|---|---|---|---|---|---|---|
| People's Republic of Congo | Odzala | 0° 48' N, 14° 52' E | 300–600 | — | — | — | Griffiths, 1972; Fay and Agnagna, 1992; Nishihara, 1995; Magliocca et al., 1999 |
| | Mboukou | 0° 56' N, 17° 25' E | — | — | — | — | |
| | Motaba | 2° 45' N, 17° 15' E | — | — | — | — | |
| | Ndoki | 2° 20' N, 16° 19' E | — | 1430 | — | — | |
| | Mbomo | 1° 00' N, 14° 35' E | — | — | — | — | |
| | Garabinzam | 1° 37' N, 13° 28' E | — | — | — | — | |
| | Ouesso[b] | 1° 34' N, 16° 02' E | — | 1596 | — | — | |
| Central African Republic | Dzanga/Sangha | 2° 50' N, 16° 28' E | 400–500 | 1400 | 18.0/31.0 | — | Remis, 1997a; Goldsmith (1999) recorded 1952 mm of rainfall in 1994 at Bai Hokou |
| | Bai Hokou | 2° 56' N, 16° 16' E | — | 1430 | 19.5/27.7 | — | |
| | Bayanga | | 20–300 | 1300 | — | — | |
| Mayombe/Cabinda | Tshela | 4° 59' S, 12° 57' E | — | — | — | — | Griffiths, 1972; Piton et al., 1979; Dowsett and Dowsett-Lemaire, 1991; rainfall increases eastward with increasing elevation |
| | Kakamoeka | 4° 08' S, 12° 03' E | — | — | — | — | |
| | Onkassa | 5° 18' S, 12° 52' E | — | — | — | — | |
| | Pointe-Noire[b] | 4° 48' S, 11° 51' E | 50 | 1231 | 22.5/29.3 | 25.9 | |
| | Cabinda[b] | 5° 33' S, 12° 11' E | 20 | 1331 | 22.0/28.0 | — | |
| | Luki[b] | 5° 37' S, 13° 06' E | 350 | 1136 | 20.0/28.0 | — | |

[a] Area elevation ranges are those known to be inhabited by gorillas.
[b] Weather stations in proximity to gorilla habitats.
[c] Griffiths (1972) and Letouzey (1968) report 2976 mm and 2800 mm average rainfall respectively for the city of Campo ~30 km west of the reserve.

ocean, respectively (Fig. 18.1). Northwards, between 4° and 5° N latitude, the western equatorial forest gives way to woodlands and grassland mosaics. In the west, the northern limit of this forest coincides roughly with the Sanaga River.

A small isolated population of western gorillas, however, is found in the Cross River watershed ranging as far as 6° 30′ N latitude. This population is approximately 260 km northwest of the western equatorial forest and the Sanaga River. Cross River gorillas (*G. g. diehli*) inhabit low-lying and submontane forests at elevations of 150–1600 m within a geographic range of approximately 750 km$^2$ (Sarmiento and Oates, 2000). It is the most westerly gorilla population and the one farthest from the equator.

### Eastern gorillas (Gorilla beringei)

Eastern gorillas are distributed in forests east of the great bend in the Congo River (Fig. 18.2). They cover a geographic area of approximately 112 000 km$^2$ (Butynski, 2001) at elevations of 650–4000 m (Table 18.1). This area is bounded to the west by the Lualaba River and to the east by the eastern edge of the Albertine rift, its lakes, and the associated edge of a forest–savanna mosaic (Fig. 18.2). The Lindi and upper Luama Rivers approximate the northern and southern boundaries of this area, respectively. With the exception of Virunga and Tshiaberimu gorillas, all eastern gorillas range at elevations below 2600 m (Table 18.1). Eastern gorillas are not found much farther north than the equator, or farther south than 4° 20′ S latitude (Schaller, 1963; personal observation) displaying a much narrower latitudinal range than western gorillas.

Dryer climates in east Africa, especially north of the equator (Kortlandt, 1983, 1995), do not fully explain the narrower latitudinal range of eastern gorillas. It is not known why eastern gorillas do not range northwards into rainforests between the equator and the Uele River, and westwards into forests between the Congo and Uele Rivers, nor why they are not found in the forests flanking the right (east) bank of the Lualaba River. With the exception of Virunga (*G. b. beringei*) and Bwindi gorillas, absence of gorillas from forests on the eastern edge of the Albertine rift, from the Ruwenzoris to the eastern shore of Lake Tanganyika, is also unexplained. Verifying from which areas they are indeed absent and understanding why this is so may reveal habitat variables adverse to gorilla survival and/or processes associated with their disappearance.

### Local distribution

Within their respective geographic ranges, both eastern and western gorillas are unevenly distributed. Populations may be separated by as much as 50 km or as little as 3 km and are typically concentrated along rivers, forest boundaries, abandoned villages and/or roads where there is an abundance of herbaceous vegetation onwhich gorillas feed (Schaller, 1963; Carroll, 1988; Fay and Agnagna, 1992; White *et al.*, 1995). These habitats include glades or swamps within or adjacent to forests, and grassy meadows and clearings within high-altitude forests. Most known gorilla populations live at densities of 0.5–1.5 animals/km², but densities as high as 10.6 animals/km² and as low as 0.05 animals/km² have been reported for western gorillas (Table 18.2). Densities above or near 10 animals/km² have been reported in forests with Marantaceae undergrowth (Bermejo, 1995), in secondary forests along logging roads (Carroll, 1988), and in swampy clearings with a high salt concentration (Magliocca *et al.*, 1999). Very low gorilla densities are usually associated with human hunting (Agnagna *et al.*, 1991; Dowsett and Dowsett-Lemaire, 1991; personal observation).

Many gorilla populations within both eastern and western groups are at present (geographically) discontinuous (Schaller, 1963; Sarmiento and Butynski, 1996; Hall *et al.*, 1998; Saltonstall *et al.*, 1998; Sarmiento and Oates, 2000; Butynski, 2001), but two populations are morphologically distinct enough from other populations in their respective group to be considered separate taxa, i.e., Virunga gorillas and Cross River gorillas (Groves, 1970; Sarmiento and Butynski, 1996; Sarmiento and Oates, 2000; Oates *et al.*, this volume). Both of these populations are small and inhabit comparatively higher altitudes in extreme gorilla habitats. Bwindi gorillas have not yet been shown to be distinct relative to other eastern gorillas, although they are allopatric and morphologically distinct from Virunga gorillas (Sarmiento *et al.*, 1996).

More precise data on gorilla distribution is necessary to test whether other populations that appear to be distinct are geographically discontinuous and thus separate taxa. As noted, total land area each gorilla taxon occupies sets upper limits on its total population number and helps determine its conservation status and needs.

### Gorilla censusing

Censusing is a principal component of any conservation initiative. It is necessary for arriving at population numbers that determine an animal's conservation status and needs. Censusing also indicates whether numbers of individuals comprising a population are stable, decreasing, or increasing, and can help in determining the degree of success or failure of any implemented conservation measure. Barring epidemics, heavy poaching, or recent environmental modifications, one census per decade after the initial baseline census is probably more than enough to monitor gorilla population numbers (T.M. Butynski, personal communication). Unfortunately, the degree of accuracy of gorilla

Table 18.2 *Area inhabited, habitat type and census methods used to arrive at population density, and numbers of eastern and western gorillas in each area*

| Area | Habitat type | Population | | Census type | Range area (km²) | References and notes |
| --- | --- | --- | --- | --- | --- | --- |
| | | Density (gorillas/km²) | Numbers | | | |
| *Eastern gorilla* | | | | | | |
| Mt. Tshiaberimu | Montane/bamboo/ericaceous | 0.65 | 16 | Total nest count | 64 | Butynski and Sarmiento, 1995, 1996; gorillas inhabit about 25 km² |
| Lubero-Mbuhi | Mid-altitude and montane forest | ? | 700 | Density estimate | 2500 | Sarmiento and Butynski, 1998; Kyungu and Vwirsihikya, 1999; author's notes; includes 1960 km² outside Tayna reserve |
| Bwindi-Impenetrable | Mid-altitude and montane forest | 0.91 | 300 | Total nest count | 330 | Sarmiento et al., 1996; Butynski and Kalina, 1998; gorillas do not use bamboo zone |
| Virunga | Bamboo/Hagenia–Hypericum/ericaceous | 0.74 | 320 | Total nest count | 434 | Sholley, 1990, 1991 |
| Kahuzi-Biega, upper sector | Montane/bamboo/ericaceous | 0.43 | 258 | Total nest count | 600 | Yamagiwa et al., 1993; Yamagiwa (1999) claimed these numbers were reduced by half after 1997–98, but more recent reports claim only 70 gorillas are left (Anonymous, 2000) |

| | | | | | | |
|---|---|---|---|---|---|---|
| Kahuzi-Biega, lower sector | Mid-altitude to lowland evergreen forest | 0.86 | 11 030 | Transect estimate | 12 770 | Hall et al. (1998) includes adjacent Kasese region; Yamagiwa (1983) reports 0.4 gorillas/km$^2$ for Utu |
| Itombwe forest/Sibatwe | Mid-altitude, montane and bamboo forest | 0.25 | 875 | Transect estimate | 3 500 | Butynski et al. 1996; Omari et al. 1999; author's notes; total area of Itombwe is 6000 km$^2$ |
| Walikale | Mid-altitude forest | ? | 50 | — | >100 | Author's notes |
| Masisi | Mid-altitude forest | 0.83 | 25 | Total nest count | 40 | Mwanza et al. 1988; Yamagiwa et al. 1992 |
| Maiko north | Lowland evergreen forest | 0.25 | 625 | Density estimate | 2 500 | Hart and Sikubwabo, 1994; Hart and Hall, 1996; total area of Maiko is 10 830 km$^2$ |
| Maiko south | Lowland evergreen forest | 0.24 | 24 | Density estimate | 100 | Hart and Sikubwabo, 1994 |
| *Western gorilla* | | | | | | |
| Cross River watershed | Low to mid-altitude and degraded montane forest | — | >200 | Transect estimate | 2 300 | Oates, 1998; Groves and Maisels, 1999; Sarmiento and Oates, 2000; includes 1075 km$^2$ of gorilla habitat in Mbe and Afi Mts., Cross River National Park, Nigeria and Takamanda Reserve, Cameroon |
| Southern Cameroon | | 0.12 | 6000 | Density estimate | 52 000 | Harcourt, 1996 |
| Dja Reserve | Lowland dry and inundated forests, swamps | 1.71 | 2500 | Transect estimate | 5 260 | Willamson and Usongo, 1996; Usongo, 1998a; *Raphia* and *Cyperus pandanus* swamps are the most common |

Table 18.2 (cont.)

| Area | Habitat type | Population Density (gorillas/km²) | Population Numbers | Census type | Range area (km²) | References and notes |
|---|---|---|---|---|---|---|
| Lobéké Forest | Dense and *Gilbertiodendron* forests,[a] swamps | 2.5 | 5 000? | Transect estimate | 2 000 | Stromayer and Ekobo, 1992; Meder, 1995; Usongo and Fimbel, 1995; Fay, 1997; Usongo, 1998b; Lac Lobéké (450 km²) is part of the Lobéké forest reserve |
| Boumba Bek | Dense and *Gilbertiodendron* forests,[a] swamps | 0.14–0.19 | 150–200 | Transect estimate | 2 000 | Stromayer and Ekobo, 1992 |
| Campo | Lowland dense forests (Biafran littoral) | ? | 200–300 | Density estimate | 2 712 | Calvert, 1985; Lee et al., 1988; Usongo, 1998a |
| Mongokele | Lowland dense forests (Biafran littoral) | 1.2 | — | Transect estimate | see Campo | Fay, 1997 |
| Douala-Edea | Dense forest and mangroves | ? | ? | — | 1 600 | Prescott et al. (1994) report a 1992 gorilla sighting in proximity to the reserve, and animals may still exist within the reserve |
| Nki forest | Dense and *Gilbertiodendron* forests,[a] swamps | — | — | — | 2000 | |
| Nanga Eboko | Lowland dry and inundated forests, swamps | ? | ? | — | 160 | |
| Nyong swamps | Inundated and swamp forests | — | — | — | ? | |

| | | | | | | |
|---|---|---|---|---|---|---|
| Equatorial Guinea | Lowland dense forests | 0.22 | 990 | Transect estimate | 4 500 | Gonzalez-Kirchner, 1997; Sabater Pi (1966) estimated numbers at 5000; density value is average of range between 0.71 and 0.12 and includes forests with no known gorillas |
| Gabon | | 0.18 | 35 000 | Transect estimate | ~194 500 | Tutin and Fernandez, 1984; Meder, 1995 |
| Lopé | Dense forests[a] | 0.5 | 2 500 | Transect estimate | 5 000 | White, 1992; White et al., 1995; Tutin and Fernandez (1984) estimated population of Lopé-Okanda at 1526–2057 |
| Moukalaba-Dougoua | Dense forests[a] | 0.22–0.39 | 390? | — | 1 000 | Tutin and Fernandez (1984) provide population and density estimates for Moabi and Sette Cama sectors, which include Petit Loango, Iguela, Ngove-Ndogo, Ounga plain, Moukalaba and Moukalaba-Dogoua reserves; these areas are some of the highest density gorilla habitats, i.e. thickets and secondary forest |
| Wonga-Wongue | Costal scrub, inundated and dense forest, savanna | 0.02 | 244 | — | 5 000 | Tutin and Fernandez, 1984; Sayer et al., 1992 |

Table 18.2 (cont.)

| Area | Habitat type | Population Density (gorillas/km²) | Numbers | Census type | Range area (km²) | References and notes |
|---|---|---|---|---|---|---|
| Central African Republic | | 0.26 | 9 000 | Density estimate | 35 000 | Carroll, 1988; Harcourt, 1996 |
| Dzanga-Sangha region | Gilbertiodendron and dense forests, Raphia swamps | 0.89–1.45 | 6 934 | Transect estimate | 6 000 | Carroll, 1988 |
| Ndoki National Park sector | Gilbertiodendron and dense forests, Raphia swamps | 1.1–2.0 | 1 232 | Transect estimate | 770 | Fay, 1989 |
| Dzanga National Park sector | Gilbertiodendron and dense forests, Raphia swamps | 0.84 | 380 | Transect estimate | 450 | Remis, 2000 |
| People's Republic of Congo | | 0.15 | 34 000 | Density estimate | 222 300 | Fay and Agnagna, 1992; Butynski, 2001 |
| Odzala | Dense forests,[a] swamps and savanna mosaics | 10.3 | — | Transect estimate | 2 840 | Bermejo, 1995; Fay, 1997; Magliocca et al., 1999 |
| Lake Mboukou | Dense forests and Raphia swamps | 1.2 | — | Transect estimate | — | Tutin and Fernandez, 1984; Sayer et al., 1992 |
| Motaba | Gilbertiodendron and inundated forests, Raphia swamps | 0.1 | — | Transect estimate | 1 100 | Tutin and Fernandez, 1984; Sayer et al., 1992 |
| Nouabale-Ndoki | Gilbertiodendron and dense forests, Raphia swamps | 0.2 | 774? | Transect estimate | 3 870 | Mitani, 1992; Nishihara, 1994, 1995; density after Fay and Agnagna (1992), based on area of 1490 km² calculated from coordinates |

| Location | Habitat | | | Method | Estimate | Reference |
|---|---|---|---|---|---|---|
| Mbomo | Dense and secondary forests | 0.4 | — | Transect estimate | 1 351 | Fay and Agnagna, 1992 |
| Garabinzam | Primary and mixed secondary forests | 0.5 | — | Transect estimate | — | Fay, 1997 |
| Southern Congo | Dense forest mosaic savanna and swamps | 0.05 | 2 615 | Density estimate | 52 300 | Fay, 1997 |
| Mayombe forest | Dense forest mosaic savanna and swamps | ? | — | — | 16 280 | Agnagna et al., 1991; Dowsett and Dowsett-Lemaire, 1991; author's notes; Mayombe forest covers approximately 11 000 km² in PRC, Conkouati and Dimonika Reserves, 3000 km² in Gabon, 2500 km² in Angola, Cabinda, and 350 km² in DRC, Luki Reserve, and contains the most southernly distributed gorilla populations |

[a] Marantaceae understory.

censuses varies with method(s) used and many methods lack an estimated degree of error.

### Accuracy of census methods

#### Total nest counts

If done carefully, total nest counts can have a very low error, as long as groups are tracked daily with night nests noted over a minimum of three consecutive days (Weber and Vedder, 1983; Sholley, 1990, 1991; Yamawiga *et al.*, 1993). Because each gorilla usually makes a new nest every night, the number of nests constructed each evening represents the number of gorillas. Repeated counts over three days insures that no nests are overlooked and that the gorilla group censused exhibits a constant membership (Harcourt and Groom, 1972; Yamagiwa *et al.*, 1993). Errors, however, may occur when counting infants and solitary animals. These may be either counted more than once or overlooked (Yamawiga *et al.*, 1993). Errors are more likely to underestimate than overestimate numbers (Weber and Vedder, 1983). According to Weber and Vedder (1983), Virunga gorilla total nest counts prior to their 1976–78 census failed to count 54% of solitary males and 20% of infants six months of age or less.

#### Density-based population estimates

Complete nest counts are not feasible when covering very large areas with limited manpower. Estimates on gorilla population size, based on known areas of available habitat or estimates of total area occupied (as deduced from current sightings and museum and field records), multiplied by gorilla density (arrived at from average gorilla densities in areas with known gorilla numbers) may have considerable error. Accurate gorilla densities are available for only a small percentage of gorilla populations (most of which are eastern gorillas in protected and isolated habitats; see Table 18.2), and densities are known to vary (Emlen and Schaller, 1960; cf. Harcourt and Groom, 1972; Weber and Vedder, 1983; Sholley, 1990). Moreover, habitats that supported gorillas in the past may not currently contain gorillas. Heavy poaching, as is presently occurring in the west African equatorial forests (Amman, 2000) may result in considerable overestimates of gorilla numbers when these are based on average density and available habitat. Density-based estimates, therefore, provide only a rough approximation and are not reliable as baseline data when implementing conservation measures or gauging conservation status.

#### Estimates based on transect nest counts

The alternative to density-based population estimates is transect nest counts. Less work-intensive than total nest counts, they are suitable for surveying larger areas. In this method, all nests found along a transect or at a specific distance from the transect are counted (Ghiglieri, 1984; Tutin and Fernandez, 1984). Dividing total nest number by transect area and by nest life (i.e., time in days a nest remains recognizable) arrives at gorilla density for the surveyed area. Multiplication of gorilla density by total area inhabited arrives at gorilla numbers (Tutin and Fernandez, 1984).

Unfortunately, the variables necessary to accurately estimate population numbers from transect nest counts are unknown, so that estimates based on this method are unlikely to be accurate. Principally, nest life varies according to plant material used in nest construction, climate, and season. The rate of deterioration, therefore, may not be the same for all encountered nests and this rate may differ with habitat. Variation in nest life underestimates gorilla numbers when nests deteriorate quicker on average than expected, or overestimates numbers when nests deteriorate slower than expected. During the day, when immobilized by heavy rains, gorillas frequently build day nests. During transect counts these may occasionally be confused with night nests. Moreover, gorillas may move in the night and build a second set of nests or build two nests and only sleep in one (Tutin *et al.*, 1995). All of this adds further error to transect nest counts.

In areas where gorillas and chimpanzees are sympatric, the nests of the two African apes may be confused (Tutin and Fernandez, 1984), especially when these are built in trees and poorly seen from the ground. Collection of hair or feces from nests or of feces from below nests can help in determining whether gorillas or chimpanzees built the nests. Otherwise, estimates of the number of unidentified nests belonging to gorillas can be arrived at by multiplying unidentified nest number by the ratio of gorilla to total African ape (chimpanzee and gorilla) tree nests.

The intervals at which transects must be placed from each other (transect intervals) to arrive at accurate population estimates and how this distance varies given differences in nest life are also unknown (cf. Carroll, 1988; Fay, 1989). Depending on nest life, transects placed either too close or too far from each other may be expected to overestimate or underestimate population numbers, respectively.

Empirical data on rate of nest deterioration, varying nest plant material, relative locality of nest (steep hill, swamp, height off ground, etc.), seasonality, and climatic conditions are necessary to arrive at estimates of average nest life that will make transect counts more accurate. Empirical data on frequency of nest identification at different levels of deterioration and at varying positions to the transect (Tutin and Fernandez, 1984) also need to be considered. Data on nest life collected by the same observer returning to the same group of nests repeatedly over specified periods are unlikely to accurately reflect age at which a nest would be recognizable to a first-time observer. To a returning observer the memory of having seen the nest when newly constructed and on subsequent visits thereafter is likely to prolong the time any one nest remains recognizable. Collecting data on nest-life employing returning observers probably has the effect of overestimating average nest life, and thus underestimating gorilla numbers.

Estimates of transect nest counts are also affected by the localized distribution of gorillas within their geographic ranges (Emlen and Schaller, 1960). In this regard, transects at any one time may arbitrarily pass through regions of high or low population density providing inaccurate estimates for the concerned area. Not only may transects be biased in sampling nonrepresentative percentages of different habitat types comprising any one area (Tutin *et al.*, 1995), but counts may fortuitously occur at times when the number of nests found along a transect is not representative of actual gorilla densities. Although long transects over large areas are likely to cancel out such bias (Carroll, 1988), shorter transects over smaller areas may present errors that are too great to provide useful population or density estimates. In this regard, minimum length of transects necessary to provide accurate estimates, given number of gorilla groups, average distances of their day ranges and habitat types, is unknown. Moreover, to what degree the presence of transects or roads may promote or discourage gorilla nesting and lead to biased counts is also unknown (Plumptre and Reynolds, 1997).

Computer modeling of transect nest counts based on gorilla populations with known numbers and known coordinate data for nest localities may be used to associate distance between transects with a degree of accuracy for population density and size estimates. Such models would also allow prediction of error with varying nest life, transect intervals, distribution of habitat types, and gorilla group numbers.

Satellite land imaging may in theory provide greater accuracy to transect censuses. It can correlate differences in gorilla densities with electromagnetic wave reflecting and emitting properties of vegetation, to arrive at more accurate measures of suitable habitat area. This method, however, has yet to be successfully applied.

Transect nest counts in areas with well-verified population numbers could provide an estimated degree of error for transect methods. In this case, total amount of error in both predicted and unpredicted confounding variables could be arrived at. Currently, transect nest count estimates of gorilla population numbers lack an estimated degree of error and are unlikely to provide reliable baseline data for assessing degree of threat (Harcourt, 1996). Until methods for estimating gorilla numbers can present reasonable accuracy with a known degree of error, the conservation status and needs of those gorilla taxa whose population numbers are based on these methods is not known.

## Demography and censuses

Censuses may also provide data on population structure that may be used to predict changes in population number over time and can help determine a population's viability and conservation needs. Data on (1) color and size (diameter) of feces, (2) color and morphology of hair found in nests, and (3) nest size and locality relative to other nests, collected during total or transect nest counts, can reveal details as to population structure (Schaller, 1963; Harcourt and Groom, 1972; Casimir, 1977; Weber and Vedder, 1983; Yamawiga *et al.*, 1993). These include number of young born per year, number of adult males in a group, and number of solitary males. Testing for estrogens and progesterone in urine (Czekala and Sicotte, 2000) and/or feces (Wasser, 1996; Jurke *et al.*, 2000) can reveal number of cycling females. These data bear on a population's fertility and survival rate, and may serve to predict a population's future growth or decline, all factors important to conservation. Through cross-population comparisons, these demographic data may also be used to reveal habitat variables, which may be affecting survival. While demographic data gleaned from total nest counts has an estimated degree of error, that gleaned from transect nest counts is unknown and needs to be worked out (Tutin *et al.*, 1995) to be useful for conservation.

Long-term studies in which researchers spend long periods with groups and are able to identify individuals in a population provide data on age distributions, longevity, mortality (including infant survival rate), interbirth intervals, number of offspring, and length of postnatal life stages (e.g., generation length, gestation length, age at first and last pregnancy, etc.). These data may serve to gauge more precisely whether populations are increasing, decreasing, or stable over time (Harcourt, 1994, 1995; Walsh, 1995). Relatively recent advances in DNA sequencing and ability to identify individuals based on variable number tandem repeats (VNTR, or microsatellites) in DNA (Jeffreys *et al.*, 1985; Field *et al.*, 1998; Goossens *et al.*, 2000) obtained from feces or hair, also enable total nest count censuses when repeated (at intervals from one to five years), to arrive at long-term demographic data. Collection of these types of data is otherwise possible only through long-term contact and visual identification of individuals. Presently, the meager long-term demographic data that do exist for gorillas have been gathered for Virunga gorillas (Harcourt, 1994; Watts, 1998*a*). This population is small, isolated, and unique (Schaller, 1963; Groves, 1970; Sarmiento *et al.*, 1996). It lives at the periphery of gorilla distribution in an extreme gorilla habitat, and is unlikely to be representative of other gorilla populations. Collection of demographic data for each unique gorilla taxon would provide greater accuracy to predictions on population growth and decline and would thus provide more accuracy for determining conservation needs.

## Gorilla taxonomy

Taxonomy (the science of classification) provides a foundation for most biological studies including conservation. Any observations or conclusions made on an individual, a group, or a population depend on taxonomy to determine whether these observations may apply to other individuals, groups, or populations, and to predict over what range of organisms or taxa these observations and conclusions are applicable. Within conservation, taxonomy gains additional significance, since number of organisms belonging to a taxonomic category (e.g., subspecies or species) contributes to determining appropriate conservation needs (Butynski, 2001).

Taxonomy is a hierarchical system of classification that is based on the continuity of life (i.e., generational continuity), but relies on those discontinuities that develop over geographical distance and time to define its categories (Sarmiento *et al.*, 2002). Because in practice taxonomy summarizes biological similarities and differences among organisms and populations, all biological data pertaining to an organism bear on its classification. Species is the most objective taxonomic category and serves as the basis for all taxonomy. Species are objectively arrived at based on whether or not distinct populations maintain their distinctiveness in areas of sympatry (overlapping distribution: Cracraft, 1989). Unable to be directly assessed objectively, allopatric (i.e., geographically separate) populations are classified as separate species if their phenotypic distinctiveness is comparable in magnitude to those of closely related species (within the same genus or family) which are objectively determined. Allopatric populations that are for the most part distinct, but show a level of phenotypic distinctiveness less than that seen in closely related and objectively determined species (or a level comparable to that exhibited by populations inhabiting different geographical areas within the continuous range of a species) are usually accorded subspecific status (Sarmiento *et al.*, 2002).

Currently, gorilla taxonomy is in a state of flux. Data collected in the last two decades suggest gorillas do not easily fit the single-species taxonomy proposed by Coolidge (1929) 70 years ago and revised by Groves (1970) 40 years later. Considering the phenotypic distinctiveness of eastern and western gorillas, and the fact that both of these groups can each be further broken down into a number of subspecies, a two-species taxonomy seems more appropriate (Groves, 1996, 2001; Sarmiento and Butynski, 1996).

A two-species taxonomy, however, is not without its problems, since eastern and western gorillas are allopatric populations (Coolidge, 1929; Schaller, 1963; Groves, 1970). Taxonomic consistency prescribes placing eastern and western gorillas in different species given that their morphologic differences are comparable in magnitude to those distinguishing the two currently recognized species of chimpanzees (*Pan troglodytes* and *P. paniscus*: Sarmiento *et al.*, 1995; Sarmiento and Butynski, 1996; Butynski, 2001; Sarmiento, in press). Unfortunately, the two chimpanzee species are also allopatric (Vanderbroek, 1959; Kortlandt, 1983, 1995), and not objectively determined. In this regard, there is no objectively determined species within great apes to establish a benchmark as to the magnitude of phenotypic distinctiveness corresponding to species differences (Simpson, 1961; Mayr, 1963, 1982).

Considering the role of distribution in classification, without a satisfactory taxonomic resolution, animals are best referred to and treated by a common name and a locality of origin, rather than assigned a specific binomial or trinomial that may or may not accurately summarize each population's biological distinctiveness. More than likely, there are more discontinuous gorilla populations that on morphological analysis will prove to be distinctive. These populations must be properly identified (i.e., following modern taxonomic rules) and accurately censused to determine their conservation status.

### Genetics and alpha taxonomy

Aside from recognizing that species differences imply discontinuity of gene flow, modern systematics, as founded in synthetic evolutionary theory, does not provide genetics with a role in defining mammalian taxa at lower levels (Dobzhansky, 1937). Direct relationships between magnitude of genetic, morphological, and/or behavioral differences to each other or to species or subspecies designations do not exist. Unless a specific morphological, behavioral, and/or genetic difference can predict whether or not there is genetic interchange between populations (e.g., protein specific acrosomes, size and geometry of copulatory organs, specific chromosomal inversions or translocations, etc.), no one degree of difference can be absolutely ascribed to any one taxonomic level (Cracraft *et al.*, 1998; Wu, 2000).

Whether or not useful alpha taxonomies relying on genetic differences (as have been currently reported in the literature: Ruvolo *et al.*, 1994; Garner and Ryder, 1996) can be constructed based on the principle of consistency as is done for phenotypic differences is arguable. Phenotypic differences directly correspond to differences in species recognition and selection, and are thus likely to correlate with species divergence. Genetic differences, however, may or may not reflect selection. Of the small percentage of DNA differences separating closely related species, only a small fraction is functional (genes coding for proteins). The rest of DNA is allegedly impervious to selection, and hypothesized to mutate at rates dependent on generation time (but see Templeton, 1998). As such, magnitude of DNA differences can either reflect divergent selection or total time populations have been separated. Owing to differences in selection pressures and rates of evolution, populations with a similar behavior and phenotype that have been separated for a relatively long time may conceivably show as much or more DNA differences than phenotypically and behaviorally distinct populations that have been separated relatively recently.

The use of mitochondrial DNA (mtDNA) in systematics, either for inferring relatedness or relative divergence dates, is increasingly proving problematic. Not only have nuclear inserts (Numts) been confused for mtDNA providing spurious mtDNA sequences and cladograms for gorillas (Jensen-Seaman, 2000; Greenwood and Sarmiento, in press; Clifford *et al.*, this volume), but recent reports that mtDNA may be recombining in humans, chimpanzees, and other mammals (Gyltenstein *et al.*, 1991; Eyre-Walker *et al.*, 1999; Awadalla *et al.*, 1999) seriously question its use in systematic studies (Hey, 2000). Concentrating on nuclear DNA that codes for functional proteins, and/or preferably genes underlying phenotypic differences between populations, seems the most sensible solution for making genetic and morphological data complementary and useful for resolving alpha taxonomy. Until these problems can be ironed out, genetic-based alpha taxonomies (see Garner and Ryder, 1996) have no clear implications for gorilla conservation.

### Genetics and conservation

Despite the fact that the role of genetics in taxonomy still needs to be defined, genetic studies have a host of well-defined roles in conservation. As mentioned above, genetic studies enable collection of demographic data. They can also establish sex ratios and decipher relations within and between populations. This enables collection of more detailed demographic data such as verifying which males father offspring and during which time in their life they are most likely to do so, etc. (Field *et al.*, 1998). Molecular genetics can also be used to identify present and past corridors of migration enabling identification of corridor types and of their specific habitat variables. This may reveal sex differences in the distance males and females are able to disperse, and whether or not specific corridor types preferentially filter one sex over the other. All of this information is essential for maintaining or re-establishing proper corridors that will insure breeding diversity and could prove critical to survival of endangered subspecies comprising small fragmented populations (Beier and Noss, 1998; Vucetich and Waite, 2000; Sarmiento, in press).

Genetic studies may also be used to identify past population bottlenecks (Nunney, 2000) and associate these to known climatic or habitat changes, further defining variables that have affected gorilla survival. Considering the relationship between genetic diversity and effective population size, genetic data may be used to arrive at rough estimates of population size (Crandall *et al.*, 1999; Schwartz *et al.*, 1999; Williamson and Slatkin, 1999). Depending on how genetic diversity is gauged, either as a function of segregating sites or as a proportion of nucleotide differences per sequence pair, it can provide estimates of historic or current population size, respectively (Crandall *et al.*, 1999). Contrasting the two estimates (or contrasting historic with actual population size) may provide data on recent explosions or crashes in populations (Waples, 1991; Crandall *et al.*, 1999), and/or on population fragmentation or fusion. These data are all important in helping to determine which populations are prioritized for conservation.

Finally, in small isolated populations, genetic studies may be used to test extinction risk models, most of which assume random mating (Ginzburg *et al.*, 1990; Taylor, 1994; Dushoff, 2000; Poon and Otto, 2000; Purvis *et al.*, 2000), to arrive at more realistic limits as to minimum numbers of individuals needed for population survival. This would help settle discussions as to effects small populations have on infant survival and group viability, and clarify whether minimum viable population size is a species-specific quality (Harris *et al.*, 1987; Ralls *et al.*, 1988; Berger, 1990; Thomas, 1990; Walter, 1990; Harcourt, 1991; Frankham, 1995; Nunney, 2000; Chapman *et al.*, 2001).

## Ecology and links between habitat and survival

### *Gorilla habitats*

The habitat provides food and shelter for animals. It contains an animal's competitors, predators and prey, and diseases and parasites that may infect it. For an animal to survive and reproduce, its anatomical structure and physiology must correspond to the habitat or habitats it lives in. Recognizing habitat variables necessary for gorilla survival and reproduction is thus essential to conservation.

Ambient temperature, rainfall, available sunlight (visible versus ultraviolet radiation), and atmospheric pressure (humidity and oxygen concentration) are all important determinants of vegetation types that will flourish. As a primary source of energy in the food chain, vegetation types designate specific animal life and together with terrain (geologic formations) define spaces animals move in. Latitude and elevation directly influence temperature, rainfall, available sunlight, ultraviolet radiation, and atmospheric pressure, and are thus important habitat determinants. Latitude and longitude specify locality on earth, and thus (1) geography – affecting climate, (2) geologic formations – determining soil type, terrain, and water availability, and (3) biogeographic region(s). Collectively all the above variables may define vegetation zones comprising a dominant plant or plants that may be used to characterize gorilla habitat types or ecosystems (Dandelot, 1965; Letouzey, 1968; White, 1983; White *et al.*, 1995; Fischer, 2000). As summarized in Table 18.2, gorillas are found in a number of different forest habitats, and do not show a predilection for any one type.

### Variables common to gorilla habitats

Comparisons of the range of habitats gorillas are currently found in reveals some variables bearing on their survival (Tables 18.1 and 18.2). Postulating that habitats most complementary to survival have the greatest population densities (gorilla mass per area), comparisons of gorilla densities throughout a range of habitats may more directly associate quantitative and qualitative habitat differences with carrying capacity. Differences in gorilla population growth curves or predicted curves (based on age and sex profiles) across different habitats may also be used to make such associations (see Magliocca *et al.*, 1999). These comparisons, however, present only a rough estimate of those variables important to gorilla survival. Habitat carrying capacity may oscillate over time (Chapman *et al.*, 2001) and epidemics, plagues, natural catastrophes, areas of evolutionary origin, habitat destruction, and human pressures may also affect animal densities and distribution. As such, differences in gorilla numbers may not always divulge specific habitat constants, or limits, of habitat variables directly bearing on survival. To identify these variables more detailed studies of the animal's environmental interactions are necessary. In this manner, causal relationships between habitat and gorilla survival can be arrived at, and predictions can be made as to the effect of changing any one habitat variable.

Presently, however, our knowledge of habitat variables causally linked to gorilla survival is extremely limited. As noted by Harcourt (1995), we are unable to predict the consequences of removing a single main food item from the habitat, even in the well-studied Virunga gorilla population. Ultimately, insuring gorilla survival depends on revealing causal links that can predict the consequences of habitat changes.

### Causal links

#### Diet and behavior

An animal's ability to survive and reproduce depends on its energy intake outpacing its energy output. Because diet and locomotor and feeding behaviors directly reflect these energy trade-offs, dietary and behavioral studies are likely to reveal causal links between habitat and survival. To explore these links, data on diet and behaviors must be defined in comparable terms (Sarmiento, 1985).

Behaviors may be defined grossly in terms of energy expended and in more detail through physical variables of displacement, velocity, acceleration, force, and power. The physical qualities of the anatomy corresponding to a behavior (mass and shape dimensions) specify the limits of Newtonian variables quantifying movement (Sarmiento, 1985, 1988, 1994).

Dietary studies must grossly define foods in terms of nutrients and energy provided. More detailed dietary studies must seek out correspondence between physical and chemical aspects of food items to (1) gnathic structure, (2) gastrointestinal tract physiology, and (3) nutrient and vitamin requirements. Considering that gorillas eat a wide range of foods (Schaller, 1963; Casimir, 1975; Sabater Pi, 1977; Watts, 1984; Calvert, 1985; Williamson *et al.*, 1990; Remis, 1997*b*), structural or physiological specializations to diet are unlikely to favor any one food, but are likely to be a compromise of the physical and chemical qualities of a wide range of food items.

Shape and form of food items and ease with which they may be separated from the remaining plant or soil may further correlate energy expended in harvesting with muscular power, and hand and foot segment proportions.

### Habitat structure and food distribution

The habitat structure determines the energy an animal expends to locate its food. Size, density, locality/distribution, and ease of accessibility of food items, and any difference in foods with seasonality or other climatic cycles, help to determine feeding postures, locomotor behaviors, and energy gorillas expend in gathering and processing foods. The density of food items, how quickly plant foods regenerate or come to fruit, and how often such a supply can be visited sets a minimum limit on distances gorillas must travel to find foods, and helps define the size of a group's home range (Watts, 1998*b,c*). Whether items eaten are found on the ground or on trees, and how readily accessible these are to gorillas are factors in determining the energy expended in collecting them and whether or not they serve as a viable food source.

Substrate cohesion, substrate irregularities including vegetative growth and friction coefficient of plantar, palmar, and phalangeal pads, limit how steep a gradient gorillas can climb up or descend down (Sarmiento, 1985, 1988). Plant types, density of vegetation, and the intervening spaces formed by vegetation determine where on the ground gorillas move or whether they prefer to move on vegetation. Together with watercourses, other impassable gradients help determine distances gorillas must move between feeding and/or sleeping sites.

Containing some foods gorillas eat (Schaller, 1963; Jones and Sabater Pi, 1971; Sabater Pi, 1977; Tutin and Fernandez, 1985), trees are an important component of the gorilla's habitat. The ability of gorillas to negotiate supports and intervening arboreal spaces, and the energetic costs expended in doing so, partly determine arboreal foods available to them and tree types preferentially used for resting, nesting, and feeding. The animal's vertical displacements when moving either in an arboreal or terrestrial habitat are directly related to energy expended in executing movements (Sarmiento *et al.*, 1996). Distances traveled between feeding sites are correlates of foods eaten and the energy and essential nutrients provided by such foods. When integrated, all causal links based on energy trade-offs and physical limits of gorilla structure would be expected to predict limits of environmental variables past those for which gorilla survival is no longer possible, and thus the threat of habitat changes.

### Life forms coexisting with gorillas

Life forms coexisting with gorillas may prey on gorillas or compete with them for food resources. They may directly cause or be vectors of gorilla disease, or they may exist in symbiosis, exhibiting parasitism, mutualism, or commensalism. In the latter two cases, they may either benefit gorillas directly or indirectly by helping to maintain gorilla habitats. Although predators and disease vectors may directly affect gorilla mortality, most nongorilla life-forms may be expected to affect (1) quantity, quality, and types of foods available to gorillas, (2) energy gorillas use to find and collect foods, and (3) energy gorillas can extract from foods. Their potential influence on energy trade-offs in gorillas makes some organisms important causal links between habitat and gorilla survival.

Many catarrhines overlapping with gorillas compete with them for food sources (Tutin and Fernandez, 1985), and have probably evolved sympatrically to do so. Humans, chimpanzees, mangabeys, drills, guenons, red colobus, and black and white colobus may all occur in gorilla habitats (Schouteden, 1947; Malbrant and Maclatchy, 1949; Jones and Sabater Pi, 1971; Butynski, 1984; Tutin and Fernandez, 1985; Oates, 1988; Grubb, 1990; Dowsett and Dowsett-Lemaire, 1991). Putty-nosed and blue monkeys of the *Cercopithecus nictitans/mitis* group, however, are the only catarrhines found throughout the gorilla's range. Although overlap in foods consumed by gorillas and sympatric nongorilla catarrhines in some forests are known (Tutin and Fernandez, 1985), there is no information as to what effect sympatric animals have on plant regeneration, gorilla food abundance, and/or habitat carrying capacity.

Found in all gorilla habitats along the Western Rift and in many parts of the western equatorial forest, elephants (*Loxodonta*) act both to disperse tree seeds and disturb vegetation (Alexandre, 1978; Short, 1983; Struhsaker *et al.*, 1996; Hawthorne and Parren, 2000). They thus promote fruits and secondary growth consumed by gorillas, and often form the trails gorillas move through. Depending on rate of forest destruction – a corollary of elephant number per unit area – elephants may be either beneficial or harmful to gorillas. Sympatry of gorillas and elephants in many forests suggests a beneficial relationship. Comparison of forests where gorillas and elephants coexist to those gorilla habitats where elephants were recently extirpated may bring some insights into this relationship.

Much less is known of other mammals or vertebrates that share gorilla habitats. Because all these animals are affected by and effect habitat changes, and some may directly affect gorilla mortality, a rise or fall in their numbers may enable predictions to be made on gorilla habitat carrying capacity and ultimately signal threats to gorilla survival.

Disease-causing viruses, bacteria, parasites, or parasite eggs may be ingested by gorillas with foods eaten and are all correlates of diet. Association of bacteria and parasites with specific habitats, animal vectors, and/or food types can predict when and where gorillas may be susceptible to disease or parasite infestations, types of habitat disturbances predisposing these, and whether benefits of living in these habitats and ingesting tainted food items outweighs costs. Aside from parasite identification in gorilla gastrointestinal tracts from the better-known populations (Fossey, 1983; Kalema, 1995; Ashford *et al.*, 1996), much of the work on parasites and their costs to the animals still needs to be done. Generally, parasites and disease rob energy from organisms, reducing available energy from foods eaten (Hochberg *et al.*, 1992) and/or increasing an organism's energy cost (Lochmiller and Deerenberg, 2000). Because infant immune systems are not fully developed, diseases and parasites are probably important determinants of infant mortality directly impacting on population viability.

Parasites or diseases carried by humans or domesticated animals bordering small isolated gorilla forests may be especially harmful (Butynski, 2001). In such cases, infestations could proceed relatively rapidly attacking a gorilla population from many focal points. Water-borne parasites, bacteria, or viruses may have already impacted on gorilla distribution in this manner. Notably, small isolated gorilla forests along the Western Rift and elsewhere have watershed areas at higher elevations than surrounding human habitation (Sarmiento and Oates, 2000; personal observation). This suggests that forests with water sources at elevations below human habitation and open to human contamination are unsuitable gorilla habitat. As experienced by Gombe chimpanzees (Goodall, 1986; Wolfe *et al.*, 1998) and elsewhere (Kortlandt, 1996), polio or other waterborne diseases may have been fatal to gorillas in the past, and may account for their present distribution.

Insect plagues have potential to devastate vegetation including gorilla food sources. Agricultural crops not native to gorilla habitats could serve as a foothold for exotic insects with no local natural predators. Small isolated gorilla forests surrounded by agricultural land, given their proportionately smaller ratio of area versus length of perimeter (edge), seem especially susceptible. In these small forests, other biotic invasions (Elton, 1958) of either plants or animals may also pose grave risks to gorilla populations. Inability to predict which exotics may take hold, disperse, and pose survival threats requires a long-term strategy of vigilance (Mack *et al.*, 2000).

All the above causal relationships are likely to be of consequence in predicting threats to gorilla survival. Unfortunately, given the multitude of links it is both difficult and time-consuming to reveal all possible threats, or to predict which is more likely to endanger gorillas. This knowledge is probably best gained through experience and constant vigilance of gorillas and their habitats over time, especially of small populations that are expected to be more susceptible to habitat changes.

### Humans and gorillas

Many human groups have traditionally hunted gorillas and their sympatric life forms. The effect hunting has had on these animals and whether at some level it is a sustainable activity and/or one that in the long run promotes animal diversity is unknown. The long archaeological record of hunting in Africa (Kuman *et al.*, 1999; de Heinzelin *et al.*, 2000) suggests that humans may have served as an important check on numbers of large mammals especially elephants. Such a check may have slowed or impeded destruction of closed habitats. Although a burgeoning human population has made hunting a nonsustainable activity (Robinson and Boomer, 1999), limited animal culling may be necessary to maintain gorilla habitats, especially in areas where natural predators or hunters no longer exist. Likewise, human use of nonanimal resources in gorilla habitats may also be important to habitat maintenance. Notably eastern gorillas are often found close to human settlements where forest disturbances promote the secondary vegetation they feed on (Schaller, 1963).

Currently, tourists and researchers are visiting some of the small and unique gorilla populations on a daily basis. Although problems with gorilla tourism have been discussed (Butynski and Kalina, 1998; Butynski, 2000, 2001; Plumptre *et al.*, this volume), only recently are some of these problems being studied (Goldsmith, 2000). Given tourism's unknown effect on gorilla survival, it is odd that international conservation organizations promote tourism on one or possibly two critically endangered gorilla taxa, i.e., Virunga and Bwindi gorillas, as a panacea for gorilla conservation.

### Conclusions

As reviewed above, gorilla conservation consists mainly of two parts, (1) assessing the degree to which gorillas are threatened and (2) insuring gorilla survival.

The first part deals mainly with taxonomy and geographic distribution, and depends on arriving at accurate numbers for each distinctive gorilla population. Ultimately conservationists must assess population viability based on gorilla numbers and available habitats (Harcourt, 1996; Butynski, 2001). The second part entails protection of gorillas and their habitats, and depends on revealing causal links between habitat and gorilla survival that can predict or timely recognize threats to either. It is based mainly on ecology, but touches upon many branches of biology.

As regards the first part, there is a pressing need to carefully map out gorilla distribution, investigate geographic discontinuities among populations, and establish which populations are distinctive and merit separate taxonomic status. There is an urgency to develop census methods for large areas which can provide (1) accurate estimates of gorilla numbers with a known degree of error, and (2) demographic data that can predict future changes in gorilla numbers. Advancements in biochemistry, genetics, computer modeling, and satellite land imaging can provide greater accuracy to census methods, and help collect demographic data that is otherwise unavailable. Long-term genealogical and life-history data on small isolated gorilla populations may serve to accurately evaluate threats arising from dwindling population numbers and suggest possible solutions.

As regards the second part, it is imperative to reveal causal links between gorilla habitat and survival. Integrated multidisciplinary studies investigating diet, locomotor and feeding behaviors, food distribution, habitat structure, and the effect sympatric organisms have on gorillas and gorilla habitats need to be undertaken. The long life-span of gorillas, the large number of interacting variables, and meager a priori knowledge as to which variables are liable to change, suggests that first-hand experience over a prolonged period is probably the most practical means of recognizing causal links. Without prior knowledge of causal relationships, close monitoring of habitat variables (including both vegetation and fauna) over long periods (more than ten years), and an aggressive research policy aimed at detecting causes early and arriving at solutions before they compromise gorilla populations, is necessary to conserve gorillas. Currently, long-term studies are sorely needed especially for the small isolated gorilla populations at the limits of gorilla distribution. The small size of these populations and of forests they inhabit makes them more susceptible to diseases, plagues, biotic invasions climatic changes, and displacement by humans (see Harcourt *et al.*, 2001). Whether or not conserving these gorilla populations is a priority, their long-term study is necessary if only to collect practical knowledge needed to insure the survival of all gorillas.

The realization over the last few decades that humans have a compassionate concern for and identify with gorillas has prompted conservation organizations to use them as flagship species for fund raising. Touting "gorilla conservation" and "ecotourism" as politically correct catch-phrases, nongovernmental organizations (NGOs) backed by donor countries have used these terms as euphemisms for funneling western funds into third-world countries. Instead of gorilla conservation, these funds have gone mainly towards maintaining a western presence and building tourist infrastructure and services, with the aim of generating tourist revenues and stimulating (in terms of western economy) third-world economies. It has been the axiom of donor countries and many NGOs that conservation does not entail research, with the tacit but mistaken belief that if gorillas are able to generate funds their conservation will follow. Hopefully, this chapter dispels the myth that gorilla conservation entails generating an interest in gorillas, gorilla tourism, and fund-raising.

Gombe stands as a grim reminder as to how projects without conservation research and long-term goals despite their ability to generate money and worldwide publicity can be disastrous to animals (Goodall, 1986; Wallis and Lee, 1999). As such, it underscores the need to build great ape conservation into a predictive science.

## References

Agnagna, M., Barnes, R., and Ipandza, M. (1991). *Inventaire Préliminaire des Éléphants de Forêt au Sud du Congo*. Wildlife Conservation International. New York:

Akeley, M.L.J. (1929). Africa's great National Parks. *Natural History*, **29**, 638–650.

Akeley, M.L.J. (1931). National parks in Africa: The extension of wildlife conservation. *Science*, **74**, 584–588.

Alexandre, D.Y. (1978). Le rôle disséminateur des éléphants en forêt de Taï, Côte d'Ivoire. *Compte Rendus, La Terre et la Vie*, **32**, 47–72.

Amman, K. (2000). Exploring the bushmeat trade. In *Bushmeat: Africa's Conservation Crisis*, ed. K. Amman, pp. 16–27. London: World Society for the Protection of Animals.

Anonymous (2000). Situation in the Kahuzi-Biega National Park. *Gorilla Journal*, **20**, 3–4.

Ashford, R.W., Lawson, H., and Butynski, T.M. (1996). Patterns of intestinal parasitism in the mountain gorilla *Gorilla gorilla beringei* in the Bwindi-Impenetrable Forest, Uganda. *Journal of Zoology, London*, **239**, 507–514.

Awadalla P., Eyre-Walker, A., and Smith, J.M. (1999). Linkage disequilibrium and recombination in hominid mitochondrial DNA. *Science*, **286**, 2524–2525.

Beier, P. and Noss, R.F. (1998). Do habitat corridors provide connectivity? *Conservation Biology*, **12**, 1241–1252.

Berger, J. (1990). Persistence of different-sized populations: An empirical assessment of rapid extinctions in bighorn sheep. *Conservation Biology*, **4**, 91–98.

Bermejo, M. (1995). *Recensement des gorilles et chimpanzés du Parc National d'Ozala*, Report to ECOFAC-CONGO, AGRECO/CTFT.

Bourlière, F. and Versucheren, J. (1960). *L'Ecologie des Ongulés du Parc National Albert*, Vol. 1. Brussels: Institut des Parcs Nationaux du Congo Belge.

Butynski, T.M. (1984). *Ecological Survey of the Impenetrable (Bwindi) Forest, Uganda, and Recommendations for its Conservation*. New York: Wildlife Conservation International.

Butynski, T.M. (2000). Peer review. *The Sciences*, **40**, 46.

Butynski, T.M. (2001). Africa's great apes. In *Great Apes and Humans: The Ethics of Coexistence*, eds. B. Beck, T. Stoinski, M. Hutchins, T.L. Maple, B.G. Norton, A. Rowan, E.F. Stevens, and A. Arluke, pp. 3–56. Washington, D.C.: Smithsonian Institution Press.

Butynski, T.M. and Kalina, J. (1998). Gorilla tourism: A critical look. In *Conservation of Biological Resources*, eds. E.J. Milner-Gulland and R. Mace, pp. 294–313. Oxford: Blackwell Scientific Publications.

Butynski, T.M. and Sarmiento, E.E. (1995). Gorilla census on Mt. Tshiaberimu: Preliminary report. *Gorilla Journal*, **1**, 11.

Butynski, T.M. and Sarmiento, E.E. (1996). The gorillas of Mt. Tshiaberimu, Zaïre. *Gorilla Conservation News*, **10**, 14–15.

Butynski, T.M., Hart, J.A., and Omari, I. (1996). Preliminary report on a survey of the southern Itombwe Massif. *Gorilla Journal*, **13**, 13–17.

Calvert, J.J. (1985). Food selection by western gorillas (*G. g. gorilla*) in relation to food chemistry. *Oecologia*, **65**, 236–246.

Carroll, R.W. (1988). Relative density, range extension, and conservation potential of the lowland gorilla (*Gorilla gorilla gorilla*) in the Dzanga-Sangha region of southwestern Central African Republic. *Mammalia*, **52**, 309–323.

Casimir, M.T. (1975). Feeding ecology and nutrition of an eastern gorilla group in the Mt. Kahuzi region (République du Zaïre). *Folia Primatologica*, **24**, 81–136.

Casimir, M.T. (1977). An analysis of gorilla nesting sites of the Mt. Kahuzi region (Zaïre). *Folia Primatolica*, **32**, 290–308.

Chapman, A.P., Brook, B.W., Clutton-Brock, T.H., Grenfell, B.T., and Frankham, R. (2001). Population viability analyses on a cycling population: A cautionary tale. *Biological Conservation*, **97**, 61–69.

Coolidge, H.J. (1929). A revision of the genus *Gorilla*. *Memoirs of the Museum of Comparative, Zoology, Harvard*, **50**, 291–381.

Cracraft, J. (1989). Speciation and its ontology. In *Speciation and its Consequences*, eds. D. Otte and J.A. Endler, pp. 28–59. Sunderland, MA: Sinauer Associates.

Cracraft, J., Feinstein, J., Vaughn, J., and Helm-Bychowski, K. (1998). Sorting out tigers (*Panthera tigris*): Mitochondrial sequences, nuclear inserts, systematics and conservation genetics. *Animal Conservation*, **1**, 139–150.

Crandall, K.A., Posada, D., and Vasco, D. (1999). Effective population sizes: Missing measures and missing concepts. *Animal Conservation*, **2**, 317–319.

Czekala, N. and Sicotte, P. (2000). Reproductive monitoring of free-ranging female mountain gorillas by urinary hormonal analysis. *American Journal of Primatology*, **51**, 209–215.

Dandelot, P. (1965). Distribution de quelques espèces de Cercopithecoidea en relation avec les zones de végétation de l'Afrique. *Zoologica Africana*, **1(1)**, 167–176.

464     *Esteban E. Sarmiento*

de Heinzelin, J., Clark, D., Schick, K.D., and Gilbert, W.H. (2000). *The Acheulean and the Plio-Pleistocene Deposits of the Middle Awash Valley, Ethiopia.* Tervuren, Belgium: Musée Royal de L'Afrique Central.

Dobzhansky, T. (1937). *Genetics and the Origin of Species.* New York: Columbia University Press.

Dowsett, R.J. and Dowsett-Lemaire, F. (1991). *Flore et Fauna du Bassin du Kouilou (Congo) et leur Exploitation.* Conoco: Tauraco Press.

Dushoff, J. (2000). Carrying capacity and demographic stochasticity: Scaling behavior of the stochastic logistic model. *Theoretical Population Biology,* **57**, 59–65.

Elton, C.S. (1958). *The Ecology of Invasions by Animals and Plants.* London: Methuen.

Emlen, J.T. and Schaller, G. (1960). Distribution and status of the mountain gorilla (*Gorilla gorilla beringei*) – 1959. *Zoologica,* **45**, 41–52.

Eyre-Walker, A., Smith, N.H., and Smith, J.M. (1999). How clonal are human mitochondria? *Proceedings of the Royal Society of London, Series B,* **266**, 477–483.

Fay, J.M. (1989). Partial completion of a census of the western lowland gorilla (*Gorilla g. gorilla* (Savage and Wyman)) in southwestern Central African Republic. *Mammalia,* **53**, 203–215.

Fay, J.M. (1997). The ecology, social organization, populations, habitat and history of the western lowland gorilla (*Gorilla g. gorilla* Savage and Wyman, 1847). PhD thesis, Washington University, St. Louis, MO.

Fay, J.M. and Agnagna, M. (1992). Census of gorillas in Northern Republic of Congo. *American Journal of Primatology,* **27**, 275–284.

Field, D., Chemnick, L., Robbins, M., Garner, K., and Ryder, O. (1998). Paternity determination in captive lowland gorillas and wild mountain gorillas by microsatellite analysis. *Primates,* **39**, 199–209.

Fischer, E. (2000). Flora and vegetation of the afromontane region in central and east Africa. *Bonner Zoologisches Monographien,* **46**, 121–131.

Fossey, D. (1983). *Gorillas in the Mist.* Boston, MA: Houghton Mifflin.

Frankham, R. (1995). Inbreeding and extinction: A threshold effect. *Conservation Biology,* **9**, 792–799.

Garner, K.J. and Ryder, O.A. (1996). Mitochondrial DNA diversity in gorillas. *Molecular Phylogenetics and Evolution,* **6**, 39–48.

Ghiglieri, M.P. (1984). *The Chimpanzees of Kibale Forest: A Field Study of Ecology and Social Structure.* New York: Columbia University Press.

Ginzburg, L.R., Ferson, S., and Akcakaya, H.R. (1990). Reconstructibility of density dependence and the conservative assessment of extinction risks. *Conservation Biology,* **4**, 63–70.

Goldsmith, M.L. (1999). Ecological constraints on the foraging effort of western gorillas (*Gorilla gorilla gorilla*) at Bai Hoku, Central African Republic. *International Journal of Primatology,* **20**, 1–23.

Goldsmith, M.L. (2000). Effects of ecotourism on the behavioral ecology of Bwindi gorillas, Uganda: Preliminary results. *American Journal of Physical Anthropology, Supplement* **30**, 161.

Gonzalez-Kirchner, J.P. (1997). Census of western lowland gorilla population in Rio Muni region, Equatorial Guinea. *Folia Zoologica,* **46**, 15–22.

Goodall, J. (1986). *The Chimpanzees of Gombe: Patterns of Behavior*. Cambridge, MA: Harvard University Press.

Goossens, B., Latour, S., Vidal, C., Jamart, A., Ancrenaz, M., and Bruford, M.W. (2000). Twenty new microsatellite loci for use with hair and fecal samples in the chimpanzee (*Pan troglodytes troglodytes*). *Folia Primatologica*, **71**, 177–180.

Greenwood, A. and Sarmiento, E.E. (in press). MtDNA, phylogenies and the problems caused by numts. *African Primates*.

Griffiths J.F. (1972). *World Survey of Climatology*, Vol. 10, *Climates of Africa*. New York: Elsevier.

Groves, C.P. (1970). Population systematics of the gorilla. *Journal of Zoology, London*, **161**, 287–300.

Groves, C.P. (1996). Do we need to update the taxonomy of gorillas? *Gorilla Journal*, **12**, 3–4.

Groves, C.P. (2001). *Primate Taxonomy*. Washington, D.C.: Smithsonian Institution Press.

Groves, J. and Maisels, F. (1999). Report on the large mammal fauna of the Takamanda Forest Reserve, South West Province, Cameroon, with special emphasis on the gorilla population. Unpublished report to WWF, Cameroon.

Grubb, P. (1990). Primate geography in the Afro-tropical forest biome. In *Vertebrates in the Tropics*, eds. G. Peters and R. Hutterer, pp. 187–214. Bonn: Museum Alexander Koenig.

Gyltenstein, U., Wharton, D., Jossefson, A., and Wilson, A.C. (1991). Parental inheritance of mitochondrial DNA in mice. *Nature*, **352**, 255–257.

Hall, J.S., White, L.J.T., Inogwabini, B.-I., Omari, I., Morland, H.S., Williamson, E.A., Saltonstall, K., Walsh, P., Sikubwabo, C., Bonny, D., Kiswele, K.P., Veder, A., and Freeman, K. (1998). Survey of Grauer's gorillas (*Gorilla gorilla graueri*) and eastern chimpanzees (*Pan troglodytes schweinfurthii*) in the Kahuzi-Biega National Park lowland sector and adjacent forest in Eastern Democratic Republic of Congo. *International Journal of Primatology*, **19**, 207–235.

Harcourt, S. (1991). Endangered species. *Nature*, **354**, 10.

Harcourt, A.H. (1994). Population viability estimates: Theory and practice for a wild gorilla population. *Conservation Biology*, **9**, 134–142.

Harcourt, A.H. (1995). PVA in theory and practice. *Conservation Biology*, **9**, 707–708.

Harcourt, A.H. (1996). Is the gorilla a threatened species? How should we judge? *Biological Conservation*, **75**, 165–176.

Harcourt, A.H. and Groom, A.F. (1972). Gorilla census. *Oryx*, **11**, 355–356.

Harcourt, A.H., Kineman, J., Campbell, G., Yamagiwa, J., and Redmond, I. (1983). Conservation and the Virunga gorilla population. *African Journal of Ecology*, **21**, 139–142.

Harcourt, A.H., Parks, S.A., and Woodroffe, R. (2001). Human density as an influence on species/area relationships: Double jeopardy for small African reserves? *Biodiversity and Conservation*, **10**, 1011–1020.

Harris, R.B., Maguire, L.A., and Shaffer, M.L. (1987). Sample sizes for minimum viable population estimation. *Conservation Biology*, **1**, 72–76.

Hart, J.A. and Hall, J.S. (1996). Status of eastern Zaïre's forest parks and reserves. *Conservation Biology*, **10**, 316–327.

Hart, J.A. and Sikubwabo, C. (1994). *Exploration of the Maiko National Park of Zaïre 1989–1992*. Wildlife Conservation Society Working Paper no. 2. New York: Wildlife Conservation Society.

Hawthorne, W.D. and Parren, M.P.E. (2000). How important are forest elephants to the survival of woody plant species in Upper Guinean forests? *Journal of Tropical Ecology*, **16**, 133–150.

Hey, J. (2000). Human mitochondral DNA recombination: Can it be true? *Trends in Ecology and Evolution*, **15**, (5) 181–182.

Hilton-Taylor, C. (2000). *2000 Red List of Threatened Animals*. Gland, Switzerland: IUCN.

Hochberg, M.E., Michalakis, Y., and De Meerus, T. (1992). Parasitism as a constraint on the rate of life-history evolution. *Journal of Evolutionary Biology*, **5**, 491–504.

Jeffreys, A.J., Wilson, V., and Thein, S.L. (1985). Hypervariable minisatellite regions in human DNA. *Nature*, **314**, 67–73.

Jensen-Seaman, M.I. (2000). Evolutionary genetics of gorillas. PhD thesis, Yale University, New Haven, CT.

Jones, C. and Sabater Pi, J. (1971). Comparative ecology of *Gorilla gorilla* (Savage and Wyman) and *Pan troglodytes* (Blumenbach) in Rio Muni, West Africa. *Biblioteca Primatologica*, **13**, 1–96.

Jurke, M.H., Hagey, L.R., Jurke, S., and Czekala, N.M. (2000). Monitoring hormones in urine and feces of captive bonobos (*Pan paniscus*). *Primates*, **41**, 311–339.

Kalema, G. (1995). Epidemiology of the intestinal parasite burden of mountain gorillas, *Gorilla gorilla beringei*, in Bwindi Impenetrable National Park, southwest Uganda. *Zebra Foundation Newsletter*, **1995**, 19–33.

Kortlandt, A. (1983). Marginal habitats of chimpanzees. *Journal of Human Evolution*, **12**, 231–278.

Kortlandt, A. (1995). A survey of the geographical range, habitats and conservation of the pygmy chimpanzee (*Pan paniscus*): An ecological perspective. *Primate Conservation*, **16**, 21–36.

Kortlandt, A. (1996). An epidemic of limb paresis (polio?) among the chimpanzee population at Beni (Zaïre) in 1964, possibly transmitted by humans. *Pan Africa News*, **3**, 9–10.

Kuman, K., Inbar, M., and Clarke, R.J. (1999). Paleoenviroments and cultural sequence of the Florisbad Middle Stone Age hominid site, South Africa. *Journal of Archaeological Sciences*, **26**, 140–145.

Kyungu, J.C. and Vwirsihikya, K. (1999). The new gorilla reserve at Mbuhi. *Gorilla Journal*, **19**, 8–9.

Lee, P.C., Thornback, J., and Bennett, E.L. (1988). *Threatened Primates of Africa: The IUCN Red Data Book*. Gland, Switzerland: IUCN.

Letouzey, R. (1968). *Étude Phytogéographique du Cameroon*. Paris: Éditions Lechevalier.

Lochmiller, R.I. and Deerenberg, C. (2000). Trade-offs in evolutionary immunology: Just what is the cost of immunity? *Oikos*, **88**, 87–89.

Mack, R.N., Simberloff, D., Lonsdale, W.M., Evans, H., Clout, M., and Bazzaz, F.A. (2000). Biotic invasions: Causes, epidemiology, global consequences, and control. *Ecological Applications*, **10**, 689–710.

Magliocca, F., Querouil, S., and Gautier-Hion, A. (1999). Population structure and group composition of western lowland gorillas in north-western Republic of Congo. *American Journal of Primatology*, **48**, 1–14.

Malbrant, R. and Maclatchy, A. (1949). *Faune de L'Equateur Africain Français*, Vol. 2, *Mammifères*. Paris: Paul Lechevalier.

Matschie, P. (1904). Bemerkungen über die Gattung *Gorilla*. *Sitzungsberichte des Gesellschaft naturforschender Freunde, Berlin*, **1904**, 45–53.

Mayr, E. (1963). *Animal Species and Evolution*. Cambridge, MA: Harvard University Press.

Mayr, E. (1982). *The Growth of Biological Thought, Diversity, Evolution and Inheritance*. Cambridge, MA: Harvard University Press.

Meder, A. (1995). Regenwälder und Gorillas in Gabun. *Gorilla Journal*, **11**, 16–18.

Mitani, M. (1992). Preliminary results of the studies on wild western lowland gorillas and other sympatric diurnal primates in Ndoki forest, Northern Congo. In *Topics in Primatology*, Vol. 2, *Behavior, Ecology and Conservation*, eds. N. Itoigawa, Y. Sugiyama, G.P. Sackett, and R.K.R. Thompson, pp. 215–224. Tokyo: Tokyo University Press.

Mwanza, N., Maruhashi, T., Yumoto, T., and Yamagiwa, J. (1988). Conservation of eastern lowland gorillas in the Masisi region, Zaïre. *Primate Conservation*, **9**, 111–114.

Nishihara, T. (1994). Population density and group organization of gorillas (*Gorilla gorilla gorilla*) in the Nouabalé-Ndoki National Park. *Journal of African Studies*, **44**, 29–45.

Nishihara, T. (1995). Feeding ecology of western lowland gorillas in the Nouabale-Ndoki National Park, Congo. *Primates*, **36**, 151–168.

Nunney, L. (2000). The limits to knowledge in conservation genetics. In *Evolutionary Biology*, eds. M.T Clegg, R.J. MacIntyre, and M. Hecht, pp. 179–194. New York: Plenum Press.

Oates, J.F. (1988). The distribution of *Cercopithecus* monkeys in the West African forests. In *A Primate Radiation: Evolutionary Biology of the African Guenons*, eds. A. Gautier-Hion, F. Bourlière, J.P. Gautier, and J. Kingdon, pp. 79–103. Cambridge, U.K.: Cambrige University Press.

Oates, J.F. (1998). The gorilla population in the Nigeria–Cameroon border region. *Gorilla Conservation News*, **12**, 3–6.

Omari, I., Hart, J.A., Butynski, T.M., Birashirwa, R., Upoki, A., M'Keyo, Y., Bengana, F., Bashonga, M., and Bagurubumwe, N. (1999). The Itombwe massif, Democratic Republic of Congo: Biological surveys and conservation with an emphasis on Grauer's gorilla and birds endemic to the Albertine rift. *Oryx*, **33**, 301–322.

Piton, B., Pointeau, J.H., and Wauthy, B. (1979). Données hydroclimatiques à Point-Noire (Congo) 1953–1977 (+1978 et 1979). Paris: ORSTOM.

Plumptre, A.J. and Reynolds, V. (1997). Nesting behavior of chimpanzees: Implications for censuses. *International Journal of Primatology*, **18**, 475–485.

Poon, A. and Otto, S.P. (2000). Compensating for our load of mutations: Freezing the meltdown of small populations. *Evolution*, **54**, 1467–1479.

468     *Esteban E. Sarmiento*

Prescott, J., Rapley, W.A., and Joseph, M.M. (1994). Status and conservation of chimpanzee and gorilla in Cameroon. *Primate Conservation*, **14–15**, 7–12.

Purvis, A., Gittlemam, J.L., Cowlishaw, G., and Mace, G.M. (2000). Predicting extinction risk in declining species. *Proceedings of the Royal Society of London, Series B*, **267**, 1947–1952.

Ralls, K., Ballou, J.D., and Templeton, A. (1988). Estimates of lethal equivalents and the cost of inbreeding in mammals. *Conservation Biology*, **2**, 185–193.

Remis, M.J. (1997a). Ranging and grouping patterns of a western lowland gorilla group at Bai Hokou, Central African Republic. *American Journal of Primatology*, **43**, 111–133.

Remis, M.J. (1997b). Western lowland gorillas (*Gorilla gorilla gorilla*) as seasonal frugivores: Use of variable resources. *American Journal of Primatology*, **43**, 87–104.

Remis, M.J. (2000). Preliminary assessment of the impacts of human activities on gorillas *Gorilla gorilla gorilla* and other wildlife at Dzanga-Sangha reserve, Central African Republic. *Oryx*, **34**, 56–65.

Robinson, J.G. and Boomer, R.E. (1999). Towards wildlife management in tropical forests. *Journal of Wildlife Management*, **63**, 1–13.

Rogers, M.E., Maisels, F., Williamson, E.A., Fernandez, M., and Tutin, C.E.G. (1990). Gorilla diet in the Lopé Reserve, Gabon: A nutritional analysis. *Oecologia*, **84**, 326–339.

Ruvolo, M., Pan, D., Zehr, S., Disotell, T., and von Dornum, M. (1994). Gene trees and hominoid phylogeny. *Proceedings of the National Academy of Sciences, U.S.A.*, **91**, 8900–8904.

Sabater Pi, J. (1966) Gorilla attacks against humans in Rio Muni, West Africa. *Journal of Mammalogy*, **47**, 123–124.

Sabater Pi, J. (1977). Contribution to the study of alimentation of lowland gorillas in the Natural State, in Rio Muni: Republic of Equatorial Guinea (West Africa). *Primates*, **18**, 183–204.

Saltonstall, K., Amato, G. & Powell, J. (1998). Mitochondrial DNA variability in Grauer's gorillas of Kahuzi-Biega National Park. *Journal of Heredity*, **89**, 129–135.

Sarmiento, E.E. (1985). Functional differences in the skeleton of wild and captive orang-utans and their adaptive significance. PhD thesis, New York University, New York.

Sarmiento, E.E. (1988). Anatomy of the hominoid wrist joints, its evolutionary and functional implications. *International Journal of Primatology*, **9**, 281–345.

Sarmiento, E.E. (1994). Terrestrial traits in the hands and feet of gorillas. *American Museum Novitates*, **3091**, 1–56.

Sarmiento, E.E. (in press). Cross River gorillas: The most endangered gorilla subspecies. *Primate Conservation*.

Sarmiento, E.E. and Butynski, T.M. (1996). Present problems in gorilla taxonomy. *Gorilla Journal*, **12**, 5–7.

Sarmiento, E.E. and Butynski, T.M. (1998). Preliminary report of the Alimbongo, Bingi, Lutunguru survey. *Gorilla Conservation News*, **12**, 12–13.

Sarmiento, E.E. and Oates, J. (2000). The Cross River gorillas: A distinct subspecies *Gorilla gorilla diehli* Matschie 1904. *American Museum Novitates*, **3304**, 1–55.

Sarmiento, E.E., Butynski, T., and Kalina, J. (1995). Taxonomic status of the gorillas of the Bwindi-Impenetrable Forest, Uganda. *Primate Conservation*, **16**, 40–43.

Sarmiento, E.E., Butynski, T., and Kalina, J. (1996). Gorillas of Bwindi-Impenetrable Forest and the Virunga Volcanoes: Taxonomic implications of morphological and ecological differences. *American Journal of Primatology*, **40**, 1–21.

Sarmiento, E.E., Stiner E., and Mowbray K. (2002). Morphology based systematics and problems with hominoid systematics. *New Anatomist*, **269**, 50–66.

Savage, T.S. and Wyman, J. (1847). Notice of the external characters and habits of *Troglodytes gorilla*: A new species of orang from the Gaboon river with osteology of the same. *Boston Journal of Natural History*, **5**, 417–441.

Sayer, J.A., Harcourt, C.S., and Collins, M.N. (1992). *The Conservation Atlas of Tropical Forests: Africa*. New York: Simon & Schuster.

Schaller, G.B. (1963). *The Mountain Gorilla: Ecology and Behavior*. Chicago, IL: University of Chicago Press.

Schouteden, H. (1947). *De Zoogdieren van Belgisch-Congo en van Ruanda-Urundi*, Vols. 1–3. Tervuren, Belgium: Musée Royal du Congo Belge, Zoologie.

Schwartz, M.K., Tallmon, D.A., and Luikart, G. (1999). Using genetics to estimate the size of wild populations: Many methods, much potential, uncertain utility. *Animal Conservation*, **2**, 321–323.

Sholley, C.R. (1990). 1989 Census of mountain gorillas in the Virungas of central Africa. Report to Dian Fossey Gorilla Fund/Morris Animal Foundation, Englewood, CO.

Sholley, C.R. (1991). Conserving gorillas in the midst of guerillas. In *American Association of Zoological Parks and Aquariums, Annual Conference Proceedings*, pp. 30–37.

Short, J. (1983). Density and seasonal movements of forest elephants (*Loxodonta africana cyclotis* Matschie) in Bia National Park, Ghana. *African Journal of Ecology*, **21**, 175–184.

Simpson, G.G. (1961). *Principles of Animal Taxonomy*. New York: Columbia University Press.

Stromayer, A.K. and Ekobo, A. (1992). The distribution and number of forest-dwelling elephants in extreme south-eastern Cameroon. *Pachyderm*, **15**, 9–14.

Struhsaker, T.T., Lwanga, J.S., and Kasenene, J.M. (1996). Elephants, selective logging and forest regeneration in the Kibale forest Uganda. *Journal of Tropical Ecology*, **12**, 45–64.

Taylor, B.L. (1994). The reliability of using population viability analysis for risk classification of species. *Conservation Biology*, **9**, 551–558.

Templeton, A.R. (1998). The role of molecular genetics in speciation studies. In *Molecular Approaches to Ecology and Evolution*, eds. R. De Salle and B. Schierwater, pp. 131–156. Boston, MA: Birkhauser Press.

Thomas, C.D. (1990). What do real population dynamics tell us about minimum population size? *Conservation Biology*, **4**, 324–327.

Tutin, C.E.G. and Fernandez, M. (1984). Nationwide census of gorilla (*Gorilla g. gorilla*) and chimpanzee (*Pan t. troglodytes*) populations in Gabon. *American Journal of Primatology*, **6**, 313–334.

Tutin, C.E.G. and Fernandez, M. (1985). Foods consumed by sympatric populations of *Gorilla g. gorilla* and *Pan t. troglodytes* in Gabon: Some preliminary data. *International Journal of Primatology*, **6**, 27–43.

Tutin, C.E.G. and Fernandez, M. (1993). Relationships between minimum temperatures and fruit production in some tropical forest trees in Gabon. *Journal of Tropical Ecology*, **9**, 241–248.

Tutin, C.E.G., Parnell, R.J., White, L.J.T., and Fernandez, M. (1995). Nest building by lowland gorillas in Lopé Reserve, Gabon: Environmental influences and implications for censusing. *International Journal of Primatology*, **16**, 53–76.

Usongo, L. (1998*a*). Conservation status of primates in Cameroon. *Primate Conservation*, **18**, 59–65.

Usongo, L. (1998*b*). Conservation status of primates in the proposed Lobéké Forest Reserve, southeast Cameroon. *Primate Conservation*, **18**, 66–68.

Usongo, L. and Fimbel, C. (1995). Preliminary survey of arboreal primates in Lobéké Forest Reserve, southeast Cameroon. *African Primates*, **1**(2), 46–48.

Vanderbroek, G. (1959). Notes écologiques sur les anthropoides africains. *Annales de la Société Royale Zoologique de Belgique*, **89**, 203–211.

Vedder, A.L. (1989). Feeding ecology and conservation of the mountain gorilla (*Gorilla gorilla beringei*). PhD thesis, University of Wisconsin, Madison, WI.

Voysey, B.C., McDonald, K.E., Rodgers, M.E., Tutin, C.E.G., and Parnell, R.J. (1999). Gorillas and seed dispersal in the Lopé Reserve, Gabon. II: Survival and growth of seedlings. *Journal of Tropical Ecology*, **15**, 39–60.

Vucetich, J.A. and Waite, T.A. (2000). Is one migrant per generation sufficient for the genetic management of fluctuating populations? *Animal Conservation*, **3**, 261–266.

Wallis, J. and Lee, D.R. (1999). Primate conservation: The prevention of disease transmission. *International Journal of Primatology*, **20**, 803–826.

Walsh, P.D. (1995). PVA in theory and practice. *Conservation Biology*, **9**, 704–705.

Walter, H.S. (1990). Small viable population: The red-tailed hawk of Socorro Island. *Conservation Biology*, **4**, 441–443.

Waples, R. (1991). Genetic methods for estimating the effective population size from temporal changes in allele frequencies. *Genetics*, **152**, 755–781.

Wasser, S.K. (1996). Reproductive control in wild baboons measured by fecal steroids. *Biological Reproduction*, **55**, 393–399.

Watt, A. (1956). *Working Plan for Mgahinga Central Forest Reserve Kigezi District, Uganda*. Kampala: Forest Department.

Watts, D.P. (1984). Composition and variability of mountain gorilla diets in the central Virungas. *American Journal of Primatology*, **7**, 323–356.

Watts, D.P. (1998*a*). Seasonality in the ecology and life histories of mountain gorillas (*Gorilla gorilla beringei*). *International Journal of Primatology*, **19**, 929–948.

Watts, D.P. (1998*b*). Long-term habitat use by mountain gorillas (*Gorilla gorilla beringei*). I: Consistency, variation, and home range size and stability. *International Journal of Primatology*, **19**, 651–680.

Watts, D.P. (1998*c*). Long-term habitat use by mountain gorillas (*Gorilla gorilla beringei*). II: Reuse of foraging areas in relation to resource abundance, quality and depletion. *International Journal of Primatology*, **19**, 681–702.

Weber, A.W. and Vedder, A. (1983). Population dynamics of the Virunga gorillas: 1959–1978. *Biological Conservation*, **26**, 341–366.

White, F. (1983). *The Vegetation of Africa*. Paris: UNESCO.

White, L.J.T. (1992). Vegetation history and logging disturbance: Effects on rainforest mammals in the Lopé Reseserve, Gabon. PhD thesis, University of Edinburgh, Edinburgh, U.K.

White, L.J.T., Rogers, M.E., Tutin, C.E.G., Williamson, E.A., and Fernandez, M. (1995). Herbaceous vegetation in different forest types in the Lopé Reserve, Gabon: Implications for keystone food availability. *African Journal of Ecology*, **33**, 124–141.

Williamson, E.A., Tutin, C.E.G., Rogers, M.E., and Fernandez, M. (1990). Composition of the diet of lowland gorillas at Lopé in Gabon. *American Journal of Primatology*, **21**, 265–277.

Williamson, E.G. and Slatkin, M. (1999). Using maximum likelihood to estimate population size from temporal changes in allele frequencies. *Genetics*, **152**, 755–781.

Williamson, L. & Usongo, L. (1996). Survey of gorillas *Gorilla gorilla* and chimpanzees *Pan troglodytes* in the Réserve de Faune du Dja, Cameroon. *African Primates*, **2**(2), 67–72.

Wolfe, N.D., Escalante, A.A., Karesh, W.B., Kilbourn, A., Spielman, A., and Lal, A. (1998). Wild primate populations in emerging infectious disease research: The missing link? *Emerging Infectious Diseases*, **4**, 149–158.

Wu, C. (2000). Genetics of species differentiation. In *Evolutionary Biology*, eds. M.T. Clegg and M. Hecht, pp. 239–248. New York: Plenum Press.

Yamagiwa, J. (1983). Diachronic changes in two eastern lowland gorilla groups (*Gorilla gorilla graueri*) in the Mt. Kahuzi region, Zaïre. *Primates*, **24**, 174–183.

Yamagiwa, J. (1999). Slaughter of gorillas in the Kahuzi-Biega park. *Gorilla Journal*, **19**, 4–6.

Yamagiwa, J., Mwanza, N., Yumoto, T., and Maruhashi, I. (1992). Travel distances and food habits of eastern lowland gorillas: A comparative analysis. In *Topics in Primatology*, vol. 2, *Behavior, Ecology and Conservation*, eds. N. Itoigawa, Y. Sugiyama, G.P. Sackett, and R.K.R. Thompson, pp. 267–281. Tokyo: Tokyo University Press.

Yamagiwa, J., Mwanza, N., Spangenberg, A., Maruhashi, T., Yumoto, T., Fischer, A., and Steinhauer-Burkart, B. (1993). A census of eastern lowland gorillas *Gorilla gorilla graueri* in Kahuzi Biega National Park, with reference to mountain gorillas *G. g. beringei* in the Virunga region, Zaïre. *Biological Conservation*, **64**, 83–89.

# 19 *The Cross River gorilla: Natural history and status of a neglected and critically endangered subspecies*

JOHN F. OATES, KELLEY L. McFARLAND, JACQUELINE
L. GROVES, RICHARD A. BERGL, JOSHUA M. LINDER, AND
TODD R. DISOTELL

In this chapter we summarize information on the morphology, genetics, and natural history of the West African gorilla population inhabiting the forests on the Nigeria–Cameroon border at the northern headwaters of the Cross River, a region at the western and northern limits of the species' range. A recent morphological analysis of skeletal specimens from this population has shown that they are sufficiently distinct from other western gorillas to justify being classified as the subspecies *Gorilla gorilla diehli*, a taxonomic name originally applied to them in the early twentieth century (Sarmiento and Oates, 2000). Just as the distinctiveness of the Cross River gorillas is being appreciated, their continued survival is in jeopardy. Recent surveys suggest that approximately 250 probably remain, concentrated in nine or more isolated hill areas. Because these gorillas are still hunted for their meat and parts of their habitat are under threat, they are one of Africa's most endangered primate taxa. After reviewing data on the status of the Cross River gorillas, our chapter ends by discussing some options for improving their prospects for survival.

In addition to the literature, the information we summarize derives from our own research: Field surveys in Nigeria (by JFO and KLM) and Cameroon (by JLG); an ecological study of a subpopulation inhabiting Afi Mountain Wildlife Sanctuary, Nigeria (by KLM); and the sequencing of mtDNA extracted from hairs of Nigerian gorillas shed into sleeping nests (by RAB and JML in the laboratory of TRD). All our results should be regarded as preliminary. Field surveys (attempting to clarify the distribution and status of the remaining population), and the analysis of ecological and genetic data are still in progress.

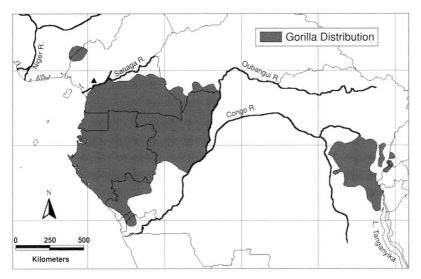

Fig. 19.1. Distribution of *Gorilla gorilla* (adapted from Butynski, 2001). Cross River gorillas occur only on the Nigeria–Cameroon border (upper left). The triangle (▲) indicates the position of a set of gorilla nest sites seen in 2001 in the Yabassi region, north of the Sanaga River by R.J. Dowsett and F. Dowsett-Lemaire (personal communication).

## Geographical range

All gorilla distribution maps show the isolated nature of the Cross River gorilla population, restricted to a small area on the edge of the Cameroon Highlands, north of Mamfe (Fig. 19.1). Until very recently this population was thought to be separated by about 250 km from the main population of western gorillas, which occurs south of the Sanaga River. However, in 2001, nests identified as belonging to gorillas were observed in the hills east of Yingui in the Yabassi region, north of the Sanaga, and about 200 km south of the most southerly section of the Cross River population (R.J. Dowsett and F. Dowsett-Lemaire, personal communication). By contrast, the Virunga Volcanoes gorilla population is separated by only about 30 km from the population in the Bwindi-Impenetrable Forest National Park, and by about 100 km from the population in the highlands of Kahuzi-Biega.

Within their range, the occurrence of Cross River gorillas has been confirmed in the Mbe Mountains and the Forest Reserves of Afi River, Boshi Extension, and Okwangwo of Nigeria's Cross River State, and in the Takamanda and Mone River Forest Reserves, and the Mbulu Forest, of Cameroon's South West Province (March, 1957; Critchley, 1968; Harcourt *et al.*, 1989; Oates *et al.*,

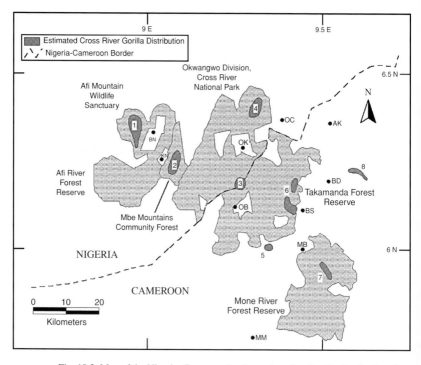

Fig. 19.2. Map of the Nigeria–Cameroon border region, showing the distribution of Cross River gorillas (dark shading) in relation to reserves and protected areas (light shading). Gorilla subpopulations are numbered as follows: (1) Afi Mountain, (2) Mbe Mountains, (3) Obonyi-Okwa Hills, (4) Boshi Extension area, Cross River National Park Okwangwo Division, (5) Takpe Hill, (6) Basho Hills, (7) Mone River Forest Reserve, (8) Upper Mbulu Forest. Towns and villages are keyed as follows: AK, Akwaya; BD, Badshama; BN, Buanchor; BS, Basho 1; KN, Kanyang; MB, Mbu; MM, Mamfe; OB, Obonyi 1; OC, Obudu Cattle Ranch; OK, Okwa 1.

1990; JLG, unpublished data) (Fig. 19.2). In 1991, Boshi Extension and Okwangwo were incorporated into the Okwangwo Division of Cross River National Park.

## Taxonomy and cranial morphology

The Cross River gorillas first became known to science as long ago as 1904 when Matschie described them as a new species, *Gorilla diehli*, based on a study of nine skulls (four adult males, two adult females, one juvenile male) collected for the Berlin Museum by Herr Diehl from Dakbe (= Takpe), Oboni (= Obonyi), and Basho in German-administered Kamerun, close to the border with Nigeria. Matschie (1904) noted in his description of this species such features as an absolutely short skull and molar row, and a low and broad nuchal plane compared to 44 other western gorillas and one Virunga gorilla. The collecting localities are villages that lie within, or close to, today's Takamanda Forest Reserve in Cameroon (Fig. 19.2). Subsequently, Rothschild (1904, 1908) reduced *diehli* to a subspecies of *Gorilla gorilla*, and Coolidge (1929) – in a wholesale simplification of gorilla taxonomy – lumped all West African gorillas including *diehli* into the subspecies *Gorilla gorilla gorilla*, and all eastern gorillas into the subspecies *G. g. beringei*. However, Coolidge did note that, based on his measurements of male skulls, the Kivu (= Virunga) and western Cameroons (= West Cameroon and Nigeria) populations were possibly separable from other eastern and western gorillas, respectively.

No further thorough review of gorilla systematics occurred until the major craniometric studies by Colin Groves in the 1960s (Groves, 1967, 1970). Following these studies, Groves (1970) recognized eastern lowland gorillas as a separate subspecies, *G. g. graueri*, but retained a single western subspecies. This arrangement has been widely followed until the present day.

Groves relied particularly on calculations of mean squared generalized distance (Mahalanobis $D^2$), a statistic that he found grouped gorilla skull measurements into three clusters, "between which the morphological contrasts were greater than those obtained between the constituent populations of each" (Groves, 1970:297). In terms of the $D^2$ statistic, the Cross River samples (which Groves referred to as "Nigerian") were the most distinctive of western gorillas. Recently, Stumpf and Fleagle reanalyzed Groves's original data, correcting for size differences among the skulls; this reanalysis not only found the Cross River gorillas to be the most divergent of western gorilla populations, but also found the $D^2$ difference between Cross River and other western gorillas to be comparable to the difference between Virunga and eastern lowland gorillas (Stumpf *et al.*, 1998, this volume). This analysis did not consider gorillas from Bwindi and Kahuzi-Biega, localities from which Groves had very limited samples.

Sarmiento and Oates (2000) have independently analyzed measurements made on the skulls of 36 male and 29 female Cross River gorillas in museum collections in Berlin, London, and New York, as well as a skull from Afi in Nigeria. Compared with the same measurements on the skulls of other western gorillas, the Cross River sample has significantly smaller values for a range of dimensions (including, among others, skull length, vault length, cheek-tooth surface area, and biglenoid diameter); all female Cross River gorillas, all females of other western gorillas, and all male Cross River gorillas could be correctly assigned to their respective populations by a discriminant analysis based on 11 such measurements, although two male western lowland gorillas were assigned to Cross River. Taken together, these studies strongly argue for again recognizing the Cross River gorillas as the subspecies *Gorilla gorilla diehli*, a name already formally revived by Sarmiento and Oates (2000) and by Groves (2001, this volume).

## Genetics

Mitochondrial DNA control region (D-loop) sequences also suggest significant divergence between Cross River and other western gorillas. Shed hairs were collected from gorilla night nests throughout the Nigerian portion of the Cross River gorilla range. Hair samples were collected with flame sterilized forceps and/or latex gloves to minimize contamination (fortunately, the gorilla D-loop is readily differentiable from that of *Pan* or *Homo* due to the presence of a large deletion, so sample contamination with human cells is not a large analytical problem). Samples were placed in glassine envelopes, with desiccant when possible, to prevent fungal growth and DNA degradation. Samples were stored at ambient temperature until transported to the New York University molecular anthropology laboratory, where they were stored at −20 °C. DNA was extracted from hair samples using the Allen–Vigilant buffer digest method for hair digestion (Allen *et al.*, 1998; Vigilant, 1999). Only one hair was used per extraction to prevent combining the DNA of two individuals. The hypervariable region 1 of the mitochondrial D-loop was amplified via the polymerase chain reaction (PCR) using a combination of custom (GDFOREX: 5'-ACTATTCTCTGTTCTTTCATGGGGAGAC-3') and published oligonucleotide primers (Goldberg, 1996; Jensen-Seaman, 2000). A nested polymerase chain reaction (PCR) protocol was employed, optimized for amplification from low-quality DNA sources (typical of extractions from shed hair). Purified PCR product was cycle-sequenced with the BigDye-DNA sequencing kit (Applied Biosystems, Foster City, CA) and sequenced on an Applied Biosystems ABI 310 automated sequencer. Sequence data were corrected and aligned by hand. Sequence data were compared to published sequences of known geographic origin (Garner and Ryder 1996; GenBank: C. Roos, unpublished data) and analyzed using PAUP (Swofford, 1998) and MacClade (Maddison and Maddison, 2000).

Our preliminary analysis of the sequence data supports the presence of a deep east–west split within the gorilla clade as a whole, as reported by Ruvolo *et al.* (1994) and Garner and Ryder (1996), and confirmed by Jensen-Seamen *et al.* and Clifford *et al.* (this volume). Within the western group, a comparison of the D-loop sequences we have obtained from Nigerian gorillas with published sequences from other populations suggests that Cross River gorillas form a clade distinct from other western gorillas. Maximum likelihood (Fig. 19.3), branch-and-bound using maximum parsimony, and neighbor-joining analysis all

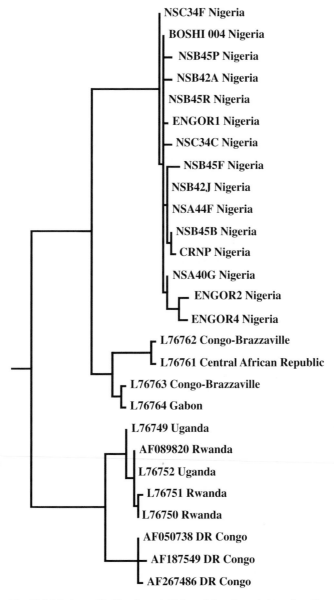

Fig. 19.3. Maximum likelihood tree (with branch lengths scaled to reflect Tamura–Nei distances) (Swofford, 1998) of Cross River gorilla D-loop sequences (top portion of tree, labeled "Nigeria") and sequences from other gorilla populations. Data from other populations are from Garner and Ryder (1996) and GenBank (C. Roos, unpublished data).

produced trees supporting this division. Based on multiple pairwise comparisons of Tamura–Nei branch lengths, the Cross River–western gorilla split is approximately 40% more divergent than that between populations traditionally classified as *G. g. beringei* and *G. g. graueri*.

Our ability to draw firm conclusions from the molecular data is limited, however. Up to this time, we have obtained 15 sequences from the Cross River population. Additional sequence data have been generated, but were determined to be nuclear insertions of mtDNA (Numts) based on comparisons with other putative Numts (see Jensen-Seaman, 2000). Our tree, generated using D-loop data, only resolves clearly if these insertions are taken into account. Furthermore, the comparative database of published gorilla D-loop sequences is severely limited; few western gorilla sequences of known geographic origin are available, with all others coming from captive individuals. Sequence data for the eastern clade are similarly limited.

Although the molecular data must be interpreted with caution, they suggest a situation similar to that seen in chimpanzees. Based on mtDNA sequence data, it has been suggested that chimpanzees in the Nigeria–Cameroon border region belong to a form, *Pan troglodytes vellerosus*, distinct from other western chimpanzees (Gonder *et al.*, 1997; Gagneux *et al.*, 1999; Gonder, 2000), and that south of the Sanaga River *P. t. vellerosus* be replaced by *P. t. troglodytes*.

If the gorilla mtDNA sequence differences we have found accurately reflect phylogenetic history, then either the Cross River population deserves subspecific status, or all the eastern populations should be collapsed into a single subspecies, *G. g. beringei*. Additional sequence data from the mitochondrial genome and from independent nuclear loci are needed from both the Cross River and other western groups in order to more accurately evaluate this hypothesis from a genetic standpoint. However, the preliminary genetic data we present here are consistent with the morphological evidence, and together provide support for the proposal to place the Cross River gorillas in their own subspecies.

### Ecology

Cross River gorillas have been recorded over an altitudinal range from below 200 m in the lower-lying parts of Takamanda in Cameroon to about 1500 m on the edge of the Obudu Plateau in Nigeria. Since 1990 the great majority of gorilla signs in this region have been found in hill areas above 400 m.

As would be expected from the northerly location, the climate in the range of the Cross River gorilla is highly seasonal, with a prolonged dry season. In parts of the range there can be less than 100 mm of rainfall per month for a five-month period (November–March) (Fig. 19.4). By the middle of this long dry season, grass fires are common around the gorillas' forest habitat, and the forest itself sometimes burns.

The lowland forests in the gorillas' range are semi-deciduous, containing such upper-canopy tree species as *Berlinia bracteosa, Brachystegia nigerica, Cylicodiscus gabunensis, Lophira alata,* and *Piptadeniastrum africanum.* At around 700 m the forest structure changes; in this submontane zone large mahoganies and *Santiria trimera* are often common. Above 1000 m there appear such characteristically montane tree species as *Cephaelis mannii* and *Podocarpus milanjianus,* and the stature of the forest is noticeably lower. At the highest elevations there is low-stature montane forest and the trees bear abundant epiphytes. On the Obudu Plateau itself (1500–1800 m) much of the original forest has been converted to grassland by many years of cultivation, burning, and cattle-grazing. Current evidence is not sufficient to show whether the gorillas have distinct preferences for certain forest types in their range. Their present restriction to elevations above 400 m may result mainly from the fact that there is greater hunting pressure from humans in the lowlands (see below).

The highly seasonal nature of the Cross River gorillas' habitat suggests that the animals may face greater seasonal changes in the abundance of preferred foods (and particularly fruit) than do other gorillas. Until KLM's study, however, only anecdotal information was available on their diet to examine this possibility. For instance, Collier (Anonymous, 1934) noted that Cross River gorillas ate the fruits of *Musanga* and *Dacryodes* (= *Pachylobus*), the young leaves of *Musanga,* the pith of *Dracaena,* and the leaves, leaf bases, and stalks of the terrestrial herbs *Palisota, Aframomum, Costus,* and *Coleus.* In Takamanda, Critchley (1968) recorded gorillas eating the fruits of the large woody climber *Saba* (= *Landolphia*) *florida* and the leaf stems of the terrestrial aroid *Anubias,* while Fay (1987) recorded evidence of feeding on the leaf bases of *Palisota.*

McFarland's data on diet were gathered from an unhabituated group at Afi Mountain studied over a 32-month period between February 1996 and June 1999. Feces were collected at nest sites, and feeding signs were noted along trails followed by the gorilla group. The feeding trails were paced for distance and at each feeding site, the location, species and part consumed, amount eaten, and habitat type were recorded. Fecal samples collected from nest sites and along feeding trails were weighed and washed; non-fruit vegetation (fiber, bark, and leaf fragments) was then separated from fruit parts (seeds and skins), and each portion was weighed separately. Finally, seeds were separated by species, and counted and weighed.

Table 19.1 summarizes the data collected at Afi on the number of species eaten each month from December 1997 to November 1998. Despite the Afi

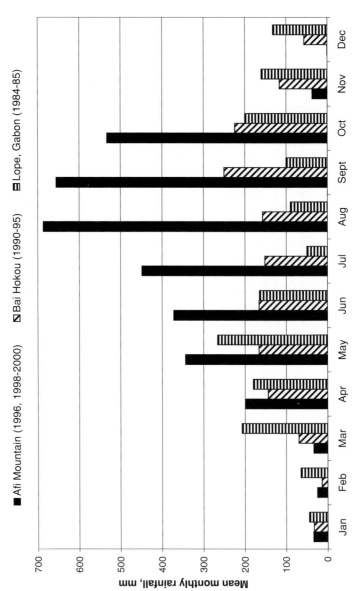

Fig. 19.4. Monthly rainfall at Afi Mountain, Nigeria, compared with two other western gorilla study sites. Bai Hokou and Lopé data based on graphs in Goldsmith (1996), Remis (1994), and Williamson (1988).

Table 19.1. *Number of plant species found per month in the diet of the gorilla study group at Afi Mountain, using two different methods (data collected by KLM)*

| | Data from fecal analysis | | | | Data from feeding trails | | | |
|---|---|---|---|---|---|---|---|---|
| Month | Number of days sampled | Number of fecal samples collected | Total number of fruit species in feces | Mean number of fruit species per sample | Number of days sampled | Number of km sampled | Number of feeding sites[a] | Number of species |
| Dec 1997 | 10 | 97 | 12 | 1.2 | 7 | 5.2 | 179 | 27 |
| Jan 1998 | 9 | 161 | 18 | 1.2 | 9 | 9.5 | 144 | 38 |
| Feb 1998 | 9 | 126 | 22 | 4.3 | 12 | 18.8 | 154 | 43 |
| Mar 1998 | 9 | 62 | 24 | 4.7 | 10 | 8.5 | 84 | 23 |
| Apr 1998 | 12 | 160 | 31 | 4.1 | 16 | 16.3 | 376 | 45 |
| May 1998 | 14 | 151 | 13 | 3.3 | 14 | 17.2 | 291 | 30 |
| June 1998 | 8 | 100 | 21 | 2.9 | 8 | 10.5 | 177 | 24 |
| July 1998 | 11 | 172 | 22 | 3.0 | 12 | 11.8 | 250 | 25 |
| Aug 1998 | 15 | 151 | 16 | 1.6 | 17 | 12.9 | 518 | 42 |
| Sept 1998 | 10 | 96 | 8 | 1.3 | 13 | 5.5 | 174 | 25 |
| Oct 1998 | 9 | 74 | 13 | 1.6 | 13 | 8.4 | 311 | 41 |
| Nov 1998 | 5 | 47 | 6 | 0.8 | 7 | 3.7 | 179 | 23 |

[a]One feeding site consists of feeding remains at least 2 m from any other feeding remains.

gorillas' relatively high-altitude habitat (400–1300 m), their diet was found to be as diverse as that of other western gorillas. Gorillas at Afi consumed at least 190 different food items (plant species plus part) from 125 identified species (21 herb species, 33 vines and climbers, 9 shrubs, and 62 trees), together with fruits from an additional 41 unidentified species. By contrast, gorillas at Bai Hokou, Central African Republic (CAR) (463 m altitude) consumed 230 items from 129 species (Remis, 1997*a*); at Nouabalé-Ndoki, Congo (300–400 m altitude), 182 items from 152 species (Nishihara, 1995); at Lopé, Gabon (100–700 m altitude), 182 items from 134 species (Williamson *et al.*, 1990; Tutin *et al.*, 1994); and at Itebero, Democratic Republic of Congo (DRC) (600–1300 m altitude), 75 species (Yamagiwa *et al.*, 1992). While gorillas are known to consume invertebrates at many other lowland gorilla sites (Williamson *et al.*, 1990; Nishihara and Kuroda, 1991; Yamagiwa *et al.*, 1991; Remis, 1997*a*), there was no evidence of invertebrate consumption by the Afi study group.

Fruit remains were found in 90.2% of all fecal samples at Afi. This is a somewhat lower percentage than at other western gorilla study sites, where 95.8–99.7% of fruit remains have been recorded (Rogers *et al.*, 1988; Goldsmith, 1996; Remis, 1997*a*; Doran and NcNeilage, 1999). However, the number of fruit species found in Afi fecal samples (at least 83) is similar to other sites: CAR, >77 species (Remis, 1994); Congo, 115 species (Nishihara, 1995); Gabon, ≥87 species (Williamson *et al.*, 1990); DRC, >24 species (Yamagiwa *et al.*, 1992). The mean number of fruit species eaten per month at Afi (17.2, range 6–31) was actually higher than at Bai Hokou (15.7: Goldsmith, 1996) or Lopé (14.2, range 9–18: Tutin and Fernandez, 1993). However, the mean number of fruit species per fecal sample was lower at Afi (2.6) than at Bai Hokou (3.5: Goldsmith, 1996) or Lopé (3.2: Tutin and Fernandez, 1993).

The Afi study group showed strong seasonal changes in diet, with more non-fruit vegetation being eaten when fruit was scarce. Table 19.2 summarizes the results from two methods of estimating the relative abundance of different food types in the diet. Based on mean percentage of total fecal weight per sample, a strong seasonal pattern is apparent with at least five months of relatively low abundance of fruit in the diet (<10%: August–September, November–January). A phenological study of 397–890 trees of 56–135 species monitored monthly in 1998–2000 found that fruit availability at Afi was significantly lower during August–January when, on average, only 8.2% of trees had fruit (both unripe and ripe; $t = 5.43$, df $= 20$, $p < 0.001$). These months do not strictly coincide with the dry season, unlike the pattern seen in other gorilla studies (Williamson, 1988; Tutin *et al.*, 1994; Goldsmith, 1996; Remis, 1997*a*); at Afi, August–October is the latter part of the rainy season.

Based on percentage of remains at feeding sites, the gorillas ate terrestrial herbs, and the bark and leaves of climbers and trees more frequently during

Table 19.2. *Relative abundance of different food types in the diet of the Afi Mountain study group, using two different methods (data collected by KLM)*

| | Mean percentage of fecal weight per sample | | Percentage of feeding sites[a] | | | | |
|---|---|---|---|---|---|---|---|
| Month | Non-fruit vegetation | Fruit | Herb pith | Woody bark/pith | Tree bark | Leaves | Fruit |
| Dec 1997 | 97.5 | 2.5 | 65.4 | 18.4 | 2.2 | 15.6 | 2.2 |
| Jan 1998 | 95.3 | 4.7 | 53.5 | 22.2 | 9.0 | 7.6 | 10.4 |
| Feb 1998 | 45.0 | 55.0 | 37.7 | 25.3 | 3.9 | 15.6 | 26.6 |
| Mar 1998 | 45.7 | 54.3 | 44.0 | 26.2 | 4.8 | 7.1 | 27.4 |
| Apr 1998 | 36.9 | 63.1 | 44.1 | 17.6 | 1.1 | 21.8 | 29.3 |
| May 1998 | 29.6 | 70.4 | 45.4 | 6.2 | 0.3 | 32.3 | 23.0 |
| June 1998 | 60.8 | 39.2 | 72.3 | 4.5 | 5.6 | 24.3 | 14.7 |
| July 1998 | 50.2 | 49.8 | 54.8 | 10.0 | 4.0 | 15.2 | 26.8 |
| Aug 1998 | 90.7 | 9.3 | 66.2 | 18.0 | 10.0 | 32.8 | 3.9 |
| Sept 1998 | 97.1 | 2.9 | 66.7 | 19.0 | 5.7 | 31.0 | 1.1 |
| Oct 1998 | 77.0 | 23.0 | 56.9 | 22.2 | 7.7 | 21.9 | 8.0 |
| Nov 1998 | 99.3 | 0.7 | 75.4 | 8.9 | 3.9 | 16.2 | 2.2 |

[a] More than one food category was often recorded at a single feeding site.

the months of fruit scarcity, although the pattern is less strong than the one shown by the fecal weight data. This is most likely due to the fact that herb- and bark-feeding sites are more easily detected than fruit-feeding sites, so that the frequencies of herb- and bark-feeding are likely to be overestimated by this method. Table 19.3 lists 14 food species that were eaten more frequently by Afi gorillas during fruit-scarce months than when fruit was abundant. Many of these species are available throughout the year, but are more sought out by the gorillas when fruit is scarce. At Lopé, two similar foods (*Milicia excelsa* bark, and the pith of *Marantochloa* species) are also eaten more frequently during the fruit-scarce dry-season months (Rogers *et al.*, 1994; Tutin *et al.*, 1997). *Morus mesozygia*, an important bark and leaf food resource at Afi, is in the same family (Moraceae) as *Milicia*.

Two herb species that appear to be unique to the Cross River gorilla habitat are important foods during fruit-scarce months. *Amorphophallus difformis* (Araceae) was the second most frequently eaten herb and, when available, appeared to be preferred over *Aframomum* spp. (Zingiberaceae), the most frequently eaten herb (when both occur in the same area, *Aframomum* is often ignored). *Amorphophallus* is the only seasonally available herb (it was only observed at Afi in the wet season) and has not been reported in other gorilla habitats. Large amounts of the pith and leaves of *Stylochiton* (Araceae)

Table 19.3. *Foods eaten more frequently by Afi gorillas during fruit-scarce months, based on evidence from feeding sites (data collected by KLM)*

| Food category and species | Percentage of all records in a given food category[a] | Percentage of records for each species that occurred during fruit-scarce months[b] |
|---|---|---|
| Herb pith | | |
| *Marantochloa leucantha* | 1.7 | 73.3 |
| *Palisota* sp. | 10.0 | 69.8 |
| *Stylochiton* sp. | 5.9 | 88.2 |
| Species G-1 | 1.7 | 100 |
| *Amorphophallus difformis* | 28.5 | 51.5 |
| Woody pith | | |
| Species C-5 | 6.2 | 92.8 |
| Species S-1 | 2.4 | 90.9 |
| Tree bark | | |
| *Pterocarpus osun* | 11.0 | 68.8 |
| Species T-11 | 4.1 | 100 |
| *Morus mesozygia* | 10.3 | 66.7 |
| *Milicia excelsa* | 28.3 | 48.8 |
| *Newbouldia laevis* | 2.1 | 66.7 |
| Species T-40 | 4.1 | 100 |
| *Grewia mollis* | 10.3 | 100 |
| Leaves | | |
| *Stylochiton* sp. | 12.8 | 91.6 |
| *Morus mesozygia* | 2.5 | 81.3 |
| *Milicia excelsa* | 5.7 | 72.9 |
| Species T-40 | 3.1 | 65.0 |
| *Amorphophallus difformis* | 25.0 | 38.9 |

[a] For example, *Marantochloa leucantha* was recorded at 1.7% of feeding sites in which herb pith was eaten.
[b] For example, 73.3% of feeding sites in which *Marantochloa leucantha* was recorded occurred during fruit-scarce months (August–September and November–January).

were also eaten during the fruit-scarce wet-season months of August and September; this herb has also not been reported in the diet of other gorilla populations.

We are not yet sure whether the unusual ecological conditions that the Cross River gorillas face produce any unusual patterns of ranging behavior and social organization. The Afi study group ranged over an area of approximately 32 km$^2$. In general, they moved on a north–south axis, covering a major portion of their range over a three-month period. They tended to remain in one area for a period of up to several weeks, and then traveled long distances ($\geq 3$ km) to another area. Preliminary results from a survey of the subpopulation in the Boshi Extension forests of Nigeria by Ernest Nwufoh (personal communication to JFO, 2000) suggest similar long-range movements.

### Grouping patterns

Most of the reported sleeping-nest clusters of Cross River gorillas indicate that the animals live in relatively small groups. The great majority of nest clusters found in Nigeria contain eight or fewer nests (Obot *et al.*, 1997; Nwufoh, 1999; Sarmiento and Oates, 2000). In Cameroon, JLG recorded 87 nest clusters between January 1998 and April 2001. Average nest-cluster size was 3.6 (range 1–23), and only three clusters had more than eight nests.

An exception to this pattern of small nest groups is Afi Mountain. Oates *et al.* (1990) found clusters of up to 19 nests on Afi Mountain, and although KLM found very variable numbers of nests at sleeping sites in her study area at Afi ($n = 140$ nest sites, range 8–38), most nest sites contained at least 14 nests, with a modal size of 18. Afi group size is high compared to western lowland gorilla study sites where groups with $\geq 18$ individuals are rare: Lopé, $n = 8$ groups, median = 10 individuals, range 4–16 (Tutin *et al.*, 1992); Bai Hokou, $n = 163$ nest sites, mean = 8, maximum = 15 (Remis, 1994); $n = 248$ nest sites, mean = 9, range 3–18 (Goldsmith, 1996). However, Bermejo (1997) reported large groups of 20 and 32 individuals at Lossi, Congo ($n = 7$ groups, mean = 14 individuals).

This variation in nest-group size at Afi may indicate that more than one group was using the study area, or, as observed at other gorilla study sites, that some fission–fusion pattern of grouping (or subgrouping) was occurring (Olejniczak, 1996; Remis, 1997*b*; Doran and McNeilage, 1998; Goldsmith, 1999). The latter hypothesis is supported by the observation of variation in nest-group sizes at consecutive nest sites; KLM found differences of up to 11 nests from one night's nest site to the next. In addition, results from a broad-area population census of Afi Mountain found fresh nests only in one limited area, suggesting that there is only one group on the mountain. Our tentative conclusion is that the variation in nest-group size observed at Afi is best explained as resulting from the presence of one group of 18–25 individuals (probably with two adult males) that sometimes split into smaller parties, and sometimes nested in the same location on consecutive nights. Relatively large groups with more than one silverback have been observed at Lopé (Tutin *et al.*, 1992), Bai Hokou (Remis, 1997*b*), and Mbeli Bai (Olejniczak, 1996). One relatively clear instance of group splitting was recorded at Afi. On two consecutive nights, one site with 22 individual nests was followed by two sites of 10 and 12 individual nests, 300 m apart. Such flexibility in grouping could be a response to patterns of food availability. However, more in-depth analysis of our data and further research will be needed before any firmer conclusions can be drawn.

## Population status

Although there are many indications that the population of Cross River gorillas is very small and declining, it has proved difficult to census the animals accurately for several reasons. Because only small numbers survive, direct sightings of the gorillas are very rare, and nest sites occur at a low frequency. In addition, most of the areas the gorillas inhabit have dry, stony soil and a sparse understory, so the animals often do not leave obvious traces of their travel paths. Finally, these gorillas appear to be presently concentrated in the most rugged hill areas of their former range, probably in large part as a way of avoiding hunters. These areas are difficult for researchers to survey; in particular, straight-line transect censuses are very time-consuming and physically demanding in this terrain, and sometimes impossible to conduct.

### *Present distribution*

While studies in the late 1980s suggested that a few Cross River gorillas still occasionally used the edges of the Obudu Plateau, above the Boshi Extension Forest Reserve (Harris *et al.*, 1987; Harcourt *et al.*, 1989), and the Makone River valley in the lowland of Takamanda (Fay, 1987), our own surveys since 1990 have found definite evidence (nest, dung, and direct sightings) only in the following locations (Oates *et al.*, 1990; McFarland, 1994; Groves and Maisels, 1999; JLG, unpublished data) (Fig. 19.2):

- Afi Mountain above Buanchor, in the northwest corner of the Afi River Forest Reserve, Nigeria. The area currently inhabited by gorillas is approximately 35 km$^2$.
- The Mbe Mountains above Kanyang; this is community-controlled land, proposed for annexation to the Cross River National Park. The area occupied by gorillas is approximately 30 km$^2$.
- The dissected escarpment on the western face of the Obudu Plateau, including the valleys of the Mache, Asache, and Enyimayi Rivers, in the Okwangwo Division of Cross River National Park (formerly part of the Boshi Extension Forest Reserve). This area covers 25–30 km$^2$.
- The hills straddling the Nigeria–Cameroon border, between the Oyi and Manyu Rivers, south of Okwa in the Cross River National Park, and north

of Obonyi 1 in the Takamanda Forest Reserve. This is also an area of about 25–30 km$^2$.

- The hills lying within the Takamanda Forest Reserve to the west of Basho 1 and Basho 2, and to the east of the Makone River. This is an area of about 10 km$^2$.
- A highland area south east of Takpe village, just outside the Takamanda Forest Reserve, covering approximately 4 km$^2$.
- The hills of the Mone River Forest Reserve, between the villages of Mbu and Amebeshu. This locality was confirmed only in 2001; surveys are continuing in Mone to establish the distribution of gorillas in the reserve.
- In the Upper Mbulu Forest, in the hills between Badshama and Ashunda, about 15 km east of Takamanda Forest Reserve.
- In the western Bamenda Highlands, between Kenshi (6° 06′ N, 9° 43′ E) and Njikwa (6° 10′ N, 9° 50′ E). These gorillas were only located late in 2001 and their distribution is still being assessed (this region is not included in Fig. 19.2).

This evidence indicates that Cross River gorillas are scattered across nine or more hill areas, separated from each other by lowlands where no firm evidence of gorillas has been found since 1987. We suspect, however, that wandering males may travel away from the hills, and therefore provide some opportunities for gene flow. The sighting by Fay (1987) west of the Makone River in the central lowlands of Takamanda was of a single blackback male. People from Takamanda village itself also report seeing an adult male gorilla in the lowland forest area north of the village (JLG and T. Sunderland, unpublished data).

## Numbers

March (1957) guessed at a population of 100–200 gorillas in the Boshi–Okwangwo–Takamanda area as a whole (not realizing that gorillas also lived at Mbe and Afi), while Harcourt *et al.* (1989) estimated a population in Nigeria alone of 100–300, with 150 as their best guess. From field surveys we have conducted since 1990, including multiple recensusing of most of the Nigerian population, we estimate that no more than 35 individuals probably survive within each of the eight areas in Nigeria and Cameroon in which we have found definite gorilla signs; several of these areas almost certainly support fewer than 30 gorillas. We will consider these areas individually, and our analysis is summarized in Table 19.4.

Harcourt *et al.* (1988, 1989) estimated a population of about 50 individuals on Afi Mountain in early 1988, based in part on finding four nest clusters along

Table 19.4. *Estimated numbers of Cross River gorillas surviving at different localities in 2001*

| Locality | Estimated area inhabited (km²) | Estimated numbers |
|---|---|---|
| Afi Mountain Wildlife Sanctuary | 35 | 25–30 |
| Mbe Mountains | 30 | 24–32 |
| Cross River National Park, Boshi Extention forests | 25–30 | 20 |
| Cross River National Park, Okwa Hills and Takamanda, Obonyi Hills | 25–30 | 20–30 |
| Takamanda, Basho Hills | 10 | 15–20 |
| Takpe Hill, near Takamanda | 4 | 10–15 |
| Other areas of Takamanda | ?[a] | ?25[b] |
| Mone River Forest Reserve | ?[a] | ?25[b] |
| Upper Mbulu Forest and Bamenda Highlands | ?[a] | ?50[b] |
| Total | 140–200 | 204–247 |

[a]Unknown.

[b]Preliminary estimates based on reconnaissance surveys; further surveys in progress.

40 km of trail, and calculating a density of 1–1.5 gorillas/km² in an estimated 40 km² area occupied by gorillas. Oates *et al.* (1990) estimated a population of 40 gorillas at Afi, in 30 km², while the long-term observations by KLM suggest that in 1998–9 there were no more than 25 gorillas inhabiting an area of about 35 km² on the mountain. However, we do not think it likely that the gorilla population size at Afi has shrunk from 50 to 25 over an 11-year period (few gorillas appear to have been killed at Afi in this time). Rather, the change probably reflects an increase in the accuracy of the population census.

Similarly, Harcourt *et al.* (1988, 1989) estimated that a further 50 gorillas might inhabit the Mbe Mountains, where they sighted one group and six nest sites along 60 km of trail, and where they estimated that 40 km² were available to gorillas. Further surveys by Oates *et al.* (1990) reduced this to an estimate of 30 gorillas. After a thorough survey of the Mbe Mountains in 1995–7, Obot *et al.* (1997) estimated a gorilla population size of 24–32 individuals; although they state that this population occupied an area of 40 km², their map suggests an area of only 25–30 km².

In the Boshi Extension forests of the Obudu Plateau escarpment (an area not fully surveyed by Harcourt *et al.*), Oates *et al.* (1990) estimated a total population of 20–30 individuals. Further surveys by Nwufoh (1999) in the same area in 1998–9 also suggested a population of about 20 weaned individuals in these forests.

In 1968, Critchley estimated a total population of 25–50 gorillas in Takamanda. During her 1998–9 surveys of Takamanda, JLG found nests only in two highland areas of the reserve; she estimated that if gorillas occurred at roughly the same density in all highland areas, then the total number of weaned gorillas would be about 140. However, because a third highland area located near a village held no evidence of gorillas, this estimate was adjusted to take into account the relative proximity of different highland areas to villages; the adjusted estimate was that approximately 100 individuals remained in Takamanda (Groves and Maisels, 1999). Given what we have learned about gorilla densities from intensive, repeated surveys in Nigeria, it is probable that even this is an overestimate, and that 75–100 gorillas remain today in Takamanda Forest Reserve and in immediately adjacent hill areas such as Takpe.

Reconnaissance surveys undertaken by JLG in early 2001 in the Mone Forest Reserve and contiguous Mbulu Forest in Cameroon found a small number of gorilla nests in each locality. In late 2001, JLG, RAB, and JML also confirmed the presence of at least two gorilla groups in the western Bamenda Highlands, just outside Kenshi. Although there is considerable highland habitat in these forests, there are also many villages. Further studies are planned to obtain estimates of population density and distribution.

In general, as more data are gathered from more thorough field surveys, a picture emerges of smaller numbers of animals living at each locality than had been estimated by earlier and more superficial surveys. Our current estimate is that the repeatedly censused areas of Afi, Mbe, and Boshi Extension in Nigeria support a total of 70–90 gorillas. There might be at least another 150 animals in Cameroon, including the subpopulation straddling the Nigeria–Cameroon border north of Obonyi, and those in Mone, Mbulu, and the western Bamenda Highlands.

### Threats and trends

Hunting animals for meat is probably the greatest threat to the survival of Cross River gorillas, as it is for many other great ape populations in western and central Africa (for more on this "bushmeat" trade, see Bowen-Jones and Pendry, 1999 and Plumptre *et al.*, this volume). In the late 1980s, Harcourt *et al.* (1989) estimated that a minimum of 15 gorillas were being killed each year in Nigerian forests alone. Since that time hunting pressure on the animals has eased, at least in Nigeria (in part due to the increased national and international attention they have received), but it has not ceased. There is evidence that one gorilla was killed in the Mbe Mountains in 1991 (Stewart, 1992) and that two were killed in the Afi Mountains in 1993 (McFarland, 1994). Hunters interviewed by JFO in 1998 said that they knew of gorillas being killed in the Boshi Extension area in 1995 and in early 1996 (Oates, 1998), and one of KLM's

Afi study group was shot by a hunter in October 1998 (McFarland, 1998). A gorilla with a wound probably caused by a gunshot was reported to have been seen in the Boshi Extension forest in 1998, and a Cross River National Park ranger reported one gorilla as having been killed by an Okwangwo hunter in the same year (Oates, 1999*a*). In villages in and around Takamanda, JLG found 10 skulls of gorillas that were reported to have been killed in the 1990–8 period (Groves and Maisels, 1999). Taken together, these reports suggest an average of at least two Cross River gorillas killed each year during 1990–8 .

Hunting is not the only threat to the Cross River gorillas; their habitat is also being eroded by fire and agriculture. Because the area inhabited by this subspecies has such a long dry season, the forest can become very susceptible to fire by the late dry season. Fires started outside the forest reserves and national park by farmers, hunters, and pastoralists during this season often sweep through the forest, burning understory plants and sometimes destroying trees. Farms have encroached on parts of the reserves and Park and are eroding the remaining forest in the Mbe Mountains. At this time, the habitats of each of the remaining Nigerian and Takamanda gorilla subpopulations are at least tenuously connected by forest, but fire and farming are threatening to sever these connections, especially between Afi and Mbe.

### Conservation efforts and options

As of the year 2001, Cross River gorillas enjoy somewhat better protection in Nigeria than in Cameroon. As we mentioned earlier in this chapter, the former Boshi Extension and Okwangwo Forest Reserves (each of which is used by a gorilla subpopulation) became components of the Okwangwo Division of Cross River National Park in 1991. Since 1990 there has been a proposal to incorporate the community-controlled forests of the Mbe Mountains into the same national park (Oates *et al.*, 1990), and in May 2000 the Afi Mountain area was gazetted by the Cross River State Government as a wildlife sanctuary. Takamanda and Mone in Cameroon still have the status of forest reserves, a status which provides no special protection for their fauna beyond national laws (rarely enforced) regulating the hunting of endangered species such as gorillas. However, a joint project between the Cameroon Ministry of the Environment and Forests and the German development agency GTZ is currently studying the status of the forests between Mamfe and Akwaya, including Takamanda and Mone; this is likely to lead to improvements in the protected status of at least some areas of the forests.

Tropical rainforests are notorious for, among other things, their "paper parks": protected areas that appear on maps but which have no real protection on the ground (see also Plumptre *et al.*, this volume). Cross River National Park has suffered to some extent from this problem. The 640 km² of the Park's Okwangwo Division was the location from 1991 to 1998 of an integrated conservation-and-development project (ICDP) managed by the World Wide Fund for Nature. This project supported a range of educational and agricultural development efforts around the Park, but placed a relatively low emphasis on serious policing within the Park, so that until recently the hunting of most animals continued at a high level (Oates, 1999*a*,*b*). Since the ICDP came to an end there have been signs of a more serious commitment by the Park authorities to protection efforts. Local people have shown a lack of co-operation with park officials, however, because the ICDP raised their expectations of development assistance beyond what was actually delivered (Oates, 1999*b*). For instance, the communities controlling the Mbe Mountains have resisted attempts to incorporate that portion of gorilla habitat into the Park, holding out for compensation or development assistance that they have been led to expect as a result of the presence of the ICDP, while people in the Okwangwo enclave have impeded ranger patrols because of lack of progress in a resettlement scheme.

The smaller-scale conservation project at Afi Mountain has been more successful. The recent gazetting of the wildlife sanctuary there builds on efforts by the nongovernmental organization Pandrillus and by KLM since the early 1990s. From 1996, KLM's ecological research project employed a permanent team of gorilla trackers on Afi Mountain, and since 1999 that team has been integrated into a larger conservation effort supported by a consortium of conservationists and scientists who are working in co-operation with the state forestry commission (Suter and Oates, 2000). By monitoring the movements of the gorillas on the mountain, and reporting to the government on poaching activities and encroachments, the trackers have managed to reduce the hunting of gorillas and other endangered primates (such as chimpanzees and drills). Efforts are now under way to build a long-term conservation program at Afi, which will include a protection system, a community education and awareness-raising program, and continuing field research.

The Afi experience conforms to a lesson learned from many sites in the African rainforest (including, for example, Taï in Côte d'Ivoire, Kibale in Uganda, Karisoke in Rwanda, Lopé in Gabon, and Tiwai in Sierra Leone): that the presence of a long-term research program, based at a permanent field station or camp, makes a strong positive contribution to conservation. Because of their inherent interest in both the animals they are studying and their environment, field researchers are usually strongly committed to conservation; and because they spend so much time in the animals' environment, they often develop a good understanding of real-life conservation problems.

Based on such experiences, we are endeavoring to establish long-term field research projects on as many subpopulations of Cross River gorillas as possible, and to integrate this research with conservation efforts. For instance, there is no evidence of any gorillas having been killed in Takamanda since 1998, almost certainly because of heightened awareness produced by JLG's field activities. We are therefore trying to create a more permanent research presence at Takamanda, while encouraging the upgrading of its protected status.

Given the small size and fragmented distribution of the remaining population of Cross River gorillas, together with the continuing growth of the dense human population around its last habitats, this subspecies is now critically endangered. Indeed, it could be argued that even improved conservation efforts will be too late to save Cross River gorillas from extinction. Each subpopulation appears to be very small and, if there is little gene flow between them, all could eventually die out from small-population effects even if rigorously protected (Gilpin and Soulé, 1986). However, geography and historical accounts suggest these subpopulations have been relatively isolated for a long time, yet they have persisted. Most of the localities from which the gorillas were collected or reported in the first half of the twentieth century are still inhabited by the animals, despite intense pressures. It is likely that wandering males have contributed to low levels of gene flow between the population isolates and, given the long generation time of gorillas, demographic and population-genetic factors are probably much less immediate threats to population persistence than hunting and habitat destruction (Harcourt, 1995). Although the population is currently fragmented, there are still forested connections between most of the surviving subpopulations, and the total area of available habitat across their range is clearly sufficient to support a much larger number of animals. Therefore, Cross River gorillas are not unequivocally beyond rescue.

If, in spite of efforts to increase protection, the subpopulations do not expand significantly and are not able to maintain adequate gene flow, active management interventions (such as the translocation of individuals) may eventually have to be considered to counter the genetic and demographic problems that could arise in the long term from small population size. However, given the serious potential for disrupting already fragile populations, and all the practical difficulties attendant on gorilla translocation, such measures should be considered only as a last resort.

We acknowledge that, in the long run, conservation of animals like the Cross River gorilla and of its habitat will be difficult in the absence of major changes in the culture and economy of West Africa. The gorillas of Cameroon and Nigeria have at least been fortunate in not suffering from the side effects of the civil wars that have increased threats to many eastern gorilla populations (Plumptre *et al.*, this volume), and there have recently been some promising political changes in Nigeria that have shown benefits for conservation. Since the return of civilian rule to Nigeria in 1999, the government of Cross River State has shown a stronger commitment to conservation. Not only has the government supported the creation of the Afi Mountain Wildlife Sanctuary, but it has also begun to make plans for a conservation education center on the Obudu Plateau, a center that would give special emphasis to the Cross River gorilla. Such initiatives, along with continued long-term field research, deserve every encouragement.

### Acknowledgments

We are most grateful to the government agencies who authorized our field work: in Nigeria, the management of Cross River National Park, and the Cross River State Forestry Commission; in Cameroon, the Ministry of Environment and Forests. Research was supported by grants from the Wildlife Conservation Society, the L.S.B. Leakey Foundation, Primate Conservation Incorporated, the Margot Marsh Biodiversity Foundation, the Whitley Foundation, WWF Cameroon, the Columbus Zoo, the National Science Foundation (SBR-9506892 to Disotell), and the Professional Staff Congress of the City University of New York. Field work was made possible through the efforts of many colleagues and assistants, including particularly: Liza Gadsby, Peter Jenkins and the staff of Pandrillus; Ernest Nwufoh and the staff of Primates Preservation Group; Livinius Abang, Raymond Abang, Martin Achu, Albert Ekinde, Dominic Elabi, Peter Eshin, Charles Ewa, Milton Nchinda, Francis Ndim, Rita Nyiam, Donatus Nyiamson, Dennis Osang, and Fidelis Osang. We thank Ryan Raaum for providing invaluable help with laboratory analyses, and JLG is grateful to Fiona Maisels and David Hill for their guidance.

### References

Allen, M., Engstrom, A.S., Meyers, S., Handt, O., Saldeen, T., von Haeseler, A., Pääbo, S., and Gyllensten, U. (1998). Mitochondrial DNA sequencing of shed hairs and saliva on robbery caps: Sensitivity and matching probabilities. *Journal of Forensic Science*, **43**, 453–464.

Anonymous (1934). Notes on gorilla. *Nigerian Field*, **3**, 92–102.

Bermejo, M. (1997). Study of western lowland gorillas in the Lossi Forest of North Congo and a pilot gorilla tourism plan. *Gorilla Conservation News*, **11**, 6–7.

Bowen-Jones, E. and Pendry, S. (1999). The threat to primates and other mammals from the bushmeat trade in Africa, and how this threat could be diminished. *Oryx*, **33**, 233–246.

Butynski, T.M. (2001). Africa's great apes. In *Great Apes and Humans: The Ethics of Coexistence*, eds. B. Beck, T. Stoinski, M. Hutchins, T.L. Maple, B.G. Norton, A. Rowan, E.F. Stevens, and A. Arluke, pp. 3–56. Washington, D.C.: Smithsonian Institution Press.

Critchley, W.R. (1968). Final report on Takamanda gorilla survey. Unpublished report to Winston Churchill Memorial Trust, London.

Coolidge, H.J., Jr. (1929). A revision of the genus *Gorilla*. *Memoirs of the Museum of Comparative Zoology, Harvard*, **50**, 291–381.

Doran, D.M. and McNeilage, A. (1998). Gorilla ecology and behavior. *Evolutionary Anthropology*, **6**, 120–131.

Doran, D. and McNeilage, A. (1999). Diet of western lowland gorillas in south-west Central African Republic: Implications for subspecific variation in gorilla grouping and ranging patterns. *American Journal of Physical Anthropology, Supplement* **28**, 121.

Fay, J.M. (1987). Report on the participation of J. Michael Fay in the Takamanda Gorilla Survey project (May 1–May 20 1967) funded by the World Wildlife Fund-U.S. Unpublished report to WWF-U.S., Washington, D.C.

Gagneux, P., Wills, C., Gerloff, U., Tautz, D., Morin, P.A., Boesch, C., Fruth, B., Hohmann, G., Ryder, O.A., and Woodruff, D.S. (1999). Mitochondrial sequences show diverse evolutionary histories of African hominoids. *Proceedings of the National Academy of Sciences U.S.A.*, **96**, 5077–5082.

Garner, K.J. and Ryder, O.A. (1996). Mitochondrial DNA diversity in gorillas. *Molecular Phylogenetics and Evolution*, **6**, 39–48.

Gilpin, M.E. and Soulé, M.E. (1986). Minimum viable populations: Processes of species extinction. In *Conservation Biology: The Science of Scarcity and Diversity*, ed. M.E. Soulé, pp. 19–34. Sunderland, MA: Sinauer Associates.

Goldberg, T.L. (1996). Genetics and biogeography of eastern African chimpanzees (*Pan troglodytes schweinfurthii*). PhD thesis, Harvard University, Cambridge, MA.

Goldsmith, M.L. (1996). Ecological influences on the ranging and grouping behavior of the western lowland gorillas at Bai Hokou, CAR. PhD thesis, State University of New York, Stony Brook, NY.

Goldsmith, M.L. (1999). Ecological constraints on the foraging effort of western gorillas (*Gorilla gorilla gorilla*) at Bai Hokou, Central African Republic. *International Journal of Primatology*, **20**, 1–23.

Gonder, M.K. (2000). Evolutionary genetics of chimpanzees (*Pan troglodytes*) in Nigeria and Cameroon. PhD thesis, City University of New York.

Gonder, M.K., Oates, J.F., Disotell, T.R., Forstner, M.R.J., Morales, J.C., and Melnick, D.J. (1997). A new west African chimpanzee subspecies? *Nature*, **388**, 337.

Groves, C.P. (1967). Ecology and taxonomy of the gorilla. *Nature*, **213**, 890–893.

Groves, C.P. (1970). Population systematics of the gorilla. *Journal of Zoology, London*, **161**, 287–300.

Groves, C.P. (2001). *Primate Taxonomy*. Washington, D.C.: Smithsonian Institution Press.

Groves, J. and Maisels, F. (1999). Report on the large mammal fauna of the Takamanda Forest Reserve, South West Province, Cameroon, with special emphasis on the gorilla population. Unpublished report to WWF-Cameroon, Yaoundé.

Harcourt, A.H. (1995). Population viability estimates: Theory and practice for a wild gorilla population. *Conservation Biology*, **9**, 134–142.

Harcourt, A.H., Stewart, K.J., and Inaharo, I.M. (1988). Nigeria's gorillas: A survey and recommendations. Unpublished report to Nigerian Conservation Foundation, Lagos.

Harcourt, A.H., Stewart, K.J., and Inaharo, I.M. (1989). Gorilla quest in Nigeria. *Oryx*, **23**, 7–13.

Harris, D., Fay, J.M., and MacDonald, N. (1987). Report of gorillas from Nigeria. *Primate Conservation*, **8**, 40.

Jensen-Seaman, M.I. (2000). Evolutionary genetics of gorillas. PhD thesis, Yale University, New Haven, CT.

Maddison, D.R. and Maddison, W.P. (2000). *MacClade*, v. 4.0. Sunderland, MA: Sinauer Associates.

March, E.W. (1957). Gorillas of Eastern Nigeria. *Oryx*, **4**, 30–34.

Matschie, P. (1904). Bemerkungen über die Gattung *Gorilla*. *Sitzungsberichte des Gesellschaft naturforschender Freunde, Berlin*, **1904**, 45–53.

McFarland, K. (1994). Update on gorillas in Cross River State, Nigeria. *Gorilla Conservation News*, **8**, 13–14.

McFarland, K. (1998). Gorilla killed in Cross River State, Nigeria. *Gorilla Journal*, **17**, 14–15.

Nishihara, T. (1995). Feeding ecology of western lowland gorillas in the Nouabalé-Ndoki National Park, Congo. *Primates*, **36**, 151–168.

Nishihara, T. and Kuroda, S. (1991). Soil-scratching behaviour by western lowland gorillas. *Folia Primatologica*, **57**, 48–51.

Nwufoh, E.I. (1999). Survey of gorillas in Boshi Extension portion of the Cross River National Park. Unpublished report submitted to WWF-CRNP project, Obudu, Nigeria, and to Wildlife Preservation Trust International, Philadelphia, PA.

Oates, J.F. (1998). The gorilla population in the Nigeria–Cameroon border region. *Gorilla Conservation News*, **12**, 3–6.

Oates, J.F. (1999*a*). Update on Nigeria. *Gorilla Conservation News*, **13**, 6.

Oates, J.F. (1999*b*). *Myth and Reality in the Rain Forest: How Conservation Strategies are Failing in West Africa*. Berkeley, CA: University of California Press.

Oates, J.F., White, D., Gadsby, E.L., and Bisong, P.O. (1990). Conservation of gorillas and other species. In *Cross River National Park (Okwangwo Division): Plan for Developing the Park and Its Support Zone*, Appendix 1, eds. J.O. Caldecott, J.F. Oates, and H.J. Ruitenbeek. Godalming, U.K.: WWF-UK.

Obot, E., Barker, J., Edet, C., Ogar, G., and Nwufoh, E. (1997). *Status of gorilla* (G. *gorilla gorilla*) *populations in Cross River National Park and Mbe Mountain*, Cross River National Park (Okwangwo Division) Technical Report no. 3. Obudu, Nigeria.

Olejniczak, C. (1996). Update on the Mbeli Bai gorilla study, Nouabale-Ndoki National Park, northern Congo. *Gorilla Conservation News*, **10**, 5–8.

Remis, M.J. (1994). Feeding ecology and positional behavior of western lowland gorilla (*Gorilla gorilla gorilla*) in the Central African Republic. PhD thesis, Yale University, New Haven, CT.

Remis, M.J. (1997*a*). Western lowland gorilla (*Gorilla gorilla gorilla*) as seasonal frugivores: Use of variable resources. *American Journal of Primatology*, **43**, 87–109.

Remis, M.J. (1997*b*). Ranging and grouping patterns of a western lowland gorilla group at Bai Hokou, Central African Republic. *American Journal of Primatology*, **43**, 111–133.

Rogers, M.E., Williamson, E.A., Tutin, C.E.G., and Fernandez, M. (1988). Effects of the dry season on gorilla diet in Gabon. *Primate Report*, **22**, 25–33.

Rogers, M.E., Tutin, C.E.G., Williamson, E.A., Parnell, R.J., Voysey, B.C., and Fernandez, M. (1994). Seasonal feeding on bark by gorillas: An unexpected keystone food? In *Current Primatology*, Vol. 1, *Ecology and Evolution*, eds. B. Thierry, J.R. Anderson, J.J. Roeder, and N. Herrenschmidt, pp. 37–43. Strasbourg, France: Université Louis Pasteur.

Rothschild, W. (1904). Notes on anthropoid apes. *Proceedings of the Zoological Society of London*, **1904**(2), 413–440.

Rothschild, W. (1908). Note on *Gorilla gorilla diehli* Matschie. *Novitates Zoologicae*, **15**, 391–392.

Ruvolo, M., Pan, D., Zehr, S., Goldberg, T, Disotell, T., and von Dornum, M. (1994). Gene trees and hominid phylogeny. *Proceedings of the National Academy of Sciences U.S.A.*, **91**, 8900–8904.

Sarmiento, E. and Oates, J.F. (2000). The Cross River gorillas: A distinct subspecies *Gorilla gorilla diehli* Matschie 1904. *American Museum Novitates*, **3304**, 1–55.

Stewart, K.J. (1992). Gorilla conservation in Nigeria, 1991. *Gorilla Conservation News*, **6**, 11.

Stumpf, R.M., Fleagle, J.G., Jungers, W.L., Oates, J.F., and Groves, C.P. (1998). Morphological distinctiveness of Nigerian gorilla crania. *American Journal of Physical Anthropology, Supplement*, **26**, 213.

Suter, J. and Oates, J.F. (2000). Sanctuary in Nigeria for possible fourth subspecies of gorilla. *Oryx*, **34**, 71.

Swofford, D.L. (1998). *PAUP\*: Phylogenetic Analysis Using Parsimony (\*and Other Methods)*, v.4. Sunderland, MA: Sinauer Associates.

Tutin, C.E.G. and Fernandez, M. (1993). Faecal analysis as a method of describing diets of apes: Examples from sympatric gorillas and chimpanzees at Lopé, Gabon. *Tropics*, **2**, 189–197.

Tutin, C.E.G., Fernandez, M., Rogers, M.E., and Williamson, E.A. (1992). A preliminary analysis of the social structure of lowland gorillas in the Lopé Reserve, Gabon. In *Topics in Primatology*, Vol. 2, *Behavior, Ecology and Conservation*, eds. N. Itoigawa, Y. Sugiyama, G. P. Sackett, and R.K.R. Thompson, pp. 245–254. Tokyo: University of Tokyo Press.

Tutin, C.E.G., White, L.J.T., Williamson, E.A., Fernandez, M., and McPherson, G. (1994). List of plant species identified in the northern part of Lopé Reserve, Gabon. *Tropics*, **3**, 249–276.

Tutin, C.E.G., Ham, R.M., White, L.J.T., and Harrison, M.J.S. (1997). The primate community of the Lopé Reserve, Gabon: Diets, responses to fruit scarcity, and effects on biomass. *American Journal of Primatology*, **42**, 1–24.

Vigilant, L. (1999). An evaluation of techniques for the extraction and amplification of DNA from naturally shed hairs. *Biological Chemistry*, **380**, 1329–1331.

Williamson, E.A. (1988). Behavioural ecology of western lowland gorillas in Gabon. PhD thesis, University of Stirling, U.K.

Williamson, E.A., Tutin, C.E.G., Rogers, M.E., and Fernandez, M. (1990). Composition of the diet of lowland gorillas at Lopé in Gabon. *American Journal of Primatology*, **21**, 265–277.

Yamagiwa, J., Mwanza, N., Yumoto, T., and Maruhashi, T. (1991). Ant eating by eastern lowland gorillas. *Primates*, **32**, 247–253.

Yamagiwa, J., Mwanza, N., Yumoto, T., and Maruhashi, T. (1992). Travel distances and food habits of eastern lowland gorillas: A comparative analysis. In *Topics in Primatology*, Vol. 2, *Behavior, Ecology and Conservation*, ed. N. Itoigawa, Y. Sugiyama, G.P. Sackett, and R.K.R. Thompson, pp. 267–281. Tokyo: University of Tokyo Press.

# Afterword

MICHELE L. GOLDSMITH AND ANDREA B. TAYLOR

Once, in the Central African Republic, a BaAka pygmy working with one of us (MLG) left a "precious" pack of matches on a log near a nesting site. "Wanga, you left your matches behind." His response, "The gorillas have been without fire for too long – it is not fair." The human-like quality of gorillas is one reason we are compelled to study them. By examining all aspects of their biology, we begin to understand them and, in so doing, gain insight into ourselves.

In this multidisciplinary approach to gorilla biology, the contributing authors have advanced our knowledge of these apes. Some have yielded novel insights, while others have provided greater clarity with more detailed and comprehensive analyses. Yet, with each insight, additional questions are generated, and challenges remain.

As has been pointed out by several contributors, though the use of molecular genetics to address questions of phylogeny and taxonomy is not new, we are still in the formative stages in terms of our ability to interpret genetic variation from molecular data and draw biologically meaningful distinctions. This is reflected in the fact that debates over gorilla taxonomy, as presented both here and elsewhere, center at least as much on methods as on findings. One need only compare the results and interpretations of molecular studies that rely on different (and sometimes even identical!) genetic markers to see that this is so.

Which genetic markers are phylogenetically informative? What sample sizes are needed to achieve reliable results? How do we recognize homoplasy? And how does variation in mutation rates impact on the information derived from a given site at a given locus? Morphologists have been grappling with these same (or very nearly the same) questions for decades. Clearly, we are still faced with the challenge of interpreting biological variation, whether morphological, genetic, or in some other form. Some may view molecular genetics as "replacing" the use of morphology, but the difficulty of interpreting variation is inherently the same; trading one approach for the other will not resolve this.

Despite findings of morphological and genetic differences among local populations of gorillas, and evidence of tremendous behavioral flexibility among gorilla groups, the greatest degree of difference, based on all lines of evidence presented thus far, is still between western and eastern groups, upholding what Coolidge concluded nearly a century ago. Yet we are plainly witnessing a trend

498

towards splitting long-standing subspecies of gorillas into distinct species, and restructuring populations long subsumed within existing subspecies into separate subspecies. Though differences in morphology and behavior have been used to bolster arguments in favor of these changes, much of this taxonomic splitting appears largely to correspond to genetic differences reflected in mtDNA, one of only a handful of genetic markers that have been studied to date.

Taxonomy should be periodically re-evaluated. But the question of whether gorillas comprise one species or two, or whether the *Pan troglodytes–P. paniscus* split is the standard against which gorilla taxonomy should be evaluated, ought not to take precedence over empirical and robust assessments of all relevant lines of evidence that factor into the definition and recognition of species. This is what Coolidge attempted to do (limited as he was to morphology), when he characterized the patterning and degree of variation in gorillas based on 213 skulls collected from all lowland and mountain gorilla sites known at the time. And this is likely the reason that his taxonomy, with minor modification, has been retained for so long. What has been conspicuously absent in the current taxonomic debates surrounding gorillas (and other primate taxa), and this is not unimportant, are the relevant discussions of what constitutes a species, how species are defined in nature, in extant forms as well as in fossils, and how we recognize and identify subspecies. Future progress in understanding gorilla diversity depends upon addressing these two issues – species recognition *and* taxonomic classification – hand in hand.

While molecular genetics appears to be providing much of the evidentiary basis for taxonomic revisions, the impetus for these modifications seems to be driven, at least in part, by conservation. The conservation status of a population is based on the actual numbers of individuals that can be counted (either practically or theoretically). Therefore, an increase or decrease in animal numbers will affect how we estimate the relative threat of extinction. The mountain gorillas provide an excellent case in point. By separating Bwindi gorillas from the rest of the mountain gorilla population, we are left with 350 Virunga gorillas and only 290 Bwindi gorillas. This certainly indicates a more dismal situation, and stresses the need for prompt action, than if both populations were members of a single subspecies. Nigerian gorillas, with somewhere between 100 and 200 individuals, provide a similar example.

A pressing reason for the reappraisal of gorilla biology may be the dire status of gorilla populations in the wild, but there is no necessary link between taxonomy and conservation. The status of populations in the wild should arguably be based on their population size and geographic isolation. What is important about the Virunga and Bwindi mountain gorillas (or the Nigerian gorillas, for that matter) is not whether they comprise one subspecies or two, but the fact that they live in small and fragmented habitats geographically isolated from one another. Conservation of endangered animals is best served when populations are evaluated based on the factors that critically influence their status in the wild, on a case-by-case basis, not on their Latin names.

It has become impossible to discuss the great apes without attention to conservation. One has only to pick a volume from the bookshelf on virtually any primate, but especially an ape, that has been published in the last decade or so, to realize how routine is the inclusion of a "foreword" or "afterword" on conservation. The study of conservation, and early conservation attempts by primatologists, were not, in general, well received by the scientific community as a whole. Many believed conservation had no place in biology or anthropology. Apparently, this is no longer the case. We were certainly surprised, as well as heartened, to find that each of the four authors who contributed introductory perspectives felt the need to highlight the importance of gorilla conservation. Thus, while an awareness of the importance of conservation is not new, its acceptability as an integral part of primate science and biology is certainly a welcome change and it comes not a moment too soon. The need for involvement is clear when we consider that since these chapters were written, we have lost thousands of eastern lowland gorillas in the Democratic Republic of Congo. These losses are mostly due to coltan mining, which continues at an unsustainable rate due to instability caused by warfare. Few of us have heard of this ore, though all of us use it in our cell phones and laptops.

Gorillas have long captivated us. At first we feared them, as great white hunters shared their tales of the vicious man-eaters they saw in the jungles of Africa. Hollywood, and its invention of "King Kong", a ferocious yet sensitive beast, spurred the fascination on. But our image of this creature changed with the discoveries of Dian Fossey, and we have been in awe of the humanness of these gentle giants ever since. If we are truly interested in gorilla biology as a whole, and not simply in one piece of their biology, then gorilla conservation should be of great importance to all biological anthropologists – whether in the laboratory or in the forest. Only with their survival will we have the opportunity to further our exploration, both of gorillas and of ourselves.

# Index

501